Progress in Nonequilibrium Greens Functions IV 2009

Journal of Physics: Conference Series Volume 220

Glasgow, United Kingdom
17-21 August 2009

ISBN: 978-1-61738-580-3
ISSN: 1742-6588

Printed from e-media with permission by:

Curran Associates, Inc.
57 Morehouse Lane
Red Hook, NY 12571

Some format issues inherent in the e-media version may also appear in this print version.

Printed by Curran Associates, Inc. (2010)

For permission requests, please contact the Institute of Physics at the address below.

Institute of Physics
Dirac House, Temple Back
Bristol BS1 6BE UK

Phone: 44 1 17 929 7481
Fax: 44 1 17 920 0979

techtracking@iop.org

Additional copies of this publication are available from:

Curran Associates, Inc.
57 Morehouse Lane
Red Hook, NY 12571 USA
Phone: 845-758-0400
Fax: 845-758-2634
Email: curran@proceedings.com
Web: www.proceedings.com

TABLE OF CONTENTS

PLASMAS, EXCITONS, BARYONS

QUANTUM TRANSPORT

FINITE SYSTEMS AND QUANTUM TRANSPORT

NUMERICAL CONCEPTS

PREFACE

This is the fourth volume[1] of articles on the theory of Nonequilibrium Green's functions (NEGF) and their modern application in various fields such as plasma physics, semiconductor physics, molecular electronics and high energy physics. It contains 23 articles written by experts in many-body theory and quantum transport who summarize recent progress in their respective area of research.

The articles are based on talks given at the interdisciplinary conference *Progress in Nonequilibrium Green's functions IV* which was held 17–21 August 2009 at the University of Glasgow, Scotland. This conference continues the tradition of the previous meetings which started in 1999 and which aimed at an informal exchange across field boundaries. The previous meetings and the earlier proceedings proved to be very stimulating not only for young researchers but also for experienced scientists, and we are convinced that this fourth volume will be as successful as the previous ones. As before, this volume includes only extended review-type papers which are written in a way that they are understandable to a broad interdisciplinary audience.

All papers published in this volume of *Journal of Physics: Conference Series* have been peer reviewed through processes administrated by the Editors assuring highest scientific standards. In the review process some papers were substantially revised and improved and some were rejected.

This conference would not have been possible without the remarkable work of the local organizing team around John Barker and Scott Roy and the generous financial support from the University of Glasgow and the Deutsche Forschungsgemeinschaft via SFB-Transregio 24.

Michael Bonitz and Karsten Balzer
Kiel, February 2010

[1] The first two volumes are *Progress in Nonequilibrium Green's functions*, M Bonitz (ed) and *Progress in Nonequilibrium Green's functions II*, M Bonitz and D Semkat (eds), which were published by World Scientific (Singapore), in 2000 and 2003, respectively (ISBN 981-02-4218-2 and ISBN 981-238-271-2). The third volume *Progress in Nonequilibrium Green's functions III* M Bonitz and A Filinov (eds) was published in *Journal of Physics: Conference Series*, volume 35 (2006).

Conference program and participants

Workshop participants

Carl-Olof Almbladh, Lund University, Sweden
Frithjof Anders, Technische Universität Dortmund, Germany
Karsten Balzer, Kiel University, Germany
John Barker, University of Glasgow, Scotland
Sebastian Bauch, Kiel University, Germany
Gordon Baym, University of Illinois at Urbana-Champaign, USA
Michael Bonitz, Kiel University, Germany
Andrew Brown, University of Glasgow, Scotland
Pawel Danielewicz, Michigan State University, USA
Barry Dunietz, University of Michigan, USA
Holger Fehske, Greifswald University, Germany
David Ferry, Arizona State University, USA
Dirk Gericke, University of Warwick, United Kingdom
Matthew Gilbert, University of Illinois at Urbana-Champaign, USA
Maria Hellgren, Lund University, Sweden
David Hochstuhl, Kiel University, Germany
Frank Jahnke, University Bremen, Germany
Antti-Pekka Jauho, Technical University of Denmark, and Helsinki University of Technology
Kimmo Kainulainen, University of Jyväskylä, Finland
Andele Kalvova, Charles University, Prague, Czech Republic
Peter Kleinert, Paul-Drude Institute, Germany
Wolf-Dietrich Kraeft, Rostock University, Germany
Dietrich Kremp, Rostock University, Germany
Antonio Martinez, University of Glasgow, Scotland
John McGrady, University of Glasgow, Scotland
Petri Myohanen, University of Jyväskylä, Finland
Marc Puig von Friesen, Lund University, Sweden
Felix Richter, Rostock University, Germany
Arnau Rios, University of Surrey, United Kingdom
Scott Roy, University of Glasgow, Scotland
Tomi Ruokola, Technical University of Denmark
Stefano Sanvito, Trinity College, Dublin, Ireland
Sebastian Schmitt, Technische Universität Dortmund, Germany
Peter Schmitteckert, Karlsruhe Institute of Technology, Germany

Dirk Semkat, Rostock University, Germany
Vaclav Spicka, Charles University, Prague, Czech Republic
Adrian Stan, University of Jyväskylä, Finland
Gianluca Stefanucci, University of Rome, Italy
Anna-Maija Uimonen, University of Jyväskylä, Finland
Kris van Houcke, Ghent University, Belgium
Robert van Leeuwen, University of Jyväskylä, Finland
Claudio Verdozzi, Lund University, Sweden
Jan Vorberger, University of Warwick, United Kingdom
Jeremy Watling, University of Glasgow, Scotland

List of Talks at the conference

(R)—Review Talk; (I)—Invited Talk; (C)—Contributed Talk.

F. Anders
University of Dortmund
(I) *Quantum transport through nano-devices: A scattering-states numerical renormalization group approach to open quantum systems*

K. Balzer, D. Hochstuhl, S. Bauch, and M. Bonitz
Christian-Albrechts-Universität Kiel
(I) *Efficient grid-based method in non-equilibrium Green's function calculations for inhomogeneous systems*

J. Barker
University of Glasgow
(I) *NEGF quantum transport phase space formalism with compact support*

S. Bauch, D. Hochstuhl, K. Balzer, and M. Bonitz
Christian-Albrechts-Universität Kiel
(C) *Quantum kinetic approach to time-resolved photoionization of multi-electron atoms*

G. Baym
University of Illinois at Urbana-Champaign
(R) *Connections between quark-gluon plasmas and cold atom physics*

P. Danielewicz[1], and A. Rios[2]
[1]Michigan State University, [2]University of Surrey
(I) *Towards Application of Green's Functions to Nonequilibrium Processes in Nuclear Systems*

B. Dunietz
University of Michigan
(I) *A spectroscopic analysis of electron transport dynamics through molecular junctions*

D.K. Ferry[1], L. Huang[1], R. Akis[1], Y.-C. Lai[1], J.P. Bird[2], N. Aoki[3], Y. Ujie[3], S. Motooka[3], T. Morimoto[3], Y. Ochiai[3], R. Brunner[4], R. Meisels[4], and F. Kuchar[4]
[1]Arizona State University, [2]University at Buffalo, [3]Chiba University, [4]Montanuniversität Leoben
(R) *Open Quantum Dots in Bilayer Graphene: Study of Conductance Fluctuations with Recursive Green's Functions*

D. Gericke, K. Wünsch, and J. Vorberger

University of Warwick
(I) *Structural Properties of Warm Dense Matter*

C. Gies, S. Ritter, P. Gartner, and F. Jahnke
University of Bremen
(I) *Photon statistics of semiconductor microcavity lasers and its connection to carrier correlations*

M. Gilbert
University of Illinois at Urbana-Champaign
(I) *Probing the existence and the stability of the Superfluid Phase of Bilayer Graphene*

D. Hochstuhl, S. Bauch, K. Balzer, and M. Bonitz
Christian-Albrechts-Universität Kiel
(C) *Non-equilibrium Green function approach to photoionization processes in atoms and molecules*

A.-P. Jauho
Technical University of Denmark, and Helsinki University of Technology
(I) *Ab initio calculations of phonons and vibrations in nonequilibrium nanowire*

K. Kainulainen
University of Jyväskylä
(I) *Extended Quasiparticle Approximation and nonlocal coherence*

P. Kleinert
Paul-Drude-Institut für Festkörperelektronik
(C) *Unified theory of double-time quantum transport and quantum diffusion*

W.-D. Kraeft
University of Rostock
(C) *Thermodynamics of Strongly Coupled Plasmas*

D. Kremp, D. Semkat, and K.Henneberger
Universität Rostock
(I) *BEC-BCS transition in excitonic systems*

A. Martinez, N. Seoane, A. Brown, J. Barker, and A. Asenov
University of Glasgow
(I) *Dopants and Roughness induced Resonances in thin Si Nanowire transistors: A self-consistent NEGF-Poisson study*

J. McGrady, and V. Georgiev
University of Glasgow
(I) *NEGF for molecular chemistry: a users perspective*

P. Myöhänen[1] A. Stan[1], G. Stefanucci[2,3], and R. van Leeuwen[1,3]
[1]University of Jyväskylä, [2]Università di Roma, [3]European Theoretical Spectroscopy Facility (ETSF)
(I) *Kadanoff-Baym approach to quantum transport through interacting nanoscale systems: From the transient to the steady-state regime*

M. Puig von Friesen, C. Verdozzi, and C.-O. Almbladh
Lund University
(I) *Off-equilibrium dynamics of Hubbard Clusters*

F. Richter, D. Semkat, and K. Henneberger

University of Rostock
(C) *The photon Green's function for bounded media: Splitting property and nonequilibrium radiation laws*

A. Rios[1], P. Danielewicz[2], and B. Barker[2]
[1]University of Surrey, [2]Michigan State University
(C) *Time-dependent Green's Functions Approach to Nuclear Reactions*

T. Ruokola[1], T. Ojanen[2], and A.-P. Jauho[1,3]
[1]Helsinki University of Technology, [2]Freie Universität Berlin, [3]Technical University of Denmark
(C) *Radiative heat transport in quantum circuits*

S. Sanvito
Trinity College, Dublin
(I) *Ab initio electronic transport in macro-molecules from first principles*

P. Schmitteckert
Intitute of Technology Karlsruhe
(I) *Non-equilibrium transport calculations within DMRG*

P. Schmitteckert
Karlsruhe Institute of Technology
(C) *Calculating Green's function from finite systems*

G. Stefanucci
University of Rome
(C) *Time-dependent quantum transport in nanoscopic Josephson junctions: A NEGF approach*

A.-M. Uimonen, P. Myohanen, A. Stan, and R. van Leeuwen
University of Jyväskylä
(C) *Kadanoff-Baym approach to density and current bistability in quantum transport*

K. van Houcke[1], and N. Prokofev[2]
[1]Universiteit Gent, [2] University of Massachusetts
(I) *Diagrammatic Monte Carlo: what happens to the sign-problem?*

B. Velicky, A. Kalvova, and V. Spicka
Charles University of Prague
(I) *Fast transients in mesoscopic systems*

J. Vorberger[1], D.O. Gericke[1], Th. Bornath[2], and M. Schlanges[3]
[1]University of Warwick, [2]Universität Rostock, [3]E.M. Arndt Universität Greifswald
(C) *Energy Transfer Rates and Relaxation in Dense Two-Temperature Plasmas*

List of Posters

A. Alvermann[1], D.M. Edwards[2], and H. Fehske[1]
[1]Ernst-Moritz-Arndt Universität, [2]Imperial College
Calculation of Drude weight for a correlated quantum transport process

S. Bauch, K. Balzer, and M. Bonitz
Christian-Albrechts-Universität Kiel
Quantum Breathing Mode for Fermions and Bosons at Arbitrary Coupling

V.S. Filinov[1], M. Bonitz[2], Y.B. Ivanov, V.V. Skokov, P.R. Levashov[1] and V.E. Fortov[1]
[1]Russian Academy of Sciences, [2]Christian-Albrechts-Universität Kiel, [3]Gesellschaft für

Schwerionenforschung, Darmstadt, [4]Russian Research Center "Kurchatov Institute", Moscow, [5]Joint Institute for Nuclear Research, Dubna
Equation of state of strongly coupled quark-gluon plasma—Path integral Monte Carlo results

M. Heimsoth, D. Hochstuhl, and M. Bonitz
Kiel University
Time-dependent unrestricted Hartree-Fock approximation for strongly correlated mesoscopic Bose systems

T. Koch[1], J. Loos[2], A. Alvermann[1], and H. Fehske[1]
[1]Ernst-Moritz-Arndt Universität, Greifswald, [2]Academy of Sciences of the Czech Republic, Prague
Transport through a vibrating quantum dot: Polaronic effects

P. Myohanen[1], A.-M. Uimonen[1], G. Stefanucci[2], and R. van Leeuwen[1]
[1]University of Jyväskylä, [2]University of Rome
Time-dependent quantum transport with the Kadanoff-Baym equations

A. Rios
University of Surrey
Self-Consistent Green's Functions in equilibrium: the ladder approximation and nuclear matter

S. Schmitt
Technische Universität Dortmund
Single impurity Anderson model out of equilibrium—a comparitive study

D. Semkat, F. Richter, D. Kremp, G. Manzke, W.-D. Kraeft, and K. Henneberger
Universität Rostock,
Ionization equilibrium in electron-hole plasmas: Mott effect vs. Bose-Einstein condensation

C. Verdozzi
Lund University
TDDFT and the Hubbard model

All papers published in this volume of *Journal of Physics: Conference Series* have been peer reviewed through processes administered by the proceedings Editors. Reviews were conducted by expert referees to the professional and scientific standards expected of a proceedings journal published by IOP Publishing.

Progress in Nonequilibrium Green's Functions IV IOP Publishing
Journal of Physics: Conference Series **220** (2010) 012001 doi:10.1088/1742-6596/220/1/012001

Structural Properties of Warm Dense Matter

D.O. Gericke, K. Wünsch, A. Grinenko and J. Vorberger
Centre for Fusion, Space and Astrophysics, Department of Physics, University of Warwick, Coventry CV4 7AL, UK

E-mail: D.Gericke@warwick.ac.uk

Abstract. We investigate the structure in warm dense matter by *ab initio* simulations and classical fluid equations. This comparison yields valuable information on the system properties such as effective ion-ion interactions and charge states. Extensions to systems with multiple ion species are discussed and the limits of reduced descriptions are demonstrated. The structural information is then used to predict the signal strength in x-ray scattering experiments which, in turn, can be used as diagnostics for warm dense fluids. Here, we need to consider the inelastic feature due to the dynamic response of free electrons as well as the elastic (Rayleigh) peak related to the static ion structure. In a last example, we apply *ab initio* simulations to calculate the ionic structure in highly excited graphite close to melting.

1. Introduction

Warm dense matter (WDM) is a state with densities comparable to solids and temperatures of a few electron volts. These conditions naturally occur, e.g., in the crusts of neutron stars, the main volume of brown dwarfs, in the interior of giant gas planets [1, 2, 3, 4] as well as the center of earth-like planets [5]. In the laboratory, WDM states are created, at least as a transient state, during the interaction of intense lasers or particle beams with solids [6]. Special interest is thus directed to the properties of WDM when designing targets for inertial confinement fusion. In indirectly driven targets, WDM states are found in the converters, the walls of the hohlraum, and the hydrogen core during the compression phase [7, 8]. Due to reduced shock-heating of the hydrogen, WDM is particularly prominent in the pre-compressed core of targets for fast ignition, a more energy efficient inertial confinement fusion scheme with higher gains [9, 10].

In the region of WDM, the correlation, the electron Fermi and the effective ionization energies are all of the same order of magnitude making it intrinsically a strongly correlated quantum system. The challenges for a theoretical description are therefore similar to those arising when considering such diverse systems as ultra-cold atoms in traps, electrons in high-temperature superconductors and nuclear matter [11]. In contrast to the other systems, WDM has two distinct species: the strongly coupled but classical ions and the partially degenerate electrons which can be affected by correlations as well. The strength of the ion-ion correlations and the electron degeneracy can be estimated by the parameters [12]

$$\Gamma_{ii} = \frac{Z_i^2 e^2}{a_i k_B T_i} \qquad \text{and} \qquad n_e \Lambda_e^3 = n_e \left(\frac{2\pi\hbar^2}{m_e k_B T_e} \right)^{3/2} , \tag{1}$$

where $a_i = (3/4\pi n_i)^{1/3}$ is the mean inter-ionic distance. The classical coupling parameter Γ gives the ratio of the mean potential to the kinetic energies. Of course, a similar parameter can be defined for the electrons, but one has to replace the classical thermal energy, that is $k_B T$, by the appropriate energy of the quantum system which becomes the Fermi energy for highly degenerate systems. Due to high electron densities and temperatures, the electrons are however mainly weakly to moderately coupled in WDM.

Whereas degeneracy in weakly coupled systems can be easily treated, strongly coupled systems represent a challenge for a theoretical description. The definitions (1) demonstrate that strong coupling may be created by either low temperatures or high enough particle densities. An extreme example for the first approach are ultra-cold plasmas [13, 14, 15] which are created by laser-ionization of ultra-cold gases in magneto-optical traps. As the laser energy considerably heats the electrons, one can describe the electron background as a nondegenerate and weakly coupled fluid [16]. The ions, however, show significant correlation effects: the laser ionization suddenly changes the potential energy landscape in the system and, as a result, the ions build a structure according to the new forces. This releases large amounts of potential energy which increases the ion temperature by orders of magnitude to states with $1 < \Gamma < 5$ [17, 18, 19]. Further laser cooling of the ions can even increase the coupling enough to create Coulomb crystals [20] similar to those known in dusty plasmas [21, 22].

Although the effects of strong coupling can be studied in detail in ultra-cold and dusty plasmas, the ions in such systems can be mainly described as classical one component plasmas (OCP) and lack the combination of partially degenerate electrons and strongly coupled ions. This mutual influence of quantum behavior and strong interactions is only found in the region of WDM with its high densities and moderate temperatures. Again, the degenerate electron fluid alone may be sufficiently described by the random phase approximation (RPA) or its extensions considering weak binary collisions [12]. Moreover, effective approaches based on integral equations [23, 24] exist to describe strongly coupled plasmas in the classical domain. For Coulomb systems with coupling strength up to $\Gamma = 100$, these approximate models have been shown to yield results similar to direct simulations [25]. However, integral equations as well as molecular dynamics and Monte Carlo simulations treat systems with known interactions. In WDM, the effective ion-ion interactions and the electronic response have to be self-consistently determined as both strongly influence each other. This fact becomes particularly clear when considering the charge state of the ions: due to the strong interactions with free electrons and other ions, thermally stable bound states can vanish and move into the continuum (as the conduction electrons in metals); moreover, a precise distinction between bound and free states might not be possible as the binding and correlation energies are often of the same order and, thus, the bound states become highly resonant.

The demand for highly accurate theoretical description of WDM can nowadays not be met by an approach based on Green's functions. We therefore apply density functional molecular dynamics (DFT-MD) simulations as an alternative method. DFT-MD fully meets the demands for a theoretical description of WDM as the electrons are treated with density functional theory (full quantum mechanics) and the ion positions are moved under the influence of forces that include the direct ion-ion as well as the electron-ion contributions. The latter are, of course, determined self-consistently from the DFT electron configuration. We then use comparisons with more approximate classical hypernetted chain (HNC) calculations to obtain valuable information on ion-ion interactions and the charge states. Both methods are applied to lithium, beryllium and CH as these materials were recently investigated experimentally in the WDM region [26, 27, 28, 29, 30]. We find that most materials can be described well by a linearly screened Coulomb potential which can be improved by an additional repulsive potential. For beryllium, we find a stable charge state of $Z = 2$ which is consistent with measurements and can also rule out suggested quantum-potentials. Extensions to mixtures show that an approximative treatment based on one component calculations [31] works for weakly coupled ions, but fails for strong coupling. Finally, we apply the DFT simulation technique to highly excited carbon in the solid phase.

2. Determining the Ion Structure in WDM

We will use a combination of *ab initio* simulations and integral equations derived in fluid theory (HNC approach) to obtain information on the ionic structure, the effective ion-ion interaction and the electron configuration in WDM. While the full quantum simulations serve as a benchmark, the HNC calculations are matched with the results to determine the effective interactions. Such reduced models are urgently needed for practical applications as the DFT-MD simulations are computationally very expensive. For instance, for plasma diagnostics by means of x-ray scattering [32, 33, 34], one needs the ion structure for many parameters as the density and temperature are obtained by matching the experimental spectrum by changing the input density and temperature in the theoretical calculation. Such a diagnostics method would not be applicable if DFT-MD simulations with run times of days/weeks would be used for every possible set of system parameters.

2.1. DFT-MD simulations

We apply Born-Oppenheimer DFT-MD simulations, that is DFT for the electrons and classical MD for the ions where the electronic configuration is updated at every step. In more detail: the Hohenberg-Kohn theorem [35, 36] proves that the electron ground state density contains sufficient information to fully determine the electron subsystem and the electron wave functions are not needed. This freedom is used by the Kohn-Sham ansatz that connects the fully interacting electron system with its noninteracting counterpart by requesting that the electron density is kept [37]. Of course, this requires that the noninteracting electrons are considered in an additional external potential which is still to be determined. The electron density can then be derived from the generalized Schrödinger equation for single electrons

$$\left[-\frac{1}{2}\nabla^2 + V^\sigma_{\mathrm{KS}}(r)\right]\ \psi^\sigma_i(r) = \varepsilon^\sigma_i\ \psi^\sigma_i(r)\,. \tag{2}$$

The Kohn-Sham potential V_{KS} can be decomposed as

$$V^\sigma_{KS}(r) = V_{\mathrm{ext}}(r) + V_{\mathrm{H}}[n] + V^\sigma_{\mathrm{xc}}[n^\uparrow, n^\downarrow]\,, \tag{3}$$

where the first term is the real external potential created by the ions, the second term is the Hartree field and the last contribution is the exchange-correlation potential. As this last term contains the full many-body physics, it is not fully known and, thus, is the only point where the DFT calculations become approximate. However, one can derive powerful approximations for the exchange-correlation potential V^σ_{xc}. The easiest is the local density approximation [36] which is further improved by the generalized gradient approximation. The latter is used for the presented results by applying the Perdew-Burke-Ernzerhof type [38] for the exchange-correlation potential.

The original DFT is a pure ground state calculation and thus not applicable in WDM as thermal and Fermi energies are comparable. Thus, the Fermi distribution deviates from a step function and a $T=0$ treatment is not applicable. Temperature effects in the electron component can be included by using the Mermin functional [39] that populates the energy eigenstates by the real temperature-dependent Fermi function.

For our DFT-MD simulations we use the VASP package [40, 41, 42] that implements density functional theory and combines it with a MD solver for the ions: for a given ion configuration, the Kohn-Sham equations are solved and the energy levels are then populated according to the electron Fermi distribution. From the given electron configuration, electron-ion forces are derived and the ions are moved in a classical MD step. Then the loop is started again and a

Journal of Physics: Conference Series **220** (2010) 012001 doi:10.1088/1742-6596/220/1/012001

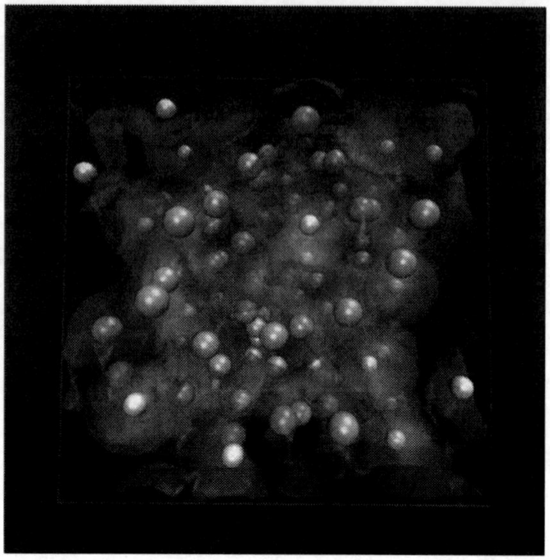

Figure 1. Results from density functional molecular dynamics (DFT-MD) simulations of hydrogen with a density of $\varrho = 3.7\,\mathrm{gcm}^{-3}$ and a temperature of $T = 4000\,\mathrm{K}$. Shown are the hydrogen nuclei (cyan spheres) and an iso-surface of the electron density (red). Here, hydrogen is pressure ionized and, therefore, is in an atomic-metallic phase.

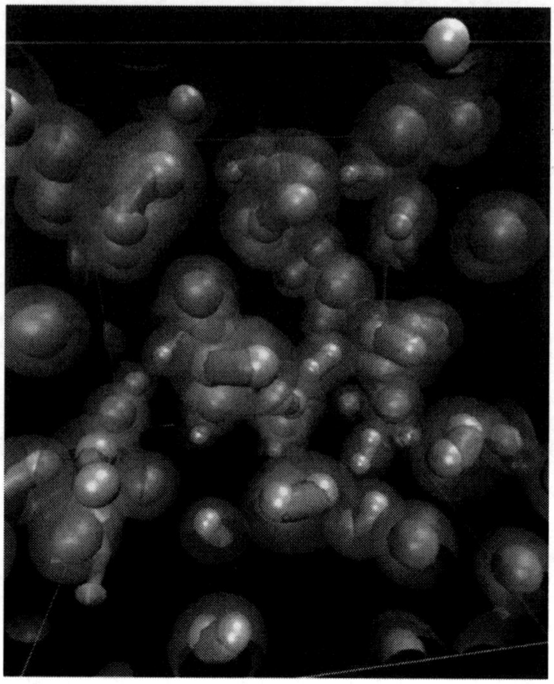

Figure 2. Same as left figure, but for a lower density of $\varrho = 0.2\,\mathrm{gcm}^{-3}$ and with an additional iso-surface shown in grey. Clearly, hydrogen is here in the molecular phase.

new electron density is calculated for the new ion configuration. It is important to notice that the electrons are considered to instantaneously react on the new ion field and that the DFT calculations are done at every MD step (Born-Oppenheimer DFT-MD).

Initially, the ions are not in equilibrium and one has to allow for an equilibration phase before extracting the physical properties. During this time, the ions build up correlation energy which, depending on their initial configuration, can considerably change their temperature. The ion temperature is then controlled via a Nose-Hoover thermostat [43]. Combined with the Mermin functional, we can thus consider both electron and ion distributions at a given finite temperature.

We have performed the simulations for 128 or 256 nuclei while treating all electrons as valence. For the fluids considered, we can use the Gamma point only to sample the Brillouin zone. The VASP package uses a plane wave expansion to solve the Kohn-Sham equations (2). To include all bound states and fully sample the Fermi distribution, we use a rather high cutoff energy of 800 eV. This requires roughly five times the number of bands as for ground state calculations which makes the calculations computationally even more expensive and also limits the temperature that can be considered. The ions are moved with a time step of 0.4 fs to fully resolve the motion under strong fields from other ions. When extracting the structural information, we average over more than 1000 different configurations (ion steps) to counter the bad statistics due to the small systems size.

Figures 1 and 2 show snap shots from DFT-MD simulations of hydrogen at two densities. Although the information that can be extracted from such pictures is rather limited, one can already observe that the two systems are in quite different states: while the electron density in the first figure fills the whole space, the electrons are localized around the ions for the second case and the ions are paired here. We can therefore conclude that Fig. 1 shows hydrogen in the metallic phase whereas hydrogen is a molecular fluid for the conditions of Fig. 2.

2.2. *HNC calculations*

To extract more information from the DFT-MD simulations, we compare the results with data from hypernetted chain (HNC) calculations [24]. Important quantities are the ionic pair distribution $g_{ii}(r)$ and the ion-ion structure factor $S_{ii}(k)$. As the HNC scheme uses a given ion-ion pair potential, the comparison with DFT-MD simulations assesses the quality of the potential used.

HNC calculations are based on the Ornstein-Zernike relation that defines the direct correlation functions $c(r)$ as a functional of the total correlation function $h(r)=g(r)-1$. Considering multi-component systems, we have in Fourier space [23]

$$h_{ab}(k) = c_{ab}(k) + \sum_c n_c \, c_{ac}(k) \, h_{cb}(k) \,. \tag{4}$$

As this equation is a definition, it can be considered to be exact. However, we need a closure relation to fully determine $h(r)$ and $c(r)$ which can be obtained by a cluster expansion [44, 45]

$$g_{ab}(r) = \exp[\, -\beta V_{ab}(r) + h_{ab}(r) - c_{ab}(r) + B_{ab}(r)] \,. \tag{5}$$

Here, $B(r)$ is the bridge function that is not exactly known. Therefore, different approximations to the closure relation (5) have been tested against direct MD or MC simulations (see, e.g., Ref. [25]). It turns out that the HNC approximation which neglects the bridge function, that is $B(r)=0$, yields quite good results for Coulomb-like systems. Thus, we adopt this approach in the following. An even faster approach is given by the mean spherical approximation (MSA) for charged hard spheres which can be solved analytically [46, 47]. Interestingly, this yields also rather good results for Coulomb systems up to high coupling strength [48].

Figure 3 shows an example for the pair distribution $g(r)$ in a mixture of two ion species where all particles interact via pure Coulomb forces (no screening). Typical for the structure in strongly coupled systems are the correlation hole at small separations and the following maxima and minima. As the aluminium ions are more highly charged, they show a more pronounced

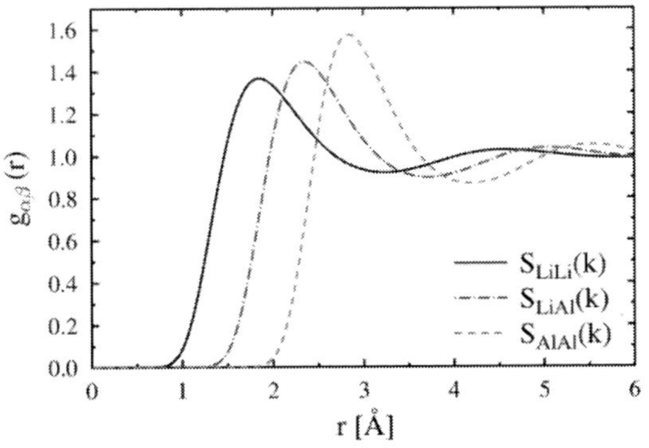

Figure 3. Partial pair distributions for a mixture of aluminum and lithium ions with $n_{\rm Al}=n_{\rm Li}=7.6\times10^{22}\,{\rm cm}^{-3}$ and $T=1\,{\rm eV}$. The ion charge states are $Z_{\rm Al}=3$ and $Z_{\rm Li}=1$.

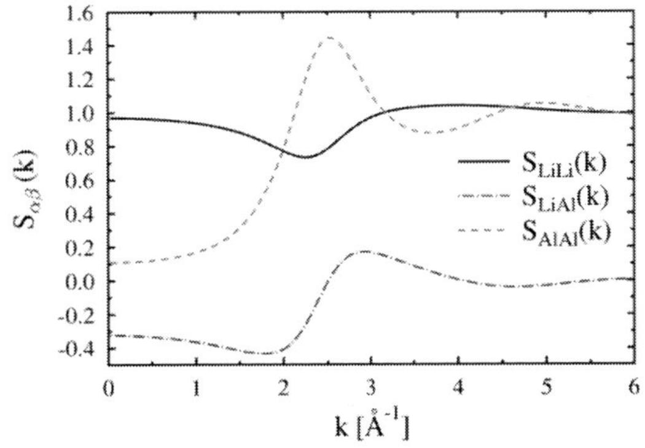

Figure 4. Partial structure factors for the same system as in Fig. 3. Note that the cross term $S_{\rm AlLi}$ lacks unity in the definition and, thus, has a different behaviour at large k.

structure. This component is also mainly responsible for the Al-Li and the Li-Li structure as the aluminium ions imprint their configuration on the lithium ions [49].

Figure 4 shows the partial ion-ion structure factors defined as

$$S_{ab}(\mathbf{k}) = \delta_{ab} + \sqrt{n_a n_b} \int d\mathbf{r} \, [g_{ab}(r) - 1] \exp(i\mathbf{k}\,\mathbf{r}) \tag{6}$$

for the same case. Again, the Al-Al term shows the most structure. In contrast to the pair distribution, the Li-Li structure factor appears almost without structure. Here, we observe a competing effect of correlations and screening: although pure Coulomb interactions are applied the (nonlinear!) screening of the Coulomb forces by the other ion species is self-consistently included in the HNC scheme. This becomes particularly pronounced at small k. Indeed, the structure factor at $k=0$ directly shows the strength of screening as an unscreened one component plasmas has $S_{ii}(0) = 0$. Thus, the value of $S_{\mathrm{LiLi}}(k)$ at small k indicates strong screening not weak direct forces. The quite different behaviour of $S_{\mathrm{AlLi}}(k)$ results mainly from the fact that its definition does not include unity as the Kronecker symbol in Eq. (6) vanishes for $a \neq b$.

3. Quantum Potentials, Effective Ion-Ion Interactions and Ion Charge States

First we evaluate the ability of the multi-component HNC approach to describe electron-ion systems. Then we use the comparison of the ionic structure obtained by DFT-MD and HNC to extract information on the effective ion-ion interaction and the electronic configuration.

3.1. Testing Electron-Ion Quantum Potentials

The HNC equations (4) and (5) are formulated for multi-component systems. They could also be used to model electron-ion systems and thus go beyond linear screening. However, the HNC scheme is a classical approach and neglects the quantum nature of the electrons. Technically, this results in too high electron densities near the ion core and too high electron-electron distributions for degenerate systems.

These issues might be resolved by the use of quantum pseudo-potentials for the electron-ion and electron-electron interactions that mimic the quantum behaviour in classical calculations. Different forms of such potentials have been suggested or derived [50, 51]. The application of such potentials is usually limited to weakly coupled and weakly degenerate plasmas making their application in calculations for WDM highly questionable. However, recent experimental results from x-ray Thomson scattering on warm dense beryllium [28] suggested that the ultra-weak Klimontovich-Kraeft (KK) potential (see Ref. [49] for its definition) is applicable here. Here, the results from HNC calculations using the KK potential should be compared to DFT-MD data to justify or falsify such a statement.

Figure 5 shows a comparison of the DFT-MD data with HNC results using the KK potential and different charge states of the ions. Clearly, the HNC results cannot match the DFT-MD data even for the wide range of ion charge states considered. The ions always seem to be too strongly correlated which results from the fact that the KK potential is very weak and thus screening of the ion-ion potential is underestimated [49]. Although it is less pronounced, the comparison for the ionic structure factors in Fig. 6 indicates the same. Similar statements can be made for other quantum pseudo-potentials as well [52]. We thus can conclude that indeed a classical treatment using the KK potential cannot be applied for the WDM condition considered here and the low intensities of the elastic scattering feature reported in Ref. [28] must have a different reason than the ion structure.

3.2. Extracting the Ion Charge State from DFT-MD Simulations

In contrast to HNC scheme using the KK potential, the DFT-MD data for warm dense beryllium and other materials can be matched by HNC calculations that treat only the ions explicitly and consider the effect of the electrons by using a linearly screened Coulomb potential for the ion-ion interaction (Yukawa model) [52]. Even better agreement is obtained if a short range repulsion, that mimics the interactions of the fully occupied shell of core electrons, is added.

We can now adopt this model and match the DFT-MD results by using different charge states for the ions. Fig. 7 shows this comparison for the case of solid-density beryllium tested experimentally [27, 28]. We find much better agreement than in Fig. 5 for all charge states considered. As expected, the best match is here found for $Z=2$ that is the stable helium-like configuration. Thus, we have here a case with a well-defined charge state as the ionization energy of the 1s electrons is too large to be overcome by either temperature or correlations and the 2s shell is fully ionized as in cold beryllium. Of course, one can also compare the electronic configuration around an ion directly with the bound state density for isolated atoms. For the case presented here, this gives also a charge state of $Z=2$ [53].

3.3. Ion Structure in Moderately to Strongly Coupled Multi-Component Plasmas

Both DFT-MD and HNC are methods that can be naturally applied to systems with multiple ion species. The only problem that arises is purely technical as the computational demands strongly increase with the number of species considered; in particular, for the DFT-MD simulations. Nevertheless, partial ion-ion structure factors for multi-component plasmas are often calculated from one component results using the relation [54]

$$S_{ab}(k) = \delta_{ab} + \frac{\sqrt{n_a n_b}}{n_a + n_b} \frac{Z_a Z_b}{Z_f^2} \left[S_{ii}^{\mathrm{1comp}}(k) - 1 \right]. \tag{7}$$

Here, n_a and Z_a are the density and the charge state of the ion species a, respectively, $Z_f = \sum_a n_a Z_a / \sum_b n_b$ is the average ion charge and $S_{ii}^{\mathrm{1comp}}(k)$ denotes the ion-ion structure factor for a one component system of particles with an average charge Z_f. It can be shown that the relation (7) is exact within RPA [55]. Deviations can thus only arise for moderately or strongly coupled plasmas.

The range of applicability of formula (7) is investigated in Fig. 8 by comparison with full multi-component calculations. Both calculations were done using the HNC scheme for ions that interact via statically screened Coulomb forces (screening is determined by the electron

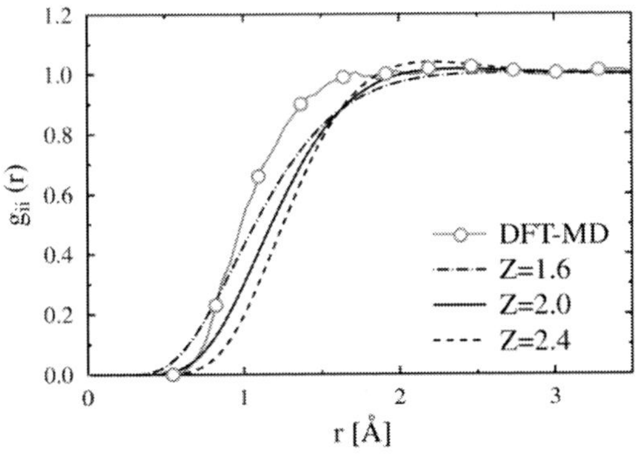

Figure 5. Ion-ion pair distribution in solid-density beryllium with $T=12\,\mathrm{eV}$. Shown are results from DFT-MD and multi-component HNC calculations using the KK-potential for the electron-ion interaction.

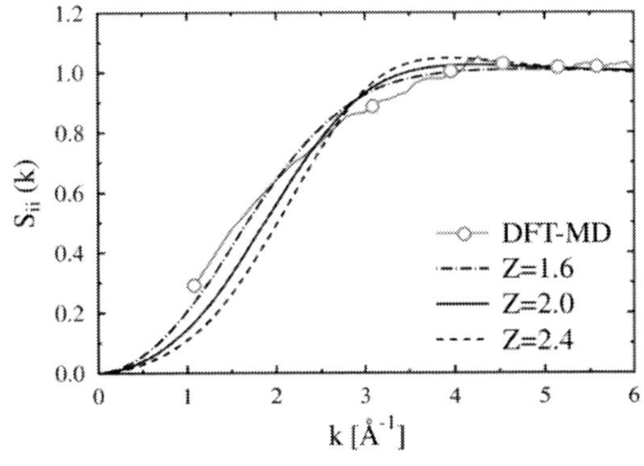

Figure 6. Comparison of the ionic structure factor from DFT-MD and multi-component HNC for beryllium as in Fig. 5. Note that the small k-behavior reflects differences for large distances in real space.

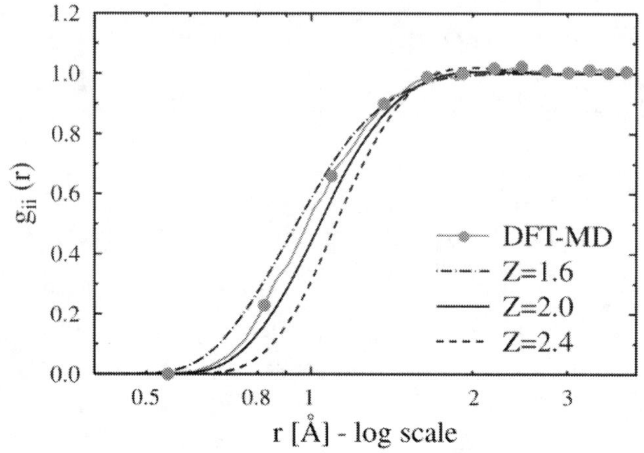

Figure 7. Comparison of HNC and DFT-MD results of the ion-ion structure for beryllium with the same parameters as in Fig. 5. The HNC calculations consider the ions only and use a screened Coulomb potential plus a strong short range repulsion as an effective interaction.

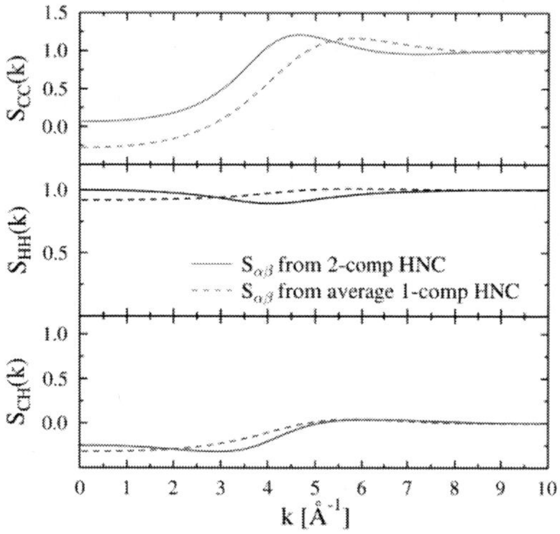

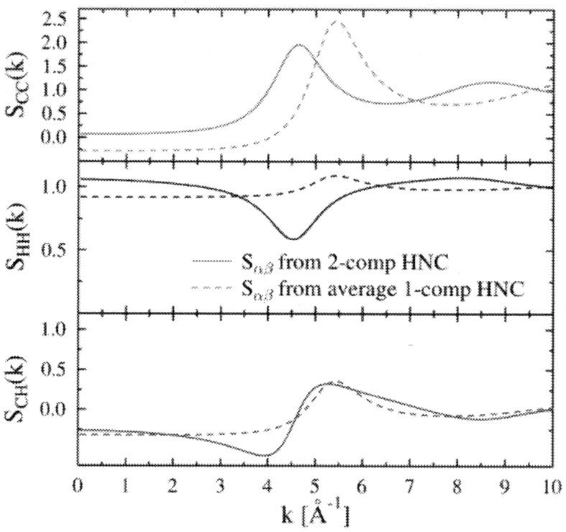

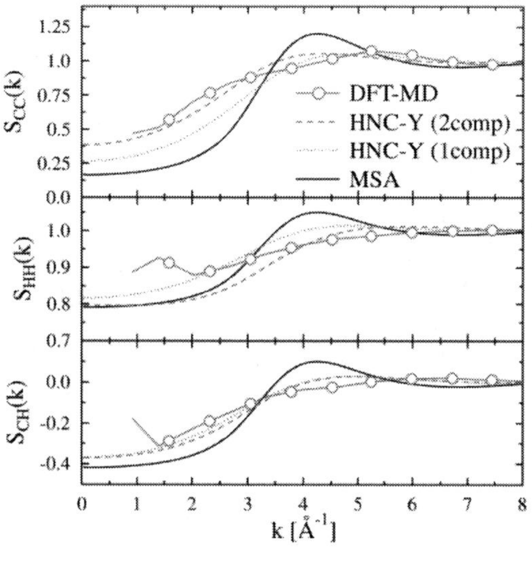

Figure 8. Partial structure factors of a CH plasma with different coupling strength which were calculated with the full two-component HNC approach or from HNC structure factors for one component plasmas via formula (7). The upper panels show data for a mass density of $\rho = 4.96\,\mathrm{gcm}^{-3}$ and a temperature of $T = 10\,\mathrm{eV}$ (left panel) and $T = 1\,\mathrm{eV}$ (right). The charge states of the ions are assumed to be $Z_\mathrm{H} = 1$ and $Z_\mathrm{C} = 3$ which results in coupling parameters of $\Gamma = 1.9$ and $\Gamma = 19$, respectively. The lower left panel shows results for the conditions measured in shock-driven plastics as presented in Ref. [30].

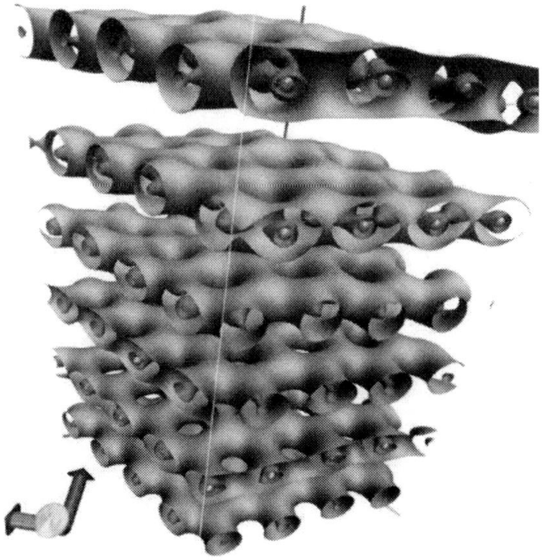

Figure 9. Snap shot from a DFT-MD simulation of carbon with normal density and $T = 0.2\,\mathrm{eV}$. The brown and red areas label iso-surfaces of the electron density related to the L and K shell electrons, respectively. The cyan spheres are the carbon nuclei.

properties as usual). Clearly, we have increasing deviations of the two approaches if the coupling strength is increased. This makes the application of the approximate formula in the WDM region highly questionable. The lower panel also displays a line labelled 'MSA' which uses the mean spherical approximation to calculate the one component structure factor. The differences to the HNC(1comp) data reflect the deviating predictions of MSA and HNC [48].

3.4. Ion Structure of Highly Excited Solids

The warm dense matter region not only covers strongly coupled fluids but its definition often also applies for solids close to melting. DFT-MD simulations can be used in this parameter region without problems (indeed, DFT-MD was invented for condensed matter) whereas the HNC scheme is restricted to fluids as it uses radial symmetric functions.

Figure 9 shows a snap shot for warm graphite at normal density. One can clearly observe that the graphite layers are still intact and that the electron configuration is also very close to cold graphite, i.e., we have a small ionization degree. More precise information can be obtained

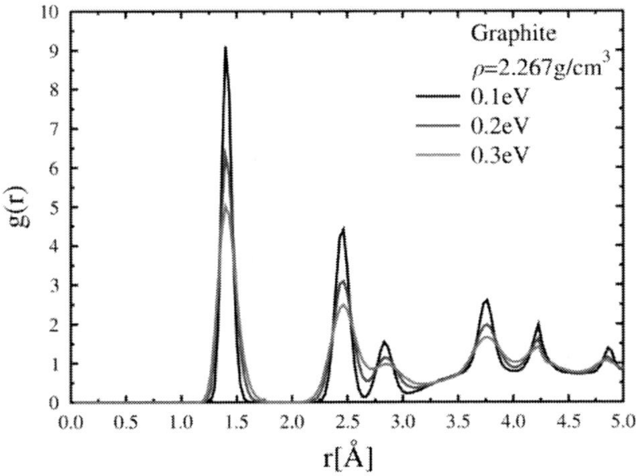

Figure 10. Ionic pair distribution for graphite at normal density and for different temperatures. These data were extracted from DFT-MD simulations. The lines correspond to different temperatures where the ion-ion pair distribution $g(r)$ is, of course, more strongly peaked at lower temperatures.

from condensed quantities such as the pair distribution shown in Fig. 10. All lines show the known order in graphite where the strong first peak is associated with the nearest neighbors in the plane and the next to ions in the planes above/below. At these graphs one can study the temperature effects on the ion structure close to melting as the peaks reduce in height and broaden with increasing temperature. This signature can be used as temperature diagnostics and also indicates melting as the well-defined spacing between peaks becomes more evenly spaced in the fluid.

4. Application to X-Ray Scattering as Plasma Diagnostics

The diagnostics of WDM states by means of x-ray scattering is one of the main applications for the structure calculations in the previous section. As for weakly coupled plasmas, the intensity of the radiation scattered under the angle θ is given by

$$P_s(\theta,\omega)\,d\Omega d\omega = \frac{P_0 r_0^2}{4\pi A} N S_{ee}^{tot}(k,\omega)\,\left[1+\cos^2\theta\right]\,d\Omega d\omega\,, \tag{8}$$

where $k=(4\pi E_0/hc)\sin(\theta/2)$ and $\omega=(E_o-E_s)/\hbar$ are the photon's changes in wave number and frequency, respectively. P_0 is the incident flux, r_0 is the classical electron radius, N the number of electrons in the scattering volume and A the area through which the flux goes. The last term is a geometrical factor that accounts for the fact that the incoming photons are unpolarized as in many laser-driven experiments. The many-body properties, that determine the scattering intensity, are contained in the total electron structure factor S_{ee}^{tot}.

Although only the electrons contribute to light scattering, the electron structure factor S_{ee}^{tot} must be calculated for a fully correlated electron-ion systems as the ion configuration strongly influences the electron properties. Moreover, both bound and free electrons must be included as the x-ray energy can drive internal excitations as well as bound-free transitions. Bound electrons also dominate elastic scattering in many WDM experiments [56, 27, 57, 30]. According to Chihara [58, 59] the total electron structure factor for the partially ionized materials under consideration can be decomposed into

$$S_{ee}^{tot}(k,\omega) = \left[f(k)+q(k)\right]^2 S_{ii}(k,\omega) + Z_f S_{ee}^0(k,\omega) + S_{bf}(k,\omega)\,. \tag{9}$$

Here, the first term considers the contribution of electrons moving with the ions, that is, $f(k)$ is the atomic or ionic form factor and $q(k)$ is the density of the screening cloud in Fourier space. For light elements, $f(k)$ can be described by hydrogen-like wave functions using the nuclear charge as a fitting parameter [33, 60]. The standard approach for $q(k)$ is to use it in linear response [31] which is supported by the fact that the ion structure is also described well in this approximation (see section 3). The dynamic ion-ion structure factor can often be used in its static form $S_{ii}(k,\omega)=S_{ii}(k)\delta(\omega)$ as laser-based x-ray sources are spectrally too broad to resolve the ion dynamics. Due to spectral resolution of the spectrometers used, this elastic contribution appears as a wide peak around $\omega=0$ (Rayleigh feature). However, future experiments applying x-ray lasers may resolve the dynamics of the ion response yielding enhanced information on strong coupling effects and the mutual interactions of electrons and ions [61].

The second term in Eq. (9) described the dynamic response of the free electrons, that is, it is just the dynamic structure factor of a free electron gas [12]. This can be mainly described within RPA; more advanced approximations include the effect of electron-ion collisions, e.g., via a Mermin ansatz [62]. However, the determination of the collision frequency in WDM is an issue that is not fully resolved yet (see, e.g., Refs. [63, 64, 65, 66] for recent developments). The last remaining term determining the total electron structure factor describes contributions of internal excitations of the target electrons. For light elements from hydrogen to beryllium, these are of minor importance.

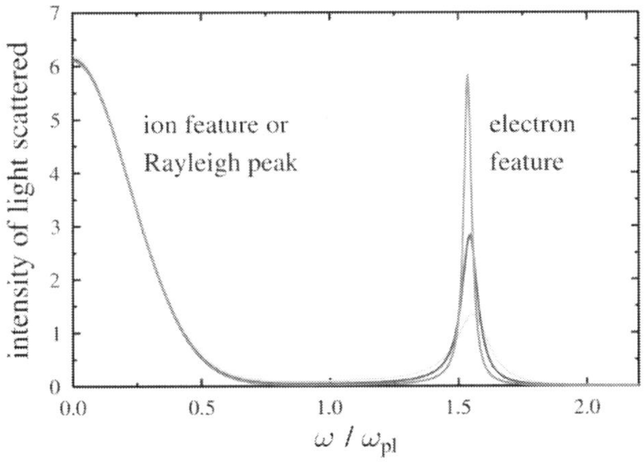

Figure 11. Synthetic scattering spectrum for 2.96 keV x-rays that experience a momentum transfer of $k = 10^{10}\,\mathrm{m}^{-1}$ when scattered on beryllium at solid density and $T = 12\,\mathrm{eV}$ [28]. The Rayleigh peak has been artificially broadened by a Gaussian to mimic the limited experimental detector response. The electron feature is calculated for three different values of the collision frequency (see text).

Figure 11 shows an artificial scattering spectrum for warm dense beryllium. The two distinct features are related to the first and second terms of Eq. (9). Such model spectra are then fit to the experimentally measured data by changing temperature, density ionization degree. The best fit is then considered to yield the conditions investigated. Experimental spectra are also blurred due to different k values and the detector response. We considered this effects by using a Gaussian for the ion feature that has the same weight as the first term in Eq. (9).

To obtain the data shown, we kept the first term constant and calculated the electron response with different collision frequencies as this quantity is not well understood. We used collision frequencies in Born approximation as defined in Ref. [63] and also frequencies that were artificially reduced (narrowest peak) or enhanced (broadest peak) by a factor of two to demonstrate the effect of the uncertainty in calculated collision frequencies. RPA data are not shown as these will give a high, very narrow peak. Clearly, the broadening of the plasmon peak plays an important role when interpreting the data. Thus, x-ray scattering might be also used to investigate the damping mechanism in WDM.

As both dynamic contributions are either reasonably well described or negligible, the structural information contained in the first term of Eq. (9) constitutes the major theoretical challenge for the description of x-ray scattering in WDM. The previous sections have given a state of the art description for this quantity based on classical integral equation and quantum simulations as DFT-MD from simple metals such as beryllium to multi-component fluids as plastics and highly excited solids.

Acknowledgements

We thankfully acknowledge fruitful discussions with Dr. Gregori (Oxford), Dr. Glenzer (Livermore) and Prof. Kraeft (Rostock) as well as financial support from EPSRC.

References

[1] Guillot T 1999 *Science* **286**, 72
[2] Guillot T *et al.* 2004 in *Jupiter*, Eds: Bagenal F *et al.* (Cambridge University Press)
[3] Remington B A , Drake P R and Ryutov D D 2006 *Rev. Mod. Phys.* **78** 755
[4] Van Horn H M 1991 *Science* **252** 384
[5] Vocadlo L, Alfe D, Gillan M J, Wood I G, Brodholt J P and Price G D 2003 *Nature* **424**, 536
[6] Koenig M *et al.* 2005 *Plasma Phys. Contr. Fusion* **47**, B441
[7] Lindl J D 1998 *Inertial Confinement Fusion* (Springer Verlag, New York)
[8] Atzeni S and Meyer-ter-Vehn J 2004 *The Physics of Inertial Fusion* (Clarendon Press, Oxford, 2004).
[9] Tabak M *et al.* 2005 *Phys. Plasma* **12**, 057305

[10] Key M H 2007 *Phys. Plasmas* **14**, 055502
[11] National Research Council 2003 *Frontiers in High Energy Density Physics: the X-Games of Contemporary Science* (National Academies Press, Washington, D.C.)
[12] Kremp D, Schlanges M, and Kraeft W-D 2005 *Quantum Statistics of Nonideal Plasmas* (Springer-Verlag, Berlin, Heidelberg)
[13] Killian T C, Kulin S, Bergeson S D, Orozco L A, Orzel C and Rolston S L 1999 *Phys. Rev. Lett.* **83** 4776
[14] Killian T C 2007 *Science* **316** 705
[15] Killian T C, Pattard T, Pohl T and Rost J M 2007 *Phys. Reports* **449** 77
[16] Gupta P, Laha S, Simien C E, Gao H, Castro J, Killian T C and Pohl T 2007 *Phys. Rev. Lett.* **99** 75005
[17] Gericke D O and Murillo M S 2003 *Contrib. Plasma Phys.* **43** 298
[18] Gericke D O, Murillo M S, Semkat D, Bonitz M and Kremp D 2003 *J. Phys. A* **36** 6087
[19] Simien C E *et al.* 2004 *Phys. Rev. Lett.* **92** 143001
[20] Pohl T, Pattard T and Rost J M 2004 *Phys. Rev. Lett.* **92** 155003
[21] Morfill G E, Thomas M H, Konopka U and Zuzic M 1999 *Phys. Plasmas* **6** 1769
[22] Bonitz M *et al.* 2006 *Phys. Rev. Lett.* **96** 075001
[23] Ornstein L and Zernike F 1914 *Proc. Akad. Sci. (Amsterdam)* **17** 793
[24] Hansen J P and McDonald I R 1990 *Theory of Simple Liquids* (London: Academic Press)
[25] Baus M and Hansen J P 1980 *Phys. Reports* **59** 1
[26] Garcia Saiz E *et al.* 2008 *Nature Physics* **4** 940
[27] Glenzer S H *et al.* 2003 *Phys. Rev. Lett.* **90** 175002
[28] Glenzer S H *et al.* 2007 Phys. Rev. Lett. **98** 065002
[29] Kritcher A L *et al.* 2008 *Science* **322** 69
[30] Barbrel B *et al.* 2009 *Phys. Rev. Lett.* **102** 165004
[31] Gregori G, Ravasio A, Höll A, Glenzer S H and Rose S J 2007 *High Energy Density Physics* **3**, 99
[32] Landen O L *et al.* 2001 *J. Quant. Spectrosc. Radiat. Transf.* **71**, 465
[33] Gregori G, Glenzer S H, Rozmus W, Lee R W and Landen O L 2003 *Phys. Rev. E* **67** 026412
[34] Glenzer S H and Redmer R 2009 *Rev. Mod. Phys.* **81**, 1625
[35] Hohenberg H and Kohn W 1964 *Phys. Rev.* **136** B864
[36] Martin R M 2004 *Electronic Structure, Basic Theory and Practical Methods* (University Press, Cambridge)
[37] Kohn W and Sham L J 1965 *Phys. Rev.* **140** A1133
[38] Perdew J P, Burke K and Ernzerhof M 1996 *Phys. Rev. Lett.* **77**, 3865
[39] Mermin N D 1965 *Phys. Rev.* **137** A1441
[40] Kresse G and Hafner J 1993 *Phys. Rev. B* **47** 558
[41] Kresse G and Hafner J 1994 *Phys. Rev. B* **49** 14251
[42] Kresse G and Furthmüller J 1996 *Phys. Rev. B* **54** 11169
[43] Hoover W G 1985 *Phys. Rev. A* **31**, 1695
[44] Van Leuwen J M J, Groenveld J and J. DeBoer J 1959 *Physica* **25** 792
[45] DeBoer J, Van Leuwen J M J and Groeneveld J 1964 *Physica* **30** 2265
[46] Palmer R G and Weeks J D 1973 *J. Chem. Phys.* **58** 4171
[47] Singh H B and Holz A 1983 *Phys. Rev. A* **28** 1108
[48] Wünsch K, Vorberger J, Gregori G and Gericke D O 2009 *J. Phys. A* **42** 214053
[49] Wünsch K, Hilse P, Schlanges M and Gericke D O 2008 *Phys. Rev. E* **77** 056404
[50] Y. Klimontovich and W.-D. Kraeft, High. Temp. Phys. (USSR) **12**, 212 (1974).
[51] C. Deutsch, Phys. Lett. **60A**, 317 (1977).
[52] Wünsch K, Vorberger J and Gericke D O 2009 *Phys. Rev. E* **79**, 010201
[53] Vorberger J *et al.* (submitted to Phys. Rev. Lett.)
[54] Gregori G *et al.* 2006 *J. Quant. Spectrosc. Radiat. Transf.* **99** 225
[55] Totsuji H 1981 *Phys. Rev. A* **24** 1077
[56] Riley D, Woolsey N C, McSherry D, Weaver I, Djaoui A and Nardi E 2000 *Phys. Rev. Lett.* **84** 1704
[57] Ravasio A *et al.* 2007 *Phys. Rev. Lett.* **99** 135006
[58] Chihara J 1987 J. Phys. F: Met. Phys. **17** 295
[59] Chihara J 2000 J. Phys.: Condens. Matter **12** 231
[60] Wünsch K, Vorberger J, Schlanges M and Gericke D O 2008 J. Phys. (Conf. Series) **112** 032077
[61] Gregori G and Gericke D O 2009 Phys. Plasmas **16** 056306
[62] Mermin N D 1970 *Phys. Rev. B* **1** 2362
[63] Reinholz H, Redmer R, Röpke G and Wierling A 2000 *Phys. Rev. E* **62** 5648
[64] Bornath Th, Schlanges M, Hilse P and Kremp D 2001 *Phys. Rev. E* **64** 026414
[65] Grinenko A and Gericke D O 2009 J. Phys. A **42**, 214039
[66] Grinenko A and Gericke D O 2009 Phys. Rev. Lett. **103** 065005

Progress in Nonequilibrium Green's Functions IV IOP Publishing
Journal of Physics: Conference Series **220** (2010) 012002 doi:10.1088/1742-6596/220/1/012002

Energy and Temperature Relaxation described by Nonequilibrium Green's Functions

J. Vorberger[1], D.O. Gericke[1], Th. Bornath[2], M. Schlanges[3]

[1] Centre for Fusion, Space & Astrophysics, University of Warwick, Coventry, UK
[2] Institut für Physik, Universität Rostock, Rostock, Germany
[3] Institut für Physik, E.-M.-Arndt Universität Greifswald, Greifswald, Germany

E-mail: j.vorberger@warwick.ac.uk

Abstract. A quantum kinetic approach is presented to investigate the energy relaxation of dense strongly coupled two-temperature plasmas. We derive a balance equation for the mean total energy of a plasma species including a quite general expression for the transfer rate. An approximation scheme is used leading to an expression of the transfer rates for systems with coupled modes relevant for the warm dense matter regime. The theory is then applied to dense beryllium plasmas under conditions such as realized in recent experiments. Special attention is paid to the influence of correlation and quantum effects on the relaxation process.

1. Introduction

The recent developments of new methods for the creation and probing of warm dense matter and strongly coupled plasmas have brought an impressive progress in the experimental characterization of high energy density matter. For instance, x-ray scattering allows not only the measurement of equation of state (EOS) data, but also to obtain structural, dynamic, and collective properties of matter [1, 2, 3, 4, 5]. For the creation of such high energy density states in the laboratory, one relies on dynamic experiments using intense particle or laser beams to compress and heat the sample. In any case, this leads to the production of highly nonequilibrium states first. Although in most experiments not of primary interest, such nonequilibrium states deserve attention for several reasons. The first one is a practical one for EOS measurements and the like, as the time delay between pump and probe pulses needs to be greater than the relaxation time. The second reason is given by the interest to study the processes that lead from a highly nonequilibrium state towards equilibrium. During the first and shortest stage of a few femtoseconds, equilibrium distributions are established within the electron and ion subsystems. Ionisation and recombination is finished shortly thereafter. On a picosecond timescale, the process of temperature relaxation equals electron and ion temperatures. Apart from the following hydrodynamic motion of the substance, temperature relaxation takes the longest time of all relaxation processes and is also experimentally accessible.

An intense discussion about its order of magnitude started when Dharma-wardana & Perrot [6] found considerably smaller energy transfer rates as predicted by classical Landau-Spitzer (LS) theory [7, 8] and early computer simulations [9] if coupled electron-ion modes were included. Indeed, experiments investigating dense plasmas, found strong indications of longer relaxation times, thus smaller energy transfer rates [11, 12]. One of the shortcomings of the LS approach was addressed applying a quantum approach for binary collisions which yields however even larger rates [13, 14, 15]. Furthermore, considering independent collective modes in the subsystems within the Fermi-Golden-Rule approach, rates close to the LS results were obtained [16]. Therefore, collective modes seem to be the only candidate to explain the lower electron-ion energy transfer in dense strongly coupled plasmas.

However, a rigorous description of temperature relaxation requires not only appropriate transfer rates but has also to involve the interplay of all contributions to the internal energy. In addition to the kinetic parts, correlations and exchange energies [17, 18, 19], and the ionisation kinetics including excitations [20, 21, 22] can be important.

Here, we assume ionisation, recombination, and charge transfer processes to be completed and concentrate on the electron-ion energy relaxation. We use a quantum kinetic description which enables us to rigorously derive the balance equations and expressions for the transfer rates within an appropriate approximation scheme. The influence of coupled mode effects on temperature relaxation are still under discussion (see Refs. [6, 23, 24, 25, 26, 27]). We therefore re-derive the coupled mode energy transfer rate based on the density operator formalism. Together with adequate expressions for the internal energies/heat capacities, this enables us to determine the time evolution of electronic and ionic temperature in a nonideal system with ionic and electronic correlations and exchange. We demonstrate the effect of the nonideality contributions and the transfer rates on the evolution of the temperatures, on the relaxation time, and on the final temperature.

2. The energy balance equation

We consider a spatially homogeneous nonequilibrium plasma. To investigate the relaxation of the plasma towards thermodynamic equilibrium, balance equations can be used. With the single particle Hamiltonian H_a and the two-body potential V_{ab}, the mean kinetic and the mean potential energy of species "a" can be defined by

$$\langle K_a(t)\rangle = \mathrm{Tr}_1\{H_a\rho_a(t)\}\,, \quad \langle V_a(t)\rangle = \frac{1}{2}\sum_b \mathrm{Tr}_{1,2}\{V_{ab}\,\rho_{ab}(t)\}, \tag{1}$$

where ρ_a and ρ_{ab} are the one- and two-particle density operators, respectively. Note that definition (1) implies the symmetric distribution of electron-ion contributions to the internal energy into the electron and ion subsystems. The reduced density operators are determined by the following equations of motion which are the first two equations of the quantum version of the BBGKY hierarchy (time variables dropped) [28]

$$i\hbar\frac{\partial}{\partial t}\rho_a = [H_a,\rho_a] + \sum_b \mathrm{Tr}_2[V_{ab},\rho_{ab}]\,, \tag{2}$$

$$i\hbar\frac{\partial}{\partial t}\rho_{ab} = [H_{ab},\rho_{ab}] + \sum_c \mathrm{Tr}_3[(V_{ac}+V_{bc}),\rho_{abc}]\,. \tag{3}$$

The square brackets denote the commutator $[A,B] = AB - BA$.

Using the first equation, one can find the following expression for the change of the mean kinetic energy over time [19]

$$\frac{\partial}{\partial t}\langle K_a\rangle = \frac{1}{2}\sum_b \mathrm{Tr}_{1,2}\frac{1}{i\hbar}\{(H_a+H_b)[V_{ab},\rho_{ab}]\} + \frac{1}{2}\sum_b \mathrm{Tr}_{1,2}\frac{1}{i\hbar}\{(H_a-H_b)[V_{ab},\rho_{ab}]\}\,. \tag{4}$$

The first term on the r.h.s. of Eq. (4) can be transformed using Eq. (3). Then one can show easily that it is equal to the time derivative of the mean potential energy of species a, i.e. equation (4) can be cast into [19]

$$\frac{\partial}{\partial t}\langle K_a\rangle + \frac{\partial}{\partial t}\langle V_a\rangle = \sum_b Z_{ab}\,, \tag{5}$$

where Z_{ab} defines the energy transfer rate between species a and b

$$Z_{ab} \;=\; \frac{1}{2}\mathrm{Tr}_{1,2}\frac{1}{i\hbar}\{(H_a - H_b)[V_{ab}, \rho_{ab}]\}\,. \tag{6}$$

Note that only terms with $b \neq a$ give non-zero contributions to the total rate and that $Z_{ab} = -Z_{ba}$. Summation over species a in Eq. (5) then gives

$$\sum_a \frac{\mathrm{d}}{\mathrm{d}t}\langle K_a\rangle + \sum_a \frac{\mathrm{d}}{\mathrm{d}t}\langle V_a\rangle = \sum_{a,b} Z_{ab} = 0\,, \tag{7}$$

which expresses the conservation of the total energy.

3. Energy transfer rate

We write the transfer rate in the form

$$Z_{ab} = \frac{1}{2}\mathrm{Tr}_{1,2}\frac{1}{i\hbar}\{(H_a - H_b)[V_{ab}, \rho_a\rho_b + \rho_{ab}^{\mathrm{corr}}]\}\,, \tag{8}$$

where $\rho_{ab}^{\mathrm{corr}}$ is the correlation part of the two-particle density operator ρ_{ab}. Our goal is to derive an expression for the energy transfer rate for dense nonequilibrium plasmas where collective phenomena can play an important role. For this purpose, it is useful to apply the following relation ($b \neq a$)

$$\langle 12|\rho_{ab}^{\mathrm{corr}}(t)|2'1'\rangle = i\hbar L_{ab}^{<}(11't, 22't')_{|t=t'}\,. \tag{9}$$

Here, $L_{ab}^{<}$ is the correlation function of density fluctuations given by

$$i\hbar L_{ab}^{<}(11't, 22't') = \langle\psi_b^\dagger(2',t')\psi_b(2,t')\psi_a^\dagger(1',t)\psi_a(1,t)\rangle - \langle\psi_b^\dagger(2',t')\psi_b(2,t')\rangle\langle\psi_a^\dagger(1',t)\psi_a(1,t)\rangle\,, \tag{10}$$

with $\psi_a^\dagger$ and ψ_a being the creation and annihilation operators of species a, respectively. Spin variables will be suppressed for simplicity. With Eq. (9) and using the equations of motion for the function $L_{ab}^{<}(t,t')$, one can derive the following expression for the energy transfer rate for spatially homogeneous systems [29],

$$Z_{ab}(t) = -2\,\mathcal{V}\,\mathrm{Im}\int\frac{d^3q}{(2\pi\hbar)^3}\int_0^\infty\frac{d\omega}{2\pi}\,\omega\,V_{ab}(q)\,i\hbar L_{ab}^{<}(\mathbf{q};\omega,t)\,, \tag{11}$$

where $\mathcal{V}$ denotes the volume, and $L_{ab}^{<}(\mathbf{q};\omega,t)$ is the Fourier transform of $L_{ab}^{<}(\mathbf{r}_1t_1, \mathbf{r}_2t_2)$ with respect to the difference variables $\tau = t_1 - t_2$ and $\mathbf{r} = \mathbf{r}_1 - \mathbf{r}_2$.

Expression (11) allows for the determination of the energy transfer rate applying appropriate approximation schemes for the correlation function of density fluctuations. For this purpose, we start from the equation for the function L_{ab} defined on the Keldysh time contour [30, 31]

$$L_{ab}(t,t') = \Pi_{ab}(t,t') + \sum_{c,d}\int_C d\bar{t}\,\Pi_{ac}(t,\bar{t})\,V_{cd}\,L_{db}(\bar{t},t')\,. \tag{12}$$

Here, Π_{ab} is the polarization function, V_{ab} is the Coulomb potential, and the symbol $\int_C$ stands for the integration along the contour. For simplicity, only the time variables were given explicitly. Applying the well-known techniques originally introduced by Keldysh, the equation for the correlation function $L^<_{ab}$ can be obtained from Eq. (12) [30, 31]. Furthermore, the equations for the retarded and advanced functions follow easily.

Now, we will specify our considerations on a two-component plasma consisting of electrons with density n_e and ions of density n_i. As a first approximation we assume $\Pi_{ab} = \delta_{ab}\Pi_a$ where Π_a is considered to be the polarization function of the subsystem of species a, i.e. contributions from the other species are neglected. Then a system of equations follows for the functions L_{ee}, L_{ei}, L_{ie}, and L_{ii} defined on the contour.

For further considerations, it is useful to introduce the auxiliary functions $\mathcal{L}_e$ and $\mathcal{L}_i$, defined by

$$\mathcal{L}_a = \Pi_a + \Pi_a V_{aa} \mathcal{L}_a , \quad (a = e, i) , \tag{13}$$

which describe the pure electron and ion subsystems, respectively. Then, the equation for L_{ei} takes the form

$$\begin{aligned} L_{ei} &= \mathcal{L}_e V_{ei} \mathcal{L}_i + \mathcal{L}_e V_{ei} \mathcal{L}_i V_{ie} L_{ei} \\ &= \mathcal{P}_{ei} + \mathcal{P}_{ei} V_{ei} L_{ei} , \end{aligned} \tag{14}$$

with $\mathcal{P}_{ei} = \mathcal{L}_e V_{ei} \mathcal{L}_i$. Here, the equations are written in a compact form where the variables are suppressed.

The equation for the function $L^<_{ei}$ for physical times can be obtained using the Langreth-Wilkins rules [32]. In this equation, we introduce difference and sum variables, and we account for only the terms of lowest order in the gradient expansion [31]. Finally, after Fourier transformation with respect to the difference variables we get the following expression for the electron-ion correlation function of density fluctuations [29]

$$L^<_{ei}(\mathbf{q};\omega,t) = \frac{\mathcal{P}^<_{ei}(\mathbf{q};\omega,t)}{\left|1 - \mathcal{L}^R_e(\mathbf{q};\omega,t)\, V_{ei}(q)\, \mathcal{L}^R_i(\mathbf{q};\omega,t) V_{ie}(q)\right|^2} , \tag{15}$$

where $\mathcal{P}^<_{ei} = \mathcal{L}^<_e V_{ei} \mathcal{L}^A_i + \mathcal{L}^R_e V_{ei} \mathcal{L}^<_i$ with $\mathcal{L}^{R/A}_a$ being the retarded and advanced response functions of the subsystems $(a = e, i)$.

In the following, a nonequilibrium plasma state is described corresponding to a situation where the electron and ion subsystems can be considered to be in local equilibrium with the electron temperature T_e and the ion temperature T_i. The correlation functions of the subsystems, $\mathcal{L}^<_a$, can then be expressed in terms of spectral functions $\mathcal{A}_a$ and the Bose function as follows

$$i\mathcal{L}^<_a(\mathbf{q};\omega,t) = \mathcal{A}_a(\mathbf{q};\omega,T_a) n_B(\omega/T_a), \qquad \mathcal{A}_a(\mathbf{q};\omega,T_e) = -2\mathrm{Im}\mathcal{L}^R_a(\mathbf{q};\omega,T_a) , \tag{16}$$

and $n_B(\omega/T_a) = 1/[\exp(\hbar\omega/k_B T_a) - 1]$. Inserting the expression (15) with (16) into the general expression (11) for the energy transfer rate, we arrive at

$$\begin{aligned} Z_{ei}(t) &= -\mathcal{V} \int \frac{d^3q}{(2\pi\hbar)^3} \int_0^\infty \frac{d\omega}{2\pi}\, \hbar\omega\, [V_{ei}(q)]^2 \\ &\times \frac{\mathcal{A}_e(\mathbf{q};\omega,t)\, \mathcal{A}_i(\mathbf{q};\omega,t) \Big[n_B(\omega/T_e) - n_B(\omega/T_i)\Big]}{\left|1 - \mathcal{L}^R_e(\mathbf{q};\omega,t)\, V_{ei}(q)\, \mathcal{L}^R_i(\mathbf{q};\omega,t) V_{ie}(q)\right|^2} . \end{aligned} \tag{17}$$

Expression (17) has the same structure as that given by Dharma-wardana and Perrot [6] for systems with coupled modes. If the denominator is set equal to unity, which corresponds to the approximation $L_{ei}^{<} = \mathcal{P}_{ei}^{<}$, the expression reduces to the Fermi-Golden-Rule formula valid for the regime of weak coupling between electrons and ions [29, 6]. A detailed discussion of coupled mode effects compared to the Fermi-Golden-Rule approach is given in [27].

The key quantities in the transfer rate (17) are the density response functions of the subsystems. Strong correlations in the subsystems can be accounted for by local field corrections (LFC). Using static local field corrections, the retarded density response function can be written as

$$\mathcal{L}_a^R(\mathbf{q};\omega,t) = \frac{\Pi_a^0(\mathbf{q};\omega,t)}{1 - V_{aa}(q)\left[1 - G_a(\mathbf{q},t)\right]\Pi_a^0(\mathbf{q};\omega,t)} \,. \tag{18}$$

Π_a^0 is the retarded polarization function in random phase approximation (RPA)

$$\Pi_a^0(\mathbf{q};\omega,t) = \int \frac{d\mathbf{p}}{(2\pi\hbar)^3} \frac{f_a(\mathbf{p},t) - f_a(\mathbf{p}+\mathbf{q},t)}{\hbar\omega + \epsilon_a(\mathbf{p}) - \epsilon_a(\mathbf{p}+\mathbf{q}) + i\varepsilon} \,, \tag{19}$$

where f_a is the Fermi function for the subsystem of species a being in local equilibrium with temperature T_a. The local field correction factor G_a is given by [33]

$$G_a(\mathbf{q},t) = 1 - \frac{k_B T_a}{n_a V_{aa}(q)} \left(\frac{1}{S_{aa}(\mathbf{q},t)} - 1 \right) \,, \tag{20}$$

with $S_{aa}(\mathbf{q},t)$ $(a = e, i)$ being the static structure factors of the electron and ion subsystems. For numerical calculations, it is sometimes convenient to use the coupled mode formula (17) in another form given in terms of the dielectric functions. For the derivation of this formula, we refer to Refs. [27, 29].

4. Energy relaxation, temperature relaxation, and quasi equation of state

In order to study temperature relaxation, a connection between temperature and internal energy needs to be introduced into Eq. (5). This can formally be done via heat capacities and leads for a two component electron ion system to two coupled differential equation for the temporal evolution of the temperatures of the species

$$\frac{\partial T_e}{\partial t} = \frac{Z_{ei}(T_e,T_i)}{\Delta} \left(\frac{\partial U_e}{\partial T_i} + \frac{\partial U_i}{\partial T_i} \right) \,, \tag{21}$$

$$\frac{\partial T_i}{\partial t} = -\frac{Z_{ei}(T_e,T_i)}{\Delta} \left(\frac{\partial U_e}{\partial T_e} + \frac{\partial U_i}{\partial T_e} \right) \,, \tag{22}$$

$$\Delta = \frac{\partial U_e}{\partial T_e}\frac{\partial U_i}{\partial T_i} - \frac{\partial U_e}{\partial T_i}\frac{\partial U_i}{\partial T_e} \,. \tag{23}$$

The internal energies of the subsystems are defined via $U_a = \langle K_a \rangle + \langle V_{aa} \rangle + 1/2\langle V_{ab} \rangle$ with $\{a,b\} = \{e,i\}$ $(a \neq b,\ \langle V_{ab} \rangle = \langle V_{ba} \rangle)$.

The internal energies for the electron and ion subsystems have to be calculated taking into account different electron and ion temperatures. This presents some challenges in the general case (see e.g. discussions in Refs. [10, 6]) and can only fully be overcome by taking one step back describing the whole process on the basis of kinetic equations.

Here, we need a theory that can handle electron degeneracy and strong ion coupling. So far, no unified approach that includes electrons and ions on the same basis is able to achieve this. For our purpose, we therefore use a perturbation expansion in the electron coupling to approximate the electron-electron internal energy contributions. This part is

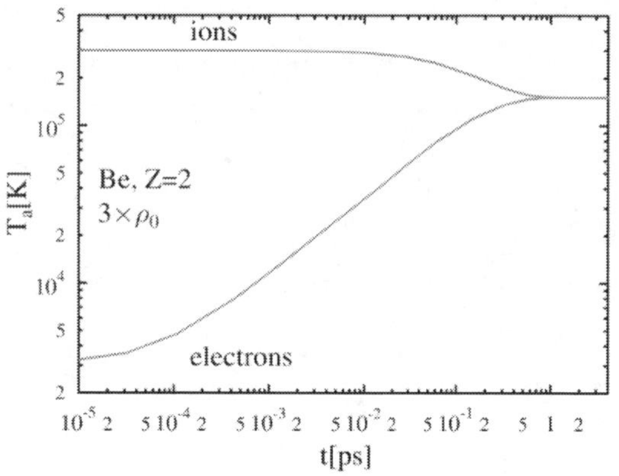

Figure 1. Nonideal temperature relaxation in a shock produced beryllium plasma at threefold solid density $\rho_0 = 1.848 g/cm^3$. The ion charge state is $Z = 2$. The conditions are the same as in a recent experiment [36]. The energy transfer rate is in coupled mode approximation including local field effects for the ions. Heat capacities as described in Eqs. (4) & (25).

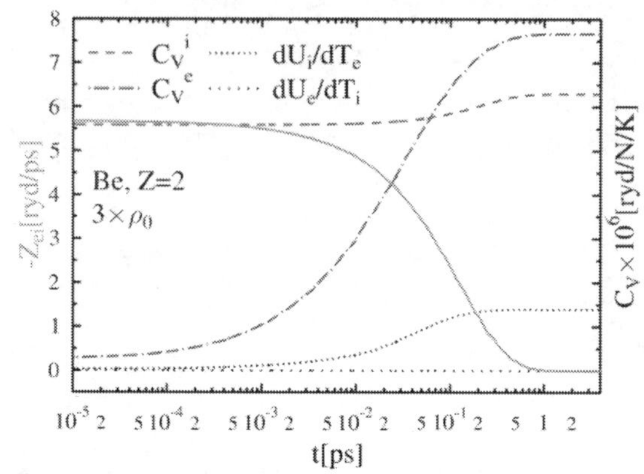

Figure 2. Energy transfer rate and heat capacities for the temperature relaxation presented in Fig. 1. The red solid line with the left ordinate gives the energy transfer rate. The heat capacities are plotted in blue (normal terms) and black (cross terms). The heat capacities include the factor of 2 from the densities: $n_e = 2 \cdot n_i$.

independent of the ion temperature and given in terms of Feynman diagrams to be [34, 35, 31]

$$U_{ee} = U_{ee}^{id} + \quad (24)$$

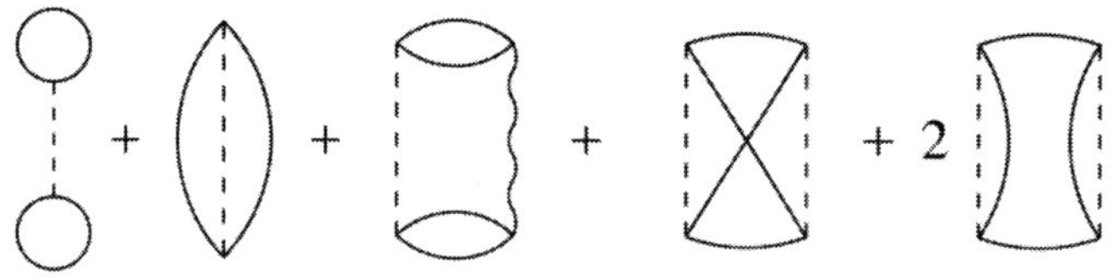

The terms considered are from left to right the Hartree, Hartree-Fock, Montroll-Ward, and normal and anomal e^4 exchanges terms. This approximation is sufficient for most laser produced plasmas with hot electrons and also for shock produced plasmas with degenerate electrons. In both cases, the electron heat capacities resulting are very small compared to the ion heat capacities. In the first case this is due to the high coupling of the ions, in the second case due to the degeneracy of the electrons. The pure ion and electron-ion contributions can be evaluated using the effective one component plasma model [37]. The ions interact via an electronically linearly screened potential. Using the hypernetted chain approximation the internal energy can be written as

$$U_{ii}^{corr} + U_{ei} = \frac{n_i}{2} \int \frac{d\mathbf{k}}{(2\pi\hbar)^3} \left[\phi_{ii}^{eff}(k) - \phi_{ii}(k)\right] + \frac{n_i^2}{2} \int d\mathbf{r} [g_{ii}(r, T_i, T_e, n_e) - 1] \phi_{ii}^{eff}(r, T_e, n_e) \,. \quad (25)$$

The pure Coulomb potential is denoted by ϕ, whereas ϕ^{eff} is the effective interaction. The cross term can be approximately evaluated by subtracting the contribution that arises in this equation when using a pure Coulomb potential. It turns out that writing the internal energy as in Eq. (25) means to give the electron-ion contribution the electron temperature. There exist further arguments that Eq. (25) misses terms with temperature derivatives [38]. However, for the physical systems under consideration here, we estimate that those terms are of minor importance.

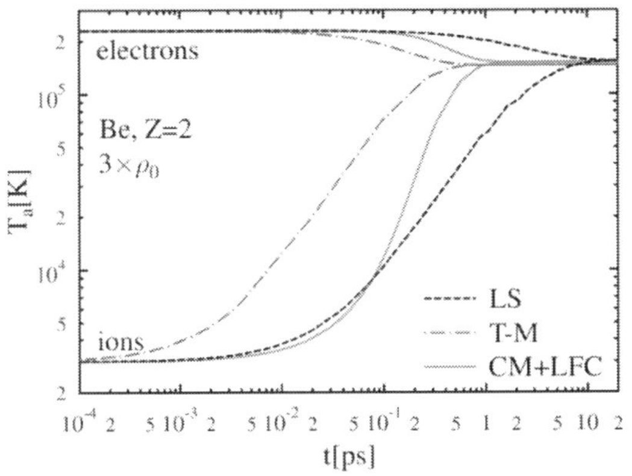

Figure 3. Nonideal temperature relaxation in a laser-heated beryllium with threefold solid density $\rho_0 = 1.848\, g/cm^3$. The ion charge state is $Z = 2$. The curve labelled LS displays a classical, ideal relaxation. Its final temperature is 154000 K. T-M uses ideal quantum relaxation and a T-Matrix energy transfer rate yielding $T(\infty) = 146000\,\mathrm{K}$. The red curve includes a coupled mode energy transfer rate and nonideal contributions to the heat capacity; the final temperature is here 151000 K.

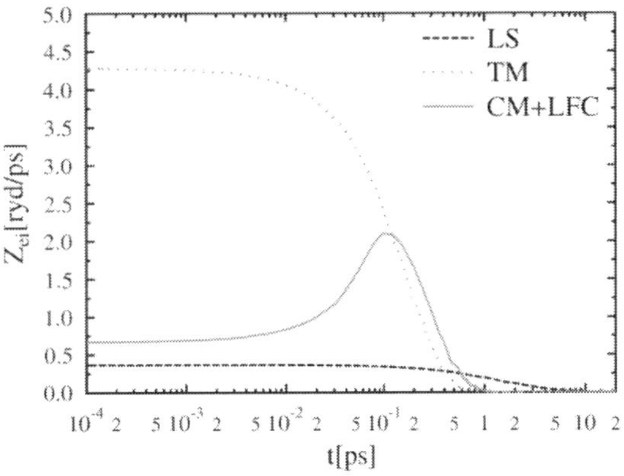

Figure 4. Energy transfer rates for the temperature relaxation presented in Fig. 3. TM and CM rates merge when the temperature difference is small so that the condition for ion acoustic modes is not fulfilled anymore ($T_i < 0.27\,T_e$) [27]. For small times, CM effects lower the rate. The LS rate (with the Coulomb logarithm chosen here) is so low because the approximations made break down [27].

5. Temperature relaxation in compressed Beryllium

Beryllium is of particular interest due to its use in inertial fusion and related experiments [39, 3, 36]. We therefore study here two examples for temperature relaxation in dense compressed beryllium. Both cases are chosen with conditions as in the experiment - threefold compression, final temperature about 13 eV, charge state $Z = 2$ for the ions. Figure 1 shows the resulting equilibration from a shock produced state as in Ref. [36]. The initial temperatures are results of the nonideal relaxation under the boundary condition that the final temperature is 13 eV. This choice of initial temperatures would have resulted in a final temperature of 8.7 eV in case of an ideal classical relaxation and in a final temperature of 12.7 eV for an ideal quantum relaxation. Quantum effects for the electrons are very important as they decrease the electron heat capacity to almost zero for small times (see Fig. 2). The ideal relaxation using the quantum heat capacity for the electrons is already very close to the full result, because the correlation contributions to the heat capacities of electrons and ions partially cancel. The pure electron-electron heat capacity is always very small as electron degeneracy is very high. However, as temperature differences between ions and electrons become smaller, the electron-ion contributions grow faster than the single species terms. Since the electron-ion internal energy gets distributed evenly between electrons and ions, the electron subsystem receives considerable heat capacity to balance the massive ion-ion correlation heat capacity. Contributions to the heat capacity from $\partial U_a/\partial T_b$ ($a \neq b$) terms are small compared to the full heat capacities. The final temperature is reached after 1.7 ps. This value is sensitive to the choice of energy transfer rate. In general, coupled mode energy transfer rates predict slower relaxation than T-Matrix or Fermi Golden Rule transfer rates. Not much can be said about Landau-Spitzer energy transfer rates as their behavior strongly depends on the choice of Coulomb logarithm. For an extensive discussion see Refs. [27, 29].

An overview of the relaxation using different energy transfer rates is given in Fig. 3. In this case, we study the opposite case to the shock produced plasma of Fig. 1. The hypothetical case of Fig. 3 assumes a laser produced and compressed state of beryllium. The initial temperatures are still the same but since the electrons are hot and the ions cold on the outset, the final temperature is different (13.3 eV). The electronic heat capacity is still dominated by quantum effects, although the higher temperature increases the ideal electron heat capacity to 88% of the classical value. In Fig. 3, the T-Matrix relaxation is fastest and the Landau-Spitzer relaxation slowest. The influence of the heat capacities (correlation contributions) on the relaxation time is small compared to the influence of the energy transfer rates. In general, correlations in the heat capacities slow relaxation down and quantum effects (which lower heat capacities) speed relaxation up. The coupled mode energy transfer rates are the smallest rates up to date which can be justified quantum statistically. It was therefore predicted that temperature relaxation using coupled mode rates gives relaxation times much longer than previously obtained. However, as Figs. 3 & 4 show, the influence of the coupled mode rate is not so much about the relaxation time as about the time evolution of the relaxation. There are two important factors: Relaxation is driven by temperature difference - the bigger the temperature difference the bigger the energy transfer. In Fig. 3 for instance, one has to observe that the temperature difference in the coupled mode case after about 0.1 ps is an order of magnitude larger than in the T-Matrix case. This drives a faster relaxation. The second point to make is that the coupled mode effect ceases to exist for small temperature differences which gives additional increase to the energy transfer. The relaxation time obtained are 1 ps for T-Matrix rates, 1.7 ps for coupled mode rates, and 16 ps for the Landau-Spitzer calculation. The significance of the Landau-Spitzer result can be doubted as the classical approximation scheme breaks.

6. Summary & Outlook

Relaxation times in the order of hundreds of picoseconds as seen in experiment [11, 12] cannot be found with the theoretical methods presented in this paper. Neither the coupled mode energy transfer rate nor the inclusion of correlation and quantum effects in the heat capacities slow down the relaxation process on a necessary scale. An improvement of our theory is possible by inclusion of bound states in the energy transfer rates. It will furthermore be necessary to put the quasi equation of state theory on a firm theoretical foundation.

Acknowledgements

This work was supported by Engineering and Physical Sciences Research Council of the UK (grant EP/D062837) and the Deutsche Forschungsgemeinschaft (SFB 652).

References

[1] Riley D, Woolsey N C, McSherry D, Weaver I, Djaoui A and Nardi E 2000 *Phys. Rev. Lett.* **84** 1704
[2] Ravasio A *et al.* 2007 *Phys. Rev. Lett.* **99** 135006
[3] Glenzer S H *et al.* 2007 *Phys. Rev. Lett.* **98** 065002
[4] Garćia Saiz E *et al.* 2008 *Nature Physics* **4** 940
[5] Kritcher A L *et al.* 2008 *Science* **322** 69
[6] Dharma-wardana M W C and Perrot F 1998 *Phys. Rev. E* **58** 3705
[7] Landau L D 1936 *Phys. Z. Sowjetunion* **10** 154
[8] Spitzer L 1962 *Physics of Fully Ionized Gases* (New York: Interscience)
[9] Hansen J P and McDonald I R 1983 *Phys. Lett. A* **97** 42
[10] Boercker D B and More R M 1986 *Phys. Rev. A* **33** 1859
[11] Celliers P, Ng A, Xu G and Forsman A 1992 *Phys. Rev. Lett.* **68** 2305
[12] Ng A, Celliers P, Xu G and Forsman A 1995 *Phys. Rev. E* **52** 4299
[13] Gericke D O, Murillo M S and Schlanges M 2002 *Phys. Rev. E* **65** 036418
[14] Gericke D O, Murillo M S and Schlanges M 2002 *Laser and Part. Beams* **20** 543

[15] Note that the break-down of the classical collision theory first yields too low than negative Coulomb logarithms which translates into too low or negative energy transfer rates in the Landau-Spitzer theory.
[16] Hazak G, Zinamon Z, Rosenfeld Y and Dharma-wardana M W C 2001 *Phys. Rev. E* **64** 066411
[17] Gericke D O and Murillo M S 2004 *Proceedings Int. Conf. on Inertial Fusion Science and Applications* ed Hogan W (Monterey)
[18] Gericke D O, Murillo M S, Semkat D, Bonitz M and Kremp D 2003 *J. Phys. A* **36** 6087
[19] Gericke D O, Bornath T and Schlanges M 2006 *J. Phys. A* **39** 4739
[20] Ohde T, Bonitz M, Bornath T, Kremp D and Schlanges M 1996 *Phys. Plasmas* **3** 1241
[21] Bornath T, Schlanges M and Prenzel R 1998 *Phys. Plasmas* **5** 1485
[22] Gericke D O, Grubert G K, Bornath T and Schlanges M 2006 *J. Phys. A* **39** 4727
[23] Dharma-wardana M W C 2001 *Phys. Rev. E* **64** 035401(R)
[24] Daligault J and Mozyrsky D 2007 *Phys. Rev. E* **75** 026402
[25] Daligault J and Mozyrsky D 2008 *High Energy Density Physics* **4** 58
[26] Gregori G and Gericke D O 2008 *Europhys. Lett.* **83** 15002
[27] Vorberger J and Gericke D O 2009 *Phys. Plasmas* **16** 082702
[28] Kraeft W D, Kremp D, Ebeling W and Röpke G 1986 *Quantum Statistics of Charged Particle Systems* (Berlin: Akademie-Verlag)
[29] Vorberger J, Gericke D O, Bornath T and Schlanges M 2010 *Phys. Rev. E* **81** 046404
[30] Keldysh V L 1964 *Zh. Eksp. Teor. Fiz.* **47** 1515
[31] Kremp D, Schlanges M and Kraeft W D 2005 *Quantum Statistics of Nonideal Plasmas* (Berlin, Heidelberg: Springer-Verlag)
[32] Langreth D C and Wilkins J W 1972 *Phys. Rev. B* **6** 3189
[33] Ichimaru S 1982 *Rev. Mod. Phys.* **54** 1017
[34] Kraeft W D, Schlanges M, Vorberger J and DeWitt H E 2002 *Phys. Rev. E* **66** 046405
[35] Vorberger J, Schlanges M and Kraeft W D 2004 *Phys. Rev. E* **69** 046407
[36] Lee H J *et al.* 2009 *Phys. Rev. Lett.* **102** 115001
[37] Hansen J P and McDonald I R 1986 *Theory of Simple Liquids* (Amsterdam: Elsevier)
[38] The effective one component plasma model provides the Helmholtz free energy F which usually gives the internal energy via $U = F - T\partial F/\partial T$. In a two temperature system, it is unclear which temperature to choose.
[39] Glenzer S H *et al.* 2003 *Phys. Rev. Lett.* **90** 175002

Dynamics of strongly correlated ions in a partially ionized quantum plasma

P. Ludwig, M. Bonitz, and H. Kählert

Institut für Theoretische Physik und Astrophysik, Christian-Albrechts Universität zu Kiel, 24098 Kiel, Germany

J.W. Dufty

Department of Physics, University of Florida, Gainesville, FL 32611, USA

Abstract. A scheme which allows to compute the dynamics of strongly correlated classical ions embedded into a partially ionized quantum plasma by first principles molecular dynamics is presented. The dynamically screened dust approach of Joyce and Lampe [Phys. Rev. Lett. **88**, 095006 (2002] is generalized to quantum systems. The electrons are treated fully quantum-mechanically taking into account their dynamical screening of the ion-ion interaction in linear response on the basis of an extended Mermin formula. The scheme allows to include the effect of the electron dynamics, electron streaming, wake effects and electron magnetization.

1. Introduction

Strong correlation effects in ensembles of charged particles are of high importance in many fields of physics, including dense plasmas, e.g. [1, 2], the electron-hole plasma or electrons in quantum dots [3], for a recent overview see [4] and references therein. The present paper is devoted to a strongly coupled mass-asymmetric two-component plasma where only the heavy component (typically the ions) is strongly coupled whereas the light component (e.g. electrons) is weakly coupled. Such a situation is expected in various compact astrophysical systems including white dwarf stars or the crust of neutron stars where one expects ion crystallization [5]. Another relevant situation are dense laboratory plasmas produced by laser or ion beam irradiation of matter where initially again a strongly coupled ion plasma is formed – so-called "warm dense matter", for a recent overview see the article by Gericke et al. in this volume [6]. A sketch of the density-temperature plane in this parameter range is shown in Fig. 1.

The parameters of interest for the present work are indicated by the shaded area in Fig. 1. There the ions (electrons) are classical (quantum) characterized by a degeneracy parameter $\chi_a = n_a\Lambda_a^3$ smaller (larger) than unity, with the DeBroglie wave length $\Lambda_a = h/(2\pi m_a k_B T_a)^{1/2}$. Our main interest is to describe strong ion coupling characterized by $\Gamma_i = q_i^2/(\bar{r}_i k_B T_i) > 1$, whereas the quantum coupling parameter (Brueckner parameter) of the electrons is small, $r_{se} = \bar{r}_e/a_B \ll 1$. In recent years there have been numerous attempts to develop simulations for such systems, based on path integral Monte Carlo (PIMC) [7, 8], quantum molecular dynamics (QMD) [9, 10] or the Wigner formalism (WMD) [11]. The best dynamical results so far have been obtained from QMD, however, these simulations involve a density functional treatment of the electrons which is very time-consuming. Moreover, current versions usually neglect the

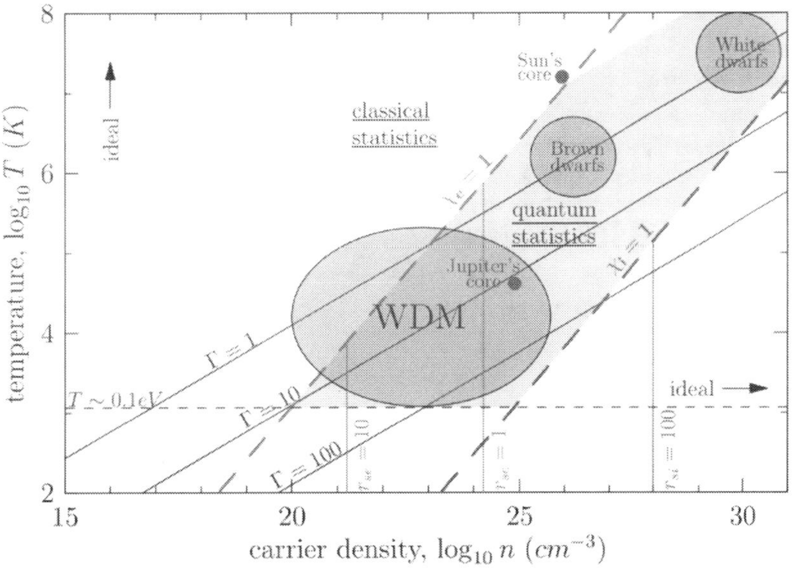

Figure 1. Phase diagram of a two-component plasma of electrons and singly charged ions in thermodynamic equilibrium with a few astrophysical examples (WDM denotes "warm dense matter"). The line $\chi_e = 1$ [$\chi_i = 1$] separates the region of classical (upper left) and quantum (lower right) behavior of the electrons [ions]. Also, the lines of constant classical (Γ) and quantum (r_s) coupling strength are shown. The present model applies to the shaded area.

dynamics of the electron subsystem. However, for dense electron-ion plasmas dynamics of the electrons, in particular dynamical screening, is expected to be important.

The approach of this paper includes these effects. It takes advantage of the weak electron coupling in the plasma phase. At the same time the ion dynamics are treated exactly, via semiclassical molecular dynamics. The model and approximations leading to effective Newtonian equations for the ions are introduced in Sec. 2. In Sec. 3 we derive the effective ion-ion pair potential which its dynamically screened by the electrons. The Mermin dielectric function is recalled in Sec. 4. A detailed derivation in the frame of quantum kinetic theory is given in the appendix. Results are summarized in Sec. 5.

2. Basic equations and approximations

The nonequilibrium dynamics of the ions in a partially ionized plasma consisting of electrons (e), ions (i), and neutrals (n) can be described by the reduced density operator for the ions, given by [12, 13]

$$i\hbar\frac{\partial \widehat{F}_i(1,..,N_i)}{\partial t} - \left[\widehat{H}_i(1,..,N_i), \widehat{F}_i(1,..,N_i)\right] = \sum_{b=e,n}\sum_{k=1}^{N_i} n_b Tr_{2_b}\left[\widehat{V}_{ib}(k,2_b), \widehat{F}_{ib}(1,..,N_i;2_b)\right], \quad (1)$$

where $b = e, n$, and $\widehat{F}_{ib}(1,..,N_i;2_b)$ is the joint density operator for N_i ions and one particle of type b. $\widehat{H}_i(1,..,N_i)$ is the hamiltonian for the ions in the presence of an external field $(\phi, \mathbf{A})$ and their mutual interactions

$$\widehat{H}_i(1,..,N_i) = \sum_{k=1}^{N_i}\left[\frac{1}{2m_k}\left(\frac{\hbar}{i}\nabla_k - \frac{e_i}{c}\mathbf{A}(\mathbf{r}_k,t)\right)^2 + e\phi\left(\mathbf{r}_k,t\right)\right] + \frac{1}{2}\sum_{k\neq\ell}^{N_i} V_{ii}(|\mathbf{r}_k - \mathbf{r}_\ell|), \quad (2)$$

where $\mathbf{A}$ and ϕ denote the vector and scalar potential of an external electromagnetic field, and $\widehat{V}_{ib}$ is the operator of binary interactions between an ion and particle of type b. Quantum exchange effects of the ions will be irrelevant and are neglected in (1).

We now start to simplify this equation and to determine effective Newtonian equations of motion for the ions. Since the electrons are assumed strongly degenerate and weakly coupled, also the electron-ion interaction on the r.h.s. of (1) is weak and $\widehat{F}_{ie}(1,..,N_i;2_e) \simeq \widehat{F}_i(1,..,N_i)\widehat{F}_e(2_e)$. In this approximation the electron contribution to the r.h.s. becomes $n_eTr_{2_e}\left[\widehat{V}_{ie}(k,2_e),\widehat{F}_{ib}(1,..,N_i;2_e)\right] \simeq \left[\widehat{W}_{ie}(k),\widehat{F}_i(1,..,N_i)\right]$, where we defined the mean field (Hartree potential) for the ions which is created by the electrons, $\widehat{W}_{ie}(k) \equiv n_eTr_{2_e}\widehat{V}_{ie}(k,2_e)\widehat{F}_e(2_e)$. This gives rise to an effective single particle Hamiltonian for each ion $\widehat{\widehat{H}}_i(1) = \widehat{H}_i(1) + \widehat{W}_{ie}(1)$ [here $\widehat{H}_i(1)$ denotes the expression under the first sum in Eq. (2)] so that equation (1) becomes

$$i\hbar\frac{\partial\widehat{F}_i(1,..,N_i)}{\partial t} - \left[\widehat{\widehat{H}}_i(1,..,N_i),\widehat{F}_i(1,..,N_i)\right] = \sum_{k=1}^{N_i} n_nTr_{2_n}\left[\widehat{V}_{in}(k,2_n),\widehat{F}_{in}(1,..,N_i;2_n)\right], \quad (3)$$

where $\widehat{\widehat{H}}_i(1,..,N_i)$ is given as in (2) except with the single particle hamiltonians $\widehat{\widehat{H}}_i(k)$ replacing $\widehat{H}_i(1)$. The l.h.s describes the effection ion dynamics driven by ion-ion and mean field ion-electron interactions, and the r.h.s. is due to the interactions with the neutrals.

Since the coupling between ions and neutrals is expected to be much weaker than between ions, an approximate treatment of the neutrals can be applied. In the following, we assume that the neutrals are close to thermodynamic equilibrium. Then the collision integral between ions and neutrals, I_{in}, can be treated using a Fokker-Planck type integral with an effective ion-neutral collision frequency ν_{in} and a diffusion coefficient D_n. The two coefficients are used as input parameters to our model. Next we take into account that the ions are classical (we briefly discuss the inclusion of quantum corrections in Sec. 5) and perform the classical limit in Eq. (3). This leads to a classical kinetic equation for the phase space distribution and possible quantum corrections [12], which is equivalent to (its characteristics are) the one set of the following Newton's equations

$$\frac{d\mathbf{p}_k}{dt} = \left(-\frac{dW_{ie}(\mathbf{r})}{d\mathbf{r}} + e_i\mathbf{E}(\mathbf{r},t) + \frac{e_i}{c}[\mathbf{v}_k \times \mathbf{B}(\mathbf{r},t)]\right)\Big|_{\mathbf{r}=\mathbf{r}_k} + \mathbf{F}^{i,n}(\mathbf{r}_k,\mathbf{p}_k,t), \quad k = 1,\ldots,N_i, \quad (4)$$

$$\frac{d\mathbf{p}_k}{dt} = -\frac{d}{d\mathbf{r}}\left(V_i^{\text{eff}}(\mathbf{r},t) + e_i\phi^{ext}(\mathbf{r},t)\right)\Big|_{\mathbf{r}=\mathbf{r}_k} + \mathbf{F}^{i,n}(\mathbf{r}_k,\mathbf{p}_k,t), \quad k = 1,\ldots,N_i, \quad (5)$$

The first line corresponds to the general case of an electromagnetic field $(\mathbf{E},\mathbf{B})$ produced by external sources and the ions which follows from the solution of Maxwell's equations. The potential W_{ie} produced by the electrons has been separated for clarity. The second line corresponds to the case of a pure electrostatic field which is composed by an external potential ϕ^{ext} plus the total effective ion-ion interaction energy $V_i^{\text{eff}}(\mathbf{r},t) = e_i\sum_{l=1}^{N_i}\phi_l(\mathbf{r},t)$. Here ϕ_l is the electrostatic potential produced by ion "l" in the partially ionized plasma which will be computed in Sec. 3. Finally, $\mathbf{F}^{i,n}(\mathbf{r}_k,\mathbf{p}_k)$ describes the force acting on ion "k" from the neutral particles. Due to the weak coupling between ions and neutrals in the plasma this effect is expected to be of minor importance. Starting from the Fokker-Planck form of the collision term I_{in} the effect of the neutrals in the classical equations of motion will be a combination of a friction force plus a random force y, i.e., $\mathbf{F}^{i,n}(\mathbf{r},\mathbf{p},t) = -\nu_{in}\mathbf{p} + \mathbf{y}(t)$ where the amplitude of the random force will be proportional to $\sqrt{D_n}$.

In Eqs. (4), (5) we have fully retained ion-ion correlations and fluctuations and are, thus, able to treat strong correlation effects in principle exactly. The quality of the solutions thereby depends on the quality of the potential energy $V_i^{\rm eff}$ which is affected (screened) by the electrons. Of particular interest for us is to include nonequilibrium situations in warm dense matter such as streaming electrons and ions, particle beams and so on. Then a full dynamic treatment of the pair interaction is crucial which we discuss now.

3. Dynamically screened ion-ion pair potential

Let us now compute the potential of a single ion moving relative to the electrons, taking into account the dielectric properties of the (unmagnetized) plasma. We start with the case of a classical charged particle, e.g. [14], and then generalize the result to quantum particles.

The Poisson equation for a polarizable medium reads ($n(\mathbf{r})$ is the number density of charged particles and n_k its Fourier component)

$$\mathrm{div}\mathbf{D}(\mathbf{r},t) = 4\pi e_a n_a(\mathbf{r},t), \tag{6}$$

which, in Fourier space, becomes $i\mathbf{k}\cdot\mathbf{D}_k(t) = 4\pi e_a n_{ak}(t)$. The electrostatic potential ϕ created by the charge density on the right is $\mathbf{E}(\mathbf{r}) = -\nabla\phi(\mathbf{r})$, corresponding, in Fourier space, to $\mathbf{E}_k = -i\mathbf{k}\cdot\phi_k$. Together with the electrodynamic definition of the dielectric tensor, $D_{k,i} = \sum_j \epsilon_{k,ij}E_{k,j}, \quad i,j = 1,2,3$, and Eq. (6) we obtain

$$\phi_k(\omega) = \frac{4\pi e_a n_{ak}(\omega)}{\sum_{ij} k_i k_j \epsilon_{k,ij}(\omega)}. \tag{7}$$

Consider now the case of a classical point charge "a" with initial position $\mathbf{r}(0) = \mathbf{r}_{0a}$, moving with constant velocity $\mathbf{v}_a$ (relative to the carriers creating the dielectric function). Then

$$n_a(\mathbf{r},t) = \delta[\mathbf{r} - \mathbf{r}_{0a} - \mathbf{v}_a t], \tag{8}$$

with the Fourier representation

$$n_{ak}(\omega) = 2\pi e^{-i\mathbf{k}\mathbf{r}_0}\delta[\omega - \mathbf{k}\mathbf{v}_a]. \tag{9}$$

Inserting this result into (7) and performing the back transform we obtain

$$\phi(\mathbf{r},t;\mathbf{v}_a) = e_a \int \frac{d^3k}{2\pi^2} \frac{e^{i\mathbf{k}(\mathbf{r}-[\mathbf{r}_{0a}+\mathbf{v}_a t])}}{\sum_{ij} k_i k_j \epsilon_{k,ij}(\mathbf{k}\mathbf{v}_a)}. \tag{10}$$

For the special case of an isotropic medium, ϵ_{ij} has only two independent components. Concentrating on longitudinal plasma oscillations we can replace $k_i k_j\,\epsilon_{k,ij} \to k^2\epsilon_k$, and the potential (10) and its Fourier transform become

$$\phi_k(\omega) = \frac{4\pi e_a n_{ak}(\omega)}{k^2\epsilon_k(\omega)}, \tag{11a}$$

$$\phi(\mathbf{r},t;\mathbf{v}_a) = \int \frac{d^3k}{2\pi^2}\frac{e_a}{k^2}\frac{e^{i\mathbf{k}(\mathbf{r}-[\mathbf{r}_{0a}+\mathbf{v}_a t])}}{\epsilon_k(\mathbf{k}\mathbf{v}_a)}. \tag{11b}$$

This result may be immediately generalized to the case of many particles. Indeed, due to linearity of Maxwell's equations, the resulting total potential is simply the sum of all potentials of the type (10), i.e. $\phi^{tot}(\mathbf{r}) = \sum_{a=1}^{N}\phi(\mathbf{r}-\mathbf{r}_{0a};\mathbf{v}_a)$. From this we obtain the total potential energy entering Newton's equations (5)

$$V_i^{\rm eff}(\mathbf{r},t) = \sum_{a=1}^{N_i} e_a\phi(\mathbf{r}-\mathbf{r}_{0a};\mathbf{v}_a). \tag{12}$$

We underline that this is an exact result fully including the ion-ion interaction (assuming point ions) in the presence of a partially ionized plasma the properties of which are included in the dynamic dielectric function $\epsilon_k(\omega)$. The result is applicable to equlibrium and can be extended to weakly nonequilibrium plasmas in the sense of a local approximation with a time-dependent distribution function $f(t)$ where $\epsilon_k(\omega) \to \epsilon_k(\omega, t) \approx \epsilon_k(\omega, [f(t)])$ [12]. In particular, the case of streaming electrons with a macroscopic velocity $\mathbf{u}_e$ is trivially included by replacing on the r.h.s. of Eq. (11b) $\mathbf{v}_a \to \mathbf{v}_a - \mathbf{u}_e$. In the special case of absence of streaming, the static limit of the dielectric function is obtained, $\epsilon_k \to (k^2 + \kappa^2)/k^2$ and the pair potential of the ions becomes the isotropic Yukawa potential, $V_{ii}(r) \to e_i^2 e^{-\kappa r}/r$. In contrast, in the case of streaming charged particles, the potential is anisotropic and non-monotonic and exhibits wake effects.

The result (12) equally applies to classical and quantum electrons and its quality is fully determined by the accuracy of the dielectric function. We, therefore, now turn to the computation of the quantum electronic dielectric function. Here we will assume that linear response (weak external field) can be applied and include collision effects in relaxation time approximation within the Mermin model [16].

4. Quantum Dielectric function

The Fourier transform of the dielectric function has, in linear response, the following general form [12]

$$\epsilon_q(\hat{\omega}) = 1 - V_q \Pi_q(\hat{\omega}), \tag{13}$$

where ϵ_q is a retarded (causal) function and is complex, with $\hat{\omega} = \omega + i\nu$, where ν is the total collision frequency. Further, $\Pi_q(\hat{\omega})$ denotes the longitudinal retarded polarization function. The simplest approximation for Π is the collisionless limit – the random phase approximation (RPA) or Lindhard polarization

$$\Pi_0(q, \hat{\omega}) = 2 \int \frac{d^3Q}{(2\pi)^3} \frac{f^{(0)}_{-,\alpha} - f^{(0)}_{+,\alpha}}{\hbar\hat{\omega} - \hbar^2 \mathbf{Q} \cdot \mathbf{q}/m_\alpha}, \tag{14}$$

where $\nu \to +0$. $f^{(0)}_{\pm,\alpha}$ denotes a Fermi distribution at the argument $\mathbf{Q} \pm \mathbf{q}/2$. An improved result which takes into account electronic correlations (collisions) in a way that sum rules are fulfilled was presented for classical systems by Rostoker and Rosenbluth [17]. The generalization to quantum systems is due to Mermin [16], for a derivation see the appendix,

$$\Pi^M(q, \hat{\omega}) = \frac{\Pi_0(q, \hat{\omega})}{1 + i\hbar\tilde{\Pi}_{\nu 0}(q, \hat{\omega})}. \tag{15}$$

Further improvements of the Mermin result have been considered by various groups. Röpke et al. have derived a Mermin-type expression which, besides particle conservation contains energy conservation [19]. However, they found that the effect was small. Another modification by this group was to include a frequency dependent collision frequency into the relaxation time collision integral [20]. Finally, we mention that a selfconsistent nonequilibrium calculation within Nonequilibrrium Green's functions which fully included sum rule preservation has been recently performed [21].

4.1. Dynamical screening and wake effects

The main motivation to include dynmical screening of the interaction between heavy particles in a two-component plasma is its importance for nonequilbrium situations. One such case is the existing of streaming light particles which causes wake effects which have a dramatic effect on the arrangement of heavy particles, e.g. dust particles in a complex plasma. This was discussed in detail by Joyce and Lampe, cf. [22] and references therein. Wake effects in a quantum plasma have also been considered by one group [23] who found an important influence on stopping power of ions in a polarizable medium. However, these were only single particle effects.

5. Discussion

A new model for the simulation of dense quantum plasmas including dynamical screening of the electrons, partial ionization and strong ion correlations has been developed. It is particularly important for high density low-temperature plasmas, in situations where the ions form liquid or solid-like structures and when the electrons are in nonequilibrium. Typical situations are streaming electrons or plasma instabilities due to fast electrons or electromagnetic fields. The proposed simulation scheme is based on classical molecular dynamics simulations where the dynamical screening effects are incorporated within a linear response approach into the screening of the ion-ion pair interactions. For the screening an extension beyond the (collisionless) RPA model has been used which is due to Mermin and a strict derivation within quantum kinetic theory has been given putting the original result of Ref. [16] on solid ground and critically assessing its scope of applicability. Correlations could also be included via other approximations such as the quasilocalized charge approximation [24].

We mention that in the present version of our model the charge state of the ions is not selfconsistently computed but is used as an input. Generalizations along this line are straightforward. Also, the co-existence of several charge states maybe included via a Saha equation. The present discussion was for point-like ions. It maybe directly extended to ions with an internal structure where the pair interaction deviates from pure Coulomb repulsion, e.g. [25].

Let us briefly comment on the question how to include quantum effects in the pair interaction. This could be of importance for astrophysical applications when the ion degeneracy parameter χ_i approaches one. In this case an approximate treatment of ion degeneracy effects is sufficient which can be included in the derivation of the classical equations of motion from the quantum equation (3). As a result the ion pair potential is modified at small distances due to quantum effects which remove the divergence. This problem was first studied by Kelbg and numerous other authors. For a discussion of the Kelbg potential and improvements, see e.g. [26]. In using the expression (11b) we assumed pointlike ions and neglected any finite ion extension. This can be corrected in the final expression by replacing the Fourier transform of the Coulomb potential, i.e. the factor $1/k^2$ in Eq. (11b), by the Fourier transform of the improved Kelbg potential [26] or of any other appropriate quantum potential.

Acknowledgements

We acknowledge stimulating discussions with G. Joyce. This work is supported by the Deutsche Forschungsgemeinschaft via grant LU 1586/1 and by the NSF/DOE Partnership in Basic Plasma Science and Engineering under the Department of Energy award DE-FG02-07ER54946.

Appendix: Kinetic derivation of the Mermin polarization

In this appendix we give a rigorous and general derivation of the Mermin polarization function (14) from quantum kinetic theory, which has not been given before. We start with the collisionless case which leads to the well-known RPA result. It is presented here to set up the notation and subsequently generalize it to include collisions. The first hierarchy equation for a multi-component system reads, in mean-field approximation, cf. Eq. (1) with $N = 1$,

$$i\hbar\frac{\partial f_{k,k',\alpha}}{\partial t} - (\epsilon_{k,\alpha} - \epsilon_{k',\alpha})f_{k,k',\alpha} - \frac{\mathcal{V}}{N_a}\langle k|[U^{\text{eff}}_{1,\alpha}, F_{1,\alpha}]|k'\rangle = 0, \tag{16}$$

with the operator of the effective potential and the mean field potential

$$U^{\rm eff}_{1,\alpha} = U_{1,\alpha} + W_{1,\alpha}, \tag{17}$$

$$W_{1,\alpha} = \sum_\beta n_\beta {\rm Tr}_2 V_{1\alpha,2\beta} F_{2\beta}. \tag{18}$$

The normalization conditions for the density operator and the distribution function are

$$N_\alpha \langle k|F_{1,\alpha}|k'\rangle = \mathcal{V} f_{k,k',\alpha}, \tag{19}$$

$$N_\alpha = \frac{N_\alpha}{\mathcal{V}} {\rm Tr}_1 F_{1,\alpha}(t) = 2\sum_k f_{kk,\alpha}(t), \tag{20}$$

where $\mathcal{V}$ is the volume and the kinetic energy of particle species "α" in momentum state $|\mathbf{k}\rangle$ is given by $\epsilon_{k,\alpha} = \frac{\hbar^2}{2m_\alpha}k^2$. Eq. (16) is the nonlinear quantum Vlasov equation (time-dependent Hartree equation).

Here we will concentrate on an unmagnetized plasma, so the external potential U is due to an electrostatic potential and $U^{\rm eff}_{1,\alpha} = e_\alpha(\phi_{ext} + \phi_{ind})$, where the induced potential is the solution of Poisson's equation, $\Delta\phi_{ind} = -4\pi\rho_{ind}$ with the plasma particles acting as source. As in the classical case ϕ_{ind} functionally depends on the particle density and thus, on the distribution function, giving rise to a nonlinear kinetic equation.

We start with a plasma without external fields, $U \equiv 0$. Then the plasma will be homogeneous and all contributions to the induced potential cancel, due to charge neutrality, $\phi_{ind} \equiv 0$. We expect that the plasma will be in a (local) thermodynamic equilibrium state which, for fermions, e.g. electrons, is given by the Fermi function[1]

$$f_{k,k',\alpha} = f^{(0)}_{k,k',\alpha} = f^{(0)}_{k,\alpha}\delta_{k,k'} = f^{EQ}_{k,\alpha}\delta_{k,k'} \tag{21}$$

$$f^{EQ}_{k,\alpha} = \left[e^{\beta(\epsilon_{k\alpha}-\mu_\alpha)} + 1\right]^{-1}, \tag{22}$$

where $\beta = 1/k_BT$ and $\mu_\alpha(n_\alpha, T)$ is the chemical potential, and the momentum delta function arises due to spatial homogeneity.

We now assume that a weak electrostatic potential ϕ_{ext} is turned on, perturbing the equilibrium state. Weakness of the potential allows us to linearize the kinetic equation and solve it with the linear response ansatz [quantities of first order in ϕ_{ext} are denoted by the superscript "(1)"]

$$U^{\rm eff}_\alpha(\mathbf{r},t) = U^{\rm eff(1)}_\alpha(\mathbf{r},t), \quad \lim_{t\to-\infty} U^{\rm eff}_\alpha(\mathbf{r},t) = 0, \tag{23}$$

$$f_{k,k',\alpha}(t) = f^{(0)}_{k,k',\alpha} + f^{(1)}_{k,k',\alpha}(t), \tag{24}$$

$$\lim_{t\to-\infty} f_{k,k',\alpha}(t) = f^{(0)}_{k,k',\alpha}.$$

Using this ansatz we can now evaluate the matrix elements of the commutator in Eq. (16). Since the potential is of first order, for the density operator the zeroth order gives the dominant contribution which is diagonal,

$$\langle k|[U^{\rm eff}_{1,\alpha}, F_{1,\alpha}]|k'\rangle \approx \langle k|[U^{\rm eff(1)}_{1,\alpha}, F^{(0)}_{1,\alpha}]|k'\rangle \tag{25}$$

$$= U^{\rm eff(1)}_{k,k',\alpha}\frac{N_a}{\mathcal{V}}\left(f^{(0)}_{k',\alpha} - f^{(0)}_{k,\alpha}\right). \tag{26}$$

[1] The only remaining contribution in Eq. (16) is the time derivative, i.e. $\frac{\partial f_{k,k',\alpha}}{\partial t} = 0$. Therefore, as in the classical case, any stationary distribution function can be used. The choice of the equilibrium distribution is based on experience. Only in the case of collisions, relaxation processes (scattering) will lead to a unique selfconsistent stationary solution.

Inserting this result into the kinetic equation (16) and taking into account that $f^{(0)}_{k,k'\alpha}$ is time-independent and diagonal, we obtain a closed equation for $f^{(1)}_{k,k',\alpha}(t)$

$$i\hbar\frac{\partial f^{(1)}_{k,k',\alpha}}{\partial t} - (\epsilon_{k,\alpha} - \epsilon_{k',\alpha})f^{(1)}_{k,k',\alpha} - U^{\mathrm{eff}(1)}_{k,k',\alpha}\left(f^{(0)}_{k',\alpha} - f^{(0)}_{k,\alpha}\right) = 0, \tag{27}$$

which is already a linear equation. Expanding the field into a Fourier integral of monochromatic oscillations[2]

$$U_{k,k',\alpha}(t) = \int d\omega\, \tilde{U}_{k,k',\alpha}(\omega)\, e^{-i(\omega+i\delta)t}, \tag{28}$$

the same expansion will also apply to $f^{(1)}_{k,k',\alpha}$ and $U^{\mathrm{eff}(1)}_{k,k',\alpha}$.

Fourier transforming Eq. (27) and cancelling the common exponent yields the result for the perturbation of the distribution,

$$\tilde{f}^{(1)}_{k,k',\alpha}(\omega) = \frac{f^{(0)}_{k',\alpha} - f^{(0)}_{k,\alpha}}{\hbar(\omega+i\delta) - (\epsilon_{k,\alpha} - \epsilon_{k',\alpha})}\, \tilde{U}^{\mathrm{eff}(1)}_{k,k',\alpha}(\omega). \tag{29}$$

This is the general result for the linear perturbation to the distribtution function, and the time-dependent expression follows by a back transform. In deriving (29) we have made no assumptions on the space dependence of the external perturbation U.

In the following we consider an exciting field which is purely periodic in space without any macroscopic spatial modulation, which can be expanded in a Fourier series $U_\alpha(\mathbf{r}) = \sum_q U_{q\alpha}e^{i\mathbf{qr}}$. Now introduce center of mass and relative momenta by (all quantities are vectors)

$$\begin{aligned} Q &= \frac{k+k'}{2}, \qquad q = k - k', \quad \text{or, vice versa,} \\ k &= Q + \frac{q}{2}, \qquad k' = Q - \frac{q}{2}. \end{aligned} \tag{30}$$

While the center of mass variable Q is related to spatial inhomegeneities, the relative momentum q is directly related to small scale spatially periodic modulations. Thus, in the present case, there will be no dependence on Q, i.e.

$$U_{k,k',\alpha} \equiv U_{Q+\frac{q}{2},Q-\frac{q}{2},\alpha} \longrightarrow U_q \tag{31}$$

and similarly for $\tilde{U}^{\mathrm{eff}(1)}_{k,k',\alpha}(\omega)$. In contrast, in the unperturbed quantities ϵ_k and $f^{(0)}_{k,\alpha}$ the momentum arguments k and k' remain. Therefore, to shorten the notation, in the following we will denote $\mathbf{Q}\pm\mathbf{q}/2 \to \pm$. Via ϵ_k and $f^{(0)}_{k,\alpha}$, the dependence on Q remains also in the perturbed distribution function. Thus, we can rewrite the result (29) for a single Fourier component q of a periodic monochromatic excitation

$$\tilde{f}^{(1)}_{+,-,\alpha}(\omega) = \frac{f^{(0)}_{-,\alpha} - f^{(0)}_{+,\alpha}}{\hbar(\omega+i\delta) - (\epsilon_{+,\alpha} - \epsilon_{-,\alpha})}\, \tilde{U}^{\mathrm{eff}(1)}_{q,\alpha}(\omega). \tag{32}$$

This is still not an explicit result for $\tilde{f}^{(1)}$ because the function also appears in the effective potential. To make further progress we consider the density disturbance and compute its Fourier components

$$\tilde{n}^{(1)}_{q,\alpha}(\omega) \;=\; 2\int \frac{d^3Q}{(2\pi)^3}\, \tilde{f}^{(1)}_{Q+\frac{q}{2},Q-\frac{q}{2},\alpha}(\omega), \tag{33}$$

[2] δ is a small positive constant which assures causality, i.e. $\lim_{t\to-\infty} U_{k,k'} = 0$.

and the prefactor 2 accounts for the spin summation. Below, we will also need the current density which is calculated in similar way

$$\tilde{\mathbf{j}}^{(1)}_{q,\alpha}(\omega) = 2\int \frac{d^3Q}{(2\pi)^3}\frac{\hbar\mathbf{Q}}{m_\alpha}\tilde{f}^{(1)}_{Q+\frac{q}{2},Q-\frac{q}{2},\alpha}(\omega). \tag{34}$$

Particle and current density are connected via the continuity equation

$$\frac{\partial n(\mathbf{r},t)}{\partial t} + \operatorname{div}\mathbf{j}(\mathbf{r},t) = 0. \tag{35}$$

Since the unperturbed expressions are time and space independent we can replace the total density and current density by the perturbed functions. The Fourier transform of this expression then reads

$$\omega\,\tilde{n}^{(1)}_{q,\alpha}(\omega) = \mathbf{q}\cdot\tilde{\mathbf{j}}^{(1)}_{q,\alpha}(\omega). \tag{36}$$

We can now explicitly compute the density perturbation by inserting the distribution function (32) into Eq. (33) and obtain

$$\tilde{n}^{(1)}_{q,\alpha}(\omega) = \Pi^{\mathrm{RPA}}_{0\alpha}(q,\hat{\omega})\,\tilde{U}^{\mathrm{eff}(1)}_{q,\alpha}(\omega), \tag{37}$$

$$\Pi^{\mathrm{RPA}}_{n\alpha}(q,\hat{\omega}) = 2\int\frac{d^3Q}{(2\pi)^3}\left(\frac{\hbar\mathbf{Q}}{m_\alpha}\right)^n \frac{f^{(0)}_{-,\alpha}-f^{(0)}_{+,\alpha}}{\hbar\hat{\omega}-[\epsilon_{+,\alpha}-\epsilon_{-,\alpha}]}, \tag{38}$$

where we defined $\hat{\omega} = \omega + i\delta$. Here $\Pi^{\mathrm{RPA}}_{n\alpha}$ denotes the n-th moment of the polarization function $\Pi^{\mathrm{RPA}}_{0\alpha}$.

The simplest way to obtain the longitudinal quantum dielectric function is to use the continuity equation and relation (13) which is valid for classical *and* quantum plasmas. Computing the total electric charge density perturbation

$$\tilde{\rho}^{(1)}_q(\omega) = \sum_\alpha q_\alpha \tilde{n}^{(1)}_{q,\alpha}(\omega), \tag{39}$$

and using the relation of the effective potential to the total electrostatic potential

$$\tilde{U}^{\mathrm{eff}(1)}_{q,\alpha}(\omega) = q_\alpha\tilde{\phi}_q(\omega), \tag{40}$$

we only need to insert (39) with the result (37) into (11a) and obtain

$$\epsilon^{\mathrm{l,RPA}}(\mathbf{k},\hat{\omega}) = 1 - \sum_\alpha \tilde{V}_{\alpha\alpha}(k)\Pi^{\mathrm{RPA}}_{0\alpha}(\mathbf{k},\hat{\omega}). \tag{41}$$

This is the longitudinal retarded dielectric function for a collisionless quantum plasma. In using the superscript "RPA" for the dielectric function and the polarization (14) we indicated that this is the result in the so-called *random phase approximation* (RPA) which is als frequently called Lindhard polarization. Note, however, that this result was obtained by many authors independently. The first, apparently were Klimontovich and Silin [27, 28], followed by [29, 30].

We now proceed to improve this result by including correlation effects. The quantum kinetic equation in momentum representation, with collision integral reads

$$i\hbar\frac{\partial f_{k,k',\alpha}}{\partial t} - (\epsilon_{k,\alpha}-\epsilon_{k',\alpha})f_{k,k',\alpha} - \frac{\mathcal{V}}{N_\alpha}\langle k|[U^{\mathrm{eff}}_{1,\alpha},F_{1,\alpha}]|k'\rangle = I_{k,k',\alpha}, \tag{42}$$

Prior to external perturbation, we again assume a homogeneous state with the distribution given by a diagonal matrix $f_{k,k',\alpha} = f^{(0)}_{k,k',\alpha} = f^{(0)}_{k,\alpha}\delta_{k,k'}$, and the same applies to the collision integral, $I_{k,k',\alpha} = I^{(0)}_{k,k',\alpha} = I^{(0)}_{k,\alpha}\delta_{k,k'}$. Then the kinetic equation (42) simplifies to a diagonal equation

$$i\hbar\frac{\partial f^{(0)}_{k,\alpha}}{\partial t} = I^{(0)}_{k,\alpha} \equiv I_{k,\alpha}[f^{(0)}_{k,\alpha}], \tag{43}$$

where in the collision integral the unperturbed distribution function has to be used. Following Mermin [16] we use the relaxation time approximation for the collision integral which is constructed by the ansatz (we drop all arguments)

$$I^{RTA}[f] \equiv -\frac{1}{\tau}\left(f - f^{EQ}\right), \quad f(0) = f_0. \tag{44}$$

The solution of this equation, together with the initial condition at $t = 0$, is $f(t) = f_0 e^{-t/\tau} + f^{EQ}[1 - e^{-t/\tau}]$, showing the decay of the initial state and the approach to the asymptotic state. Here, τ is the total relaxation time due to all scattering processes which has to be computed from a separate kinetic theory or taken from experiment. We use a simple static approximation where τ is frequency independent. Despite its simplicity, this approximation allows to achieve a selfconsistent relaxation of the distribution function to the equilibrium state. While we expect that f^{EQ} will be a Fermi function, we will not need the explicit form of the equilibrium distribution below.

We now again consider a weak perturbation by a longitudinal field $U_a = q_a\phi^{ext}$,

$$\begin{aligned} U^{\rm eff}_\alpha(\mathbf{r},t) &= U^{\rm eff(1)}_\alpha(\mathbf{r},t), \quad \lim_{t\to-\infty} U^{\rm eff}_\alpha(\mathbf{r},t) = 0, \\ f_{k,k',\alpha}(t) &= f^{(0)}_{k,k',\alpha} + f^{(1)}_{k,k',\alpha}(t), \\ \lim_{t\to-\infty} f_{k,k',\alpha}(t) &= f^{(0)}_{k,k',\alpha}. \end{aligned} \tag{45}$$

The equation for the perturbation of the distribution reads, in first order,

$$\begin{aligned} i\hbar\frac{\partial f^{(1)}_{k,k',\alpha}}{\partial t} - (\epsilon_{k,\alpha} - \epsilon_{k',\alpha})f^{(1)}_{k,k',\alpha} - U^{\rm eff(1)}_{k,k',\alpha}\cdot\left(f^{(0)}_{k',\alpha} - f^{(0)}_{k,\alpha}\right) &= I^{(1)}_{k,k',\alpha}, \\ \lim_{t\to-\infty} f_{k,k',\alpha}(t) &= f^{(0)}_{k,\alpha}\delta_{k,k'}. \end{aligned} \tag{46}$$

Here, $U^{\rm eff(1)}$ is again obtained by replacing $f_{k,k'}$ by $f^{(1)}_{k,k'}$, whereas $I^{(1)}$ is obtained by keeping in all appearances of the electron distribution functions only terms of first order in $f^{(1)}$ [21]. In case of the relaxation time approximation which is linear in f we just have to use $f^{(1)}_{k,k'}$. In order to find the explicit expression for $I^{(1)}$ we consider the long time limit of the system. The complete asysmptotic solution of the original kinetic equation will be the sum of the zero and first order terms, i.e. $f^{\rm EQ}$ and the asymptotic solution for $f^{(1)}_{k,k'}$ which we will denote $f^{(1)\infty}_{k,k'}$. To find this solution assume a purely periodic space dependence of U. Using center of mass and relative momenta Eq. (46) becomes

$$i\hbar\frac{\partial f^{(1)\infty}_{k,k',\alpha}}{\partial t} - (\epsilon_{+,\alpha} - \epsilon_{-,\alpha})f^{(1)\infty}_{k,k',\alpha} - U^{\rm eff(1)\infty}_{q,\alpha}\cdot\left(f^{(0)}_{-,\alpha} - f^{(0)}_{+,\alpha}\right) = I^{(1)\infty}_{k,k',\alpha}. \tag{47}$$

Since we are looking for a stationary solution, the time-dependence should vanish. Also, we expect that collisions have relaxed to zero, i.e. $I^{(1)\infty}_{k,k',\alpha} = I^{(1)}_{k,k',\alpha}[f^{(1)\infty}] = 0$, resulting in

$$\lim_{t\to\infty} f^{(1)}_{k,k',\alpha} = f^{(1)\infty}_{k,k',\alpha} = \frac{f^{(0)}_{+,\alpha} - f^{(0)}_{-,\alpha}}{\epsilon_{+,\alpha} - \epsilon_{-,\alpha}} U^{\text{eff}(1)\infty}_{q,\alpha}. \tag{48}$$

Note that also the effective potential carries the superscrtipt ∞ since it also depends on this solution.

Performing a Fourier transform with respect to time, according to Eq. (28) we obtain

$$\tilde{f}^{(1)\infty}_{k,k',\alpha} = \frac{f^{(0)}_{+,\alpha} - f^{(0)}_{-,\alpha}}{\epsilon_{+,\alpha} - \epsilon_{-,\alpha}} \tilde{U}^{\text{eff}(1)\infty}_{q,\alpha}, \qquad \omega = 0, \tag{49}$$

where, as a result of the long-time limit, this form is restricted to zero frequency.

Now we can construct the first order correction to the collision term in Eq. (46). Since $f^{(0)}$ relaxes towards f^{EQ}, in Eq. (44) only the difference of $f^{(1)}$ and $f^{(1)\infty}$ remains. Multiplying by $i\hbar$ we obtain

$$\tilde{I}^{(1)}_{k,k',\alpha}(\omega) = -\frac{i\hbar}{\tau}\left\{\tilde{f}^{(1)}_{k,k',\alpha}(\omega) - \frac{f^{(0)}_{+,\alpha} - f^{(0)}_{-,\alpha}}{\epsilon_{+,\alpha} - \epsilon_{-,\alpha}} \tilde{U}^{\text{eff}(1)\infty}_{q,\alpha}\right\}, \tag{50}$$

which can be inserted in Eq. (46) after Fourier transform to frequency space[3] yielding the solution

$$\tilde{f}^{(1)}_{k,k',\alpha}(\omega) = \left\{\tilde{U}^{\text{eff}(1)}_{q,\alpha} - i\hbar\nu\frac{\tilde{U}^{\text{eff}(1)\infty}_{q,\alpha}}{\epsilon_{+,\alpha} - \epsilon_{-,\alpha}}\right\} \frac{f^{(0)}_{-,\alpha} - f^{(0)}_{+,\alpha}}{\hbar\hat{\omega} - [\epsilon_{+,\alpha} - \epsilon_{-,\alpha}]}, \tag{51}$$

where we defined $\hat{\omega} = \omega + i\nu$ and used $\delta \to 0$ due to the existence of a finite collisional damping $\nu = \tau^{-1}$. This result is a straightforward extension of the collisionless random phase approximation, cf. Eq. (32). Scattering effects (terms proportional to ν) are contained in two places: first, the frequency in the denominator is replaced by a complex frequency and, second, there appears an additional contribution proportional to $U^{\text{eff}(1),\infty}$ in the numerator which renormalizes the Fourier component of the effective potential.

Eq. (51) is not an explicit solution for $\tilde{f}^{(1)}$ since this function also appears in the effective potential. To solve this problem and compute the dielectric function, we proceed as in the collisionless case. We first compute the Fourier component of the density disturbance according to Eq. (33) and also of the current density, using Eq. (34) and the definition (38),

$$\tilde{n}_{q,\alpha}(\omega) = \Pi^{\text{RPA}}_{0\alpha}(q,\hat{\omega})\tilde{U}^{\text{eff}(1)}_{q,\alpha}(\omega) - i\hbar\nu\, \tilde{U}^{\text{eff}(1)\infty}_{q,\alpha}\Pi_{\nu 0\alpha}(q,\hat{\omega}), \tag{52}$$

$$\tilde{\mathbf{j}}_{q,\alpha}(\omega) = \Pi^{\text{RPA}}_{1\alpha}(q,\hat{\omega})\tilde{U}^{\text{eff}(1)}_{q,\alpha}(\omega) - i\hbar\nu\, \tilde{U}^{\text{eff}(1)\infty}_{q,\alpha}\Pi_{\nu 1\alpha}(q,\hat{\omega}), \tag{53}$$

where we used the definition (41) of the RPA polarization and introduced in analogy to (38) a modified polarization function which arises from the collisions

$$\Pi_{\nu n\alpha}(q,\hat{\omega}) = 2\int\frac{d^3Q}{(2\pi)^3}\left(\frac{\hbar\mathbf{Q}}{m_\alpha}\right)^n \frac{1}{\epsilon_{+,\alpha} - \epsilon_{-,\alpha}} \times \frac{f^{(0)}_{-,\alpha} - f^{(0)}_{+,\alpha}}{\hbar\hat{\omega} - [\epsilon_{+,\alpha} - \epsilon_{-,\alpha}]}. \tag{54}$$

Eq. (52) indicates that collisions change the particle density (second term) compared to the RPA result. As a consequence of the local density conservation law also the current density has to change, cf. Eq. (53).

[3] We again assume a spatially periodic excitation.

Eqs. (52) and (53) contain the still unknown function $\tilde{U}^{\text{eff}(1)\infty}$ which we determine using the continuity equation (36). We now transform $\hbar\mathbf{q}$ times the integral $\Pi_{1\alpha}$, by adding and subtracting, under the integral, $\hbar\hat{\omega}$. Taking into account that $\epsilon_{+,\alpha} - \epsilon_{-,\alpha} = \hbar^2\mathbf{Q}\cdot\mathbf{q}/m_\alpha$, we obtain the identity

$$\hbar\mathbf{q}\cdot\Pi_{1\alpha}^{\text{RPA}}(q,\hat{\omega}) = 2\int\frac{d^3Q}{(2\pi)^3}\left(f_{-,\alpha}^{(0)} - f_{+,\alpha}^{(0)}\right) + (\hbar\omega + i\hbar\nu)\Pi_{0\alpha}^{\text{RPA}}(q,\hat{\omega}). \tag{55}$$

Assuming that the field-free distribution depends only on the modulus of the momentum, i.e. $f_{-\mathbf{k},\alpha}^{(0)} = f_{\mathbf{k},\alpha}^{(0)}$ the integrals over f_- and f_+ cancel. The same transformation is possible for the integral $\Pi_{\nu 1\alpha}$ with the result

$$\hbar\mathbf{q}\cdot\Pi_{\nu 1\alpha}(q,\hat{\omega}) = \Pi_{0\alpha}^{\text{RPA}}(q,0) + (\hbar\omega + i\hbar\nu)\Pi_{\nu 0\alpha}(q,\hat{\omega}). \tag{56}$$

Collecting the results (55) and (56) together we may rewrite Eq. (53),

$$\begin{aligned}
\hbar\mathbf{q}\cdot\tilde{\mathbf{j}}_{q,\alpha}(\omega) &= \hbar(\omega+i\nu)\Pi_{0\alpha}^{\text{RPA}}(q,\hat{\omega})\tilde{U}_{q,\alpha}^{\text{eff}(1)} \\
&\quad -i\hbar\nu\left[\Pi_{0\alpha}^{\text{RPA}}(q,0) + (\hbar\omega+i\hbar\nu)\Pi_{\nu 0\alpha}(q,\hat{\omega})\right]\tilde{U}_{q,\alpha}^{\text{eff}(1)\infty} \\
&= \hbar\omega\cdot\tilde{n}_{q,\alpha}(\omega) + \\
&\quad + i\hbar\nu\left\{\Pi_{0\alpha}^{\text{RPA}}(q,\hat{\omega})\tilde{U}_{q,\alpha}^{\text{eff}(1)} - \tilde{U}_{q,\alpha}^{\text{eff}(1)\infty}\left[\Pi_{0\alpha}^{\text{RPA}}(q,0) + i\hbar\nu\Pi_{\nu 0\alpha}(q,\hat{\omega})\right]\right\}.
\end{aligned} \tag{57}$$

Evidently, the continuity equation (36) is fulfilled if the terms on the last line (in the curley brackets) vanish which yields the required condition on $\tilde{U}^{\text{eff}(1)\infty}$

$$\tilde{U}_{q,\alpha}^{\text{eff}(1)\infty}(\omega) = \frac{\Pi_{0\alpha}^{\text{RPA}}(q,\hat{\omega})\tilde{U}_{q,\alpha}^{\text{eff}(1)}(\omega)}{\Pi_{0\alpha}^{\text{RPA}}(q,0) + i\hbar\nu\Pi_{\nu 0\alpha}(q,\hat{\omega})} = \frac{\tilde{n}_{q,\alpha}(\omega)}{\Pi_{0\alpha}^{\text{RPA}}(q,0)}. \tag{58}$$

With this result we can now eliminate $\tilde{U}^{\text{eff}(1)\infty}$ from all expressions. Inserting it into (52), we can solve for the density perturbation which – as in the collisionless case, cf. Eq. (37) – is proportional to the effective potential, however, with a modified coefficient

$$\tilde{n}_{q,\alpha}(\omega) = \Pi_\alpha^M(q,\hat{\omega})\tilde{U}_{q,\alpha}^{\text{eff}(1)}(\omega), \tag{59}$$

Using the result (59) we immediately obtain the dielectric function. Computing the total charge density to $\tilde{\rho}_q(\omega) = \tilde{\Phi}_q(\omega)\sum_\alpha q_\alpha^2\Pi_\alpha^M(q,\hat{\omega})$ and inserting the result into Eq. (11a) we obtain

$$\epsilon^{l,M}(\mathbf{k},\omega) = 1 - \sum_\alpha\tilde{V}_{\alpha\alpha}(k)\Pi_\alpha^M(\mathbf{k},\omega). \tag{60}$$

This is the longitudinal quantum dielectric function including collisions in relaxation time approximation. This result was first obtained by Mermin [16], therefore we use the superscript "M". The formal structure of this expression is the same as in the collisionless and classical cases, it is a general property of linear response theory. The different physical approximations are solely contained in the longitudinal polarization function, the Mermin polarization,

$$\Pi_\alpha^M(q,\hat{\omega}) \equiv \frac{\Pi_{0\alpha}^{\text{RPA}}(q,\hat{\omega})}{1 + i\hbar\nu\tilde{\Pi}_{\nu 0\alpha}(q,\hat{\omega})} \tag{61}$$

where we introduced the definition

$$\tilde{\Pi}_{\nu 0\alpha}(q,\hat{\omega}) \equiv \frac{\Pi_{\nu 0\alpha}(q,\hat{\omega})}{\Pi_{0\alpha}^{\text{RPA}}(q,0)} = \frac{1}{\hbar\omega}\left[\frac{\Pi_{0\alpha}^{\text{RPA}}(q,\hat{\omega})}{\Pi_{0\alpha}^{\text{RPA}}(q,0)} - 1\right]. \tag{62}$$

In the last equality we have eliminated the function $\Pi_{\nu 0\alpha}$ and expressed it in terms of the RPA polarization where use has been made of the identity $\hbar\omega\Pi_{\nu 0\alpha}(q,\hat{\omega}) = \Pi_{0\alpha}^{\rm RPA}(q,\hat{\omega}) - \Pi_{0\alpha}^{\rm RPA}(q,0)$. Obviously, the collisionless limit of the Mermin dielectric function, i.e. $\nu \to 0$ and $\hat{\omega} \to \omega + i\delta$, is given by the random phase approximation. The classical limit of Π_α^M gives the result derived from the the Bhatnagar/Gross/Krook collision integral which was first obtained by Rostoker and Rosenbluth [17].

References

[1] D.H.E. Dubin, and T.M. O'Neill, Rev. Mod. Phys. **71**, 87 (1999)
[2] M. Bonitz, P. Ludwig, H. Baumgartner, C. Henning, A. Filinov, D. Block, O. Arp, A. Piel, S. Käding, Y. Ivanov, A. Melzer, H. Fehske, V.S. Filinov, Phys. Plasmas **15**, 055704 (2008)
[3] A. V. Filinov, M. Bonitz, and Y. E. Lozovik, Phys. Rev. Lett. **86**, 3851 (2001)
[4] M. Bonitz, C. Henning, and D. Block, *Complex plasmas - a laboratory for strong correlations*, submitted to Reports on Progress in Physics (2009)
[5] The conditions of plasma crystallization have been derived in M. Bonitz, V.S. Filinov, V.E. Fortov. P.R. Levashov, and H. Fehske, Phys. Rev. Lett. **95**, 235006 (2005)
[6] D. Gericke, K. Wünsch, A. Grinenko and J. Vorberger, J. Phys.: Conf. Ser. (2010), this volume
[7] B. Militzer, and D.M. Ceperley, Phys. Rev. Lett. **85**, 1890 (2000)
[8] V. Filinov, M. Bonitz, W. Ebeling, and V.E. Fortov, Plasma Phys. Control. Fusion, **43**, 743 (2001); V. Filinov, M. Bonitz, and V.E. Fortov, JETP Lett. **72**, 245 (2000)
[9] F. Lambert, J. Clerouin, S. Mazevet, and D. Gilles, Contrib. Plasma Phys. **47**, 272 (2007)
[10] B. Holst, R. Redmer, and M.P. Desjarlais, Phys. Rev. B **77**, 184201 (2008)
[11] V. Filinov et al., Phys. Rev. B **65**, 165124 (2002); V.S. Filinov, P.R. Levashov, A.V. Botan, M. Bonitz, and V.E. Fortov, J. Phys. A **42**, 214002 (2009)
[12] M. Bonitz, *Quantum Kinetic Theory* (Teubner, Stuttgart, Leipzig, 1998)
[13] A. Mallik, K. Runge, J. Dufty, and H-P Cheng, J. Chem. Phys. **127**, 224707 (2007)
[14] A.F. Alexandrov, L.S. Bogdankevich, A.A. Rukhadse, *Principles of Plasma Electrodynamics* (Springer, Heidelberg, 1984)
[15] A. Fromm, J.W. Dufty, and M. Bonitz, Ann. Phys. (N.Y.) **323**, 3158 (2008)
[16] D. Mermin, Phys. Rev. B **1**, 2362 (1970)
[17] N. Rostoker and M.N. Rosenbluth, Phys. Fluids **3**, 1 (1960)
[18] G. Röpke, A. Selchow, A. Wierling, and H. Reinholz, Phys. Lett. A **260**, 365 (1999)
[19] A. Selchow, G. Röpke, and A. Wierling, Contrib. Plasma Phys. **42**, 43 (2002)
[20] T. Millat, A. Selchow, A. Wierling, H. Reinholz, R. Redmer, and G. Röpke, J. Phys. A **36**, 6259 (2003)
[21] N.H. Kwong, and M. Bonitz, Phys. Rev. Lett. **84**, 1768 (2000)
[22] G. Joyce, M. Lampe, and G. Ganguli, Phys. Rev. Lett **88**, 095006 (2002); M. Lampe, G. Joyce, and G. Ganguli, Phys. Plasmas **7**, 3851 (2000); IEEE Tans. Plasma Sci. **29**, 238 (2001); Phys. Scr. **T89**, 106 (2001)
[23] I. Abril, R. Garcia-Molina, C.D. Denton, J. Perez-Perez, and N.R. Arista, Phys. Rev. A **58**, 357 (1998)
[24] G. J. Kalman, K. I. Golden, Z. Donko, and P. Hartmann, J. Phys.: Conf. Ser. **11**, 254 (2005)
[25] K. Wünsch, J. Vorberger, and D. Gericke, Phys. Rev. E **79**, 010201 (2009)
[26] A. V. Filinov, V. Golubnychiy, M. Bonitz, W. Ebeling and, J.W. Dufty, Phys. Rev. E **70**, 046411 (2004)
[27] Yu. L. Klimontovich and V. P. Silin, Dokl. Akad. Nauk, **82**, 361 (1952)
[28] Yu. L. Klimontovich and V. P. Silin, ZhETF, **23**, 151 (1952)
[29] D. Bohm and D. Pines, Phys.Rev. **92**, 609 (1953)
[30] J. Lindhard, Kgl. Danske Videnskab. Selskab. Mat. Fys. Medd., **28**, 8 (1954)
[31] M.D. Barriga-Carrasco, Phys. Rev. E **73**, 026401 (2006); M.D. Barriga-Carrasco, Phys. Plasmas **15**, 033103 (2008)

Progress in Nonequilibrium Green's Functions IV IOP Publishing
Journal of Physics: Conference Series **220** (2010) 012004 doi:10.1088/1742-6596/220/1/012004

BEC–BCS transition in excitonic systems

D Kremp, D Semkat and K Henneberger
Institut für Physik, Universität Rostock, 18051 Rostock, Germany

Abstract. We consider quantum condensation in the electron–hole plasma of highly excited semiconductors. A theoretical approach applying the concept of time long range order in the framework of real-time Green's functions is presented and non-equilibrium generalizations of the basic equations of quantum condensation are derived. For the quasi-equilibrium case, we solve the coupled system of number and gap equations in ladder approximation for a statically screened Coulomb potential. The resulting phase boundary shows a smooth crossover from BEC to BCS.

1. Introduction

In this paper we consider quantum condensation in electron–hole plasmas (EHPs) in excited semiconductors. Quantum condensation occurs in Bose systems well-known as Bose–Einstein condensation (BEC) and in Fermi systems as condensation of Cooper pairs into the BCS state. In this connection the question arises whether there are systems with both types of condensation. The answer is clear: BEC as well as BCS-type condensation may be expected in Fermi systems with Bose-like bound states, e.g., in

(i) nuclear matter [1]: $p + n \rightleftarrows D^+$
(ii) spin polarized hydrogen [2]: $e + p \rightleftarrows H$
(iii) Fermi atoms in traps, currently intensively in the discussion [3, 4]: $A + A \rightleftarrows A_2$
(iv) EHPs in semiconductors [5, 6, 7]: $e + h \rightleftarrows X$

The properties of an EHP are well investigated, see, e.g., Refs. [8, 9, 10]. Of particular interest is the formation of bound states, the excitons (X). The thermodynamic equilibrium is connected with an ionization equilibrium $e + h \rightleftarrows X$ [11]. A special feature of the EHP is the screening of the Coulomb interaction and, therefore, a lowering of the coupling with increasing density. Due to various many-particle effects like self-energy and Pauli blocking one observes a lowering of the continuum edge (band gap renormalization) and, therefore, a lowering of the ionization energy [12]. Finally, at vanishing ionization energy, a breakup of the bound states at a certain density occurs referred to as *Mott effect*.

Some important properties of the EHP may be explained in the density–temperature (n–T) plane, see Fig. 1. Here, in a schematic presentation, the border of the Bose condensation of ideal Bose particles as well as curves of constant degree of ionization are drawn. Due to the special behavior of these curves, the n–T plane is divided roughly into an area without bound states, i.e., a high-density e–h liquid (where also a phase of hole crystallization can occur, cf. [13]) and an area with bound states. Here, a partially ionized plasma with a strong dominance of excitons (and, possibly, biexcitons and trions) at lower temperatures may be found.

Since the excitons behave approximately like composite Bose particles, BEC in the region $n\lambda^3 > 2.61$ [λ is the thermal de Broglie wavelength, $\lambda^2 = 2\pi\hbar^2/(mk_BT)$] may be expected if the chemical potential reaches the exciton 1s ground state energy [14, 15, 16, 17]. At higher densities, the region of BEC of excitons is, of course, limited by the Mott transition. However, we underline that the vanishing of the excitons does not imply a disappearance of the condensed phase. As proposed by Keldysh and Kopaev [18] and others [19, 20, 21], in the high-density highly degenerate e–h liquid the formation of weakly bound cooperative Cooper pairs of electrons and holes and their Bose condensation to a BCS state may be expected. The whole condensed phase is sometimes referred to as excitonic insulator [7].

The exciton gas as a system of interacting Bose-like particles implies the description as an interacting Bose gas. The latter one has been investigated in several "classical" papers [22] as well as in more recent works (see, e.g., [23]). It has further interesting physical properties like the occurrence of superfluidity [24, 25]. In the present work, however, the excitons are assumed to be noninteracting particles.

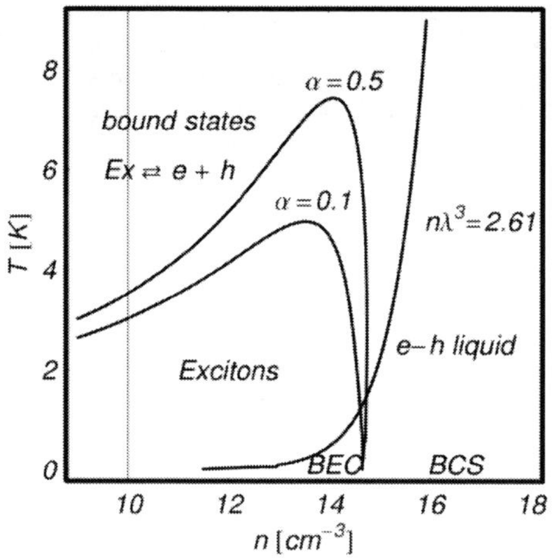

Figure 1. Example for a density–temperature plane. Parameters of germanium (Ge) are used.

An interesting problem is now the crossover from the BEC to the BCS state. It was shown by Eagles [26] and by Leggett [27] and generalized and extended to finite temperatures by Noziéres and Schmitt-Rink [5] that the crossover from the BEC to the BCS regime is smooth, despite the fact that the limits of BCS state and BEC of excitons are physically quite different. This is a general behavior of Fermi systems with Bose-like bound states and has been the subject of many papers mostly for Fermi atoms interacting via a contact potential.

While the physics of the interacting exciton gas (BEC of excitons, superfluidity etc.) is an intensely investigated field (see, e.g., [25]), for the BEC–BCS crossover in the EHP there exist only few papers. In spite of the fundamental importance of Ref. [5], the special features of Coulomb systems mentioned above are not taken into account. Moreover, a separable potential was assumed. Further investigations to the crossover problem in the EHP have been carried out in Refs. [6, 7]. The first paper is restricted to the optical behavior at the crossover and the second one considers the crossover for Coulomb systems more in detail but a simultaneous consideration of the density is missing. In recent publications [28] we elaborated a real-time Green's functions approach to quantum condensation which is the basis of the following considerations.

2. Theory and basic relations

A convenient quantity for the unified theoretical description of quantum condensation is the real-time two-particle Green's function defined on the two-time Keldysh contour by

$$G_{ab}(12, 1'2') = \frac{1}{(i\hbar)^2} \left\langle \mathrm{T}_{\mathcal{C}}\{\Psi_a(1)\Psi_b(2)\Psi_b^\dagger(2')\Psi_a^\dagger(1')\}\right\rangle \qquad (1 = \{\mathbf{r}_1, t_1, s_1^{(3)}\}) \tag{1}$$

where $\mathrm{T}_{\mathcal{C}}$ is the time ordering operator on the contour.

The two-particle Green's function describes two-particle bound and scattering states and is a solution of the Bethe–Salpeter equation (BSE),

$$\int_{\mathcal{C}} d\bar{1}\, G_a^{-1}(1,\bar{1})G_{ab}(\bar{1}2, 1'2') - i\hbar \int_{\mathcal{C}} d3d\bar{1}d\bar{2}\, G_b(2,3)W_{ab}(13,\bar{1}\bar{2})G_{ab}(\bar{1}\bar{2}, 1'2') = G_b(2,2')\delta(1-1') \tag{2}$$

Here, W_{ab} is an effective two-particle interaction. In the following, we apply the so-called V^s approximation,

$$W_{ab}(12,\bar{1}\bar{2}) = V_{ab}^s(1,2)\delta(1-\bar{1})\delta(2-\bar{2}), \tag{3}$$

where V^s is the screened Coulomb potential.

In addition we need the Dyson equation for the determination of the one particle Green's function

$$\int_{\mathcal{C}} d\bar{1} \left[G_a^{0^{-1}}(1,\bar{1}) - \Sigma_a(1,\bar{1})\right] G_a(\bar{1},1') = \delta(1-1'). \tag{4}$$

Now the important question arises: What is the signal of quantum condensation in terms of the two-particle Green's function?

It can be shown [28] that quantum condensation occurs if G_{ab} has the structure

$$G_{ab}(12,1'2') = \widehat{G}_{ab}(12,1'2') + C_{ab}(12,1'2'), \tag{5}$$

where the normal phase term $\widehat{G}_{ab}$ vanishes if the difference of all primed and all unprimed variables tends to infinity,

$$\lim_{|\{t_1,t_2\}-\{t_1',t_2'\}|\to\infty} \widehat{G}_{ab}(12,1'2') = 0 \tag{6}$$

and the *time long range order* (TLRO) contribution C_{ab} does not vanish in this limit but factorizes into the functions F_{ab} and F_{ab}^*, i.e.,

$$\lim_{|\{t_1,t_2\}-\{t_1',t_2'\}|\to\infty} C_{ab}(12,1'2') \neq 0, \qquad C_{ab}(12,1'2') = G_{ab}(12)F_{ab}^*(1'2'). \tag{7}$$

Since G_{ab} is a solution of the BSE (2), F_{ab} must be a solution of the homogeneous BSE

$$\int_{\mathcal{C}} d\bar{1}\, G_a^{-1}(1,\bar{1})F_{ab}(\bar{1}2) - \int_{\mathcal{C}} d\bar{2}\, G_b(2,\bar{2})\Delta_{ab}(1\bar{2}) = 0. \tag{8}$$

Here, the important gap function is introduced, which is fundamental for the description of the condensate,

$$\Delta_{ab}(12) = i\hbar V_{ab}^s(1,2)F_{ab}(12). \tag{9}$$

Journal of Physics: Conference Series **220** (2010) 012004 doi:10.1088/1742-6596/220/1/012004

The TLRO contribution in G_{ab} induces TLRO terms in all quantities which are connected with the two-particle Green's function, especially in the self-energy

$$\begin{aligned}\int_{\mathcal{C}} d\bar{1}\, \Sigma_a(1,\bar{1})G_a(\bar{1},1') &= \sum_b \int_{\mathcal{C}} d2\, V_{ab}(1-2)G_{ab}(12,1'2^+) \\ &= \int_{\mathcal{C}} d\bar{1}\, \Sigma_a^{\mathrm{HF}}(1,\bar{1})G_a(\bar{1},1') + \sum_b \int_{\mathcal{C}} d2\, V_{ab}^s(1,2)\Pi_{ab}(12,1'2^+) \quad (10)\end{aligned}$$

with Σ_a^{HF} being the Hartree–Fock self-energy and Π_{ab} the polarization function.

The self-energy with TLRO is given by

$$\Sigma_a(1,1') = \widehat{\Sigma}_a(1,1') + \Sigma_a^{\mathrm{LRO}}(1,1')\,, \quad (11)$$

$$\int_{\mathcal{C}} d\bar{1}\, \Sigma_a^{\mathrm{LRO}}(1,\bar{1})G_a(\bar{1},1') = \int_{\mathcal{C}} d2\, \Delta_{ab}(12)F_{ab}^*(1'2^+)\,. \quad (12)$$

Therefore, the Dyson equation on the Keldysh contour $\mathcal{C}$ including the quantum condensate takes the form

$$\begin{aligned}&\left[i\hbar\frac{\partial}{\partial t_1} + \frac{\hbar^2\nabla_1^2}{2m_a} + \epsilon_a\right] G_a(1,1') - \int_{\mathcal{C}} d\bar{1}\left[\widehat{\Sigma}_a(1,\bar{1}) + \Sigma_a^{\mathrm{LRO}}(1,\bar{1})\right] G_a(\bar{1},1') \\ &= \delta(1-1')\,, \quad (13)\end{aligned}$$

where $\epsilon_e - \epsilon_h = \epsilon_g$ is the energy gap.

From the considerations so far, we obtain the following conclusion:

TLRO occurs if the coupled system of Dyson equation (13) and homogeneous BSE (8) has a solution.

Let us consider the consequences of TLRO:

(i) Obviously, spatial *off-diagonal long range order* (ODLRO) [29, 30] as an important property of quantum condensation follows from TLRO. ODLRO is a property of the reduced density matrix,

$$\varrho_{ab}(\mathbf{r}_1\mathbf{r}_2,\mathbf{r}_1'\mathbf{r}_2';t) = \widehat{\varrho}_{ab}(\mathbf{r}_1\mathbf{r}_2,\mathbf{r}_1'\mathbf{r}_2';t) + \Phi_{ab}(\mathbf{r}_1\mathbf{r}_2,t)\Phi_{ab}^*(\mathbf{r}_1'\mathbf{r}_2',t)\,, \quad (14)$$

where Φ_{ab} is the macroscopic wave function. ODLRO follows from TLRO since the two-particle density matrix is given by

$$g_{ab}^<(t_1=t_2=t_1'=t_2') = \varrho_{ab}; \qquad \Phi(\mathbf{r}_1\mathbf{r}_2,t_1) = i\hbar F_{ab}(12)|_{t_1=t_2}\,. \quad (15)$$

(ii) In many cases it is more convenient to use instead of the two-particle Green's function the T-matrix. The T-matrix is connected with G_{ab} by the relation $T_{ab} = V_{ab}^s + V_{ab}^s G_{ab} V_{ab}^s$ and, therefore, has a TLRO term, too,

$$T_{ab}(12,1'2') = \widehat{T}_{ab}(12,1'2') + \frac{1}{i\hbar}\Delta_{ab}(12)\Delta_{ab}^*(1'2')\,. \quad (16)$$

(iii) In the equilibrium case, Kwok and Schultz [31] have shown that TLRO modifies the spectral representation as follows

$$g_{ab}^<(\omega) = A_{ab}(\omega)\,\frac{\mathcal{P}}{e^{\beta(\hbar\omega-\mu_b-\mu_a)}-1} + 2\pi\delta(\hbar\omega-\mu_b-\mu_a)\,F_{ab}\,F_{ab}^*\,, \quad (17)$$

$$T_{ab}^<(\omega) = 2\,\mathrm{Im}\,T_{ab}^{\mathrm{R}}(\omega)\frac{\mathcal{P}}{e^{\beta(\hbar\omega-\mu_b-\mu_a)}-1} + 2\pi\delta(\hbar\omega-\mu_b-\mu_a)\,\Delta_{ab}\,\Delta_{ab}^*\,. \quad (18)$$

(iv) Finally, if we take into account the expression for $\Sigma^{\rm LRO}$, Eq. (11), Dyson equation (13) and homogeneous BSE (8) are equivalent to the non-equilibrium Gorkov equations for G_a and F_{ab},

$$\left[i\hbar\frac{\partial}{\partial t_1}+\frac{\hbar^2\nabla_1^2}{2m_a}+\epsilon_a\right]G_a(1,1')-\int_C d\bar{1}\,\widehat{\Sigma}_a(1,\bar{1})G_a(\bar{1},1')-\int_C d\bar{1}\Delta_{ab}(1\bar{1})F^*_{ab}(1'\bar{1})=\delta(1-1')\,, \tag{19}$$

$$\left[i\hbar\frac{\partial}{\partial t_1'}-\frac{\hbar^2\nabla_{1'}^2}{2m_b}-\epsilon_b\right]F^*_{ab}(11')+\int_C d\bar{1}\,\widehat{\Sigma}^*_b(1',\bar{1})F^*_{ab}(1\bar{1})-\int_C d\bar{1}\Delta^*_{ab}(\bar{1}1')G_a(\bar{1},1)=0\,. \tag{20}$$

The self-energy of the normal phase will be considered in the following in screened ladder approximation [32]

$$\begin{aligned}\widehat{\Sigma}_a(1,1')=\pm i\hbar\sum_b\int_C d2d\bar{2}\,&\Big[\widehat{T}_{ab}(12,1'\bar{2})\pm\delta_{ab}\widehat{T}_{ab}(12,\bar{2}1')\\&-V^s_{ab}(1,2)G_a(1,1')G_b(2,\bar{2})V^s_{ab}(1',\bar{2})\Big]\,G_b(\bar{2},2)\,.\end{aligned} \tag{21}$$

Equations (19,20) are the basic equations for the description of the non-equilibrium behavior of a quantum many-particle system including the quantum condensate. From Eqs. (19,20), by time specialization on the Keldysh contour, there follow Kadanoff–Baym like equations for the correlation functions $F^{\gtrless}_{ab}$,

$$\begin{aligned}&\left[i\hbar\frac{\partial}{\partial t_1}+\frac{\hbar^2\nabla_1^2}{2m_a}+\epsilon_a\right]g_a^{\gtrless}(1,1')\\&-\int d\bar{1}\,\left[\widehat{\Sigma}^{\rm R}_a(1,\bar{1})g_a^{\gtrless}(\bar{1},1')+\widehat{\Sigma}^{\gtrless}_a(1,\bar{1})g^{\rm A}_a(\bar{1},1')\right]-\int d\bar{1}\Delta_{ab}(1\bar{1})F^{*\gtrless}_{ab}(1'\bar{1})=0\,,\end{aligned} \tag{22}$$

$$\begin{aligned}&\left[i\hbar\frac{\partial}{\partial t_1'}-\frac{\hbar^2\nabla_{1'}^2}{2m_b}-\epsilon_b\right]F^{*\gtrless}_{ab}(11')\\&-\int d\bar{1}\,\left[\widehat{\Sigma}^{*\rm R}_b(1',\bar{1})F^{*\gtrless}_{ab}(1\bar{1})+\widehat{\Sigma}^{*\lessgtr}_b(1',\bar{1})F^{*\rm R}_{ab}(1\bar{1})\right]-\int d\bar{1}\Delta^*_{ab}(\bar{1}1')g_a^{\lessgtr}(\bar{1},1)=0\,.\end{aligned} \tag{23}$$

The equations for the retarded and advanced Green's functions follow easily from the definition and Eqs. (22,23) as

$$\begin{aligned}&\left[i\hbar\frac{\partial}{\partial t_1}+\frac{\hbar^2\nabla_1^2}{2m_a}+\epsilon_a\right]g_a^{\rm R/A}(1,1')-\int d\bar{1}\,\widehat{\Sigma}^{\rm R/A}_a(1,\bar{1})g_a^{\rm R/A}(\bar{1},1')\\&-\int d\bar{1}\,\Delta_{ab}(1\bar{1})F^{*\rm A/R}_{ab}(1'\bar{1})=\delta(1-1')\,,\end{aligned} \tag{24}$$

$$\begin{aligned}&\left[i\hbar\frac{\partial}{\partial t_1'}-\frac{\hbar^2\nabla_{1'}^2}{2m_b}-\epsilon_b\right]F^{*\rm R/A}_{ab}(11')-\int d\bar{1}\,\widehat{\Sigma}^{*\rm A/R}_b(1',\bar{1})F^{*\rm R/A}_{ab}(1\bar{1})\\&-\int d\bar{1}\,\Delta^*_{ab}(\bar{1}1')g_a^{\rm A/R}(\bar{1},1)=0\,.\end{aligned} \tag{25}$$

From Eqs. (22–25), kinetic equations for the Wigner function (Boltzmann equation) and the macro wave function may be derived [28].

Concluding follows that the concept of TLRO is obviously very convenient for the description of quantum condensation in terms of real-time Green's functions.

3. EHP in equilibrium

In the case of thermodynamic equilibrium, because of the spectral representation, the time dependence of the gap function is simply given by

$$\Delta_{ab}(\mathbf{p},\mathbf{t}) = e^{\frac{i}{\hbar}(\mu_a+\mu_b)t}\Delta_{ab}(\mathbf{p})\,. \tag{26}$$

Therefore, it is useful to introduce the substitutions

$$F_{ab}(t,t') \Rightarrow e^{\frac{i}{\hbar}(\mu_a+\mu_b)t}F_{ab}(t-t')\,, \quad G_{ab}(t,t') \Rightarrow e^{\frac{i}{\hbar}(\mu_a+\mu_b)t}G_{ab}(t-t')\,, \tag{27}$$

where the remaining functions depend only on the time difference. Therefore, the Fourier representation can be used and the Gorkov equations are algebraic equations only,

$$\left[\hbar\omega - E_a(\mathbf{p}) - \widehat{\Sigma}_a^{\mathrm{R/A}}(\mathbf{p},\omega)\right] g_a^{\mathrm{R/A}}(\mathbf{p},\omega) - \Delta_{ab}(\mathbf{p})F_{ab}^{*\mathrm{A/R}}(-\mathbf{p},-\omega) = 1\,, \tag{28}$$
$$\left[\hbar\omega - E_b(\mathbf{p}) + \mu_a + \mu_b - \widehat{\Sigma}_b^{*\mathrm{A/R}}(-\mathbf{p},-\omega)\right] F_{ab}^{*\mathrm{R/A}}(\mathbf{p},\omega) + \Delta_{ab}^{*}(\mathbf{p})g_a^{\mathrm{A/R}}(-\mathbf{p},-\omega) \;=\; 0\,.$$

Then it is easy to find a formal solution and the spectral function

$$A_a(\mathbf{p},\omega) = \frac{\Gamma_a(\mathbf{p},\omega)}{\left(\hbar\omega - E_a(\mathbf{p}) - \mathrm{Re}\,\widehat{\Sigma}_a^{\mathrm{R}}(\mathbf{p},\omega) - \mathrm{Re}\,\Sigma_a^{\mathrm{LRO}}(\mathbf{p},\omega)\right)^2 + \left(\frac{1}{2}\Gamma_a(\mathbf{p},\omega)\right)^2} \tag{29}$$

with

$$\Sigma_a^{\mathrm{LRO}}(\mathbf{p},\omega) = \frac{|\Delta_{ab}(\mathbf{p})|^2}{\hbar\omega + E_b(\mathbf{p}) - \mu_a - \mu_b + \widehat{\Sigma}_b^{*\mathrm{R}}(\mathbf{p},\omega)} \tag{30}$$

and

$$\Gamma_a(\mathbf{p},\omega) = 2\left(\mathrm{Im}\,\widehat{\Sigma}_a^{\mathrm{R}}(\mathbf{p},\omega) + \mathrm{Im}\,\Sigma_a^{\mathrm{LRO}}(\mathbf{p},\omega)\right)\,. \tag{31}$$

In many cases, the spectral function is used in quasiparticle approximation, i.e., the damping Γ is neglected and it follows simply that

$$a_a(\mathbf{p},\omega) = 2\pi\left[|u_p|^2\;\delta\left(\hbar\omega - E_a^{+}(\mathbf{p})\right) + |v_p|^2\;\delta\left(\hbar\omega - E_a^{-}(\mathbf{p})\right)\right]\,. \tag{32}$$

This approximation is not in every case sufficient. Instead, we use an extended approximation obtained by an expansion with respect to the damping Γ [33] called *extended quasiparticle approximation*

$$\begin{aligned} A_a(\mathbf{p},\omega) \;=\; & |u_p|^2\left\{\left[1+|u_p|^2\frac{\partial}{\partial\omega}\mathrm{Re}\,\widehat{\Sigma}_a^{\mathrm{R}}(\mathbf{p},\omega)|_{\hbar\omega=E_a^+}\right]2\pi\delta(\hbar\omega - E_a^{+}(\mathbf{p})) \right.\\ & \left. -\Gamma_a(\mathbf{p},\omega)\frac{\partial}{\partial\omega}\frac{\mathcal{P}}{\hbar\omega - E_a^{+}(\mathbf{p})}\right\} \\ & +|v_p|^2\left\{\left[1+|v_p|^2\frac{\partial}{\partial\omega}\mathrm{Re}\,\widehat{\Sigma}_a^{\mathrm{R}}(\mathbf{p},\omega)|_{\hbar\omega=E_a^-}\right]2\pi\delta(\hbar\omega - E_a^{-}(\mathbf{p})) \right.\\ & \left. -\Gamma_a(\mathbf{p},\omega)\frac{\partial}{\partial\omega}\frac{\mathcal{P}}{\hbar\omega - E_a^{-}(\mathbf{p})}\right\}\,. \end{aligned} \tag{33}$$

This lengthy expression has a simple structure. The first terms in the curly brackets, proportional to delta functions, are the quasiparticle contributions with the quasiparticle energies E_a^{+-}. The renormalization terms are linearized with respect to the normal self-energy. The second, off-pole, contribution describes the interaction between the quasiparticles.

The quasiparticle dispersion for the two-component EHP is given by

$$E_a^{\pm}(\mathbf{p}) = \frac{1}{2}\left[\{\varepsilon_a(\mathbf{p},\omega) - \varepsilon_b(\mathbf{p},\omega)\} \pm \sqrt{[\varepsilon_a(\mathbf{p},\omega) + \varepsilon_b(\mathbf{p},\omega)]^2 + 4|\Delta_{ab}(\mathbf{p})|^2}\right]_{\hbar\omega = E_a^{\pm}(\mathbf{p})} \tag{34}$$

with the self-energy of the normal phase

$$\varepsilon_a(\mathbf{p},\omega) = e_a(\mathbf{p}) + \mathrm{Re}\,\widehat{\Sigma}_a^{\mathrm{R}}(\mathbf{p},\omega), \quad e_a(\mathbf{p}) = p^2/(2m_a) - \mu_a\,. \tag{35}$$

Finally, the u_p and v_p are the spectral weights with the normalization $|u_p|^2 + |v_p|^2 = 1$. They are given by the renormalization terms with respect to the TLRO self-energy

$$\left.\begin{array}{l} |u_p|^2 \\ |v_p|^2 \end{array}\right\} = \frac{1}{1 - \frac{\partial}{\partial\omega}\mathrm{Re}\,\Sigma^{\mathrm{LRO}}(\mathbf{p},\omega)\big|_{\hbar\omega = E^{\pm}(\mathbf{p})}} = \frac{1}{2}\left[1 \pm \frac{\varepsilon_a(\mathbf{p},\omega) + \varepsilon_b(\mathbf{p},\omega)}{\sqrt{[\varepsilon_a(\mathbf{p},\omega) + \varepsilon_b(\mathbf{p},\omega)]^2 + 4\,|\Delta_{ab}(\mathbf{p})|^2}}\right]_{\hbar\omega = E^{\pm}}\,. \tag{36}$$

The crucial quantity in the theory is the gap function Δ_{ab}. From the definition and the homogeneous BSE (8), approximating the dynamically screened potential by the statically screened one (denoted again by V_{ab}^s), the following integral equation for Δ_{ab} may be obtained,

$$\Delta_{ab}(\mathbf{p}) = i\hbar\int\frac{d\bar{\mathbf{p}}}{(2\pi\hbar)^3}V_{ab}^s(\mathbf{p}-\bar{\mathbf{p}})\mathcal{G}_{ab}^{\mathrm{R}}(\bar{\mathbf{p}},-\bar{\mathbf{p}};\omega)\Big|_{\hbar\omega=\mu_a+\mu_b}\Delta_{ab}(\bar{\mathbf{p}})\,, \tag{37}$$

where V_{ab}^s is the statically screened Coulomb potential. For the determination of the free pair propagator $\mathcal{G}_{ab}^{\mathrm{R}}(\omega)$ entering the gap equation

$$\mathcal{G}_{ab}^{\mathrm{R}}(\omega) = \int\frac{d\bar{\omega}}{2\pi}\int\frac{d\bar{\bar{\omega}}}{2\pi}\frac{g_a^>(\bar{\omega})g_b^>(\bar{\bar{\omega}}) - g_a^<(\bar{\omega})g_b^<(\bar{\bar{\omega}})}{\omega - \bar{\omega} - \bar{\bar{\omega}} + i\epsilon}$$

in this case it is sufficient to use the quasiparticle spectral function (32). Then the following nonlinear gap equation for an EHP is obtained ($e_{ab} = \frac{1}{2}(e_a + e_b)$, $\quad e_a = E_a - \mu_a$)

$$\Delta_{ab}(\mathbf{p}) = \frac{e^2}{\epsilon_0\epsilon_r}\int\frac{d\bar{\mathbf{p}}}{(2\pi\hbar)^3}\frac{\hbar^2}{(\mathbf{p}-\bar{\mathbf{p}})^2 + \hbar^2\kappa^2}\frac{\Delta_{ab}(\bar{\mathbf{p}})}{2\sqrt{(e_{ab}(\bar{\mathbf{p}}))^2 + |\Delta_{ab}(\bar{\mathbf{p}})|^2}}\left[f(E_a^+(\bar{\mathbf{p}})) - f(E_a^-(\bar{\mathbf{p}}))\right]\,. \tag{38}$$

A central problem in the theory of quantum condensation is the determination of the critical temperature T_c as a function of the chemical potential and the density, i.e., the phase boundary of the condensate. This boundary is determined by the vanishing of the gap. Under this condition, the gap equation is reduced to the linearized form

$$\Delta_{ab}(\mathbf{p}) = -\frac{e^2}{\epsilon_0\epsilon_r}\int\frac{d\bar{\mathbf{p}}}{(2\pi\hbar)^3}\frac{\hbar^2}{(\mathbf{p}-\bar{\mathbf{p}})^2 + \hbar^2\kappa^2}\frac{\Delta_{ab}(\bar{\mathbf{p}})}{e_{ab}(\bar{\mathbf{p}})}\left[1 - f(e_a(\bar{\mathbf{p}})) - f(e_b(\bar{\mathbf{p}}))\right]\,. \tag{39}$$

The critical temperature of quantum condensation is obtained by numerical solution of Eq. (39) (for details see Ref. [28]). The result for the model case $m_e = m_h$ is shown in Fig. 6.

We find the onset of the Bose condensation of excitons at $\mu_e + \mu_h = E_{1s} = -1$ and $T = 0$. With increasing chemical potential, the critical temperature increases and reaches a maximum.

Journal of Physics: Conference Series **220** (2010) 012004 doi:10.1088/1742-6596/220/1/012004

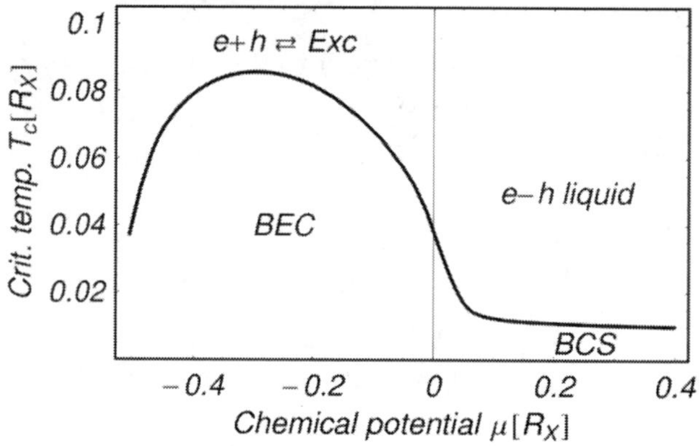

Figure 2. Critical temperature vs. chemical potential for a model semiconductor with $m_e = m_h$. Result of the solution of the gap equation under the condition of vanishing gap [28]. R_X denotes the excitonic Rydberg.

The region of Bose condensation of excitons is limited by the Mott transition and we observe a smooth crossover to the phase boundary of a BCS state.

In many cases it more convenient to know the critical temperature as a function of the density. The determination of T_c vs. density requires the coupled solution of gap and density equations,

$$n_a(\mu, T) = \int \frac{d\mathbf{p}}{(2\pi\hbar)^3} \frac{d\omega}{2\pi} A_a(\mathbf{p}, \omega)\, f_a(\omega)\,. \tag{40}$$

By inversion $\mu = \mu(n, T)$ we are able to eliminate the chemical potential from T_c.

Here it is not sufficient to use the spectral function in quasiparticle approximation. In order to take into account correlated states, especially the important bound states, A_a has to be taken in extended quasiparticle approximation, Eq. (33). However, on the phase boundary of the quantum condensate it is sufficient to consider the relation (40) for the normal phase, i.e., to use Eq. (33) with $v_p \to 0$ and $u_p \to 1$ [33, 8],

$$\begin{aligned} n_a(\mu_a, T) &= \int \frac{d\mathbf{p}}{(2\pi\hbar)^3} f(\varepsilon_a^{\mathrm{RPA}}) + \sum_{nl}(2l+1) \int \frac{d\mathbf{P}}{(2\pi\hbar)^3} n_{ab}^B\left(\frac{P^2}{2M} + E_{nl}\right) \\ &+ \int_0^\infty \frac{d\omega}{2\pi} n_{ab}^B(\omega) \mathrm{Im}\,\mathrm{Tr}_{12}\left[\frac{d}{d\omega}\mathcal{G}_{ab}^{\mathrm{R}}(\omega) T_{ab}^{\mathrm{R}}(\omega)\right]\,. \end{aligned} \tag{41}$$

Obviously, for the application of this relation we need the solution of the two-particle bound and scattering problem. Quantum mechanics determines the properties of a two-particle system from the stationary Schrödinger equation. In a dense plasma, the influence of the surrounding medium has to be taken into account, i.e., (i) influence of the Pauli blocking or phase space occupation effect, (ii) self-energy corrections to the kinetic energy, (iii) screening of the interaction between the two particles.

A solution of this problem needs a careful analysis of the Bethe–Salpeter equation for the two-time two-particle Green's function in dynamically screened ladder approximation [12, 34, 35]. The result of such investigations is an effective Schrödinger equation for the determination of the two-particle properties of the EHP following from the homogeneous BSE [12, 34, 36],

$$\left(\mathrm{H}_{ab}^0 + \mathrm{V}_{ab} - E_{\alpha\mathbf{P}}(\omega, t) + \mathrm{H}_{ab}^{pl}(\omega, t)\right) |\psi_{\alpha\mathbf{P}}(\omega, t)\rangle = 0\,. \tag{42}$$

An equivalent approach applicable in the EHP are the semiconductor Bloch equations [37, 38, 39, 40, 41]. For an electron–hole pair system, both approaches are in full agreement for zero center-of-mass momentum.

The influence of the surrounding plasma on the two-particle properties is described in Eq. (42) by the in-medium part $\mathrm{H}^{pl}_{ab}(\omega, t)$, explicitly given in Refs. [32, 34].

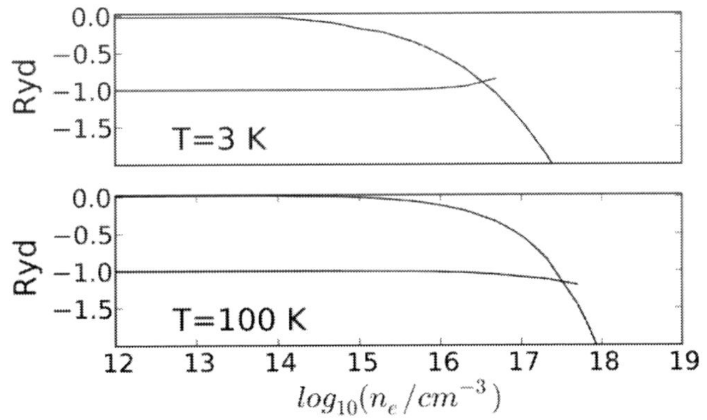

Figure 3. Two-particle spectrum vs. density: band (continuum) edge (upper curves), ground state energy E_b (lower curves) in zinc selenide (ZnSe) (in units of excitonic Rydberg) [42]

The resulting two-particle spectrum, following by extensive numerical work, is presented in Fig. 3 [42] which shows that the plasma modifies the electron–hole pair in the following way [12, 37, 38, 39, 40, 42]:

(i) There is a lowering of the band (continuum) edge

$$\Delta E_{ab} = \Sigma_a^{\mathrm{HF}} + \Sigma_b^{\mathrm{HF}} + \Delta\,, \qquad \Delta = \Delta_a + \Delta_b\,. \tag{43}$$

(ii) The exciton ground state energy is nearly independent of the density

$$E_b = E_b^0 + \delta\,. \tag{44}$$

It changes by the small quantity δ, but it shows a weaker density dependence as compared to the continuum edge. This follows from an approximative compensation of the many-particle effects [12].

(iii) The influence of the plasma leads to a lowering of the ionization energy,

$$\begin{aligned} I^{\mathrm{eff}} &= |E_b| + \Delta E_{eh} \\ &= |E_b^0| - \delta + \Delta E_{eh}\,. \end{aligned} \tag{45}$$

In statically screened approximation is $\Delta = -\kappa e^2$. In general, Δ and δ follow from the numerical solution of the effective Schrödinger equation with dynamically screened potential or, analogously, the semiconductor Bloch equations.

(iv) For $I^{\mathrm{eff}} = 0$, the bound state vanishes and merges into the scattering continuum (Fig. 3). This is usually referred to as the *Mott effect.*

Due to the lowering of the ionization energy it may be expected that the bound state part of the density has discontinuities at the critical values of the coupling parameter connected with drastic changes in the thermodynamic functions. However, one can show that the total density is a smooth function of the coupling parameter [43, 32] because there is a compensation of the discontinuities by contributions of the scattering part. By application of the higher order Levinson theorems [32], these parts may be transferred into the bound state contribution. This renormalized bound state contribution is a generalization of the Planck–Larkin sum of states to degenerate systems and can be written as [33]

$$\begin{aligned} n_{\text{bound}} &= (2s+1)^2 \int \frac{d\mathbf{P}}{(2\pi\hbar)^3} \left\{ n_B\left(\frac{P^2}{2M} + I^{eff}(\zeta_e,\zeta_h) - \zeta_e - \zeta_h\right) \right. \\ & \left. -n_B\left(\frac{P^2}{2M} - \zeta_e - \zeta_h\right) - \frac{M}{P^2} I^{eff}(\zeta_e,\zeta_h)\, n_B\left(\frac{P^2}{2M} - \zeta_e - \zeta_h\right) \right\} \end{aligned} \tag{46}$$

where n_B is the Bose function and $\zeta_a = \mu_a - \Delta_a$.

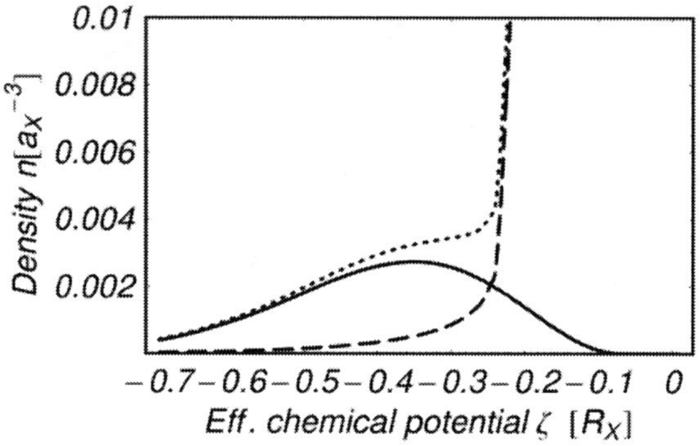

Figure 4. Electron density as a function of the effective chemical potential: total density n_e (dotted line), quasiparticle contribution n_{QP} (dashed line), bound state contribution n_{bound} (solid line) for a model semiconductor with $m_e = m_h$ [28]. a_X denotes the excitonic Bohr radius.

At lower temperatures, the remaining scattering part in Eq. (41) is small and can be neglected. Therefore, the total density can be subdivided into two parts

$$n_{\text{tot}}(\mu, T) = n_{\text{QP}}(\mu, T) + n_{\text{bound}}(\mu, T)\,. \tag{47}$$

The resulting total electron density is shown in Fig. 4 vs. the effective chemical potential. Here and in all subsequent figures, results are shown for a model semiconductor with equal masses of electrons and holes and a background dielectric constant of $\epsilon_r = 7$ corresponding approximately to the parameters of cuprous oxide (Cu_2O).

With the subdivision of the density according to Eq. (47), the degree of ionization can be defined,

$$\alpha(\mu, T) = \frac{n_{\text{QP}}(\mu, T)}{n_{\text{tot}}(\mu, T)}\,. \tag{48}$$

Isotherms of the degree of ionization as a function of the density are shown in Fig. 5. We observe a very strong increase of the degree of ionization up to $\alpha = 1$ due to the lowering of the ionization energy [11, 43]. This behavior is usually referred to as *Mott transition* as a consequence of the Mott effect. The detailed properties of α, however, are strongly modified by including exciton–exciton interaction and the possible formation of biexcitons and trions, known already from experiences from gaseous plasmas (atom–atom interaction and molecule formation) [32] and recently confirmed in first-principle numerical simulations [44].

Journal of Physics: Conference Series **220** (2010) 012004 doi:10.1088/1742-6596/220/1/012004

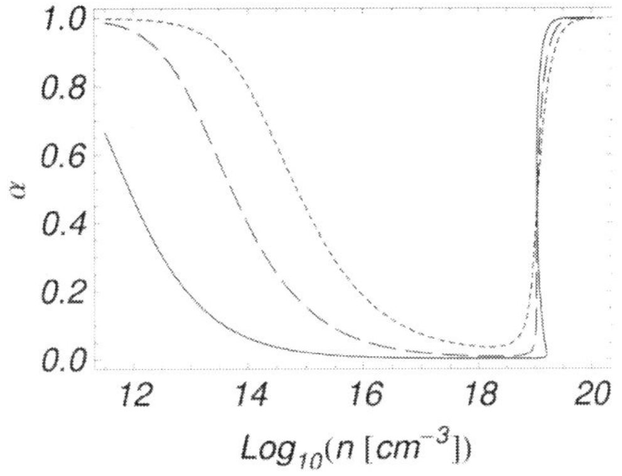

Figure 5. Degree of ionization for $T = 120$ K (solid), 160 K (dashed), and 200 K (dotted)

With the density relations (41) and (46), the critical temperature may be determined as a function of the density. The result is shown in Fig. 6. At lower densities, the critical temperature agrees with the condition for Bose condensation of ideal excitons, $n\lambda^3 \geq 2.61$, as a consequence of the neglect of the exciton–exciton interaction in our model. With increasing density, we observe a strong deviation from the ideal behavior due to the lowering of the ionization energy and further a smooth crossover to the BCS regime.

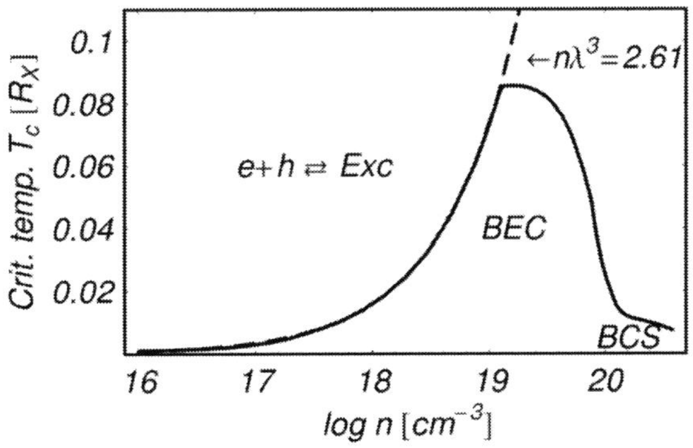

Figure 6. Phase boundary of quantum condensation vs. density, obtained by elimination of μ in $T_{crit}(\mu)$ [28]

This behavior is confirmed in Fig. 7 where the critical temperature is given in units of the Fermi energy vs. inverse Fermi momentum. This curve is qualitatively in agreement with corresponding curves of the crossover of Fermi atom systems in traps [4].

4. Conclusion

Summarizing our investigations, we have presented a description of quantum condensation in the EHP of highly excited semiconductors.

In the framework of real-time Green's functions applying the concept of time long range order for the two-particle Green's function we have derived nonequilibrium Gorkov equations on the Keldysh time contour. In order to consider the EHP, the theory has been specified to systems with Coulomb interaction using the screened ladder approximation.

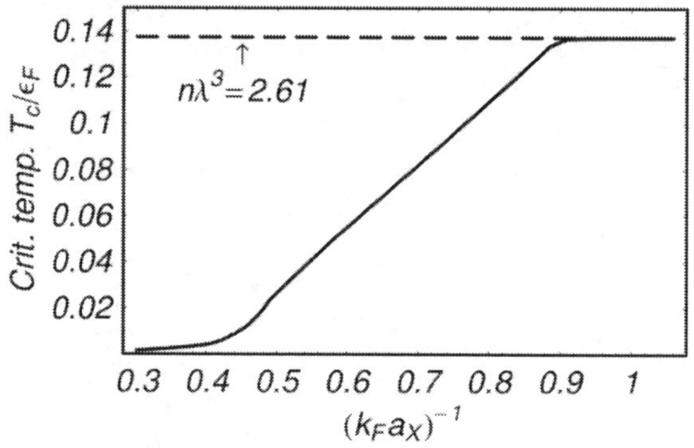

Figure 7. Critical temperature of quantum condensation vs. inverse Fermi momentum [28]

In thermodynamic quasiequilibrium, we have obtained the usual BCS scheme of equations in terms of the screened potential for the two-component case including the influence of the normal phase.

For a complete solution of the problem we solved the coupled system of gap and density equations in extended quasiparticle approximation. The analysis of the density in screened ladder approximation and the associated two-particle problem in a plasma environment shows a lowering of the ionization energy and finally a breakup of bound states with increasing density, the Mott effect. Therefore, we observe a transition from a partially ionized EHP to a high density e–h liquid connected with a change of the physical nature of the quantum condensate (BEC–BCS crossover).

The phase boundary of the quantum condensate arises from the condition of vanishing gap function. Therefore, we have determined the critical temperature as a function of the chemical potential by solution of the linearized gap equation with a screened Coulomb potential. The result shows a smooth crossover from the Bose–Einstein condensate of excitons to BCS states at high densities driven by lowering of ionization energy and Mott effect.

Acknowledgments

We would like to thank G. Manzke, F. Richter, Th. Bornath, W.-D. Kraeft (Rostock), T. Schmielau (Sheffield), and H. Fehske and F. X. Bronold (Greifswald) for many fruitful discussions. This work was supported by the Deutsche Forschungsgemeinschaft (Collaborative Research Center SFB 652).

References

[1] Röpke G 1995 in *Bose–Einstein Condensation* ed A Griffin, D W Snoke and S Stringari (Cambridge: Cambridge University Press)
[2] Fried D G *et al.* 1998 *Phys. Rev. Lett.* **81** 3811
[3] Haussmann R 1994 *Phys. Rev.* B **49** 12975
[4] Sá de Melo C A R, Randeria M and Engelbrecht J R 1993 *Phys. Rev. Lett.* **71** 3202
[5] Nozières P and Schmitt-Rink S 1985 *J. Low Temp. Phys.* **59** 195
[6] Inagaki T J and Aihara M 2002 *Phys. Rev.* B **66** 075204
[7] Bronold F X and Fehske H 2006 *Phys. Rev.* B **74** 165107
[8] Zimmermann R 1988 *Many-Particle Theory of Highly Excited Semiconductors* (Leipzig: Teubner)
[9] Haug H and Koch S W 1993 *Quantum Theory of the Optical and Electronic Properties of Semiconductors*, (World Scientific Publ.) (4th ed. 2004)
[10] Schäfer W and Wegener M 2002 *Semiconductor Optics and Transport Phenomena* (Berlin: Springer)
[11] Kraeft W D, Kilimann K and Kremp D 1975 *Phys. Status Solidi* B **72** 461

[12] Kilimann K, Kraeft W D and Kremp D 1977 *Phys. Lett.* A **61** 393; Zimmermann R, Kilimann K, Kraeft W D, Kremp D and Röpke G 1978 *Phys. Status Solidi* B **90** 175
[13] Bonitz M, Filinov V S, Fortov V E, Levashov P R and Fehske H 2005 *Phys. Rev. Lett.* **95** 235006; 2006 *J. Phys. A: Math. Gen.* **39** 4717
[14] Keldysh L V 1995 in *Bose–Einstein Condensation* ed A Griffin, D W Snoke and S Stringari (Cambridge: Cambridge University Press)
[15] Moskalenko S A 1962 *Fiz. Tverd. Tela* **4** 276
[16] Wolfe J P, Lin J L and Snoke D W 1995 in *Bose–Einstein Condensation* ed A Griffin, D W Snoke and S Stringari (Cambridge: Cambridge University Press)
[17] Hanamura E and Haug H 1977 *Phys. Rep.* **33** 209
[18] Keldysh L V and Kopaev Yu V 1965 *Sov. Phys. Solid State* **6** 2219
[19] des Cloizeaux J 1965 *J. Phys. Chem. Solids* **26** 259
[20] Halperin B I and Rice T M 1968 in *Solid State Physics* ed F Seitz, D Turnbull and H Ehrenreich (New York: Academic Press)
[21] Zimmermann R 1976 *Phys. Status Solidi* B **76** 191
[22] Bogolyubov N N 1947 *J. Phys. (USSR)* **11** 23; Beliaev S T 1958 *Zh. Eksp. Teor. Fiz.* **34** 417 [*Sov. Phys. JETP* **7** 289]; Hugenholtz N M and Pines D 1959 *Phys. Rev.* **116** 489; Nambu Y 1960 *Phys. Rev.* **117** 648
[23] Imamović-Tomasović M and Griffin A 1999 *Phys. Rev. A* **60** 494
[24] Hohenberg P C and Martin P C 1965 *Ann. Phys. (N. Y.)* **34** 291
[25] Filinov A, Bonitz M, Ludwig P and Lozovik Yu E 2006 *Phys. Status Solidi* C **3** 2457; Filinov A, Ludwig P, Bonitz M and Lozovik Yu E 2009 *J. Phys. A: Math. Gen.* **42** 214016
[26] Eagles D M 1969 *Phys. Rev.* **186** 456
[27] Leggett A J 1980 in *Modern Trends in the Theory of Condensed Matter* ed A Pekalski and R Przystawa (Berlin)
[28] Kremp D, Semkat D and Henneberger K 2008 *Phys. Rev.* B **78** 125315; Semkat D, Kremp D and Henneberger K 2009 *Phys. Status Solidi* C **6** 546
[29] Penrose G and Onsager L 1956 *Phys. Rev.* **104** 576
[30] Yang G N 1962 *Rev. Mod. Phys.* **34** 694
[31] Kwok P C and Schultz T D 1969 *J. Phys. C: Solid State Phys.* **2** 1196
[32] Kremp D, Schlanges M and Kraeft W-D 2005 *Quantum Statistics of Nonideal Plasmas* (Berlin: Springer)
[33] Kremp D, Kraeft W D and Lambert A J D 1984 *Physica* A **127** 72
[34] Bornath Th, Kremp D and Schlanges M 1999 *Phys. Rev.* E **60** 6382
[35] Haug H and Tran Thoai D B 1978 *Phys. Status Solidi* B **85** 561
[36] Schmielau T 2001 Ph.D. thesis (Universität Rostock)
[37] Manzke G, Peng Q Y, Henneberger K, Neukirch U, Hauke K, Wundke K, Gutowski J and Hommel D 1998 *Phys. Rev. Lett.* **80** 4943
[38] Nägerl J S, Stabenau B, Böhne G, Dreher S, Ulbrich R G, Manzke G and Henneberger K 2001 *Phys. Rev.* B **63** 235202
[39] Manzke G and Henneberger K 2002 *Phys. Status Solidi* B **234** 233
[40] Seemann M, Kieseling F, Stolz H, Franz R, Manzke G, Henneberger K, Passow T and Hommel D 2005 *Phys. Rev.* B **72** 075204
[41] Lorke M, Nielsen T R, Seebeck J, Gartner P and Jahnke F 2006 *Phys. Rev.* B **73** 085324
[42] Semkat D, Richter R, Kremp D, Manzke G, Kraeft W-D and Henneberger K 2009 *Phys. Rev.* B **80** 155201
[43] Ebeling W, Kraeft W D and Kremp D 1976 *Theory of Bound States and Ionization Equilibrium in Plasmas and Solids* (Berlin: Akademie–Verlag); 1979 (Moscow: Mir)
[44] Filinov V S, Fehske H, Bonitz M, Fortov V E and Levashov P 2007 *Phys. Rev. E* **75** 036401

Journal of Physics: Conference Series **220** (2010) 012005 doi:10.1088/1742-6596/220/1/012005

Mott transition in electron–hole plasmas

D Semkat, F Richter, D Kremp, G Manzke, W-D Kraeft and K Henneberger

Institut für Physik, Universität Rostock, 18051 Rostock, Germany

E-mail: `dirk.semkat@uni-rostock.de`

Abstract. In connection with the ionization equilibrium of an electron–hole plasma in a highly excited semiconductor, the influence of many-particle effects on both the chemical potentials of carriers and the Mott transition of para-excitons in zinc selenide (ZnSe) is investigated over a wide range of temperatures and carrier densities. Special attention is directed to the determination of the region where an excitonic fraction can exist and, therefore, a possible Bose–Einstein condensate can occur at low temperatures.

1. Introduction

Collective quantum phenomena have been intensely investigated in semiconductor physics for many years. Such phenomena comprise Bose–Einstein condensation (BEC) of excitons as well as condensation of Cooper-like electron–hole pairs to a BCS state. According to the theory of the ideal Bose gas, the BEC of excitons should occur if their chemical potential reaches the exciton energy, either by reduction of the temperature or increase of the density. On the other hand, excitons are bound electron–hole pairs which are subject to an ionization equilibrium $e + h \rightleftarrows X$ and, therefore, form only a subsystem of the partially ionized electron–hole plasma (EHP). Due to many-particle effects like screening of the Coulomb interaction between the carriers, their coupling weakens with increasing density leading finally to a breakup of the excitons usually referred to as Mott effect (density ionization) [1, 2, 3, 4, 5, 6]. The existence of the excitonic fraction of the EHP (and, therefore, of a possible condensate) is limited by the Mott transition from an exciton gas to a fully ionized EHP as a consequence of the Mott effect.

In this paper we investigate the influence of many-particle effects on the composition of the partially ionized EHP in zinc selenide (ZnSe) beyond simple approximations [7, 8]. Concerning the phase diagram, we concentrate on the Mott transition; our aim is not the discussion of the full phase diagram including the occurrence of biexcitons, electron–hole liquid etc.

After a brief overview over the underlying theory [9] in Sec. 2, the quasiparticle energies of the carriers and the exciton energies are analyzed in Secs. 3 and 4, respectively. Finally, in Sec. 5, the degree of ionization being a measure of the EHP composition is calculated.

2. Theory

A self-contained presentation of a quantum statistical description of thermodynamic properties of the EHP has been given in [9]. Only a brief outline shall be given here. Starting point is the

Journal of Physics: Conference Series **220** (2010) 012005 doi:10.1088/1742-6596/220/1/012005

general quantum statistical relation

$$n_a(\{\mu_c\},T) = -i\hbar g_a \int \frac{d\mathbf{k}}{(2\pi)^3} \int \frac{d\omega}{2\pi} G_a^<(\mathbf{k},\omega) = g_a \int \frac{d\mathbf{k}}{(2\pi)^3} \int \frac{d\omega}{2\pi} A_a(\mathbf{k},\omega)\, f_a(\omega)\,, \qquad (1)$$

expressing the carrier density n_a of the species a via the single-particle correlation function $G_a^<$ as a function of the chemical potentials $\{\mu_c\}$ and the temperature T. $f_a(\omega)$ is a Fermi-like distribution, and g_a denotes the band multiplicity (including spin degeneracy) of species a.

Using the spectral function in *extended quasi-particle approximation* [10] and the self-energy in screened ladder approximation, the electron density can be subdivided according to

$$\begin{aligned} n_e(\mu_e,\mu_h,\mu_X,T) &= g_e \int \frac{d\mathbf{k}}{(2\pi)^3} f_e(\epsilon_e(\mathbf{k})) + g_e \int_\Delta^\infty \frac{d\omega}{2\pi} n_{eh}^B(\omega) \mathrm{Im}\, F(\omega) \\ &\quad + g_e g_h \sum_{nl} (2l+1) \int \frac{d\mathbf{K}}{(2\pi)^3} n_X^B\left(\frac{\hbar^2 K^2}{2M}\right) \\ &= n_e^{\mathrm{QP}} + n_e^{\mathrm{scatt}} + n_X\,, \end{aligned} \qquad (2)$$

into "free" (quasiparticle n_e^{QP} and scattering n_e^{scatt}) and the bound (exciton) n_X contributions. Here, $\epsilon_e(\mathbf{k})$ is the quasiparticle energy in *random phase approximation* (RPA) including the Hartree–Fock self-energy Σ_a^{HF},

$$\epsilon_e(\mathbf{k}) = \frac{\hbar^2 k^2}{2m_e} + \mathrm{Re}\, \Sigma_e^r(\mathbf{k},\omega)\Big|_{\hbar\omega=\epsilon_e(\mathbf{k})}\,, \qquad (3)$$

n_X^B denotes the Bose distribution of ideal excitons,

$$n_X^B\left(\frac{\hbar^2 K^2}{2M}\right) = \frac{1}{\mathrm{e}^{\beta\left(\frac{\hbar^2 K^2}{2M} - \mu_X\right)} - 1}\,, \qquad (4)$$

$\beta = (k_B T)^{-1}$ and Δ represents the lowering of the band (continuum) edge. In Eq. (2), we have introduced the so-called *chemical picture*, i.e., the bound electron–hole pairs are reinterpreted as a new particle species, the excitons, where the definition of the excitonic chemical potential

$$\mu_X = \mu_e + \mu_h - E_{nl} \qquad (5)$$

is just equivalent to the well-known thermodynamic condition for the chemical (ionization) equilibrium.

In the chemical picture, the EHP is characterized by the densities of free electrons $n_e^* = n_e^{\mathrm{QP}} + n_e^{\mathrm{scatt}}$, free holes n_h^* (with $n_h^* = n_e^*$), and the density of excitons n_X (total electron density $n_e = n_e^* + n_X$). The ionization equilibrium $e + h \rightleftarrows X$ is controlled by the relation (5) which determines the composition of the system described by the degree of ionization

$$\alpha = \frac{n_e^*}{n_e} \qquad (6)$$

and plays the role of a mass-action law (MAL).

The chemical potential of the excitons μ_X is given by the inversion of

$$n_X(\mu_X,T) = g_e g_h \int \frac{d\mathbf{K}}{(2\pi)^3} \frac{1}{\mathrm{e}^{\beta\left(\frac{\hbar^2 K^2}{2M} - \mu_X\right)} - 1}\,. \qquad (7)$$

We should emphasize here, however, the limitations of the chemical picture or, more precisely, of the underlying extended quasiparticle approximation for the spectral function $A_a(\mathbf{k},\omega)$: While $A_a(\mathbf{k},\omega)$ exhibits, at lower densities, distinct bound state peaks and the pair continuum, the peaks broaden for higher densities due to the damping. Thus, a well defined distinction between bound and scattering states becomes problematic around the Mott density [11, 5, 6]. A quantitative demonstration of the effect for zinc oxide (ZnO) has been given by Klingshirn, see Fig. 24b in Ref. [12]. However, in ZnSe the excitonic linewidth is much smaller [13], as will be discussed in Sec. 4. Due to the damping of the two-particle states, the degree of ionization is not really a well defined quantity around the Mott density. It gives, however, a good qualitative measure for the state of the EHP.

The chemical potential of free carriers (neglecting scattering states) is given by

$$n_a^*(\mu_a, T) = g_a \int \frac{d\mathbf{k}}{(2\pi)^3} \frac{1}{e^{\beta(\epsilon_a(\mathbf{k})-\mu_a)}+1}\,. \tag{8}$$

An analytical determination of the chemical potential from Eq. (8) is possible at most in limiting cases (nondegenerate or highly degenerate, respectively). In the general case, knowing the RPA quasiparticle energy $\epsilon_a(\mathbf{k})$, μ_a can be obtained by numerical inversion of Eq. (8). The determination of $\epsilon_a(\mathbf{k})$ according to Eq. (3) is, however, a numerically expensive task. As an alternative to the *complete inversion* of Eq. (8), one can split the chemical potential according to [1, 14]

$$\mu_a = \mu_a^{\mathrm{id}} + \Delta\mu_a\,. \tag{9}$$

Here, μ_a^{id} is the ideal contribution, and $\Delta\mu_a$ is the correlation part. The inversion of Eq. (8) with respect to the first order in $\Delta\mu_a$ is usually referred to as *incomplete inversion* [15]. Then the ideal contribution, for arbitrary degeneracy, follows in well-known manner from

$$n_a^*(\mu_a^{\mathrm{id}}, T) = g_a \int \frac{d\mathbf{k}}{(2\pi)^3} \frac{1}{e^{\beta\left(\frac{\hbar^2k^2}{2m_a}-\mu_a^{\mathrm{id}}\right)}+1}\,. \tag{10}$$

Numerical calculations for the correlation part $\Delta\mu_a$ have been performed, e.g., in Refs. [15, 16, 17, 2].

While for a non-degenerate EHP, from Eq. (5), a Saha equation can be derived immediately [1, 18], the situation is more complicated in the case of arbitrary degeneracy. Then, the degree of ionization α has to be determined directly from Eq. (5) which reads in terms of α and n_e

$$\mu_X[(1-\alpha)n_e, T] = \mu_e[\alpha n_e, T] + \mu_h[\alpha n_e, T] - E_{nl}\,. \tag{11}$$

Equation (11) represents a form of the MAL which allows for the determination of the degree of ionization α as a function of density and temperature and, therefore, can be regarded as a generalized Saha equation. The degree of ionization is given implicitly by Eq. (11) and has to be obtained numerically [8, 9]. If an excitonic condensate is present, on the r.h.s. of Eq. (7) the density of condensed excitons occurs as an additional term, for the numerical treatment see Ref. [19].

3. Quasiparticle energies

The influence of many-particle effects on the composition of the partially ionized EHP according to the MAL [Eqs. (11) together with (8) and (7)] is contained in two quantities, namely (i) in the quasiparticle energies of the carriers $\epsilon_a(\mathbf{k})$, and (ii) in the two-particle bound state (exciton) energy E_{nl}. Both the quasiparticle energy of the carriers given by Eq. (3), and their damping Γ_a,

$$\Gamma_a(\mathbf{k},\omega) = -2\mathrm{Im}\,\Sigma_a^r(\mathbf{k},\omega) = i\left[\Sigma_a^>(\mathbf{k},\omega) - \Sigma_a^<(\mathbf{k},\omega)\right]\,, \tag{12}$$

are determined by the retarded RPA self-energy Σ_a^r which can be split into a Hartree–Fock contribution $\Sigma_a^{\rm HF}$ and a Montroll–Ward term $\Sigma_a^{\rm MW}$ which considers the dynamical screening between carriers,

$$\Sigma_a^r(\mathbf{k},\omega) = \Sigma_a^{\rm HF}(\mathbf{k}) + \Sigma_a^{\rm MW}(\mathbf{k},\omega)\,, \tag{13}$$

$$\Sigma_a^{\rm HF}(\mathbf{k}) = \int \frac{d\mathbf{k}}{(2\pi)^3} f_a(\varepsilon_a(\mathbf{q}))V_{aa}(\mathbf{k}-\mathbf{q})\,, \tag{14}$$

$$\Sigma_a^{\rm MW}(\mathbf{k},\omega) = \int \frac{d\mathbf{q}}{(2\pi)^3} \int \frac{d\omega'}{2\pi} \frac{\left[1-f^a(\varepsilon_a(\mathbf{q}))+n^B(\omega')\right]\,\hat{V}(\mathbf{k}-\mathbf{q},\omega')}{\hbar\omega-\varepsilon_a(\mathbf{q})-\hbar\omega'+i\Gamma_a(\mathbf{q})/2}\,. \tag{15}$$

The spectral function of the screened potential $\hat{V}(\mathbf{q},\omega)$ appearing in the Eqs. (19) and (15) is related to the inverse retarded dielectric function $\varepsilon^{r^{-1}}(\mathbf{q},\omega)$ via

$$\hat{V}(\mathbf{q},\omega) = 2iV_{ab}(\mathbf{q})\,\mathrm{Im}\,\varepsilon^{r^{-1}}(\mathbf{q},\omega)\,. \tag{16}$$

The function $n^B(\omega)$ represents the Bose distribution of the elementary excitations in the plasma (plasmons), and, in our calculations, we use the Lindhard dielectric function. Contributions of bound electron–hole pairs to the dielectric screening function in the sense of atomic (excitonic) polarizabilities [20] are small as compared to that of the free carriers and are neglected. This is justified if the description of the Mott transition is addressed, however, excitonic contributions to screening should be incorporated if the properties of the pure excitonic gas phase are investigated.

The Fermi distribution $f_a(\varepsilon_a(\mathbf{q}))$ appearing in Eqs. (14) and (15) contains the renormalized dispersion and the chemical potential μ_a of carriers being related to the carrier density n_a^* by Eq. (8)

$$f_a(\varepsilon_a(\mathbf{q})) = \left[e^{(\varepsilon_a(\mathbf{q})-\mu_a)/k_BT}+1\right]^{-1}\,. \tag{17}$$

In order to determine the quasiparticle energy and damping for a given carrier density and temperature, the Eqs. (3), (8), (13)–(15) have to be solved self-consistently, including the chemical potential. We start our numerical procedure with the calculation of the chemical potential $\mu_a^{\rm id}$ of free (non-correlated) carriers, cf. Eq. (10), considering only the kinetic energy part in (3) for the iteration of (8). This is used to iterate the whole system self-consistently. As a result we get the quasiparticle energy and damping which then are employed in the calculation of the carrier density of the correlated EHP with (8). In earlier papers the quasiparticle energy was approximated by the so-called Debye shift [1] or Coulomb-hole self-energy [21], given in excitonic units by $\Sigma^{\rm D} = -\kappa a_X E_X^b$ (κ – inverse screening length). We use the following parameters for ZnSe [13]: binding energy $E_X^b = 22.4$meV, Bohr radius $a_X = 3.65$nm, and effective electron and hole masses of $m_e = 0.15m_{\rm e}^0$ and $m_h = 0.86m_{\rm e}^0$, $m_{\rm e}^0$ being the free electron mass. The real part of the self-energy for electrons and holes, resulting from our calculations (QPA) is compared in Fig. 1 to two approximations. For lower carrier densities, as considered in the figure, the deviation of the Debye self-energy (static screening) is obvious (note the factor 1/2 in $\Sigma^{\rm D}$). Moreover, the influence of dynamical screening, following from the deviation of the QPA from the HF self-energies, is clearly demonstrated. One should remark that the Debye self-energy represents a rigid shift approximation, while the energy shifts in QPA and HF approximation, e.g., for the electrons, decrease for higher wave numbers. For comparison at different temperatures and densities the real part of the sum of the self-energies of electrons and holes at $k = 0$ ($\Sigma_0 = \Sigma_{e,0} + \Sigma_{h,0}$) is presented in Fig. 2. The quasiparticle energy strongly increases with the density due to the many-body effects. The increase is larger for lower temperatures. Comparing our results (QPA) to simple approximations we find out that many-body effects are overestimated by the Debye self-energy, in particular for the intermediate density region and

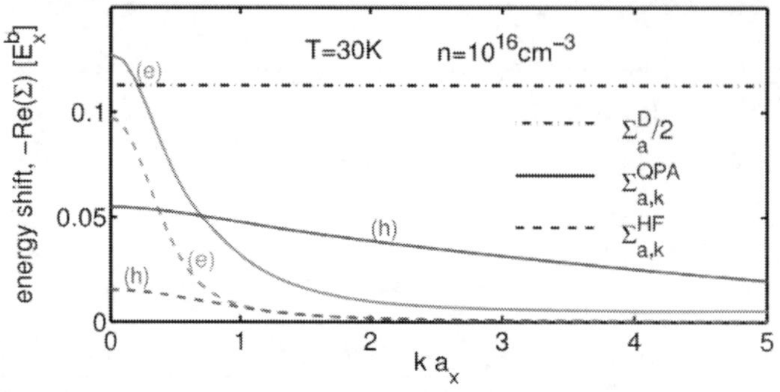

Figure 1. Real part of the self-energy for electrons (e) and holes (h) in dependence of the wave vector. Comparison of our approach QPA (solid lines) with HF self-energy (14) (dashed) and Debye shift (dash-dotted).

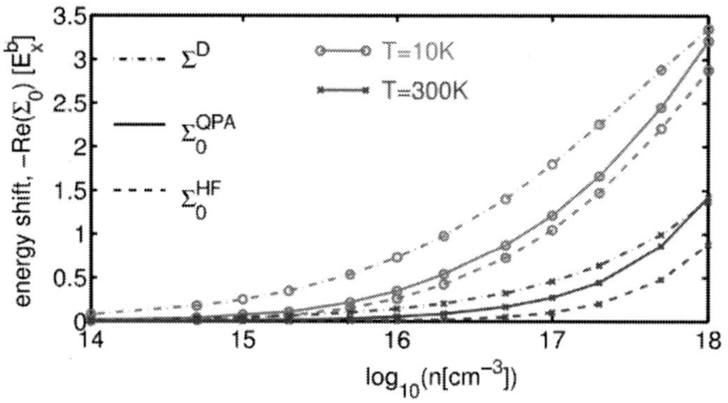

Figure 2. Real part of the sum of self-energies for electrons and holes at $k = 0$ in dependence of the density of carriers. Comparison of our approach QPA (solid lines) with HF self-energy (14) (dashed) and Debye shift (dash-dotted).

for lower temperatures. As we will show later, this is the case for those densities at which the Mott transition of excitons occurs. Only for the higher densities beyond the Mott transition, the static screening approaches the QPA. On the other side, the HF contribution to the self-energy (13) is the dominant one for low temperatures. Here, screening effects play a minor role. Often $\mathrm{Re}\,\Sigma_0$ was considered as the lowering of the band edge. At the crossing of lowered band edge and exciton binding energy, the Mott transition may be expected to occur due to the vanishing of the ionization energy. This will be discussed further in Sec. 4, and we will show that the real part of the interband self-energy (20) has to be considered instead of $\mathrm{Re}\,\Sigma_0$. Nevertheless, for $k = 0$ and at the carrier dispersion both are identical. With regard to the crossing of energy shifts and exciton binding energy E^b_X in Fig. 2, one may expect the Mott density to be much higher in QPA than in Debye approximation. This effect will become even more pronounced in the discussion of the chemical potentials in the next section.

The imaginary part of the quasiparticle energy $\Gamma_a(k)$ for electrons and holes is shown in Fig. 3 for different temperatures at a carrier density of $n = 10^{16}\mathrm{cm}^{-3}$. The results for electrons and holes differ due to their different effective masses. Just as the quasiparticle energies, the damping of the carrier states increases with the temperature. For the Mott density and at room temperature it reaches the order of the excitonic Rydberg energy. This is comparable

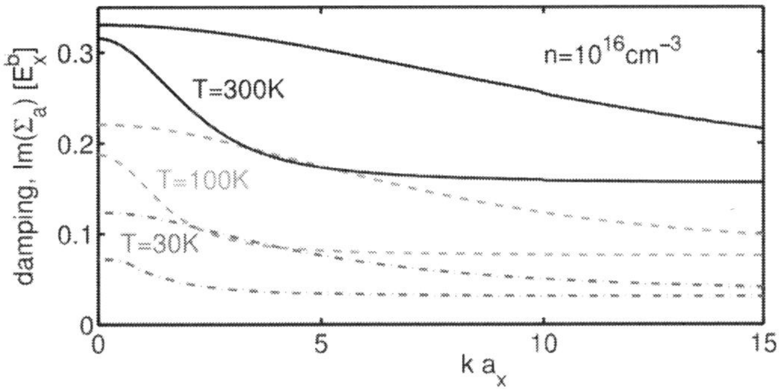

Figure 3. Imaginary part of the self-energy (damping) pairwise for electrons (lower curves) and holes (upper) in dependence of the wave vector for different temperatures.

to the results for the excitonic broadening discussed in the next section and suggests that the quasiparticle description becomes questionable at higher temperatures.

4. Exciton energies

Apart from the quasiparticle energy of the carriers, also the exciton energy contains the influence of many-particle effects on the properties of the EHP. It is well-known from both optical experiments [22] and from the theoretical point of view [3] that the exciton energy only weakly changes with increasing excitation, while the ionization is mainly generated by the lowering of the band edge, described in our approach by the quasiparticle energies of carriers. According to Ref. [3], this behavior can be described by an effective wave equation considering the influence of screening both by self-energies and by a renormalization of the Coulomb interaction.

Experimentally the influence of many-body effects in an excited semiconductor can be determined by measuring the optical spectrum detected with a weak probe pulse. A corresponding theoretical description is given by the semiconductor Bloch equations (SBE) [21]. If one considers the carriers to be in quasi-equilibrium, the carrier distributions are Fermi functions with given chemical potential and temperature, which are not affected by the weak probe pulse, and only the kinetic equation for the polarization has to be solved. In this case, the equation for the polarization $p(\mathbf{k},\omega)$, generated by the probe pulse $E(\omega)$ and coupled via the dipole matrix element d to the semiconductor, can be written in excitonic units as [23, 24, 25, 13, 26]

$$\begin{aligned}&\{\omega - \mathbf{k}^2 - \Delta^{\rm HF}(\mathbf{k}) - \Sigma^r(\mathbf{k},\omega)\}p(\mathbf{k},\omega) + \int \frac{d\mathbf{q}}{(2\pi)^3}\{N(\mathbf{k})V_{eh}(\mathbf{k}-\mathbf{q}) + \Theta(\mathbf{k},\mathbf{q},\omega)\}p(\mathbf{q},\omega)\\ &= N(\mathbf{k})\,\mathrm{d}\,E(\omega)\,. \end{aligned} \tag{18}$$

Many-body effects are contained (i) as Hartree–Fock (HF) renormalized carrier energies $\Delta^{\rm HF}(\mathbf{k})$ and Pauli blocking $N(\mathbf{k}) = 1 - f_e(\mathbf{k}) - f_h(\mathbf{k})$, and (ii) as screening effects in the renormalized interaction matrix $\Theta(\mathbf{k},\mathbf{q},\omega)$

$$\begin{aligned}\Theta(\mathbf{k},\mathbf{q},\omega) &= \Delta V^{\rm eff}(\mathbf{k},\mathbf{q},\omega) - i\Gamma(\mathbf{k},\mathbf{q},\omega)\\ &= \sum_{a\neq b}\int \frac{d\bar\omega}{2\pi}\,\frac{\left[1 - f^a(\varepsilon_a(\mathbf{q})) + n^B(\omega')\right]\,\hat V(\mathbf{k}-\mathbf{q},\omega')}{\hbar\omega - \epsilon_a(\mathbf{k}) - \epsilon_b(\mathbf{q}) - \hbar\bar\omega + i[\Gamma_a(\mathbf{k}) + \Gamma_b(\mathbf{q})]/2}\,,\end{aligned} \tag{19}$$

whose real and imaginary parts $\Delta V^{\rm eff}(\mathbf{k},\mathbf{q},\omega)$ and $\Gamma(\mathbf{k},\mathbf{q},\omega)$ are the effective interaction and the so-called off-diagonal dephasing, respectively, and in the interband self-energy $\Sigma^r(\mathbf{k},\omega)$

$$\Sigma^r(\mathbf{k},\omega) = \Delta e^{sc}(\mathbf{k},\omega) - i\Gamma(\mathbf{k},\omega) = \int \frac{d\mathbf{q}}{(2\pi)^3}\,\Theta(\mathbf{q},\mathbf{k},\omega) = \Sigma^r_e(\mathbf{k},\omega-\epsilon_h) + \Sigma^r_h(\mathbf{k},\omega-\epsilon_e)\,, \quad (20)$$

with its real and imaginary parts $\Delta e^{sc}(\mathbf{k},\omega)$ and $\Gamma(\mathbf{k},\omega)$ being the renormalized interband self-energy and the diagonal dephasing, respectively.

It is important to notice that the various many-body quantities widely compensate each other pairwise in the case of bound states. This concerns on one hand the HF energies and the Pauli blocking, and the renormalized interband self-energy $\Delta e^{sc}(\mathbf{k},\omega)$ and the effective interaction $\Delta V^{\rm eff}(\mathbf{k},\mathbf{q},\omega)$ on the other hand as well as the so-called diagonal and off-diagonal dephasing $\Gamma(\mathbf{k},\omega)$, $\Gamma(\mathbf{k},\mathbf{q},\omega)$, respectively. Moreover, there is an exact relation of the (two-particle) interband self-energy Σ^r for an energy $\hbar\omega$ to the (single-particle) carrier self-energies Σ^r_a at shifted energies $\hbar\omega - \varepsilon_b$ given in Eq. (20). The appearance of this energy shift becomes physically clear if one looks at the energies in the denominator of the effective interaction matrix $\Theta(\mathbf{k},\mathbf{q},\omega)$. It becomes resonant if the incident photon with the energy $\hbar\omega$ generates an interband transition, described by the renormalized quasiparticle energies $\varepsilon_a(\mathbf{k})+\varepsilon_b(\mathbf{q})$, accompanied by absorption or emission of a plasmon with the energy $\hbar\bar{\omega}$. The quasi-particle energies $\varepsilon_a(\mathbf{k})$ and the quasi-particle damping $\Gamma(\mathbf{k},\omega)$ are determined from Eqs. (13)–(15).

For weak probe pulses the macroscopic polarization $P(\omega)$ depends linearly on the electric field

$$P(\omega) = \int \frac{d\mathbf{k}}{(2\pi)^3}\,p(\mathbf{k},\omega) = \chi(\omega)\,E(\omega)\,, \quad (21)$$

where the susceptibility $\chi(\omega)$ characterizes the dielectric properties of the semiconductor. The experimental verification of this approach was given in several optical experiments. In previous papers [23, 24, 25, 13] we have demonstrated how the influence of many-body effects on the exciton line, e.g., carrier-induced line broadening, shift of the exciton resonance and band gap shrinkage show up in the phase and amplitude of transmitted/reflected light.

The semiconductor Bloch equation approach sketched above describes systems with a band structure in an electromagnetic field. In general, the two-particle problem in a surrounding medium needs a careful analysis of the Bethe–Salpeter equation for the two-particle Green's function. A first approach in this direction was given in Refs. [3, 4] using the dynamically screened ladder approximation. However, this result has some serious shortcomings, especially in the degenerate plasma. There, a static contribution in addition to the Hartree–Fock term, and, moreover, a division by the Pauli blocking factors $1-f_a(\mathbf{k}_1)-f_b(\mathbf{k}_2)$ occur. In a subsequent paper [11] it was shown that these shortcomings are generated by using an inappropriate retarded Green's function and by the treatment of the retardation of the screened potential by the Shindo approximation. The corrected effective Schrödinger equation was derived in Refs. [11, 27] and for the EHP in Refs. [25, 13]. For an electron–hole pair system, both approaches are in full agreement for zero center-of-mass momentum.

Numerical solutions for the pair spectrum have been given in Refs. [23, 24, 25, 13] using the SBE and in Refs. [3, 28, 29] using the effective Schrödinger equation in the nondegenerate case.

The two-particle spectrum shown in Fig. 4 exhibits the following peculiarities: (i) There is a lowering of the band (continuum) edge due to the many-body effects contained in the self-energy. (ii) The exciton ground state energy remains nearly constant up to higher densities. This follows from the approximate compensation of the many-particle effects for the bound states [3]. Moreover, at low temperatures, $\Theta(\mathbf{k},\mathbf{q},\omega)$ and, therefore, $\Sigma^r(\mathbf{k},\omega)$ are small, and Coulomb Hartree–Fock self-energy and Pauli blocking dominate. However, they nearly compensate each other at low densities, too, but lead to a weak shift for higher densities. (iii) The difference

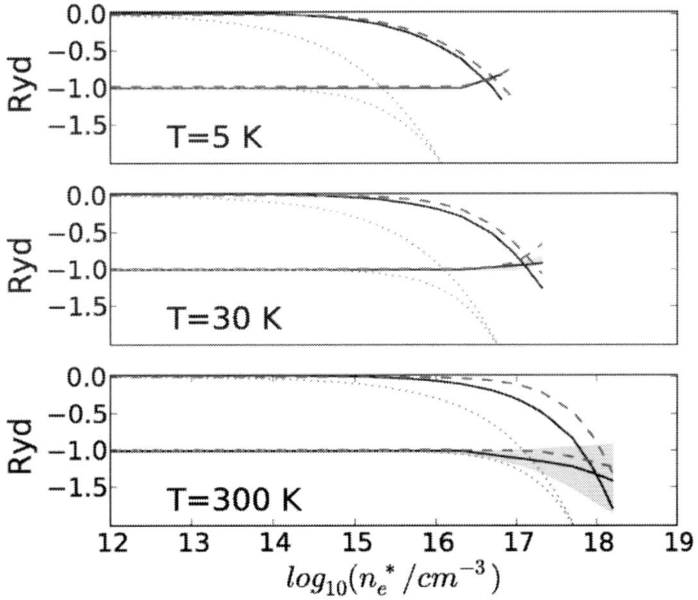

Figure 4. Two-particle spectrum vs. density in ZnSe: band edge (upper curves) and exciton ground state energy (lower curves). Solid lines: solution of the full SBE, dashed lines: Hartree–Fock approximation, dotted lines: Debye approximation. The grey area marks the width (full width at half maximum) of the exciton. Ryd denotes the excitonic Rydberg.

between band edge and exciton energy defines the effective ionization energy $I^{\rm eff}$ which is lowered with increasing density. (iv) For $I^{\rm eff} = 0$, the bound state vanishes and merges into the scattering continuum. This process is referred to as *Mott effect*, and the corresponding density is called Mott density. Above that point, the bound state does not exist, and the bound state energy is no longer a meaningful quantity. (v) Finally, a small shift of the exciton energy appears with increasing carrier density, depending on the temperature [23]: a blue shift at low temperatures turns into a red shift for higher temperatures.

While the exciton energy is shown in Fig. 4 as a well-defined line, in fact, with increasing carrier density the exciton is broadened due to screening of the Coulomb interaction between carriers. This broadening, being described in Eq. (18) by the diagonal and off-diagonal dephasing, strongly increases at higher carrier densities. At low temperatures, this broadening is small: the excitonic full width at half maximum is $\gamma = 0.012$ Ryd for 3 K and $\gamma = 0.07$ Ryd for 30 K just below the Mott density, respectively. However, at room temperature it reaches the order of the excitonic Rydberg. This is shown in Fig. 4 by the grey area. For corresponding experimental results see [30, 12].

In addition to our results from the solution of the full SBE (18), in Fig. 4 we show a comparison to two approximations: That is (i) a solution with screening effects being neglected in the SBE and only the HF energies and the Pauli blocking considered, i.e., $\Theta(\mathbf{k},\mathbf{q},\omega) = 0$ and $\Sigma^r(\mathbf{k},\omega) = 0$ (Hartree–Fock approximation). (ii) A further frequently used approximation is the static limit of $\Theta(\mathbf{k},\mathbf{q},\omega)$. It follows from Eq. (19) if $\bar{\omega}$ exceeds all other terms in the denominator. If furthermore, for $\Delta e^{sc}(\mathbf{k},\omega)$ [cf. Eq. (20)] the classical limit is applied, it leads to the Debye shift $\Sigma^{\rm D}$. Figure 4 shows that this approximation provides an inadequate description of both bound state energy and band edge even at room temperature. In particular, the Mott effect occurs at far too low densities. Static screening becomes a convenient approximation far above room temperature but completely fails at low temperatures.

5. Degree of ionization

A quantitative measure for the ionization state of the EHP is given by the degree of ionization α defined by Eq. (6). For an overview of the ionization state it is convenient to consider isolines $\alpha(n_e, T)$ = const. in the density–temperature plane. Such "ionization maps" are presented in Figs. 5–7 which differ in the approximations for the chemical potential of the carriers. All cases have in common that, with increasing temperature at fixed density, thermal ionization takes place quite smoothly, while at fixed temperature with increasing density, the Mott transition appears as an abrupt jump from $\alpha = 0$ to $\alpha = 1$.

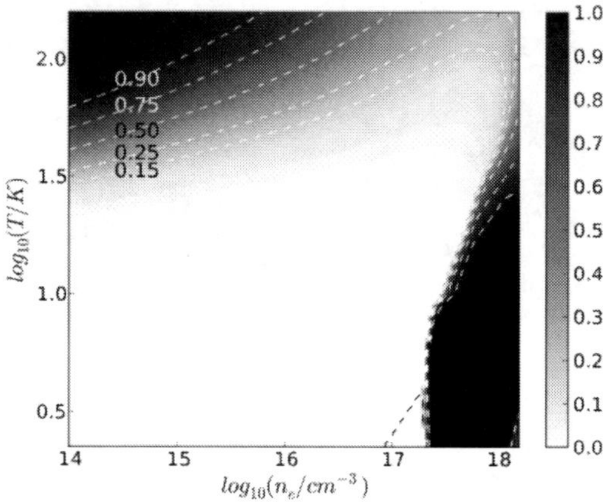

Figure 5. Isolines of the degree of ionization in the density–temperature plane for ZnSe. Solution of the MAL using self-consistent quasiparticle energies for the carriers and the SBE results for the exciton energies. The dashed triangle at the bottom borders the region where BEC of excitons is possible, cf. Eq. (22).

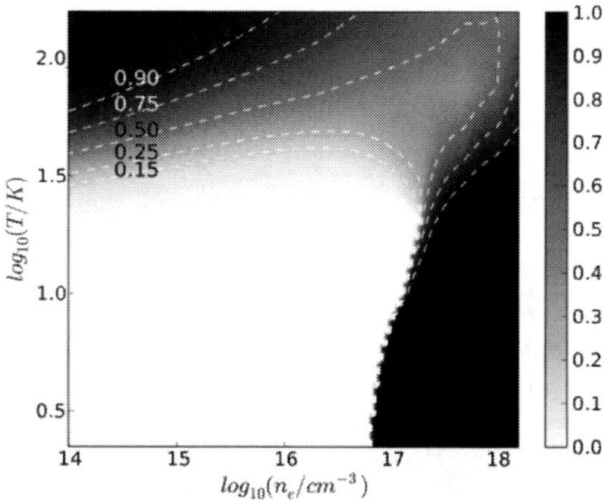

Figure 6. Same representation as in Fig. 5, but using the Padé formula from [16] for the carrier chemical potentials.

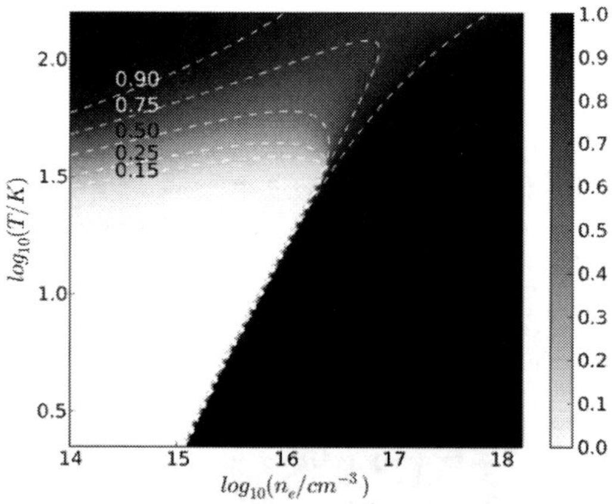

Figure 7. Same representation as in Fig. 5, but using carrier chemical potentials in Debye approximation.

However, the position as well as temperature dependence of the Mott density is entirely different. Especially the Debye approximation overestimates the Mott effect drastically so that the Mott transition occurs at more than two orders of magnitude too low densities. Already the application of the Padé formulas [16] shifts the Mott density considerably. Only our self-consistent calculations, however, exhibit a window for a possible Bose–Einstein condensation (BEC) of excitons in the depicted temperature range: Looking at the area of strong dominance of the excitons ($\alpha \approx 0$) and regarding them as noninteracting bosons, BEC may be expected under the condition

$$(1-\alpha)n_e\Lambda_X^3 \geq \zeta\,(3/2) \approx 2.61\,. \tag{22}$$

This condition is given in Fig. 5 by the triangle bordered by the dashed line at the bottom. Note that the breakup of excitons at the Mott density does not mean a disappearance of the condensate. Instead, a closer investigation shows a smooth crossover to a BCS-type condensate at high densities [31, 32, 33]. Thus, the physical nature of the condensed particles changes, i.e., we find excitons below the Mott density, and Cooper-like electron–hole pairs far above it. To which extent the survival of the excitons as resonances just above the Mott density plays a role has still to be investigated.

Acknowledgments
We would like to thank Th. Bornath and H. Stolz (Rostock) for many fruitful discussions. This work was supported by the Deutsche Forschungsgemeinschaft (Collaborative Research Center SFB 652).

References

[1] Ebeling W, Kraeft W D and Kremp D 1976 *Theory of Bound States and Ionization Equilibrium in Plasmas and Solids* (Berlin: Akademie–Verlag); 1979 (Moscow: Mir)
[2] Zimmermann R 1988 *Many-Particle Theory of Highly Excited Semiconductors* (Leipzig: Teubner)
[3] Zimmermann R, Kilimann K, Kraeft W D, Kremp D and Röpke G 1978 *Phys. Status Solidi* B **90** 175
[4] Haug H and Tran Thoai D B 1978 *Phys. Status Solidi* B **85** 561
[5] Peng Q Y, Schmielau T, Manzke G and Henneberger K 2000 *Journal of Crystal Growth*, **214-215** 810
[6] Schmielau T, Manzke G, Tamme D and Henneberger K 2000 *Phys. Status Solidi* B **221** 215
[7] Snoke D 2008 *Solid State Commun.* **146** 73
[8] Richter F, Semkat D, Kremp D and Henneberger K 2009 *Phys. Status Solidi* C **6** 532
[9] Semkat D, Richter F, Kremp D, Manzke G, Kraeft W-D and Henneberger K 2009 *Phys. Rev.* B **80** 155201
[10] Kremp D, Kraeft W D and Lambert A J D 1984 *Physica* A **127** 72
[11] Bornath Th, Kremp D and Schlanges M 1999 *Phys. Rev.* E **60** 6382
[12] Klingshirn C F 2007 *Phys. Status Solidi* B **244** 3027
[13] Seemann M, Kieseling F, Stolz H, Franz R, Manzke G, Henneberger K, Passow T and Hommel D 2005 *Phys. Rev.* B **72** 075204
[14] Kremp D, Kraeft W D and Schlanges M 1993 *Contrib. Plasma Phys.* **33** 567
[15] Stolzmann W and Kraeft W D 1979 *Ann. Phys. (Leipzig)* **36** 388
[16] Rösler M, Zimmermann R and Richert W 1984 *Phys. Status Solidi* B **121** 609
[17] Ebeling W and Richert W 1985 *Phys. Lett.* A **108** 80
[18] Kraeft W D, Kilimann K and Kremp D 1975 *Phys. Status Solidi* B **72** 461
[19] Richter F 2007 Diploma thesis, Universität Rostock (unpublished)
[20] Röpke G and Der R 1979 *Phys. Status Solidi* B **92** 501
[21] Haug H and Koch S W 1993 *Quantum Theory of the Optical and Electronic Properties of Semiconductors*, (World Scientific Publ.) (4th ed. 2004)
[22] Fehrenbach G W, Schäfer W, Treusch J and Ulbrich R G 1982 *Phys. Rev. Lett.* **49** 1281
[23] Manzke G, Peng Q Y, Henneberger K, Neukirch U, Hauke K, Wundke K, Gutowski J and Hommel D 1998 *Phys. Rev. Lett.* **80** 4943
[24] Nägerl J S, Stabenau B, Böhne G, Dreher S, Ulbrich R G, Manzke G and Henneberger K 2001 *Phys. Rev.* B **63** 235202
[25] Manzke G and Henneberger K 2002 *Phys. Status Solidi* B **234** 233
[26] Lorke M, Nielsen T R, Seebeck J, Gartner P and Jahnke F 2006 *Phys. Rev.* B **73** 085324
[27] Schmielau T 2001 Ph.D. thesis (Universität Rostock)
[28] Seidel J, Arndt S and Kraeft W D 1995 *Phys. Rev.* E **52** 5387
[29] Arndt S, Kraeft W D and Seidel J 1996 *Phys. Status Solidi* B **194** 601
[30] Neukirch U and Wundke K 1997 *Phys. Rev.* B **55** 15408
[31] Nozières P and Schmitt-Rink S 1985 *J. Low Temp. Phys.* **59** 195
[32] Bronold F X and Fehske H 2006 *Phys. Rev.* B **74** 165107
[33] Kremp D, Semkat D and Henneberger K 2008 *Phys. Rev.* B **78** 125315; Semkat D, Kremp D and Henneberger K 2009 *Phys. Status Solidi* C **6** 546

The photon Green's function for bounded media: Splitting property and nonequilibrium radiation laws

F Richter, D Semkat and K Henneberger

Institut für Physik, Universität Rostock, 18051 Rostock, Germany

E-mail: felix.richter2@uni-rostock.de

Abstract. The presence of a medium boundary has been a major obstacle for the theoretical description of the propagation, emission and absorption of light due to the loss of translational invariance. We present a nonequilibrium photon Green's function theory that is valid for bounded (i.e., spatially inhomogeneous) media systems and also yields an energy flow law which can be seen as a generalization of the Kirchhoff and Planck laws to nonequilibrium. With the help of this law, we discuss mechanisms of emission and optical signatures of quantum condensates. An important finding is that the $D^{\gtrless}$ components of the photon GF, which describe field-field fluctuations, decompose universally into two parts related to medium kinetics and external light sources. Thanks to their specific structure, the propagation of arbitrary (even nonclassical) light can be analyzed straightforwardly. These properties are used to demonstrate the energy flux and scattering of squeezed light incident on the medium.

1. The photon Green's function

1.1. Definition

The photon Green's function (PGF) [1; 2] is a nonequilibrium (real-time) Green's function for the vector potential operator $\hat{\mathbf{A}}$ of the electromagnetic field in Coulomb gauge, which obeys the Maxwell equation

$$\Box\hat{\mathbf{A}} = -\mu_0\left(\hat{\mathbf{j}}_{\text{ind}} + \mathbf{j}_{\text{ext}}\right) \tag{1}$$

with an operator-valued induced (bound) current $\hat{\mathbf{j}}_{\text{ind}}$ and a c-number external (free) current $\mathbf{j}_{\text{ext}}$.

The PGF D is defined by functional variation of Maxwell's equation for the effective field $\mathbf{A}_{\text{eff}} = \langle\hat{\mathbf{A}}\rangle$ [1; 2] on the Keldysh double time contour [3; 4]:

$$D(\underline{1},\underline{2}) = -\frac{1}{\mu_0}\frac{\delta\mathbf{A}_{\text{eff}}(\underline{1})}{\delta\mathbf{j}_{\text{ext}}(\underline{2})} \tag{2}$$

Here, $\underline{1}$ denotes variables $(\mathbf{r}_1, \underline{t}_1)$ and $\underline{t}_1$ a time on the contour. The polarization P

$$P(\underline{1},\underline{2}) = -\mu_0\frac{\delta\hat{\mathbf{j}}_{\text{ind}}(\underline{1})}{\delta\mathbf{A}_{\text{eff}}(\underline{2})} \tag{3}$$

represents the effects of the medium on the electromagnetic field and plays the role of a "photon self-energy" in the Dyson equation

$$\left(D_0^{-1}(\underline{1},\underline{2}) - P(\underline{1},\underline{2})\right) D(\underline{2},\underline{1}') = \delta_T(\underline{1},\underline{1}'), \tag{4}$$

where δ_T is the transverse delta function [5]. For notational simplicity, we restrict the discussion in this paper to isotropic media, in which D is essentially reduced to a scalar function. The generalization to anisotropic media with tensor-valued D is straightforward and has been presented in former articles referenced here.

1.2. The PGF and classical electrodynamics

As all Keldysh GF's do, D contains four different physical functions which obey a set of identity relations. With the above definitions, the "greater" and "less" components of D give the *fluctuations* of the electromagnetic field,

$$D^{>}(1,2) = D^{<}(2,1) = \frac{1}{i\hbar\mu_0}\left(\langle\hat{\mathbf{A}}(1)\hat{\mathbf{A}}(2)\rangle - \mathbf{A}_{\text{eff}}(1)\mathbf{A}_{\text{eff}}(2)\right), \tag{5}$$

while the "retarded" Green's function solves the classical wave propagation problem:

$$\mathbf{A}_{\text{eff}}(1) = -\mu_0 D^{\text{ret}}(1,2)\mathbf{j}_{\text{ext}}(2). \tag{6}$$

The electromagnetic properties of the medium here are contained in the retarded polarization function P^{ret}, which is related to the induced current and to the susceptibility χ,

$$\mathbf{j}_{\text{ind}}(1) = \frac{\partial}{\partial t_1}\mathbf{P}(1) = -\frac{1}{\mu_0}P^{\text{ret}}(1,2)\mathbf{A}_{\text{eff}}(2), \tag{7}$$

$$P^{\text{ret}}(1,2) = -\frac{1}{c^2}\frac{\partial^2\chi(1,2)}{\partial t_1 \partial t_2}. \tag{8}$$

Note that P should not be confused with the polarization field $\mathbf{P}$ appearing in the macroscopic Maxwell equations, and that the linear approximation was used. Thus, it is implied that fields are weak, and P^{ret} gives the *linear response* of the medium.

2. Bounded media problems

2.1. Spatial inhomogeneity

In the PGF approach shown here, the susceptibility χ describes fully and exactly the electromagnetic properties of the medium, for which appropriate approximations have to be employed in calculations. In nonequilibrium and bounded media systems, χ is inherently a nonlocal function: It is non-local in space (or *spatially inhomogeneous*) due to the finite size of the dipoles that create the polarization field in response to the exciting outer field, and it is non-local in time if the system is in nonequilibrium [6].

Only in bulk media in (quasi-)equilibrium, a *dispersion relation* for the wave vector of the field propagating through the system can be obtained, because due to translational invariance, the dependency reduces to differences of variables only:

$$\chi(\mathbf{r},\mathbf{r}',t,t') \quad\rightarrow\quad \chi(r-r',t-t') \quad\rightarrow\quad \chi(\mathbf{q},\omega). \tag{9}$$

As a result of such a Fourier-transformed susceptibility function, one obtains multiple branches of solutions for the dispersion relation, which are known as *polariton solutions* in semiconductor physics,

$$\mathbf{q}^2(\omega) = \frac{\omega^2}{c^2}\left(1+\chi(\mathbf{q},\omega)\right) \quad\rightarrow\quad \mathbf{q}_i(\omega). \tag{10}$$

If further the long-wavelength limit is taken ($\mathbf{q}\rightarrow 0$, since optical wavelengths are much larger than atoms), one arrives at a dielectric function reduced to frequency dispersion only, $\varepsilon(\omega) = 1+\chi(\omega)$. However, there exist a number of effects associated with medium boundaries and surfaces which cannot be represented by such a simple form [6]. On the other hand, the non-locality of the polarization function vastly complicates exact manipulations in the PGF theory.

Journal of Physics: Conference Series **220** (2010) 012006 doi:10.1088/1742-6596/220/1/012006

2.2. Vacuum polarization

In contrast to bulk systems, bounded media systems have free space or vacuum regions, where the polarization tends to zero ($P \to 0$). In these regions, the form of $D^{\rm ret}$ can be readily derived in the Fourier domain as

$$D_0^{\rm ret}(\mathbf{q},\omega) = \frac{c^2}{(\omega + {\rm i}\epsilon)^2 - c^2q^2} = D_0^{{\rm adv},*}(\mathbf{q},\omega)\,. \tag{11}$$

The addition of an infinitesimal imaginary displacement ($\epsilon \to 0$) is needed to ensure causality. With regard to the *optical theorem*, the analogue of the dissipation-fluctuation theorem in particle GF theory,

$$D^{\gtrless}(1,2) = D^{\rm ret}(1,3)P^{\gtrless}(3,4)D^{\rm adv}(4,2) + D_h^{\gtrless}(1,2)\,, \tag{12}$$

it appears necessary to add an infinitesimal contribution to P in order to ensure the validity of the theorem in the vacuum limit $P \to 0, D^{\rm ret} \to D_0^{\rm ret}$ [1; 7]:

$$P^{\gtrless}(\omega) \quad \to \quad P^{\gtrless}(\omega) \mp {\rm i}\epsilon\Theta(\pm\omega)\frac{2\omega}{c^2}\,. \tag{13}$$

This contribution, sometimes referred to as "vacuum polarization", could be justified as a homogeneous solution $D_h^{\gtrless}$ to the optical theorem, and it was known to play a decisive role for steady-state emission and absorption [1; 7] and also for the energy flow law [8; 9] to be discussed later in this article. Its interpretation, however, was not satisfying and its handling rather difficult, requiring to exploit subtle cancellation effects between infinitesimal and diverging factors.

3. The splitting property

An elegant derivation and interpretation of the "vacuum polarization" in spatially inhomogeneous systems is found with the help of the Langreth rules [10], which can be used to extract the Keldysh components of PGF products, e.g.,

$$(FG)^{\gtrless} = F^{\rm ret}G^{\gtrless} + F^{\gtrless}G^{\rm adv}\,, \tag{14}$$

$$(FG)^{\rm ret/adv} = F^{\rm ret/adv}G^{\rm ret/adv}\,. \tag{15}$$

Applying these rules to the integral form of the Dyson equation on the Keldysh contour [Eq. (4)],

$$D(\underline{1},\underline{2}) = D_0(\underline{1},\underline{2}) + D_0(\underline{1},\underline{3})P(\underline{3},\underline{4})D(\underline{4},\underline{2})\,, \tag{16}$$

leads after some rearrangements to a splitting of the field-field fluctuations into two contributions [11; 12]. After introduction of the inverse retarded dielectric function,

$$\varepsilon_T^{-1,\rm ret} = \delta + D^{\rm ret}P^{\rm ret}\,, \tag{17}$$

they can be written as

$$D^{\gtrless} = D^{\gtrless}_{\rm med} + D^{\gtrless}_{\rm vac} \qquad \left\{ \begin{array}{ll} D^{\gtrless}_{\rm med} & = D^{\rm ret}\,P^{\gtrless}\,D^{\rm adv} \\ D^{\gtrless}_{\rm vac} & = \varepsilon_T^{-1,\rm ret}\,D_0^{\gtrless}\,\varepsilon_T^{-1,\rm adv} = D_h^{\gtrless}\,. \end{array} \right. \tag{18}$$

Comparison with Eq. (12) shows that $D^{\gtrless}_{\rm med}$, the *medium-induced* contribution, is known from the bulk-matter optical theorem. It is driven by the dissipation in the medium, or more specifically, from electronic recombination or generation processes described by the rates $P^{\gtrless}$. Second, with $D^{\gtrless}_{\rm vac}$, called the *vacuum-induced* contribution, obviously the explicit form of the homogeneous solution $D^{\gtrless}_h$ is found, whose sources are now clearly identified as freely evolving fluctuations $D^{\gtrless}_0$, while medium processes do not play a role [11; 12].

Following from bare rearrangements of the Dyson equation, the splitting of the fluctuations in a bounded media system appears as a universal property of the PGF. It is valid regardless of the electromagnetic properties or geometrical shape of the medium, and spatial as well as temporal inhomogeneity are fully considered without further assumptions.

Let us have a closer look on the structure of the vacuum-induced fluctuations $D^{\gtrless}_{\rm vac}$: Their sources, the fluctuations $D^{\gtrless}_0$, are *free PGFs* defined in terms of the freely evolving vector potential $\hat{\mathbf{A}}_0$ according to Eq. (5), i.e., they are pure vacuum solutions of the Maxwell equations. These free fluctuations are renormalized in $D^{\gtrless}_{\rm vac}$ by the inverse of the retarded dielectric function $\varepsilon_T^{\rm ret}$, which describes the propagation of an incident classical free wave,

$$\Box \mathbf{A}_{\rm ext} = -\mu_0 \mathbf{j}_{\rm ext}\,, \tag{19}$$

perturbed by the presence of the medium and resulting in the effective field

$$\mathbf{A}_{\rm eff} = \varepsilon_T^{-1,\rm ret}\,\mathbf{A}_{\rm ext}\,. \tag{20}$$

Hence, the vacuum-induced fluctuations $D^{\gtrless}_{\rm vac}$ can be attributed to light coming from external sources ($\mathbf{j}_{\rm ext}$), in other words, incident light. Their propagation in presence of the medium can always be traced back to classical wave propagation, no matter what the properties of the incident light are.

It is to be noted that the splitting property does not mean a spatial decomposition of the system, since all PGFs are defined globally. Rather, it allows for a discrimination of the light fluctuations by their source together with a clear interpretation.

4. Propagation of quantized light

4.1. Mode expansion and quantized light

In the discussion above, no assumptions have been made on the properties of the incident light. Thanks to the specific structure of $D^{\gtrless}_{\rm vac}$, even the propagation of non-classical quantized light can be readily analyzed. If the freely evolving vector potential operator in the definition of $D^{\gtrless}_0$ is expanded into normal modes according to

$$\hat{\mathbf{A}}_0(\mathbf{r}) = \sum_{\lambda\mathbf{q}} \sqrt{\frac{\hbar}{2\varepsilon_0 c q V}}\,\mathbf{e}_{\lambda\mathbf{q}} \left(\hat{a}_{\lambda\mathbf{q}} e^{i\mathbf{q}\mathbf{r}} + \hat{a}^{+}_{\lambda\mathbf{q}} e^{-i\mathbf{q}\mathbf{r}} \right)\,, \tag{21}$$

where $\hat{a}_{\lambda\mathbf{q}}$ is a photon annihilation operator for mode vector $\mathbf{q}$ and polarization direction λ, one obtains a general expression with expectation values of photon operator correlations in the normal form $\langle \hat{a}^{+}\hat{a}\rangle - \langle \hat{a}^{+}\rangle \langle \hat{a}\rangle$ and in the anomalous form $\langle \hat{a}\hat{a}\rangle - \langle \hat{a}\rangle \langle \hat{a}\rangle$ [11; 13]. Their values depend on the quantum state of the light, for example

$$\langle \hat{a}^{+}\hat{a}\rangle = \left\langle 0 \left| \hat{S}^{+}_{\xi}\, \hat{a}^{+}\hat{a}\, \hat{S}_{\xi} \right| 0 \right\rangle \tag{22}$$

is to be evaluated for incident light prepared to the squeezed vacuum state [14]. This is demonstrated systematically for various states in Ref. [13]. We will see later in Eq. (29) that the normal and anomalous terms are associated with stationary and instationary energy fluxes, respectively.

Furthermore, the normal-mode expansion reveals that $D_0^{\gtrless}$ and, hence, $D_{\rm vac}^{\gtrless}$, split further according to [11; 13]

$$D_0^{\gtrless} = D_{0,\rm sp}^{\gtrless} + D_{0,\rm stim} \tag{23}$$

into a ubiquitous fixed contribution from the spontaneous ground-state fluctuations,

$$D_{0,\rm sp}^{\gtrless}(1,2) = \sum_{\lambda\mathbf{q}} \frac{c}{2\mathrm{i}Vq}\mathbf{F}(1)\otimes\mathbf{F}^*(2) \tag{24}$$

with free plane waves $\mathbf{F}_{\lambda\mathbf{q}}(\mathbf{r},t) = \mathbf{e}_{\lambda\mathbf{q}}\exp[\mathrm{i}\mathbf{qr} - \mathrm{i}cqt]$, and a contribution $D_{0,\rm stim}$ from external stimulation by a light source. For example, for incident light prepared to a Fock state with mode population number $n_{\lambda\mathbf{q}}^F$, it takes the form

$$D_{0,\rm stim}(1,2) = \sum_{\lambda\mathbf{q}} \frac{c}{2\mathrm{i}Vq} n_{\lambda\mathbf{q}}^F \mathbf{F}_{\lambda\mathbf{q}}(1)\otimes\mathbf{F}_{\lambda\mathbf{q}}^*(2)\,. \tag{25}$$

Just as the splitting into medium- and vacuum-induced contributions, the splitting into ground-state fluctuations and stimulation is very useful to analyze the various contributions separately. For both $D_{0,\rm sp}^{\gtrless}$ and $D_{0,\rm stim}$, the renormalization by the medium is formally simple. The free normal waves just have to be replaced with effective normal waves thanks to Eq. (20) [11]:

$$D_{\rm vac}^{\gtrless} = D_0^{\gtrless}(\mathbf{F}_{\lambda\mathbf{q}} \to \mathbf{A}_{\lambda\mathbf{q}})\,. \tag{26}$$

The PGF spectral function is defined as the difference between the "greater" and the "less" component of the corresponding PGF. The vacuum spectral function $\hat{D}_0$ stays a fixed universal function as it should since $D_{0,\rm stim}$ is the same for both components $D_0^{\gtrless}$. The vacuum-induced spectral function $\hat{D}_{\rm vac}$ appears as a fixed function for a given media system, i.e., it depends only on $\varepsilon_T^{\rm ret}$ [11]:

$$\hat{D}_{\rm vac}(1,2) = \sum_{\lambda\mathbf{q}} \frac{c}{2\mathrm{i}Vq}\left[\mathbf{A}_{\lambda\mathbf{q}}(1)\otimes\mathbf{A}_{\lambda\mathbf{q}}(2)^* - \mathbf{A}_{\lambda\mathbf{q}}(1)^*\otimes\mathbf{A}_{\lambda\mathbf{q}}(2)\right]\,. \tag{27}$$

4.2. Energy flow of quantized light

The energy flux vector (Poynting vector) $\hat{\mathbf{S}} = \frac{1}{\mu_0}\langle\hat{\mathbf{E}}\times\hat{\mathbf{B}}\rangle|_{\rm sym}$ [1; 15] can also be expressed by PGFs. The splitting properties (18),(23) then translate directly to $\mathbf{S}$:

$$\mathbf{S} = \mathbf{S}_{\rm sp} + \mathbf{S}_{\rm stim} + \mathbf{S}_{\rm med}\,. \tag{28}$$

The quantity $\mathbf{S}_{\rm stim}$ is just the energy flux caused by an external light source. It is equivalent to the normally-ordered energy flux $\langle : \hat{S} : \rangle$ that is measured by photodetectors, e.g., in absorption-transmission experiments [13].

We can now derive, e.g., an exact relation for the signal of nonclassical light transmitted through arbitrary media slabs [11; 13], which yields the form

$$S_{\rm stim}^t = \frac{\hbar c^2}{V}\sum_{\lambda\mathbf{q}} q_\perp |t_\mathbf{q}|^2 \left(\langle\hat{a}^+\hat{a}\rangle - |\langle\hat{a}\hat{a}\rangle|\cos\left[2q_\perp x - 2cqt + \varphi\right]\right)\,. \tag{29}$$

Both stationary and instationary parts of the signal are modulated by the classical transmission coefficient $t_\mathbf{q}$, which appears here due to the fact that the renormalization in $D_{\rm vac}^{\gtrless}$ is governed by $\varepsilon_T^{-1,\rm ret}$ which represents classical optical properties [Eq. (20)]. For slabs, the transmission coefficient is linked in an obvious way to absorptivity and reflectivity [9]: $a_\mathbf{q} = 1 - |r_\mathbf{q}|^2 - |t_\mathbf{q}|^2$. Thus, scattering of light by media is always classical, independent of the quantum state of light.

4.3. Comparison to other theories

This result arises from a quantum-kinetically exact treatment of the vector potential operator in the PGF. It is valid for arbitrarily dispersive and absorptive nonequilibrium media and fully considers the spatial inhomogeneity of bounded media problems. In contrast, recent well-developed quantum-optical approaches (compare, e.g., [14; 16; 17]) still must resort to an effective description of the medium properties, and represent ground-state fluctuations phenomenologically by noise currents, for which only recently a physical description has been established [18]. However, the present approach confirms the results of these effective theories and generalizes them to spatial inhomogeneity, nonequilibrium and oblique incidence.

5. Nonequilibrium energy flow

Further analytic evaluation of the spatially inhomogeneous nonequilibrium PGF seems out of reach without further assumptions. Let us thus leave complete universality, and consider the energy flow in a slab in the steady state according to Fig. 1.

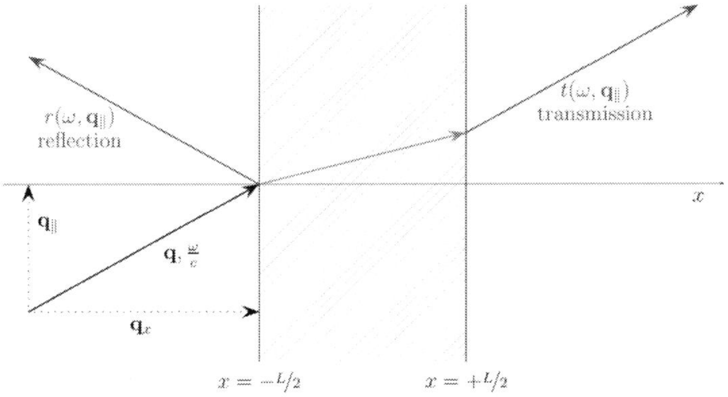

Figure 1. Slab geometry: TE-polarized light is incident in the x direction on a slab with thickness L and infinite extension in the y-z plane.

The energy conservation law (Poynting's theorem) $\frac{\partial}{\partial t}U + \operatorname{div}\mathbf{S} = -\mathbf{jE}$ [15] in this geometry reduces to

$$S(x = -L/2) - S(x = L/2) = \int \mathrm{d}x\, \mathbf{j}(x)\mathbf{E}(x)\,. \tag{30}$$

The dissipation $W = \mathbf{jE}$ can be expressed by PGFs [1; 9]

$$W(1) = \mathrm{i}\hbar \frac{\partial}{\partial t_2} \Theta(t_3 - t_1) \left[P^{>}(1,3)\, D^{<}(3,2) - P^{<}(1,3)\, D^{>}(3,2)\right]\big|_{2\to 1} \tag{31}$$

and decomposes into W_{med} and W_{vac} according to the splitting of $D^{\gtrless}$, Eq. (18). The further evaluation of these terms is far from trivial and has been presented in Refs. [8; 9]. One can show that W_{med} vanishes in steady-state slab geometry, reflecting that, e.g., any emission from a medium process must be re-absorbed in order to keep the steady state. The vacuum-induced fluctuations then are fully responsible for any energy flow in the system, and one arrives at an energy flow law [8; 9],

$$s(\mathbf{q}_\parallel, \omega) = \left[b(\mathbf{q}_\parallel, \omega) - n(\mathbf{q}_\parallel, \omega)\right] a(\mathbf{q}_\parallel, \omega)\,. \tag{32}$$

The spectrally resolved energy flux s leaving the slab surfaces is composed of an emission contribution, $s_e = ba$, and an absorption contribution, $s_a = -na$, and both are proportional to the classical absorptivity a. The prefactors b and n are distributions of internal and external excitations. Application of the splitting property and mode expansion on this theory leads to deeper insight on their properties and the interplay of light, matter and ground-state fluctuations.

5.1. Internal and external excitations in the energy flux

The ratio of the renormalized incident fluctuations and their fixed spectral function [Eq. (27)] gives, in the sense of a Kadanoff-Baym ansatz [8; 9], the distribution of photons in the incident light, which is also referred to as *external optical excitations*:

$$n^{\gtrless}(\omega, q_{\|}) = \frac{D^{\gtrless}_{\text{vac}}}{\hat{D}_{\text{vac}}}\,. \tag{33}$$

Indeed, from mode expansion follows

$$n^{\gtrless} = n^{\gtrless}_{\text{sp}} + n_{\text{stim}}\,, \qquad\qquad n^{\gtrless}_{\text{sp}} = \pm\frac{1}{2}\,, \tag{34}$$

and in a Fock state $n_{\text{stim}} \equiv n^F$. For other quantum states, n_{stim} will represent an effective mode population. Due to the definition of the spectral function, $n^> = 1 + n^<$, and $n^< = n$ is the distribution entering the above energy flow law, Eq. (32). In analogy to that, the distribution of *internal optical excitations* $b^{\gtrless}(\omega, q_{\|})$ is introduced as the ratio of (global) recombination to the spectral function of the polarization [8; 9]. We will assume here that a similar decomposition $b^{\gtrless} = b^{\gtrless}_{\text{sp}} + b_{\text{stim}}$ holds.

Re-inserting the decomposed distributions into the energy flow law leads to

$$s = 2\left(n^{>}_{\text{sp}}\, b_{\text{stim}} - b^{>}_{\text{sp}}\, n_{\text{stim}}\right)\, a\,, \tag{35}$$

a balance of absorption of incident light ($D^{\gtrless}_0 \to n_{\text{stim}}$) and emission due to medium processes ($P^{\gtrless} \to b_{\text{stim}}$). Both contributions are coupled to the ground-state fluctuations of the opposite party. This expression reduces further to

$$s = (b_{\text{stim}} - n_{\text{stim}})\, a \tag{36}$$

and results in an effectively classical, quasi-thermal picture of emission and absorption. The contributions of the ground-state fluctuations cancel out, but ignoring them in the calculations would lead to vanishing energy flow. Hence, they are indispensable for emission and absorption to occur. This result includes spontaneous and stimulated emission as well as light amplification ($a < 0$) and can also be seen as a generalization of the Planck and Kirchhoff laws to distributions in nonequilibrium [9].

5.2. Quasi-equilibrium behavior

A strong theoretical requirement arises from the Kubo–Martin–Schwinger condition [19; 20] for the polarization rates $P^{\gtrless}$: All internal (medium) excitation states must (optically) behave bosonic, regardless of the underlying fermionic particle system. In quasi-equilibrium, the distribution b will take the form of the Bose distribution

$$b \quad\to\quad b(\mu, T) = \frac{1}{\exp\left[\frac{1}{k_B T}(\hbar\omega - \mu)\right] - 1}\,, \tag{37}$$

and in full equilibrium, the Planck distribution ($\mu = 0$) is reached as a limiting case.

For the example of polaritons in the electron-hole plasma of semiconductors, the chemical potential μ in the Bose distribution gives the excitation (carrier) density, $n_e(\mu, T)$. The energy $\hbar\omega = \mu$ marks the frequency at which a crossover from amplifying ($a(\omega) < 0$) to absorbing ($a(\omega) > 0$) behavior occurs. On this basis, the emission of a semiconductor in quasi-equilibrium shall be analyzed in the next section [9; 21].

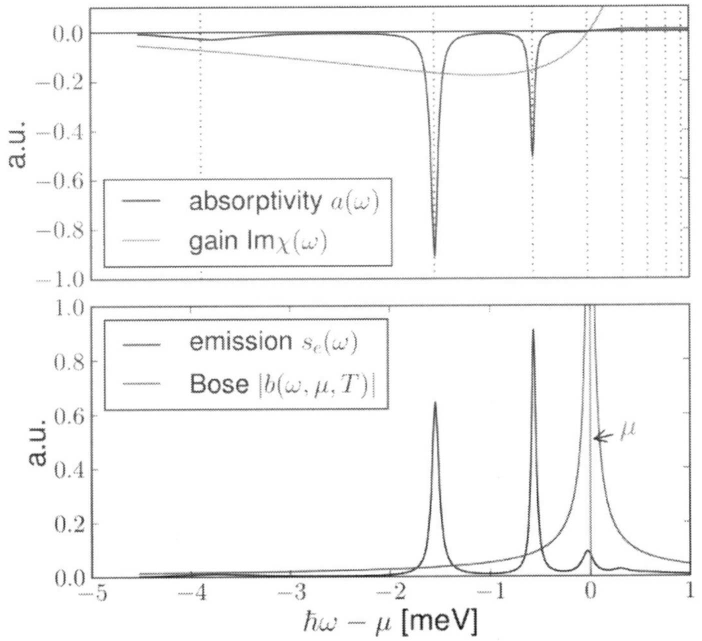

Figure 2. Absorptivity a and emission $s_e = ba$ in a 1500 nm ZnSe slab as a function of frequency relative to the chemical potential $\mu = 2807$ meV. The susceptibility χ [21; 22] was calculated for an excitation density of $n_e = 2\times10^{17}$ cm^{-3} at $T = 77$ K.

5.3. Application example: Semiconductor emission in quasi-equilibrium

In Figure 2, the emission spectrum $s_e(\omega) = b(\omega,\mu,T)\cdot a(\omega)$ perpendicular to the surface ($\mathbf{q}_\parallel = 0$) is depicted for a zinc selenide (ZnSe) slab at high excitation. The according susceptibility exhibits a strong gain region ($\mathrm{Im}\,\chi(\omega) < 0$) below the crossover energy $\mu = \hbar\omega = 2807$ meV. The absorptivity in this region, while still following the Fabry–Perot interference conditions marked by dotted vertical lines, exceeds the values in the absorbing range by orders of magnitude. The singularity of the Bose function at the crossover keeps s_e positive and shifts the weight of the Fabry–Perot resonances closer to the crossover, which is an effect of the degeneracy in the distribution of medium excitations.

6. Signatures of quantum condensation

If the temperature approaches absolute zero, the quasi-equilibrium emission will vanish in the absorbing region and reflect the gain in the amplifying region due to the behavior of the Bose distribution:

$$b(\hbar\omega > \mu, T\to 0\,\mathrm{K})\,,\to 0 \qquad\qquad b(\hbar\omega < \mu, T\to 0\,\mathrm{K}) \to -1\,. \tag{38}$$

At such low temperatures, effects of quantum condensation need to be considered. In electron-hole plasmas, quantum condensation manifests itself in the Bose–Einstein condensation of excitons at low density or as Cooper-like electron-hole pairs at higher densities, with a smooth crossover regime between both states (Fig. 3). With a recent particle Green function approach to the problem [23; 24], an anomalous contribution P_{cond} is predicted to appear in the polarization $P^{\gtrless}$ in presence of a condensate. This contribution does not affect the coherent absorption, but in the emission, an additional (background broadened) δ peak is expected at $\hbar\omega = \mu$,

$$s_{\mathrm{cond}}(\omega,\mathbf{q}_\parallel) \quad\propto\quad \delta_{\mathbf{q}_\parallel,0}\cdot\delta(\hbar\omega-\mu)\,P_{\mathrm{cond}}\int_{\mathrm{slab}} dx\,|A(x)|^2\,, \tag{39}$$

which increases for $T\to 0$, whereas the normal emission disappears [8; 9]. Its strength, given by P_{cond}, is illustrated in Fig. 4 for various temperatures. This additional peak can serve as

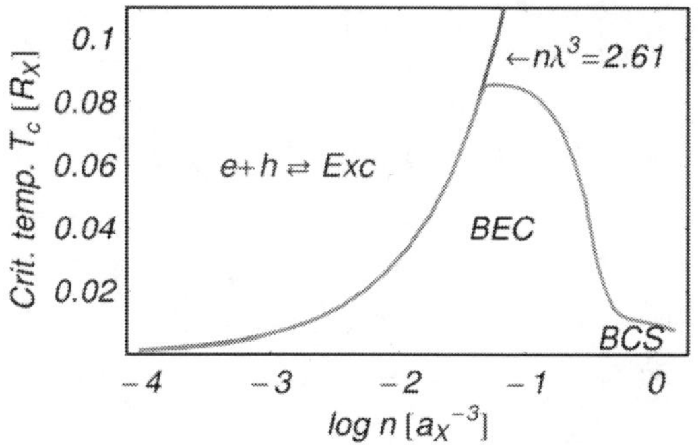

Figure 3. Phase diagram: Condensed and non-condensed phase in the electron-hole plasma of a model semiconductor, from [23; 24]. The phase boundary follows the ideal Bose condensation condition, then the condensate develops smoothly from a Bose–Einstein condensate to a BCS-like electron-hole pair state due to the breakup of the excitons (Mott effect).

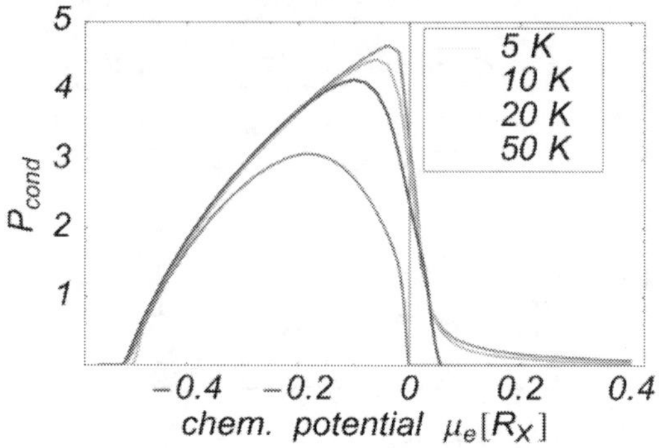

Figure 4. Strength of the condensate contribution to the polarization function at different temperatures as a function of the chemical potential of the electrons. Calculated according to the model from [23]. The crossover region around $\mu = 0$ corresponds to the Mott density.

the optical signature of a quantum condensate, since the regular emission just at this frequency vanishes $(a(\hbar\omega = \mu) \equiv 0)$ [8; 9].

7. Summary

The description of radiation in bounded media with the help of the nonequilibrium photon Green's function was presented. Thanks to the universal splitting property of the field fluctuations represented by $D^{\gtrless}$ and the specific structure of the vacuum-induced contribution, light can be discriminated by its source in this approach. The splitting into contributions from ground-state fluctuations of the vacuum field, incident light, and electronic processes in the medium translates to the energy flux vector and to the dissipation. Incident quantized light can be readily described. Regardless of its state, it is predicted to propagate classically through a medium. The results are valid for arbitrarily dispersive and absorptive media of any geometrical shape. The spatial inhomogeneity inherent to bounded media problems is fully considered.

Further analysis of energy flux and dissipation in the steady-state slab geometry leads to an energy flow law valid for nonequilibrium excitations of the outside (incident light) and the medium. Application of the splitting properties to the excitation distributions shows the influence of the ground-state fluctuations on emission and absorption. Finally, the energy flow law provides an effectively classical picture. Both emission and absorption are governed by the classical absorptivity, and the contributions of the ground-state fluctuations on the resulting energy flux cancel out.

For demonstration, the quasi-equilibrium emission of an highly excited semiconductor is discussed. In combination with a general Green function approach to quantum condensation in the electron-hole plasma, predictions can be made for the optical signature of, e.g., a Bose–Einstein condensate of excitons.

Acknowledgments
The authors would like to thank the *Deutsche Forschungsgemeinschaft* for support through *Sonderforschungsbereich 652.*

References

[1] Henneberger K and Koch S W 1995 *Microscopic theory of semiconductors: Quantum Kinetics, Confinement and Lasers* ed Koch S W (World Scientific, Singapore) pp 131–166
[2] DuBois D F 1967 *Lectures in Theoretical Physics* vol IX C ed Brittin W E (Gordon and Breach, New York) pp 469–620
[3] Keldysh L V 2003 *Progress in Nonequilibrium Green's Functions II* ed Bonitz M and Semkat D (World Scientific, Singapore) pp 4–17
[4] Keldysh L V 1964 *Zh. Eksp. Teor. Fiz.* **47** 1515 [Soviet Phys. JETP **20**, 1018 (1965)]
[5] Cohen-Tannoudji C, Dupont-Roc J and Grynberg G 1989 *Photons and atoms* (Wiley, New York)
[6] Halevi P 1992 *Spatial Dispersion in Solids and Plasma* (North-Holland, Amsterdam)
[7] Henneberger K and Koch S W 1996 *Phys. Rev. Lett.* **76** 1820
[8] Henneberger K 2008 *Phys. Status Solidi B* **246** 283
[9] Richter F, Florian M and Henneberger K 2008 *Phys. Rev. B* **78** 205114
[10] Langreth D C 1976 *Linear and Nonlinear Electron Transport in Solids* ed Devreese J T and Van Doren E (Plenum, New York)
[11] Henneberger K and Richter F 2009 *Phys. Rev. A* **80** 013807
[12] Henneberger K 2008 (*Preprint* `arXiv:0810.5058`)
[13] Richter F, Vasylyev D Y and Henneberger K 2009 (*Preprint* `arXiv:0903.5238`)
[14] Vogel W and Welsch D G 2006 *Quantum Optics* 3rd ed (Wiley-VCH, Weinheim)
[15] Richter F, Florian M and Henneberger K 2008 *Europhys. Lett.* **81** 67005
[16] Henry C H and Kazarinov R F 1996 *Rev. Mod. Phys.* **68** 801
[17] Artoni M and Loudon R 1999 *Phys. Rev. A* **59** 2279
[18] Vasylyev D, Vogel W, Henneberger K, Schmielau T and Welsch D G 2008 *Phys. Rev. A* **78** 033837
[19] Martin P C and Schwinger J 1959 *Phys. Rev.* **115** 1342
[20] Kadanoff L P and Baym G 1962 *Quantum Statistical Mechanics* (Benjamin, New York)
[21] Richter F, Manzke G and Henneberger K 2009 *Phys. Status Solidi C* **6** 528
[22] Seemann M, Kieseling F, Stolz H, Franz R, Manzke G, Henneberger K, Passow T and Hommel D 2005 *Phys. Rev. B* **72** 075204
[23] Kremp D, Semkat D and Henneberger K 2008 *Phys. Rev. B* **78** 125315
[24] Semkat D, Kremp D and Henneberger K 2009 *Phys. Status Solidi C* **6** 546

Journal of Physics: Conference Series **220** (2010) 012007 doi:10.1088/1742-6596/220/1/012007

Coherent quasiparticle approximation (cQPA) and nonlocal coherence

Matti Herranen[1,2,3], **Kimmo Kainulainen**[1,2] **and Pyry M. Rahkila**[1,2]
[1]University of Jyväskylä, Department of Physics,
P.O. Box 35 (YFL), FIN-40014 University of Jyväskylä, Finland
[2]Helsinki Institute of Physics, P.O. Box 64, FIN-00014 University of Helsinki, Finland.
[3]Institut für Theoretische Teilchenphysik und Kosmologie, RWTH Aachen University, D-52056 Aachen, Germany

E-mail: kimmo.kainulainen@jyu.fi

Abstract. We show that the dynamical Wigner functions for noninteracting fermions and bosons can have complex singularity structures with a number of new solutions accompanying the usual mass-shell dispersion relations. These new shell solutions are shown to encode the information of the quantum coherence between particles and antiparticles, left and right moving chiral states and/or between different flavour states. Analogously to the usual derivation of the Boltzmann equation, we impose this extended phase space structure on the full interacting theory. This extension of the quasiparticle approximation gives rise to a self-consistent equation of motion for a density matrix that combines the quantum mechanical coherence evolution with a well defined collision integral giving rise to decoherence. Several applications of the method are given, for example to the coherent particle production, electroweak baryogenesis and study of decoherence and thermalization.

1. Introduction

Many problems in modern particle physics and cosmology require setting up transport equations for relativistic quantum systems in out-of-equilibrium conditions, including baryogenesis [1, 2, 3, 4, 5], heavy ion collisions [6] and out-of equilibrium particle production [7]. It is straightforward to write a formal solution for the problem in the Schwinger-Keldysh method, and detailed studies have been done on *e.g.* thermalization of quantum systems [8]. It is much more difficult to find a simple enough an approximation scheme that is usable in practical applications. Here we will report a recent progress [9, 10, 11, 12, 13] towards a scheme, in which one can account at least for certain particular types of nonlocal coherence effects in the presence of hard collisions.

The key observation is that in cases with particular symmetries (the homogeneous time dependent case and stationary, planar symmetric case) the free 2-point Wigner functions have new singular shell solutions (at $k_0 = 0$ for the homogeneous system and at $k_z = 0$ in the static planar case), that carry the information about the nonlocal quantum coherence. When this shell structure is fed into the dynamical equations for the full interacting theory, they reduce to a closed set of quantum transport equations for the phase space distribution functions on the singular shells, which describe the on-shell particle numbers (mass-shells) and to a measure of the coherence (the new $k_{0,z} = 0$ shells).

Journal of Physics: Conference Series **220** (2010) 012007 doi:10.1088/1742-6596/220/1/012007

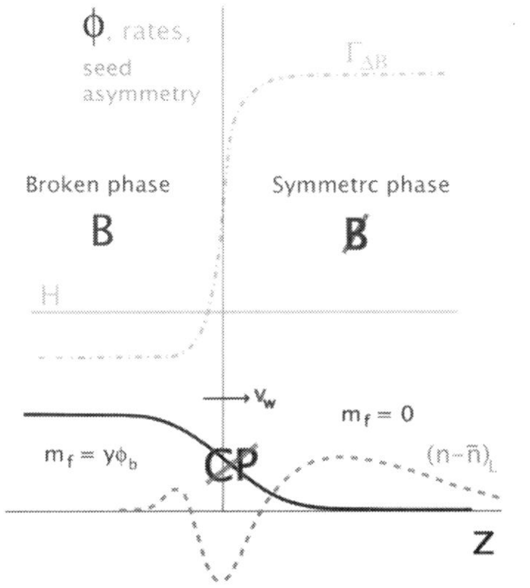

Figure 1. Shown is schematically how various quantities change over the profile of a planar phase transition front. The symmetric (massless) phase is to the right and the broken (massive) phase to the left, and the front is moving to the right with a velocity v_w. The black solid line presents the variation of the Higgs field, the gray dash-dotted line shows the baryon number violation rate $\Gamma_{\rm sph}$ and the solid gray line marks the constant expansion rate H. The thick dashed line shows the left chiral quark asymmetry.

The problem that initially led us to develop the current formalism was the electroweak baryogenesis by the CP-violating quantum reflection mechanism. We therefore briefly discuss EWBG and the quantum reflection problem in this context. First, the baryon number is broken at nonperturbative level in the electroweak theory by the electroweak anomaly which couples the baryon and lepton currents to the Chern-Simons current in the gauge sector. The anomalous B-number violation is vanishingly small in the zero temperature [14], but it becomes very fast at high temperatures. Above the *electroweak phase transition* temperature $T_c \approx 100$ GeV, where the electroweak symmetry is restored, the B-violation rate [15] $\Gamma_{\rm sph} = (20-25)\alpha_w^5 T^4 \sim 10^{-6}T^4$, is much faster rate per degree of freedom than is the expansion rate of the universe during the transition: $H \sim 10^{-16}T$. Second, the P- and CP-parities, whose breaking is necessary to give direction to the B-violation, are broken by weak interactions themselves and by the complex phases in the quark mass matrix, respectively. CP-violating phases are also abundant (and needed) in many extensions of the standard model. A third condition for any successful BG-mechanism is to have out-of-equilibrium conditions. This is much harder to establish in the electroweak setting, because typical interaction rates between particles, $\Gamma_{\rm int} \sim \alpha_w^2 T \approx 10^{-3}T$, are enormous in comparison with the expansion rate during the transition. The only way to find out-of-equilibrium conditions with a standard expansion history is that the transition is of first order, proceeding through nucleation of broken symmetry (with $\phi \neq 0$, where ϕ is the vacuum expectation value of the Higgs field) bubbles inside the symmetric $\phi = 0$ phase. In this case a new time scale, the bubble passing time, emerges: $\Delta t_b \approx l_w/v_w$, where l_w is the width of the phase transition front and v_w is the front velocity. Typically one finds $\Delta t_b \sim 10^3/T \sim \Gamma_{\rm int}^{-1}$. Because particles get their masses through the Higgs mechanism: $m \sim y\phi$, fermions change from being massless to massive excitations as they cross the phase transition front, and this sudden change is enough to push particle distributions momentarily out of equilibrium. Careful studies of the EWPT dynamics have shown that bubbles are created far apart and they grow to enormous dimensions before their coalescence completes the transition. Thus, in electroweak baryogenesis one is led to study the CP-violating, out-of-equilibrium interactions of fermions with expanding, stationary and planar phase transition walls. This setting is illustrated in Fig. 1.

The precise quantitative problem is to compute the total left chiral quark asymmetry in the front of the transition wall, as this asymmetry acts as the seed that biases the B-violating rate to create the eventual baryon asymmetry. This problem has been satisfactorily solved for the case of a very smooth phase transition front. This *semiclassical baryogenesis* formalism was first developed by use of WKB-techniques [3], and later from a more fundamental CTP-formalism [4]. However, when the wall is not particularly thick, the quantum reflection mechanism may become relevant. This problem is much harder to solve in the presence of interactions because of the inherent nonlocality of the reflection mechanism. However, it can be addressed by use of our coherent quasiparticle approximation (cQPA) scheme, and solving this problem remains one of our immediate goals. However, there are also several other interesting applications for the cQPA formalism. These include a number of homogeneous problems, such as particle production by time-varying (oscillating) background fields, *e.g.* during the preheating after inflation, or at the end of phase transitions. We have not yet completed implementing the cQPA scheme to the EWBG-program, but we have applied it to several homogeneous problems, some of which will be illustrated in the sections below.

This proceeding is organised as follows. In section 2 we introduce the basic CTP-formalism and introduce the Kadanoff-Baym equations in the mixed representation. In section 3 we derive the cQPA shell structure including the case with flavour mixing. In section 4 we find expressions for physical quantities and solve some simple collisionless problems to illustrate the physical content of the coherence solutions. In section 5 we consider the interacting theory and derive the density matrix equation with collisions for the homogeneous problems. In section 6 we extend our results for the scalar fields, and finally section 7 contains our conclusions and outlook.

2. CTP-formalism

In this chapter we very briefly review the Schwinger-Keldysh formulation of a relativistic fermionic quantum field theory. The mathematical quantity of interest is the 2-point Wightmann function:

$$iG^{<}_{\alpha\beta}(u,v) \equiv \langle \bar{\psi}_\beta(v)\psi_\alpha(u)\rangle \equiv \mathrm{Tr}\left\{\hat{\rho}\ \bar{\psi}_\beta(v)\psi_\alpha(u)\right\}. \tag{1}$$

By definition $G^<$ is an ensemble average over a quantum density operator $\hat{\rho}$, which fully describes the properties of the system. In the Schwinger-Dyson formalism one replaces the problem of finding $\hat{\rho}$ by a problem of finding all correlators defined by $\hat{\rho}$. Since $G^<$ is an "in-in"-correlator, a complex time contour has to be introduced, one convenient choice of which is shown in Fig. 2. On such a Closed Time Path (CTP) [16, 17], the basic object is a path ordered 2-point function (Dirac indices are suppressed):

$$iG_{\mathcal{C}}(u,v) = \left\langle T_{\mathcal{C}}\left[\psi(u)\bar{\psi}(v)\right]\right\rangle, \tag{2}$$

where $T_{\mathcal{C}}$ defines time ordering along the Keldysh contour $\mathcal{C}$. The correlator $G_{\mathcal{C}}(x,y)$ can be shown to obey a contour Schwinger-Dyson equation

$$G_{\mathcal{C}}(u,v) = G^0_{\mathcal{C}}(u,v) + \int_{\mathcal{C}} \mathrm{d}^4 z_1 \int_{\mathcal{C}} \mathrm{d}^4 z_2\ G^0_{\mathcal{C}}(u,z_1)\Sigma_{\mathcal{C}}(z_1,z_2)G_{\mathcal{C}}(z_2,v), \tag{3}$$

where the self energy function $\Sigma_{\mathcal{C}}$ couples $G_{\mathcal{C}}$ to an infinite hierarchy of equations for higher order Green's functions. The precise form of $\Sigma_{\mathcal{C}}$ depends on the model Lagrangian and a truncation scheme for the hierarchy. (In the weak coupling limit a natural truncation is to substitute all higher than 2-point functions by their perturbative expressions.) The complex time valued function $G_{\mathcal{C}}(x,y)$ can be shown to obey a contour Schwinger-Dyson equation The real time is regained at the expense of introducing a 2×2 matrix structure; after some manipulations one can rewrite Eq. (3) in the following matrix form:

$$G_0^{-1} \otimes G = \sigma_3\,\delta + \Sigma \otimes \sigma_3 G, \tag{4}$$

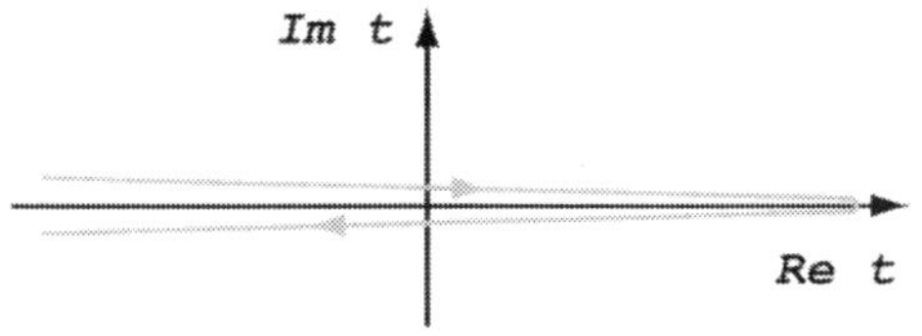

Figure 2. Schwinger-Keldysh path in complex time.

where

$$G = \begin{pmatrix} G_F & -G^< \\ G^> & G_{\bar{F}} \end{pmatrix}, \qquad \Sigma = \begin{pmatrix} \Sigma_F & -\Sigma^< \\ \Sigma^> & \Sigma_{\bar{F}}, \end{pmatrix} \tag{5}$$

and σ_3 is the usual Pauli matrix, and we defined a shorthand notation $\otimes$ for the convolution integral:

$$f \otimes g \equiv \int \mathrm{d}^4 z f(u,z) g(z,v). \tag{6}$$

Here G_F and $G_{\bar{F}}$ are the chronological (Feynman) and anti-chronological (anti-Feynman) Green's functions, respectively, and $G^<$ and $G^>$ are the Wightmann distribution functions, the former of which is just the correlator in Eq. (1). Above have also left out the labels u and v where obvious; for example $\delta \equiv \delta^4(u-v)$. Self energy functions can be calculated for example by using the 2PI effective action techniques (see eg. [18, 17, 5]). For example:

$$\Sigma^{<,>}(u,v) \equiv i \frac{\delta \Gamma_2[G]}{\delta G^{>,<}(v,u)}, \tag{7}$$

where Γ_2 is the sum of all two particle irreducible vacuum graphs of the theory. Introducing the usual retarded and advanced propagators $G^r(u,v) \equiv \theta(u^0 - v^0)(G^< + G^>)$ and $G^a(u,v) \equiv -\theta(v^0 - u^0)(G^< + G^>)$, and a similar decomposition for the self energy function Σ, equations (4) can be recast into a particularly compact from:

$$(G_0^{-1} - \Sigma^{r,a}) \otimes G^{r,a} = \delta \tag{8}$$

$$(G_0^{-1} - \Sigma^r) \otimes G^{<,>} = \Sigma^{<,>} \otimes G^a. \tag{9}$$

Equations (8) and (9) are called the pole equations and the *Kadanoff-Baym equations* [19], respectively. In general, the former will fix the spectral properties of the theory, while the latter will give the dynamical evolution, including quantum transport effects we are interested in. The KB-equation for $G^<$ can be further rewritten as

$$(G_0^{-1} - \Sigma_H) \otimes G^< - \Sigma^< \otimes G_H = \frac{1}{2}\left(\Sigma^> \otimes G^< - \Sigma^< \otimes G^>\right), \tag{10}$$

where the hermitian function $G_H \equiv \frac{1}{2}(G^r + G^a)$. The equation for $G^>$ is not needed, because $G^>$ and $G^<$ are related by the *spectral function*, $\mathcal{A} \equiv \frac{i}{2}(G^r - G^a) = \frac{i}{2}(G^> + G^<)$, and $\mathcal{A}$ and G_H are determined by the pole equations (8). The spectral function is further subject to a condition $2\mathcal{A}(t,\vec{u};t,\vec{v})\gamma^0 = \delta^3(\vec{u} - \vec{v})$, which is just the direct space version of the famous spectral sum-rule and follows directly from the canonical equal time anticommutation relation: $\{\psi(t,\vec{u}), \psi^\dagger(t,\vec{v})\} = -i\delta^3(\vec{u} - \vec{v})$.

2.1. Mixed representation

The next step, crucial to our program, is to move to the mixed representation Fourier transforming with respect to the relative coordinate $u - v$. However, it is necessary to first define more precisely the model we are interested in. With the electroweak baryogenesis and particle production problems in mind, we study a model described by the Lagrangian:

$$\mathcal{L} = i\bar{\psi}\,\partial\!\!\!/\,\psi + \bar{\psi}_L m\psi_R + \bar{\psi}_R m^*\psi_L + \mathcal{L}_{\rm int}\,, \tag{11}$$

where $m(x) = m_R(x) + im_I(x)$ is complex, possibly spacetime dependent mass and $\mathcal{L}_{\rm int}$ is the interaction part to be defined later. All nontrivial interactions responsible for coherence, will be produced by the mass term. From Eq. (11) it is easy to find that

$$G_0^{-1}(u,v) = \delta^4(u-v)(i\,\partial\!\!\!/_v - m^*(v)P_L - m(v)P_R)\,, \tag{12}$$

where $P_{L,R} = \frac{1}{2}(1 \mp \gamma^5)$ are the chiral projector matrices. Inserting the expression (12) into Eq. (10) and performing the Fourier (or Wigner) transformation we can rewrite the KB-equations in the mixed representation:

$$(k\!\!\!/ + \tfrac{i}{2}\,\partial\!\!\!/_x - \hat{m}_0 - i\hat{m}_5\gamma^5)G^< - e^{-i\Diamond}\{\Sigma_H\}\{G^<\} - e^{-i\Diamond}\{\Sigma^<\}\{G_H\} = \mathcal{C}_{\rm coll}, \tag{13}$$

where the G- and Σ functions are functions of the average coordinate $x = \frac{1}{2}(u+v)$ and the k which is the conjugate momentum to $r \equiv u - v$. For example the *Wigner* function $G^<(k,x)$ is defined as:

$$G^<(k,x) \equiv \int d^4r\, e^{ik\cdot r} G^<(x + r/2, x - r/2)\,. \tag{14}$$

The collision term is given by

$$\mathcal{C}_{\rm coll} \equiv \tfrac{1}{2}e^{-i\Diamond}\left(\{\Sigma^>\}\{G^<\} - \{\Sigma^<\}\{G^>\}\right)\,. \tag{15}$$

The $\Diamond$-operator is the following generalization of the Poisson brackets:

$$\Diamond\{f\}\{g\} = \tfrac{1}{2}\left[\partial_X f \cdot \partial_k g - \partial_k f \cdot \partial_X g\right] \tag{16}$$

and the mass operators $\hat{m}_0$ and $\hat{m}_5$ are defined as:

$$\hat{m}_{0,5}G^<(k,x) \equiv m_{R,I}(x)e^{-\frac{i}{2}\partial_x^m\cdot\partial_k^F}G^<(k,x)\,. \tag{17}$$

The KB-equation (13) together with the pole-equations (whose mixed representation forms we omit here) and the sum-rule form a complete set of equations for solving $G^<$, $\mathcal{A}$ and G_H exactly, after the interactions, a scheme to compute Σ and the mass profiles are specified. In practice these equations are way too hard to be solved in their full generality, and several approximations are needed to find a solvable set of equations. The novelty of this work is to show that a tractable approximation scheme that is general enough to treat quantum coherence simultaneously with interactions does exist. We shall now proceed to build our scheme by first constructing the full phase space structure of the free 2-point functions.

3. Free theory and the shell structure

In the collisionless case the last three terms in equation (13) are absent. However, the mass operators still contain infinite number of momentum derivatives acting on $G^<$. These gradients are controlled by space-time derivatives of equally high order acting on the mass function. So, in the regions where the mass is slowly varying we can truncate these expansions to the lowest order. Note that the spatial variations in $G^<$ are not restricted by this approximation. In this limit we find

$$(k\!\!\!/ + \tfrac{i}{2}\,\partial\!\!\!/_x - m_R - im_I\gamma^5)G^< = 0\,. \tag{18}$$

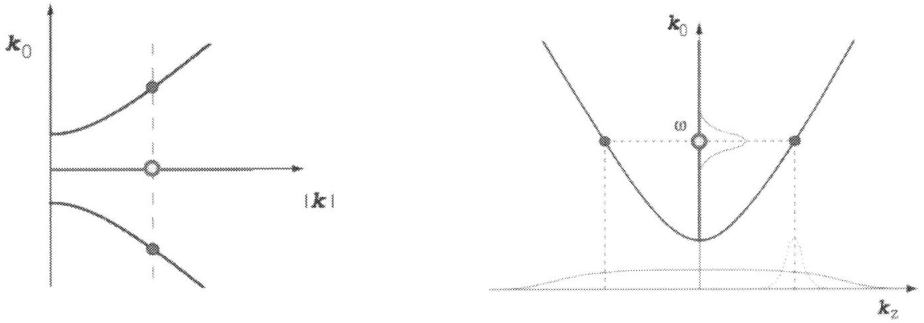

Figure 3. Dispersion relations in the case of a scalar mass function. The dark blobs show the mass-shell contributions and the circles show the corresponding coherence solution at the new $k_0 = 0$ shell in the homogeneous case (left) and at the new $k_z = 0$ shell in the planar symmetric case (right).

Even this equation is very difficult to solve with general boundary conditions. We shall therefore restrict us to two special geometries relevant for the baryogenesis and particle productions problems: the spatially homogeneous and stationary planar symmetric cases. Consider first the homogeneous case. Observing that helicity is now a good quantum number, we can use the decomposition

$$iG_h^< \gamma^0 \equiv g_h^< \otimes \tfrac{1}{2}(1 + h\hat{k}\cdot\vec{\sigma}) \tag{19}$$

in the chiral Weyl representation. When this decomposition is inserted to Eq. (18), it eventually breaks into hermitian and antihermitian 2x2-matrix equations for $g_h^<$:

$$(\mathrm{H}): \quad 2k_0 g_h^< = \{H, g_h^<\}, \qquad (\mathrm{AH}): \quad i\partial_t g_h^< = [H, g_h^<]. \tag{20}$$

where $H = -h|\mathbf{k}|\sigma_3 + m_R\sigma_1 - m_I\sigma_2$ is immediately identified as the the local Hamiltonian. The algebraic H-equation constrains the form of possible solutions. Using the Bloch representation: $g_h^< = \frac{1}{2}(g_0^h + g_i^h\sigma^i)$, it can be written as $B_{ab}g_b^h = 0$, which has nontrivial solutions only if

$$\det(B) = (k_0^2 - \mathbf{k}^2 - |m|^2)k_0^2 = 0\,. \tag{21}$$

Somewhat surprisingly, in addition to the usual mass-shell solutions there are new solutions living at shell $k_0 = 0$ (see Fig. (3)). The chiral structure of the mass-shell solution is

$$g_{h,\mathrm{m-s}}^< = 2\pi s_0\, f_{s0}^h \Big(k_0 - h|\mathbf{k}|\sigma_3 + m_R\sigma_1 - m_I\sigma_2\Big)\delta(k^2 - |m^2|)\,, \tag{22}$$

where $s_0 \equiv \mathrm{sgn}(k_0)$. This corresponds to the usual helicity eigenstate that can also be found by the use of Dirac equation. The chiral structure of the $k_0 = 0$ solutions is:

$$g_{h,0-\mathrm{s}}^< = \frac{\pi}{|\mathbf{k}|}\Big(h(m_R f_1^h - m_I f_2^h)\sigma_3 + |\mathbf{k}|(f_1^h\sigma_1 + f_2^h\sigma_2)\Big)\,\delta(k_0)\,, \tag{23}$$

Here the density functions $f_\pm^h$ and $f_{1,2}^h$ are some functions of time t and the three momentum $\mathbf{k}$, whose evolution is determined by the dynamical equation. The mass-shell functions $f_\pm^h$ are related to the usual 1-particle densities of states and the new functions $f_{1,2}^h$ describe the quantum coherence between particles and antiparticles. The complete solution to the algebraic equation is compactly written as

$$g_h^< = 2\pi\left[\sum_\pm \rho_{h\pm}\delta(k_0 \mp \omega_{\mathbf{k}}) + \rho_{h0}\delta(k_0)\right], \tag{24}$$

where $\omega_{\mathbf{k}}^2 \equiv \mathbf{k}^2 + |m^2|$ and the matrices $\rho_{h\pm}$ and ρ_0 can be read off from Eqs. (22) and (23). The full shell structure of the homogeneous model is shown in the left panel of the Fig. 3.

A similar construction can be done for the stationary, planar symmetric case [9]. In that case the spin s orthogonal to the symmetry plane is conserved and eventually, in a particular frame [9], one finds equations very similar to Eqs. (20):

$$-2sk_z g_s^< = Pg_s^< + g_s^< P^\dagger\,, \qquad is\partial_z g_s^< = Pg_s^< - g_s^< P^\dagger\,, \tag{25}$$

where $P \equiv k_0\sigma_3 - i(m_R\sigma_2 + m_I\sigma_1)$ is the local momentum operator. The algebraic equation again gives rise to a singular shell structure; only this time the mass shell is accompanied by a singular momentum shell at $k_z = 0$, where k_z is the momentum perpendicular to the planar symmetry in the static rest frame. We do not write the explicit forms for these solutions here, but they can be found in ref. [9]. The complete shell structure for the correlator in the planar symmetric case in is shown in the right panel of the Fig. 3.

The crucial issue leading to the new solutions in either of the cases discussed above is that *we did not assume translational invariance.* Indeed, if one imposes translational invariance (in time in homogeneous and in space in the planar symmetric case), the dynamical equations in (20) and (25) become algebraic and it is then easy to show that they exclude the $k_{0,z} = 0$ solutions. Another important thing to observe is that the new $k_{0,z} = 0$ solutions do not appear in the spectral function despite the fact that in the collisionless limit the equation for the spectral function coincides with Eq. (18) for $G^<$. The difference comes from the additional constraints from the spectral sum rule which are enough to kill the coherence solutions from $\mathcal{A}$. Thus, as one might expect, coherence solutions exist only as correlations between particular 1-particle phase space states in the dynamical sector of the theory.

3.1. Flavour mixing

The discussion of this section can be straightforwardly extended to the case of several mixing fields. We shall consider here only the planar symmetric case. If we replace the simple complex mass m by a general complex mass matrix M, the analysis leading to equations (25) remains intact, apart from the fact that the local momentum operator becomes a matrix: $P \equiv k_0\sigma_3 - i(M_H\sigma_2 + M_{AH}\sigma_1)$, where $M \equiv M_H + iM_{AH}$. The matrix structure makes extracting the spectral solutions more complicated. However, for a hermitian M the dispersion relation can still be expressed in a very simple form:

$$(k_z)_{ij} = \pm\frac{1}{2}\left(\sqrt{k_0^2 - m_i^2} \pm \sqrt{k_0^2 - m_j^2}\right)\,, \tag{26}$$

where $m_{i,j}$ refer to the mass eigenstates of M. This dispersion relation shows new interesting features. The diagonal entries obviously correspond to the simple DR's found above in the scalar mass case. The off-diagonal entries however correspond to the singular shells for the *flavour mixing* coherence. Interestingly, the latter DR's change continuously from one closely degenerate with mass shells and describing mixing between particles going mostly in the same direction to a shell close to $k_z = 0$ and describing the combined flavour mixing and reflection coherence between particles going on opposite directions. The complete phase space structure is shown in Fig. 4 for the case of a hermitian 2×2 mixing matrix. Clearly a very complicated set of mixing scenarios is possible with this dispersion relation complex, where one can have the two diagonal 1-particle densities coupling up to eight coherence functions. On the other hand, if the extrenous constraints are enough to restrict the mixing to, say, positive momenta only, but not enough to resolve the flavour mixing triplet (illustrated within the black circle in Fig. 4), one finds the usual flavour oscillation pattern widely studied in neutrino physics.

The complex singular shell structure illustrated in this section is the foundation of, and gives the name to our approximation scheme: the coherent quasiparticle approximation or cQPA.

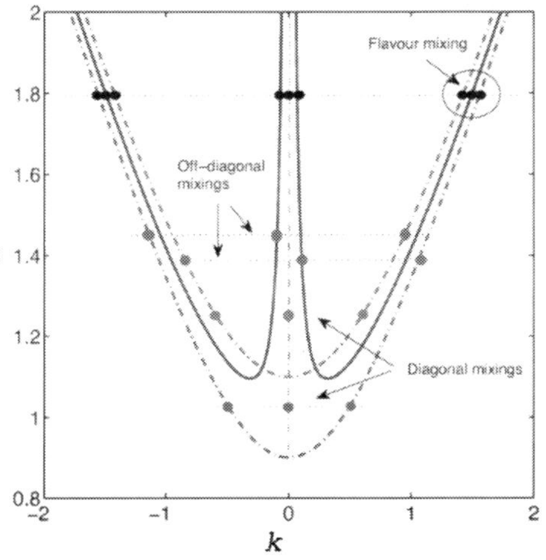

Figure 4. Shown is the complete set of dispersion relations corresponding to the case of 2×2 flavour mixing. The triplets of filled circles connected by dotted lines indicate states which at a given energy form the chiral mixing sub-complexes. The black filled circles indicate a possible flavour mixing complex when the chirality mixing is prohibited by extrenous constraints.

4. Physical density matrix

The dynamical equations in (20) and (25) look compelling. The time dependent equation (20) in particular has just the usual form of the evolution equation for a density matrix. However, they are actually meaningless because of the singular structure imposed by the algebraic constraints. That is, the $g^<_{h,s}$-matrices are mere phase space densities that cannot be identified with physical density matrices. As was already suggested in the previous section, we will have to convolve them with some k_0- and/or k_z-dependent weight functions that may quantify the prior extrenous information on the system [9]. Different types of such information functions were illustrated in the right panel of Fig. 3 by dashed lines. A *physically observable* density matrix can be formally written as

$$\rho_{\mathcal{W}}(k_0, k_z, s; z) \equiv \sum_{s'} \int \frac{\mathrm{d}k'_0}{2\pi}\frac{\mathrm{d}k'_z}{2\pi}\, \mathcal{W}(k_0, k_z, s \parallel k'_0, k'_z, s'\,;\, z)\; g^<_{s'}(k'_z, k'_0; z)\,, \tag{27}$$

where $\mathcal{W}$ encodes the prior information. Let us make clear that this description is an effective one; in fact all restrictions should be imposed on the system directly by the collision integral. For example, our knowledge that a certain neutrino beam is moving in a given direction should be described by the time-dependent correlations imposed on the system by interactions with the apparatus that generates the beam. However, it is obvious that the effective description can often be more useful in practice.

In many applications some parameters can be fixed by conservation laws while no prior information at all exists on some other parameters. For example, in our reflection problem the energy, the momentum parallel to the symmetry plane and the spin s are conserved, while we cannot assume any information on the perpendicular momentum of the state. The weight function appropriate for describing such a system then is $\mathcal{W} = 2\pi\delta(\omega - k'_0)\,\delta_{s,s'}$. Using this weight in Eq. (27) one finds that the density matrix is just the zeroth moment of the fundamental distribution with respect to k_z: $\rho_{\mathcal{W}} \equiv \rho_s(\omega, s) = \langle g^<_s \rangle$. Note that in general there is no guarantee that the equation of motion for a weighted density matrix has a simple form, but in this particular case, because the momentum operator P depends only on k_0, the collisionless equation for ρ_s indeed is just

$$is\partial_z \rho_s = P\rho_s - \rho_s P^\dagger\,. \tag{28}$$

Unlike Eq. (25), the new evolution equation (28) is well defined because the matrix ρ_s is a nonsingular object that depends only on the energy ω and spin s. A similar construction can be made in the context of a homogeneous problem, where one can fix the momentum $\mathbf{k}$ and helicity, and integrate over all energies. The resulting collisionless equation, now for a density matrix $\rho_W \equiv \rho_h(|\mathbf{k}|, h)$, is found to be the familiar one:

$$i\partial_t \rho_h = [H, \rho_h] , \tag{29}$$

where $\rho_h = \rho_{h+} + \rho_{h-} + \rho_0$ in terms of the matrices defined in equation (24).

4.1. Moment connections

From practical point of view the most important thing about the singular shell structure we have found is that it allows one to draw a one-to-one correspondence between the density matrix elements and the shell f_i-functions. Indeed, hermitian ρ matrices contain four degrees of freedom, which is exactly the number of independent shell functions. Clearly, if the coherence shell solutions were ignored two degrees of freedom would be missing, making it impossible to describe coherence phenomena. Using the Bloch-representation $\rho_h \equiv \frac{1}{2}(\langle g_0^h \rangle + \langle g_i^h \rangle \sigma^i)$, we have the following relations between $\langle g_\alpha^h \rangle$ (here the zeroth moments *w.r.t.* energy) and $f^h_{\pm,1,2}$:

$$\begin{aligned}
\langle g_0^h \rangle &= f_+^h + f_-^h \\
\langle g_1^h \rangle &= \tfrac{m_R}{\omega}(f_+^h - f_-^h) + f_1^h \\
\langle g_2^h \rangle &= -\tfrac{m_I}{\omega}(f_+^h - f_-^h) + f_2^h \\
\langle g_3^h \rangle &= -h\tfrac{|\mathbf{k}|}{\omega}(f_+^h - f_-^h) + h\Big(\tfrac{m_R}{|\mathbf{k}|} f_1^h - \tfrac{m_I}{|\mathbf{k}|} f_2^h\Big) ,
\end{aligned} \tag{30}$$

where $\omega \equiv \omega_{\mathbf{k}}$. As we shall see later, these moment-f_i function relations will allow one to rewrite the collision integral entirely in terms of the moment functions, leading to a solvable, closed system of equations for the density matrix.

4.2. Physical quantities

It is gratifying to see that the natural definition of the number density in terms of our mass-shell density functions coincides with some other nonequilibrium definitions found in the literature. Indeed, according to the usual Feynman-Stuckelberg interpretation the phase-space particle number density is identified as $n \equiv f_+$, while for (fermionic) antiparticles $\bar{n} \equiv 1 - f_-$. With these identifications, using the inverse relations of Eq. (30) we find that:

$$\begin{aligned}
n_{\mathbf{k}h} &= \frac{1}{2\omega}\Big(-h|\mathbf{k}|\langle g_3^h \rangle + m_R \langle g_1^h \rangle - m_I \langle g_2^h \rangle\Big) + \frac{1}{2}\langle g_0^h \rangle \\
\bar{n}_{\mathbf{k}h} &= \frac{1}{2\omega}\Big(-h|\mathbf{k}|\langle g_3^h \rangle + m_R \langle g_1^h \rangle - m_I \langle g_2^h \rangle\Big) - \frac{1}{2}\langle g_0^h \rangle + 1 .
\end{aligned} \tag{31}$$

Setting $\mathrm{Tr}(\rho^h) = \langle g_0^h \rangle \equiv 1$ (physically this constraint corresponds to setting the chemical potential to zero) these expressions reduce to the ones obtained in ref. [20][1], where they were derived using the solutions to a Dirac equation and a Bogolyubov transformation to diagonalize the fermionic Hamiltonian.

One can also compute the energy and the pressure densities from the diagonal elements of the energy moment tensor $\theta^{\mu\nu} = \frac{i}{4}(\bar{\Psi}\gamma^\mu \partial^\nu \Psi - \partial^\nu \bar{\Psi}\gamma^\mu \Psi) + \mu \leftrightarrow \nu$. For energy density one finds

$$\langle \mathcal{H}(x) \rangle = \langle \theta^{00}(x) \rangle = \sum_h \int \frac{\mathrm{d}^3 k}{(2\pi)^3} \omega_{\vec{k}} \left(n_{\mathbf{k}h} + \bar{n}_{\mathbf{k}h} - 1\right) , \tag{32}$$

[1] Different signs of terms involving h and m_I are due to a different sign convention in our definition of the Hermitian Wightmann function.

which, according to expectations is simply a sum of free particle and antiparticle contributions, plus the vacuum energy. For the pressure we find

$$\langle P(x)\rangle = \langle \theta^{ii}(x)\rangle = \sum_h \int \frac{d^3k}{(2\pi)^3}\frac{1}{3}\left(\frac{\mathbf{k}^2}{\omega}\left(n_{\mathbf{k}h} + \bar{n}_{\mathbf{k}h} - 1\right) - m_R f_1^h + m_I f_2^h\right). \tag{33}$$

Now, in addition to normal free particle and antiparticle terms there is an explicit contribution from the coherence shell functions $f_{1,2}$. Thus the pressure is different from the statistical one at the quantum level. However, the coherence functions $f_{1,2}$ are oscillatory with microscopic time-scales $\Delta t_{\rm osc} \sim 1/\omega$ and the classical thermodynamical pressure arises when Eq. (33) is averaged out over time-scales exceeding the quantum scale $\Delta t_{\rm osc}$.

4.3. Solutions: collisionless examples

We applied Eq. (28) to the Klein problem of reflection off a step potential wall in ref. [9]. Correct results were found for the transmission and reflection coefficients and tunneling factors when the coherence shell factors are included in Eqs. (30), while neglecting them gave only pure classical complete transmission and complete reflection solutions. Similar results were later found also for the bosonic Klein problem [10]. We also considered a reflection off a smooth mass wall, shown

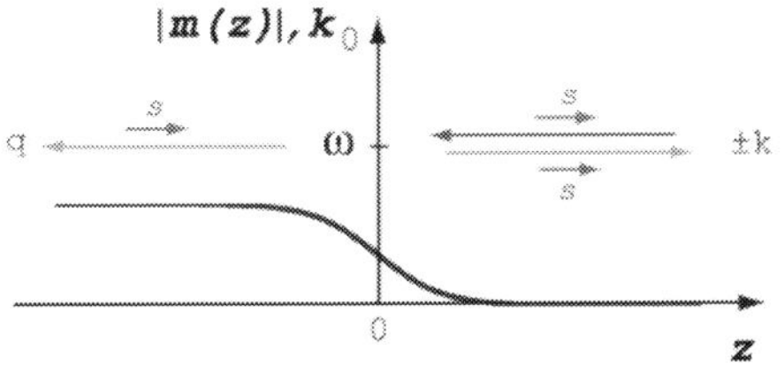

Figure 5. Reflection from a potential wall due to a spatially varying complex mass function.

in Fig. 4. This is precisely the problem of interest for the quantum reflection mechanism in the electroweak baryogenesis. Again, the integrated moment equations are exact for this case, and the moment connections are needed only asymptotically to define the reflection coefficients. The

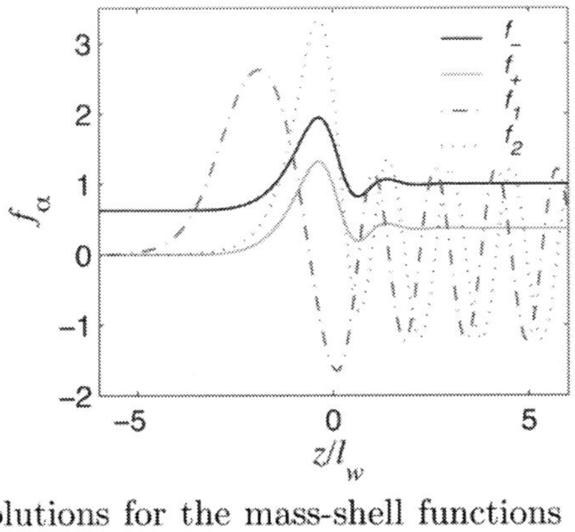

Figure 6. Shown are the mass-shell functions $f_\pm$ corresponding to the left and right moving fluxes and functions $f_{1,2}$, which encode the quantum coherence. We have taken $s = 1$ and $q/|m_{-\infty}| = 0.088$.

solutions for the mass-shell functions f_- and f_+ (here equivalent to the left and right moving currents, respectively) and the coherence functions $f_{1,2}$ are shown in figure 6. The coherence functions are oscillating in the massless phase and vanish in the massive phase. This is as expected because the boundary conditions (see Fig. 4) exclude right moving states at $z \to -\infty$. Our results for the reflection and transmission currents agree with those found earlier with use of the Dirac equation [21]. The crucial improvement over these results is to include the effects of interactions. We have not yet completed this work for the planar symmetric case. However, parallel results for the homogeneous problem do exist and they will be presented next.

5. Interacting fields

With interactions included the full Schwinger-Keldysh equations become much more complicated of course. A tractable approximation scheme can still be obtained by combining the familiar quasiparticle approximation with our zeroth gradient order treatment of the constraint equations. In this scheme we can in particular drop [10] the G_H-term in Eq. (13). Eventually we find:

$$(\not{k} + \tfrac{i}{2}\not{\partial}_x - \hat{m}_0 - i\hat{m}_5\gamma^5 - \Sigma_H)G^< = \mathcal{C}_{\rm coll}\,. \tag{34}$$

Here we are focusing only on the effects of the nontrivial mass function, so we drop the thermal corrections to the dispersion relation setting $\Sigma_H = 0$. After inserting the helicity decomposition Eq. (19), breaking the equation to the hermitian and antihermitian parts and, consistently with our cQPA scheme, dropping the collision integral in the constraint equation, which is also expanded to the lowest order in gradients[2], we find:

$$\begin{aligned} &(\mathrm{H}): \qquad 2k_0 g_h^< = \{H, g_h^<\} \\ &(\mathrm{AH}): \qquad \partial_t g_h^< = -i[H, g_h^<] - \mathcal{C}_h^- \,, \end{aligned} \tag{35}$$

where $\mathcal{C}_h^-$ is the antisymmetric part of the helicity projected collision integral. In cQPA and for an uncorrelated (for example a thermal) self-energy it can be written as [13]

$$i\mathcal{C}_h^- = \cos(\tfrac{1}{2}\partial_{k_0}^\Gamma \partial_t^g)\{\Gamma_h, g_h^<\} + i\sin(\tfrac{1}{2}\partial_{k_0}^\Gamma \partial_t^g)[\Gamma_h, g_h^<] - \{\Gamma_h, g_{\rm eq}^<\}\,, \tag{36}$$

where $\Gamma_h \equiv \frac{i}{2}(\Sigma_h^> + \Sigma_h^<)$ and we defined the hermitian helicity projected self energy function $P_h\gamma^0\Sigma^{<,>}P_h \equiv \Sigma_h^{<,>} \otimes \frac{1}{2}(1 + h\hat{k}\cdot\vec{\sigma})$. Finally $g_{\rm eq}^<$ is the equilibrium limit of the distribution $g_h^<$. We were careful to keep a subset of gradients to infinite order, because the projection onto k_0 shell of these expansions is not controlled by any small parameters (Γ or m') and need to be resummed. Of course equation (35) is equally badly defined as was our free theory evolution equation for $g^<$. It is also clear that when the integration is done, the integrated collision term will, thanks to the moment connections Eq. (30), reduce to and expression entirely expressible in terms of the density matrix components. When integration (over k_0 with a flat weight) and resummation is done, one eventually finds [13]:

$$\partial_t \rho_h = -i[H, \rho_h] - \{\Gamma_{hs}, \delta\rho_h\} - \{\Gamma_{ha}, \delta\rho_{h+} - \delta\rho_{h-}\} - \frac{1}{2\omega_{\mathbf{k}}}[\Gamma_{ha}, [H, \rho_h]]\,. \tag{37}$$

Here brackets $\{,\}$ denote an anticommutator, $\rho_h = \rho_{h+} + \rho_{h-} + \rho_0$ in terms of the matrices in Eq. (24) and $\delta\rho_i \equiv \rho_i - \rho_{i,\rm eq}$, and the collision terms $\Gamma_{h(s,a)} \equiv \frac{1}{2}(\Gamma_{h+} \pm \Gamma_{h-})$ denote symmetric and antisymmetric combinations of the mass-shell interaction strengths $\Gamma_{h\pm} \equiv \Gamma_h(k_0 = \pm\omega_{\mathbf{k}})$. Equation (37) is the sought after generalization of the standard collisionless equation of motion (29) for the density matrix to the case with decohering collisions.

We have applied the equation of motion (37) to study a system with a temporally evolving CP-violating mass function:

$$|m(t)| = \frac{1}{2}(1 + \tanh(t/\tau_w))\,, \qquad \arg[m(t)] = \frac{1}{2}\Delta\theta(1 - \tanh(t/\tau_w))\,, \tag{38}$$

where τ_w is the characteristic time of the transition. Collisions were modelled by an interaction with a thermally equilibrated fermion and scalar fields q_R and ϕ:

$$\mathcal{L}_{\rm int} = -y\,\bar{\psi}_L \phi\, q_R + h.c. \tag{39}$$

[2] Note that these are just the standard assumptions that one makes when deriving the usual Boltzmann equation from the Kadanoff-Baym equations. See *e.g.* [22].

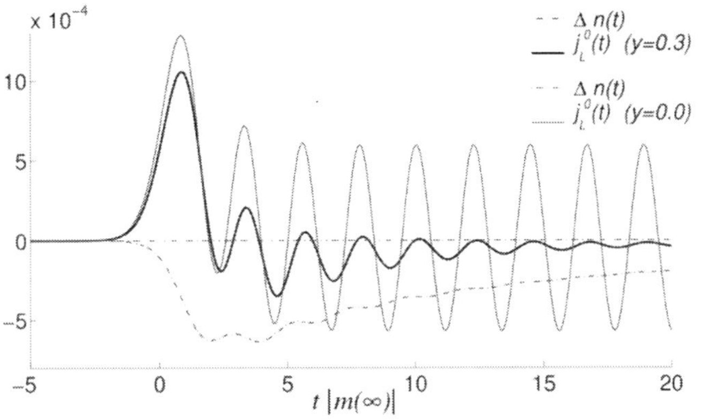

Figure 7. Shown are the total asymmetry $\Delta(n) = \sum_h (n_{\mathbf{k}h} - \bar{n}_{\mathbf{k}h})$ (see Eq. (31)) and the L-chiral current density per unit **k** volume: $j^0_{L\mathbf{k}}(t) = \sum_h (\mathrm{Tr}[\frac{1}{2}(1+\sigma_3)\rho_h] - \frac{1}{2})$ (vacuum part subtracted). The thin curves correspond to the noninteracting case and the thick curves to the interacting case.

The lowest order contribution to the self energy from this interaction is:

$$\Sigma^{<,>}(k,x) = i|y|^2 \int \frac{d^4k'}{(2\pi)^4} P_R G_q^{<,>}(k',x) P_L \Delta^{<,>}(k-k',x)\,. \tag{40}$$

Explicit analytical forms for $\Sigma^{<,>}$ can be found in ref. [10]. Numerical results for the total asymmetry and left chiral current density are shown in Fig. 7. For various parameters we used (in units of $|\mathbf{k}|$) $\tau_w = 2$, $\Delta\theta = \frac{1}{2}$, $T = 10$, $m_q = 10$, $m_\phi = 5$ and $y = 0.3$ (0 in the free case) and the initial conditions at $t \to -\infty$ were set to to thermal equilibrium: $f^h_\pm \to 1/(e^{\pm\beta\omega_k}+1)$ and $f^h_{1,2} \to 0$. In the collisionless case no total asymmetry is created and the L-chiral asymmetry oscillates with a constant amplitude after the transition. In the interacting case a temporary nonzero total asymmetry is created and both the total asymmetry and the amplitude of the oscillating L-chiral asymmetry decay as a function of time. Such a simple picture might be used to model the baryon asymmetry generation during a second order (or cross-over) electroweak phase transition, for example in models where the expansion rate of the universe is highly accelerated during the EWPT.

6. Scalar fields

Our method can also be applied to the bosonic fields with similar results. In the case of the scalar fields we wish to study the correlator: $i\Delta^<(u,v) \equiv \langle\phi(v)\phi(u)\rangle \equiv \mathrm{Tr}\{\hat{\rho}\,\phi(v)\phi(u)\}$. The derivation of the KB-equations for $\Delta^<$ proceeds analogously to the fermionic case [11]; the final result in the mixed representation is

$$\Delta_0^{-1}\Delta^< - e^{-i\diamond}\{\Pi_H\}\{\Delta^<\} - e^{-i\diamond}\{\Pi^<\}\{\Delta_H\} = \mathcal{C}_{\rm coll}\,, \tag{41}$$

with

$$\mathcal{C}_{\rm coll} = -ie^{-i\diamond}\left(\{\Gamma\}\{\Delta^<\} - \{\Pi^<\}\{\mathcal{A}_\phi\}\right)\,, \tag{42}$$

where we already used the connection $\mathcal{A}_\phi = \frac{i}{2}(\Delta^> + \Delta^<)$ and defined $\Gamma \equiv \frac{i}{2}(\Pi^> + \Pi^<)$. The self energy functions are again computable from the 2PI-generating function: $\Pi^{<,>}(u,v) \equiv -i\delta\Gamma_2[\Delta]/\delta\Delta^{>,<}(v,u)$. Of course the explicit forms of Δ_0 as well as the interactions depend on the model. Similarly to the fermionic case, we are interested in effects caused by a space-time varying mass function, so we take

$$\mathcal{L} = \tfrac{1}{2}(\partial_\mu \phi)^2 - \tfrac{1}{2}m(x)^2\phi^2 + \mathcal{L}_{\rm int}\,, \tag{43}$$

where $\mathcal{L}_{\rm int}$ is some yet to be defined interaction term. The inverse free propagator corresponding to Eq. (43) in the mixed representation then is

$$\Delta_0^{-1} \equiv k^2 - \tfrac{1}{4}\partial^2 + ik\cdot\partial - m^2 e^{-\frac{i}{2}\partial_x^m\cdot\partial_k^\Delta}\,, \tag{44}$$

where the ∂_x^m-derivative always acts on the mass term and the ∂_k^Δ-derivative to $\Delta^<$. We again consider the free theory first, and also expand the Δ_0 operator to the lowest nontrivial order in gradients. Because the operator is complex, we find two independent equations:

$$\Big(k_0^2 - \mathbf{k}^2 - m^2(t) - \tfrac{1}{4}\partial_t^2\Big) i\Delta^<(k,t) \;=\; 0 \tag{45}$$

$$k_0\partial_t i\Delta^<(k,t) \;=\; 0\,. \tag{46}$$

Unlike with fermions, the noninteracting scalar KB-equations can not be divided into algebraic constraints and dynamic equations because both equations (45-46) contain time derivatives. Yet a similar shell structure does emerge. Let us first assume that $k_0 \neq 0$. Then[3] Eq. (46) requires that $\partial_t i\Delta^< = 0$ at all times and so one must also have $\partial_t^2 i\Delta^< = 0$. Substituting this back to Eq. (45) gives

$$\left(k_0^2 - \mathbf{k}^2 - m^2(t)\right) i\Delta^<_{\rm m-s} = 0\,. \tag{47}$$

This equation has the spectral mass-shell solution parametrized by t:

$$i\Delta^<_{\rm m-s}(k_0,|\mathbf{k}|,t) = 2\pi\,{\rm sgn}(k_0) f_{s_{k_0}}(|\mathbf{k}|,t)\delta(k_0^2 - \mathbf{k}^2 - m^2(t))\,, \tag{48}$$

where $s_0 \equiv {\rm sgn}(k_0)$. However, the point $k_0 = 0$ is special: if we first set $k_0 = 0$, then equation (46) is identically satisfied, but Eq. (45) becomes $\partial_t^2\Delta^< = -4\omega_{\mathbf{k}}^2(t)\Delta^<$. For a generic time-varying mass this implies a corresponding spectral solution for $k_0 = 0$:

$$i\Delta^<_{\rm 0-s}(k_0,|\mathbf{k}|,t) = 2\pi\, f_c(|\mathbf{k}|,t)\delta(k_0)\,, \tag{49}$$

where $f_c(|\mathbf{k}|,t)$ is some real-valued function. The most complete solution for a given momentum $|\mathbf{k}|$ is the combination of the solutions (48) and (49): $\Delta^< = \Delta^<_{\rm m-s} + \Delta^<_{\rm 0-s}$. The situation is now seen to be qualitatively equivalent to the case with fermions and we interpret analogously that the new $k_0 = 0$-solution (49) describes the quantum coherence between particles and antiparticles.

6.1. Moment connection

Let us again suppose that we do not have any prior restrictions on the momentum variable $\mathbf{k}$. It is then appropriate to use the n-th moments as the weighted density functions for the problem:

$$\rho_n(|\mathbf{k}|,t) = \int \frac{{\rm d}k_0}{2\pi}\, k_0^n\, i\Delta^<(k_0,|\mathbf{k}|,t)\,. \tag{50}$$

When one computes three lowest moments using the complete shell structure of $\Delta^<$ one immediately finds:

$$\begin{aligned}\rho_0 &= \tfrac{1}{2\omega_{\mathbf{k}}}(f_+ - f_-) + f_c\\ \rho_1 &= \tfrac{1}{2}(f_+ + f_-)\\ \rho_2 &= \tfrac{\omega_{\mathbf{k}}}{2}(f_+ - f_-)\,.\end{aligned} \tag{51}$$

[3] Note that including collisions at this stage would destroy the singular phase space structure. Neglecting them here corresponds to the usual quasiparticle approximation.

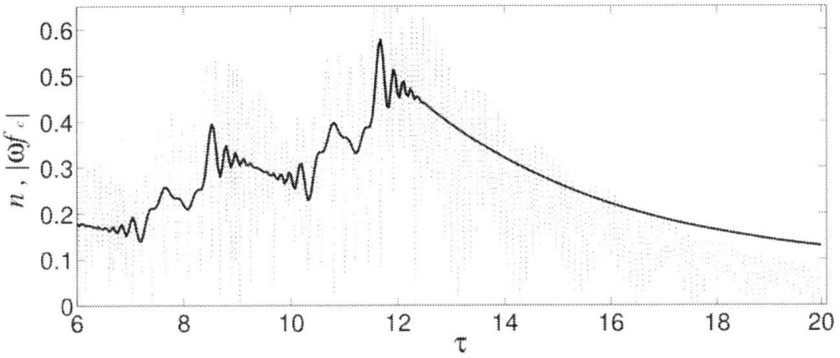

Figure 8. Shown is the number density $n_{\mathbf{k}}$ (thick solid line) and the coherence function $f_c(|\mathbf{k}|)$ (dotted line). For the driving mass function Eq. (53) we used the parameters $m_0 = T$, $A = 1.5T$ and $B = 0.1T$. For other parameters we used $|\mathbf{k}| = 0.6T$, $y = 1$, $m_\psi = 0.1T$ and $\omega_\varphi = 0.1T$, where temperature T sets the scale. At $\tau \equiv \omega_\varphi t = 0$ the system was in the adiabatic vacuum, and the driving mass term smoothly set to constant $m = 2.5T$ after $\tau > 4\pi$.

All higher moments can be trivially related to these three lowest ones. As was the case with fermions, these relations are only valid to the lowest order in mass gradients. However, when used as an *ansatz* in the complete set of equations (up to all order in gradients in the flow term), they lead to a closure and a solvable set of equations for the moment functions ρ_n.

6.2. Evolution equations with collisions

When the cQPA ansatz is inserted to the full KB-equation (41) (again dropping terms with Δ_H and Π_H), and the resulting equations are integrated over the energy, and the projections of the collision terms onto $k_0 = 0$ shell are resummed, one finds the following closed set of equations [23]:

$$\begin{aligned} \tfrac{1}{4}\partial_t^2\rho_0 + \omega_{\mathbf{k}}^2\rho_0 - \rho_2 &= -\tfrac{1}{2\omega_{\mathbf{k}}}\Gamma_m\,\partial_t\rho_0 \\ \partial_t\rho_1 &= -\tfrac{1}{\omega_{\mathbf{k}}}\Gamma_m\,(\rho_1 - \rho_{1,\mathrm{eq}}) \\ \partial_t\rho_2 - \tfrac{1}{2}\partial_t(m^2)\rho_0 &= -\tfrac{1}{\omega_{\mathbf{k}}}\Gamma_m\,(\rho_2 - \rho_{2,\mathrm{eq}})\,. \end{aligned} \qquad (52)$$

We illustrate the solution of Eqs. (52) in the case where the scalar mass term is an oscillatory function coming from a coupling of ϕ to an oscillating background field φ (the inflaton):

$$m^2(t) = |m_0 + A\cos(2\omega_\varphi t) + iB\sin(2\omega_\varphi t)|^2\,, \qquad (53)$$

Physically this situation could model the particle production after inflation (preheating). Interactions were modelled by decays and inverse decays caused by the interaction with a thermalized fermion background: $\mathcal{L}_{\mathrm{int}} = g\phi\bar{\psi}\psi$, where ψ is some generic thermalized background fermion field. As the inflaton oscillates around its minimum, ϕ-fields are excited out from the temporally varying classical background and the number density increases in each oscillation cycle. Also the magnitude of the coherence function increases at each oscillation, although the decohering effect of collisions is clearly visible between the oscillation periods. After the driving mass terms is set to constant, a phase of smooth decoherence and thermalization is observed.

7. Conclusions

We have introduced a novel approximation for the Kadanoff-Baym equations for the fermionic and bosonic 2-point functions. Our approach makes use of newly discovered extended singular phase space structure for these functions in the free theory limit in two simple geometries: the spatially homogeneous and stationary planar symmetric cases. When applied to the full KB-equations, this *coherent quasiparticle approximation* (cQPA) leads to equations of motion for a density matrix (or a set of moment functions for scalar fields), that can describe nonlocal coherent evolution of the state in the presence of hard collisions. In the homogeneous case the new shell solutions characterize the particle-antiparticle correlations and in the planar symmetric case correlations between reflecting states moving on opposite directions. Our original, and still principal motivation for developing this formalism was the desire to find a quantitative description for the electroweak baryogenesis proceeding through the CP-violating reflection mechanism. While we yet have to complete this program, we have already demonstrated the use of our formalism in a number of other interesting problems; for example the homogeneous time dependent problem of particle production in the early universe at the end of inflation or some phase transition. This formalism can be used to model homogeneous baryogenesis as well. Yet another interesting research direction is to use the cQPA to better understand the nature of the flavour coherence in the case of neutrino mixing and in particular in connection with the supernova neutrinos.

Acknowledgements
We thank the Referee for bringing the papers [24] and [25] to our attention. It will be interesting to see if any connections between these works and our formalism can be derived.

References
[1] A. Cohen, D. Kaplan, and A. Nelson, *Progress in electroweak baryogenesis*, in *43*199427 [hep-ph/9302210]; V.A. Rubakov and M.E. Shaposhnikov, Usp. Fiz. Nauk **166** (1996) 493, Phys. Usp. **39** (1996) 461.
[2] M. Joyce, T. Prokopec and N. Turok, Phys. Rev. D **53**, 2958 (1996); Phys. Rev. Lett. **75**, 1695 (1995); [Erratum-ibid. **75**, 3375 (1995)]; Phys. Rev. D **53**, 2930 (1996).
[3] J. M. Cline, M. Joyce and K. Kainulainen, JHEP **0007** (2000) 018; Phys. Lett. B **417** (1998) 79, [Err.-ibid. B **448** (1999) 321]; J. M. Cline and K. Kainulainen, Phys. Rev. Lett. **85** (2000) 5519.
[4] K. Kainulainen, T. Prokopec, M. G. Schmidt and S. Weinstock, JHEP **0106**, 031 (2001); Phys. Rev. D**66** (2002) 043502; T. Prokopec, M. G. Schmidt and S. Weinstock, Annals Phys. **314**, 208 (2004); Annals Phys. **314**, 267 (2004).
[5] T. Prokopec, M.G. Schmidt and S. Weinstock, Annals Phys. **314**, 208 (2004); Annals Phys. **314**,267 (2004).
[6] See G. Baym, these proceedings.
[7] J. H. Traschen and R. H. Brandenberger, Phys. Rev. D **42** 2491 (1990).
[8] J. Berges, S. Borsanyi, J. Serreau, Nucl. Phys. B **660**, 51 (2003).
[9] M. Herranen, K. Kainulainen and P. M. Rahkila, Nucl. Phys. B **810** (2009) 389.
[10] M. Herranen, K. Kainulainen and P. M. Rahkila, JHEP **0809** (2008) 032.
[11] M. Herranen, K. Kainulainen and P. M. Rahkila, JHEP **0905** (2009) 119.
[12] M. Herranen, K. Kainulainen and P. M. Rahkila, Nucl. Phys. A **820** (2009) 203C.
[13] M. Herranen, K. Kainulainen and P.M. Rahkila, in progress.
[14] G. 't Hooft, Phys. Rev. Lett. **37** (1976) 8.
[15] G.D. Moore and K. Rummukainen, Phys. Rev. D61 (2000) 105008; D. Bödeker, G.D. Moore and K. Rummukainen, Nucl. Phys. Proc. Suppl. 83:583-585,2000.
[16] J. S. Schwinger, J. Math. Phys. **2** 407 (1961);
L. V. Keldysh, Zh. Eksp. Teor. Fiz. **47** 1515 (1964) [Sov. Phys. JETP **20** 1018 (1965)].
[17] K. c. Chou, Z. b. Su, B. l. Hao and L. Yu, Phys. Rept. **118**, 1 (1985); E. Calzetta and B. L. Hu, Phys. Rev. D**37** (1988) 2878.
[18] J. M. Cornwall, R. Jackiw and E. Tomboulis, Phys. Rev. D **10** (1974) 2428.
[19] L. Kadanoff and G. Baym, "Quantum Statistical Mechanics" *Benjamin, New York (1962).*
[20] B. Garbrecht, T. Prokopec and M.G. Schmidt Eur. Phys. J. C38, 135 (2004).
[21] J. M. Cline, K. Kainulainen and A. P. Vischer, Phys. Rev. D **54** (1996) 2451.
[22] P. Henning, Phys. Rept. **253** (1995); G. Mahan, Phys. Reps. 145 (1987) 251.
[23] M. Herranen, PhD Dissertation, University of Jyväskylä, [arXiv:0906.3136 [hep-ph]].
[24] Y. Kluger, J. M. Eisenberg, B. Svetitsky, F. Cooper and E. Mottola, Phys. Rev. D **45** (1992) 4659.
[25] J. M. Eisenberg and G. Kaelbermann, Phys. Rev. D **37** (1988) 1197.

Journal of Physics: Conference Series **220** (2010) 012008 doi:10.1088/1742-6596/220/1/012008

Modeling transient aspects of coherence-driven electron transport

A Prociuk, H Phillips, and B D Dunietz*

The University of Michigan, Ann Arbor, Michigan 48109

E-mail: bdunietz@umich.edu

Abstract. Non-equilibrium Green's function formalism (NEGF) by employing time-dependent (TD) perturbation theory is used to solve the electronic equations of motion of model systems under potential biasing conditions. The time propagation is performed in the full frequency domain of the two time variables representation. We analyze transient aspects of the resulting conductance under effects of applied direct-current and alternating current potentials. The coherence induced response dependence on different aspects of the applied perturbation is resolved in time and analyzed using calculated TD distributions of the current operator.

1. Introduction

The study of electron and energy transport through molecular and nanoscale systems is drawing a large volume of research attention.[1, 2, 3, 4, 5, 6, 7, 8, 9, 10, 11, 12, 13, 14, 15, 16, 17, 18, 19, 20, 21, 22, 23, 24, 25, 26, 27, 28, 29, 30, 31, 32, 33, 34, 35, 36, 37, 38, 39, 40, 41, 42, 43, 44, 45] The appeal stems mainly from prospects with which nanotechnology is associated and requires understanding of the complex physics of electron transport (ET) processes at the atomic and electronic structure levels.[46, 47, 48, 49, 50, 51, 52, 53, 54, 55, 19, 20, 56] Computational modeling of the transport plays a crucial role in this research, where *insight at the most fundamental level is achieved.* While the main focus of ET studies has been the steady state, recent technological advances are promising to shift this.[57, 58] Due to these advances, the modeling of the temporal resolution of electron transport becomes essential.[59, 60, 61, 62, 63, 64, 65, 66]

In general, the study of systems, where quantum interferences affect the ET, requires resolving the time-dependent (TD) conductance. This includes the study of transient conductance, where the system still evolves to a steady state. TD studies revealed interesting quantum effects associated with photo-assisted conductances in mesoscopic systems, such as absolute negative conductance, Coulomb blockade, and Kondo effects driven by alternating-current (AC) fields.[67, 48, 49, 50, 55, 56] Other studies highlight the importance of transient effects, where the non-equilibrium nature of ET determines the switching rate and efficiency of the transport. Time-resolved THz spectroscopy was used to study carrier dynamics within nano-crystal arrays, where the inter-dots coupling has been varied by modifying the length of the surface bound ligand.[57] The importance of considering the dynamics of ET for analysis of experimental data is also demonstrated by Raman spectra of molecular-scale junctions that relate the conductance to spectroscopic changes. [68, 69, 70, 71, 72]

Journal of Physics: Conference Series **220** (2010) 012008 doi:10.1088/1742-6596/220/1/012008

The dynamics of electron transport is greatly complicated by inherent dissipative aspects. The complexity of the ET due to quantum dephasing is reflected even for model systems of non-interacting electrons in bulk-coupled systems driven by TD potentials.[73, 74, 75, 76, 77, 78, 79, 80, 81, 82, 66] Recent computational studies that analyze currents under TD applied potentials include approaches based on TD-DFT,[83, 84, 85, 59, 61, 86, 60, 87, 88, 89] density matrix equations with Floquet representation or quantum master equations,[90, 91, 92, 93, 94, 66] Floquet scattering matrix approach,[95, 96] and Keldysh non-equilibrium formalism.[97, 98, 99, 62, 100, 101, 66] Keldysh Green's function (GF)-based expansions[102, 103, 104, 105, 106] provide for a rigorous route to treat electron dynamics in biased systems[97] and are used below to analyze quantum interferences affecting the transient transport.

We are implementing a dynamical approach based on Keldysh formalism to study ET. The effect of the bulk is to broaden and shift the electronic DOS resulting with an energy distribution of the electronic density matrix ($\rho(E)$). This may involve broadened states that couple the two leads providing an efficient transport channel. The Keldysh NEGF formalism relates a coupled electronic system experiencing a TD perturbation to its thermally equilibrated initial state. In this approach, thermal coupling to the bulk is represented by the imaginary part of the time ordered contour that is used to define the electronic equations of motion (e.o.m.). Broadening effects due to the electrodes' coupling are treated consistently, where the bulk-induced manifolds of junction's states are directly propagated. The electronic density of the coupled system can be extracted from these equations.

Most descriptions of dynamic transport that are GFs-based formalisms follow the seminal work of Jauho et al.,[97] where system partitioning for the unperturbed state is employed. In this picture the system's components are each kept in thermal isolation up to switching on the perturbation. Initial state designation, as reflected from model-partitioning, however, may affect the evolution of the system to steady state. Transient effects, therefore, can reliably be treated only by a non-partitioning scheme, which is fully and consistently equilibrated. Below we employ the more consistent approach within the Keldysh formalism, where the full equilibration of the unperturbed device with the electron baths is included.[107] The importance of the partition-free approach was highlighted in several recent real time-based propagations.[59, 60, 66]

2. Approach

2.1. The Kadanoff-Baym electronic equations of motion

The central quantities of Keldysh and NEGF formalisms are the one-body Green Functions (GF)[104, 106, 108], which are functions of two space-spin and time coordinates ($x = \vec{x}, t$), with time variables expressed on the Keldysh contour (KC). The GFs are defined by pair of field operators ($\hat{\Psi}(\vec{x})$ and $\hat{\Psi}^\dagger(\vec{x}')$) at times t and t' respectively:

$$G(x,x') \equiv -i\frac{Tr[\hat{U}(t_0 - i\beta, t_0)\hat{T}_C[\hat{\Psi}_H(x)\hat{\Psi}^\dagger_H(x')]]}{Tr[\hat{U}(t_0 - i\beta, t_0)]}. \tag{1}$$

In the Heisenberg picture, these operators take the form:

$$\hat{\Psi}_H(x) = \hat{U}(t_\circ, t)\hat{\Psi}(\vec{x})\hat{U}(t, t_\circ) \tag{2}$$

where $U(t,t')$ propagates the fully time dependent hamiltonian from t' to t on the KC. For the special case of a time independent Hamiltonian, $U(t,t') = e^{-i\hat{H}(t-t')}$ and $G(x,x')$ reduces to a GF of the time difference $G^0(\vec{x},\vec{x}';t-t')$. The superscript, zero, is used to denote a GF corresponding to the Hamiltonian without a time dependent contribution. The contour ordering operator, $\hat{T}_C$, can be expanded, allowing the GF to be expressed as a sum over two types of GFs related to the relative positions of the two time variables on the contour:

$$G(x,x') = \theta(t,t')G^>(x,x') + \theta(t',t)G^<(x,x'), \tag{3}$$

where the step function, $\theta(t, t')$, is as usual zero for t appearing before t' on the Keldysh contour and one otherwise. The contour, therefore, defines 7 types of GFs that each solve a different electronic equation of motion (e.o.m.) (Kadanoff-Baym (K-B) equations). The K-B equations are related to each other through analytic continuation as set forth by the KC.[105]

We start with the K-B equation for the case of non-interacting electrons expressed by the lesser GF ($G^<$):

$$\left[i\mathbf{S}\frac{\partial}{\partial t} - \mathbf{h}(t)\right]\mathbf{G}^<(t,t') = 0 \tag{4}$$

$$-i\frac{\partial}{\partial t'}\mathbf{G}^<(t,t')\mathbf{S} - \mathbf{G}^<(t,t')\mathbf{h}(t') = 0, \tag{5}$$

where $h(t)$ is the TD electronic Hamiltonian and S is the overlap of the orbitals used in the expansion. Note that for this equation, we assume that t and t' are both on the real branch of the KC. We are neglecting electron-electron SE interaction terms. The complete one-body Hamiltonian takes the following form:

$$h(x) = h_0(\vec{x}) + v(x), \tag{6}$$

where $v(x)$ is a time dependent external potential acting on the electrons and $h_0(\vec{x})$ represents the electron kinetic energy and electron nuclear attractions. We combine equations 4-5 by rewriting in terms of the time average and correlation variables: $\bar{t} \equiv \frac{t_1+t_2}{2}$ and the $\Delta t \equiv t_1 - t_2$:

$$\frac{\partial}{\partial \bar{t}}\mathbf{G}^<(\bar{t},\Delta t) = i\left[\mathbf{G}^<(\bar{t},\Delta t)\mathbf{h}(\bar{t} - \frac{\Delta t}{2}) - \mathbf{h}(\bar{t} + \frac{\Delta t}{2})\mathbf{G}^<(\bar{t},\Delta t)\right]. \tag{7}$$

2.2. Mixed time-frequency space representation

Pure time-domain representations require using sufficiently large clusters to reliably treat conductance through a device region. In the frequency domain on the other hand, self-energy expressions can use a cluster model to effectively represent an open system. Frequency domain bulk-self energy (SE) models are used below to represent the dissipative effect of coupling to the electrodes. Earlier, we analyzed the convergence of transport calculations using bulk-coupling DFT-based SEs with their model size.[109] We introduce the frequency variable by Fourier transforming over Δt using a generalized transform operator ($\mathcal{F}$):

$$\begin{aligned}\mathcal{F}[f(\Delta t)] \equiv \int_{-\infty}^{\infty} d(\Delta t)\left[e^{i(\bar{\omega}+i\eta)\Delta t}\Theta(\Delta t) + e^{i(\bar{\omega}-i\eta)\Delta t}\Theta(-\Delta t)\right] f(\Delta t) \\ \equiv f_1(\bar{\omega}) + f_2(\bar{\omega}) \equiv f(\bar{\omega})\end{aligned} \tag{8}$$

where $\eta \to 0^+$, leads, when applied on $G^<(\bar{t},\Delta t)$, to

$$\mathbf{G}^<(\bar{t},\bar{\omega}) = \mathbf{G}_1^<(\bar{t},\bar{\omega}) + \mathbf{G}_2^<(\bar{t},\bar{\omega}), \tag{9}$$

with

$$\begin{aligned}\mathbf{G}_1^<(\bar{t},\bar{\omega}) &= \int_{-\infty}^{\infty} d(\Delta t)e^{i(\bar{\omega}+i\eta)\Delta t}\Theta(\Delta t)\mathbf{G}^<(\bar{t},\Delta t) \qquad \text{and} \\ \mathbf{G}_2^<(\bar{t},\bar{\omega}) &= -\,\mathbf{G}_1^{\dagger<}(\bar{t},\bar{\omega}).\end{aligned} \tag{10}$$

The use of the two branches, where Θ is the step function, ensures that the transformed function approaches zero as Δt approaches $\pm\infty$. The resulting transformed e.o.m takes the form:

$$\frac{\partial}{\partial \bar{t}}\mathbf{G}^<(\bar{t},\bar{\omega}) = i\left[\mathbf{G}^<(\bar{t},\bar{\omega}), \mathbf{h}_0\right] + i\int d\omega'[\mathbf{G}^<(\bar{t},\bar{\omega}+\omega')\mathbf{v}(\bar{t},\omega') - \mathbf{v}(\bar{t},\omega')\mathbf{G}^<(\bar{t},\bar{\omega}-\omega')], \tag{11}$$

where $\mathbf{v}(\bar{t},\bar{\omega})$ is the mixed representation of the time dependent contribution to the Hamiltonian, $\mathbf{v}(\bar{t})$, and can be expressed in terms of the Fourier transformed form ($\tilde{\mathbf{v}}(\omega) = \mathcal{F}[\mathbf{v}(t)]$):

$$\mathbf{v}(\bar{t},\bar{\omega}) = \frac{1}{\pi}e^{-i2\bar{\omega}\bar{t}}\int_{-\infty}^{\infty} dt e^{i(2\bar{\omega})t}\mathbf{v}(t) = \frac{1}{\pi}e^{-i2\bar{\omega}\bar{t}}\tilde{\mathbf{v}}(2\bar{\omega}). \tag{12}$$

This mixed time-frequency representation of the responding electronic system, as used below, provides substantial physical insight into the electronic response aspects of the propagated model system.[65]

2.3. Propagating the electrode-coupled system

The infinite nature of the coupled system is represented by adding bulk self-energies as shown next. Here $\bar{\mathbf{h}}_0$, $\bar{\mathbf{v}}(\bar{t},\omega')$ and $\bar{\mathbf{G}}^<(\bar{t},\bar{\omega})$ are in the AO representation and $\mathbf{S}$ is the AO overlap matrix. We parse the system's model space into a device region and two bulk regions.

$$\bar{\mathbf{h}}_0 = \begin{bmatrix} \bar{\mathbf{h}}_{0LL} & \bar{\mathbf{h}}_{0LC} & 0 \\ \bar{\mathbf{h}}_{0CL} & \bar{\mathbf{h}}_{0CC} & \bar{\mathbf{h}}_{0CR} \\ 0 & \bar{\mathbf{h}}_{0RC} & \bar{\mathbf{h}}_{0RR} \end{bmatrix} \tag{13}$$

$$\bar{\mathbf{S}} = \begin{bmatrix} \bar{\mathbf{S}}_{LL} & 0 & 0 \\ 0 & \bar{\mathbf{S}}_{CC} & 0 \\ 0 & 0 & \bar{\mathbf{S}}_{RR} \end{bmatrix} \tag{14}$$

$$\bar{\mathbf{v}}(t,\omega) = \begin{bmatrix} 0 & 0 & 0 \\ 0 & \bar{\mathbf{v}}_{CC}(t,\omega) & 0 \\ 0 & 0 & 0 \end{bmatrix} \tag{15}$$

$$\bar{\mathbf{G}}^<(\bar{t},\bar{\omega}) = \begin{bmatrix} \bar{\mathbf{G}}^<_{LL}(\bar{t},\bar{\omega}) & \bar{\mathbf{G}}^<_{LC}(\bar{t},\bar{\omega}) & 0 \\ \bar{\mathbf{G}}^<_{CL}(\bar{t},\bar{\omega}) & \bar{\mathbf{G}}^<_{CC}(\bar{t},\bar{\omega}) & \bar{\mathbf{G}}^<_{CR}(\bar{t},\bar{\omega}) \\ 0 & \bar{\mathbf{G}}^<_{RC}(\bar{t},\bar{\omega}) & \bar{\mathbf{G}}^<_{RR}(\bar{t},\bar{\omega}) \end{bmatrix} \tag{16}$$

We next solve for the bulk-coupled lesser GF in the mixed representation ($\bar{\mathbf{G}}^<_{CC}(\bar{t},\bar{\omega})$):

$$\begin{aligned} i\frac{\partial}{\partial \bar{t}}\mathbf{G}^<_{CC}(\bar{t},\bar{\omega}) &= [\mathbf{h}_{0CC}, \mathbf{G}^<_{CC}(\bar{t},\bar{\omega})] \\ &+ \sum_{I\in L,R}[\mathbf{h}_{CI}\mathbf{G}^<_{IC}(\bar{t},\bar{\omega}) - \mathbf{G}^<_{CI}(\bar{t},\bar{\omega})\mathbf{h}_{IC}] \\ &+ \int d\omega'[\mathbf{v}_{CC}(\bar{t},\omega')\mathbf{G}^<_{CC}(\bar{t},\bar{\omega}-\omega') - \mathbf{G}^<_{CC}(\bar{t},\bar{\omega}+\omega')\mathbf{v}_{CC}(\bar{t},\omega')]. \end{aligned} \tag{17}$$

$$\begin{aligned} i\frac{\partial}{\partial \bar{t}}\mathbf{G}^<_{IC}(\bar{t},\bar{\omega}) &= \mathbf{h}_{0II}\mathbf{G}^<_{IC}(\bar{t},\bar{\omega}) + \mathbf{h}_{0IC}\mathbf{G}^<_{CC}(\bar{t},\bar{\omega}) \\ &- \mathbf{G}^<_{II}(\bar{t},\bar{\omega})\mathbf{h}_{0IC} - \mathbf{G}^<_{IC}(\bar{t},\bar{\omega})\mathbf{h}_{0CC} \\ &- \int d\omega'[\mathbf{G}^<_{IC}(\bar{t},\bar{\omega}+\omega')\mathbf{v}_{CC}(\bar{t},\omega')]. \end{aligned} \tag{18}$$

where $I \in L, R$. Furthermore, given the assumptions made on the overlap matrix, we have $\mathbf{v}_{CC}(\bar{t},\omega') = \mathbf{S}_{CC}^{-1/2}\bar{\mathbf{v}}_{CC}(\bar{t},\omega')\mathbf{S}_{CC}^{-1/2}$, $\mathbf{h}_{0IJ} = \mathbf{S}_{II}^{-1/2}\bar{\mathbf{h}}_{0IJ}\mathbf{S}_{JJ}^{-1/2}$ and $\mathbf{G}^<_{IJ}(\bar{t},\bar{\omega}) = \mathbf{S}_{II}^{1/2}\bar{\mathbf{G}}^<_{IJ}(\bar{t},\bar{\omega})\mathbf{S}_{JJ}^{1/2}$ for $I, J \in L, C, R$.

We now divide the lesser GF into a sum of two parts:

$$\mathbf{G}^<(\bar{t},\bar{\omega}) \equiv \mathbf{G}^{0,<}(\bar{\omega}) + \mathbf{\Delta G}^<(\bar{t},\bar{\omega}) \tag{19}$$

where $\mathbf{G}^{0,<}(\bar{\omega})$ is the lesser GF in the absence of a TD perturbation (i.e. Hamiltonian is constant in time, see discussion proceeding equation (2)) and $\mathbf{\Delta G}^{<}(\bar{t},\bar{\omega})$ is the remainder (i.e. the part that carries all perturbation effects). This implies that

$$\frac{\partial}{\partial \bar{t}}\mathbf{G}^{0,<}(\bar{\omega}) = 0 \tag{20}$$

and therefore,

$$\frac{\partial}{\partial \bar{t}}\mathbf{G}^{<}(\bar{t},\bar{\omega}) = \frac{\partial}{\partial \bar{t}}\mathbf{\Delta G}^{<}(\bar{t},\bar{\omega}) \tag{21}$$

Using these properties and the approximations above, we can rewrite eqns. (17) and (18) in terms of $\mathbf{\Delta G}^{<}(\bar{t},\bar{\omega})$:

$$\begin{aligned} i\frac{\partial}{\partial \bar{t}}\mathbf{\Delta G}^{<}_{CC}(\bar{t},\bar{\omega}) &= \left[\mathbf{h}_{0CC}, \mathbf{\Delta G}^{<}_{CC}(\bar{t},\bar{\omega})\right] \\ &+ \sum_{I\in L,R}\left[\mathbf{h}_{CI}\mathbf{\Delta G}^{<}_{IC}(\bar{t},\bar{\omega}) - \mathbf{\Delta G}^{<}_{CI}(\bar{t},\bar{\omega})\mathbf{h}_{IC}\right] \\ &+ \int d\omega'[\mathbf{v}_{CC}(\bar{t},\omega')\mathbf{G}^{0,<}_{CC}(\bar{\omega}-\omega') - \mathbf{G}^{0,<}_{CC}(\bar{\omega}+\omega')\mathbf{v}_{CC}(\bar{t},\omega')] \\ &+ \int d\omega'[\mathbf{v}_{CC}(\bar{t},\omega')\mathbf{\Delta G}^{<}_{CC}(\bar{t},\bar{\omega}-\omega') - \mathbf{\Delta G}^{<}_{CC}(\bar{t},\bar{\omega}+\omega')\mathbf{v}_{CC}(\bar{t},\omega')], \end{aligned} \tag{22}$$

$$i\frac{\partial}{\partial \bar{t}}\mathbf{\Delta G}^{<}_{IC}(\bar{t},\bar{\omega}) = \mathbf{h}_{0II}\mathbf{\Delta G}^{<}_{IC}(\bar{t},\bar{\omega}) - \mathbf{\Delta G}^{<}_{IC}(\bar{t},\bar{\omega})\mathbf{h}_{0CC} + \mathbf{h}_{0IC}\mathbf{\Delta G}^{<}_{CC}(\bar{t},\bar{\omega}). \tag{23}$$

We integrate these equations after redefining $\mathbf{\Delta G}^{<}_{IC}(\bar{t},\bar{\omega})$ as follows:

$$\mathbf{\Delta G}^{<}_{IC}(\bar{t},\bar{\omega}) \equiv e^{-i\mathbf{h}_{0II}(\bar{t}-t_0)}\mathbf{\Delta}\mathcal{G}^{<}_{IC}(\bar{t},\bar{\omega})e^{i\mathbf{h}_{0CC}(\bar{t}-t_0)}. \tag{24}$$

Upon substitution of definition eqn. 24 into eqn. 23 we get,

$$\mathbf{\Delta G}^{<}_{IC}(\bar{t},\bar{\omega}) = \int_{-\infty}^{\bar{t}} d\bar{t}' e^{i\mathbf{h}_{0II}(\bar{t}'-\bar{t})}\mathbf{h}_{0IC}\mathbf{\Delta G}^{<}_{CC}(\bar{t},\bar{\omega})e^{-i\mathbf{h}_{0CC}(\bar{t}'-\bar{t})} \tag{25}$$

Note that the integration lower bound is set at $-\infty$ because unlike $\mathbf{v}(t)$ which starts at some finite t_0, $\mathbf{v}(\bar{t},\bar{\omega})$ is potentially infinite in extent. Equation 25 together with the property $\mathbf{\Delta G}^{<\dagger}_{IC}(\bar{t},\bar{\omega}) = -\mathbf{\Delta G}^{<}_{CI}(\bar{t},\bar{\omega})$ can be substituted into equation 22 to give

$$\begin{aligned} i\frac{\partial}{\partial \bar{t}}\mathbf{\Delta G}^{<}_{cc}(\bar{t},\bar{\omega}) &= \left[\mathbf{h}_{cc}, \mathbf{\Delta G}^{<}_{cc}(\bar{t},\bar{\omega})\right] + \int d\omega'[\mathbf{v}_{\mathbf{cc}}(\bar{t},\omega')\mathbf{G}^{<}_{cc}(\bar{t},\bar{\omega}-\omega') - \mathbf{G}^{<}_{cc}(\bar{t},\bar{\omega}+\omega')\mathbf{v}_{\mathbf{cc}}(\bar{t},\omega')] \\ &+ \int_{-\infty}^{\bar{t}} dt'[\mathbf{\Sigma}^{\mathbf{R}}(\bar{t}-t')\mathbf{\Delta G}_{\mathbf{cc}}{}^{<}(t',\bar{\omega})e^{-i\mathbf{h}_{\mathbf{cc}}(\bar{t}-t')} - e^{i\mathbf{h}_{\mathbf{cc}}(\bar{t}-t')}\mathbf{\Delta G}_{\mathbf{cc}}(t',\bar{\omega})\Sigma^{A}(t'-\bar{t})], \end{aligned} \tag{26}$$

where the self energies $\mathbf{\Sigma}^{\mathbf{R}}(t)$ and $\mathbf{\Sigma}^{\mathbf{A}}(t)$ are defined as follows:

$$\mathbf{\Sigma}^{\mathbf{R}}(t) \equiv \sum_{I\in L,R}\mathbf{h}_{0CI}\mathbf{g}^{\mathbf{R}}_{\mathbf{II}}(t)\mathbf{h}_{0IC} \tag{27}$$

$$\mathbf{\Sigma}^{\mathbf{A}}(t) \equiv \sum_{I\in L,R}\mathbf{h}_{0CI}\mathbf{g}^{\mathbf{A}}_{\mathbf{II}}(t)\mathbf{h}_{0IC} \tag{28}$$

$$\mathbf{g}^{\mathbf{R}}_{\mathbf{II}}(t) \equiv -i\Theta(t)e^{-i\mathbf{h}_{\mathbf{II}}t} \tag{29}$$

$$\mathbf{g}^{\mathbf{A}}_{\mathbf{II}}(t) \equiv i\Theta(-t)e^{-i\mathbf{h}_{\mathbf{II}}t} \tag{30}$$

with $\Theta(-t)$ the Heaviside step function.

For description of TD conductance the electronic density at the unperturbed state has to be properly evaluated as the initial conditions of the propagation. The initial $\mathbf{G}^{0,<}$ (*i.e.* the unperturbed equilibrated system) must describe well the steady state, which is thermally equilibrated with the electron baths. We remind that in the physically proper non-partitioned view[107] the different subsystems are allowed to reach thermal equilibration as one system. Namely, their coupling is not part of the turned on perturbation.[59, 60, 66]

2.4. Equilibrated Initial Conditions

The Keldysh formalism is used next to properly account for the thermal equilibration of the device with the electrodes and its affect on $G^<$. A Matsubara GF (G^M) corresponds entirely to the imaginary part of the KC and is used to describe the system at thermal equilibrium. Through KC continuation

$$G^M \rightarrow G^<(t_0, t_0) \equiv G_0^<. \tag{31}$$

In the absence of a perturbation, $\mathbf{G}^<(\bar{t}, \bar{\omega}) = \mathbf{G}^{\mathbf{0},<}(\bar{\omega}) = \mathbf{G}^{\mathbf{0},<}_{\mathbf{1}}(\bar{\omega}) - \mathbf{G}^{\mathbf{0},<\dagger}_{\mathbf{1}}(\bar{\omega})$, where

$$\mathbf{G}^{\mathbf{0},<}_{\mathbf{1}}(\bar{\omega}) = i\mathbf{G}^{\mathbf{R}}(\bar{\omega})\mathbf{G}^<_{\mathbf{0}} = -\mathbf{G}^{\mathbf{R}}(\bar{\omega})\mathbf{f}(\mathbf{h_0} - \mu I). \tag{32}$$

The Fermi-matrix $\mathbf{f}(\mathbf{A})$ is a function of a matrix and in the limit of non-interacting electrons is given by $\mathbf{f}(\mathbf{A}) = e^{-\beta\mathbf{A}}\left[\mathbf{I} + e^{-\beta\mathbf{A}}\right]^{-1}$, assuming $\mathbf{A}$ is Hermitian. This condition is satisfied for the orthogonalized basis set, where $\mathbf{h_0} = \mathbf{S}^{-1/2}\bar{\mathbf{h}}\mathbf{S}^{-1/2}$. The Fermi-matrix above provides an analytical solution for the Matsubara GF ($\mathbf{G}^{\mathbf{M}}$) in the case of non-interacting electrons.

More specifically, we evaluate the unperturbed system in the localized basis, whereas the propagation relations are all implemented in the orthogonalized representation. For a Hamiltonian of a closed system represented in a finite basis, $\mathbf{G}^{0,<}(\bar{\omega})$ is diagonal in the basis that diagonalizes $\bar{\mathbf{h}}$. This implies that $\mathbf{G}^{0,<}(\bar{\omega})$ does not introduce any off-diagonal coherences. On the other hand, an open system introduces a non-trivial $\bar{\omega}$ dependence to the Hamiltonian (through the bulk SEs). In this case, the initial guess is still guaranteed not to correspond to arbitrarily turning on the electron density of the leads at $\bar{t} = t_0$, since $\bar{\omega}$ depends on all Δt.

2.5. Propagating using frequency domain bulk self-energy

Next, the projected equation (eq. 26) is written in the integrated form, where $\mathbf{h}_{cc}$ is diagonalized becomes

$$\begin{aligned} i\Delta G^<_{cc,ij}(\bar{t}, \bar{\omega}) = \\ \int_{-\infty}^{\infty} \Theta(\bar{t} - t')e^{-i\Delta\epsilon_{ij}(\bar{t}-t')}\left\{\int d\omega'[\mathbf{v_{cc}}(t', \omega')\mathbf{G}^<_{cc}(t', \bar{\omega} - \omega') - \mathbf{G}^<_{cc}(t', \bar{\omega} + \omega')\mathbf{v_{cc}}(t', \omega')]_{ij}\right\} \\ + \int_{-\infty}^{\infty} \Theta(\bar{t} - t')e^{-i\Delta\epsilon_{ij}(\bar{t}-t')} \int_{-\infty}^{\infty} dt'' \sum_{k,l}[\Gamma_{ijkl}(t' - t'')\Delta G^<_{cc,kl}(t'', \bar{\omega})]. \end{aligned} \tag{33}$$

In equation (33), $\int_{-\infty}^{\bar{t}} dt' \rightarrow \int_{-\infty}^{\infty} dt'\Theta(\bar{t} - t')$ and the electrodes are described by a memory kernel of the form

$$\Gamma_{ijkl}(t) \equiv \Sigma^R_{ik}(t)e^{i\epsilon_j t}\delta_{lj} - \Sigma^A_{lj}(-t)e^{-i\epsilon_i t}\delta_{ik}. \tag{34}$$

We then proceed to express the e.o.m.s in the two-frequency representation by an additional Fourier transform (FT):

$$\boldsymbol{G}^<(\Delta\omega,\bar{\omega}) \equiv \int_{-\infty}^{\infty} d\bar{t} e^{i\Delta\omega\bar{t}} \boldsymbol{G}^<(\bar{t},\bar{\omega}). \tag{35}$$

Applying the Fourier transform to eqn. 33 gives

$$\begin{aligned}\Delta G^<_{CC,ij}(\Delta\omega,\bar{\omega}) &= \mathcal{G}_{ij}(\Delta\omega)\left[\tilde{\mathbf{v}}_{CC}(\Delta\omega)\mathbf{G}^{0,<}_{CC}(\bar{\omega}-\Delta\omega/2) - \mathbf{G}^{0,<}_{CC}(\bar{\omega}+\Delta\omega/2)\tilde{\mathbf{v}}_{CC}(\Delta\omega)\right]_{ij} \\ &+\frac{1}{\pi}\mathcal{G}_{ij}(\Delta\omega)\int d\omega' \left[\tilde{\mathbf{v}}_{CC}(2\omega')\boldsymbol{\Delta G}^<_{CC}(\bar{\omega}-\omega',\Delta\omega-2\omega') - \boldsymbol{\Delta G}^<_{CC}(\bar{\omega}+\omega',\Delta\omega-2\omega')\tilde{\mathbf{v}}_{CC}(2\omega')\right]_{ij} \\ &+\mathcal{G}_{ij}(\Delta\omega)\sum_{k,l}\Gamma_{ijkl}(\Delta\omega)\Delta G^<_{cc,kl}(\Delta\omega,\bar{\omega}),\end{aligned} \tag{36}$$

where

$$\Gamma_{ijkl}(\Delta\omega) \equiv \int dt e^{i\Delta\omega t}\Gamma_{ijkl}(t) = \Sigma^R_{ik}(\epsilon_j+\Delta\omega)\delta_{lj} - \Sigma^A_{lj}(\epsilon_i-\Delta\omega)\delta_{ik}. \tag{37}$$

In addition, we use, as routinely employed, a small broadening factor η in defining the propagator to enhance the stability of the solution:

$$\mathcal{G}_{ij}(\Delta\omega) = \frac{1}{\Delta\omega + i\eta - \Delta\epsilon_{ij}}. \tag{38}$$

This broadening ensures that eventually (at $t \to \infty$) the system will return to its initial equilibrium configuration. In practice, any finite grid method in the frequency domain requires an artificial broadening to resolve infinitesimally narrow peaks. The final result is

$$\begin{aligned}\sum_{k,l}\boldsymbol{\mathcal{H}}_{ijkl}(\Delta\omega)\boldsymbol{\Delta G}^<_{CC,kl}(\Delta\omega,\bar{\omega}) &= \left[\tilde{\mathbf{v}}_{CC}(\Delta\omega)\mathbf{G}^{0,<}_{CC}(\bar{\omega}-\Delta\omega/2) - \mathbf{G}^{0,<}_{CC}(\bar{\omega}+\Delta\omega/2)\tilde{\mathbf{v}}_{CC}(\Delta\omega)\right]_{ij} \\ &+\frac{1}{\pi}\int d\omega' \left[\tilde{\mathbf{v}}_{CC}(2\omega')\boldsymbol{\Delta G}^<_{CC}(\bar{\omega}-\omega',\Delta\omega-2\omega') - \boldsymbol{\Delta G}^<_{CC}(\bar{\omega}+\omega',\Delta\omega-2\omega')\tilde{\mathbf{v}}_{CC}(2\omega')\right]_{ij},\end{aligned} \tag{39}$$

where

$$\boldsymbol{\mathcal{H}}_{ijkl}(\Delta\omega) \equiv (\Delta\omega + i\eta - \Delta\epsilon_{ij})\delta_{ik}\delta_{jl} - \Gamma_{ijkl}(\Delta\omega). \tag{40}$$

The final result can be Fourier transformed back to the mixed representation to generate Wigner type information that provides important insight on the quantum mechanical effects that determine the time-dependence of the observable (*i.e.* the current operator).

We note that eqn. 39 involves a tensor of rank four that operates on a matrix (tensor of rank 2). Consequently, we employ tetradic notation to re-expresses matrices as vectors and rank four tensors as matrices. Using this notation an $n \times n$ matrix $\mathbf{A}$ with elements A_{ij} is rewritten as a vector $|A\rangle\rangle$ with elements $|A\rangle\rangle_{ni+j}$. Likewise, a tensor $\mathcal{H}$ with element $\mathcal{H}_{ijkl}$ (where i,j,k,l each spans over n elements) is rewritten as an $n^2 \times n^2$ matrix $\mathbf{H}$, with elements $\mathbf{H}_{ni+j,nk+l}$. The tetradic notation for eqn. 39 takes the form:

$$\mathbf{H}(\Delta\omega)|\boldsymbol{\Delta G}^<_{CC}(\Delta\omega,\bar{\omega})\rangle\rangle = |B^{(1)}_{CC}(\Delta\omega,\bar{\omega})\rangle\rangle + \frac{1}{\pi}\int d\omega'|B_{CC}(\omega',\Delta\omega,\bar{\omega})\rangle\rangle, \tag{41}$$

where

$$\mathbf{H}_{ni+j,nk+l}(\Delta\omega) \equiv \boldsymbol{\mathcal{H}}_{ijkl}(\Delta\omega), \tag{42}$$

$$|\boldsymbol{\Delta G}^{<}_{CC}(\Delta\omega,\bar{\omega})\rangle\rangle_{ni+j} \equiv \boldsymbol{\Delta G}^{<}_{CC,ij}(\Delta\omega,\bar{\omega}), \tag{43}$$

$$|B^{(1)}_{CC}(\Delta\omega,\bar{\omega})\rangle\rangle_{ni+j} \equiv \left[\tilde{\mathbf{v}}_{CC}(\Delta\omega)\mathbf{G}^{0,<}_{CC}(\bar{\omega}-\Delta\omega/2) - \mathbf{G}^{0,<}_{CC}(\bar{\omega}+\Delta\omega/2)\tilde{\mathbf{v}}_{CC}(\Delta\omega)\right]_{ij} \tag{44}$$

and

$$\begin{aligned} &|B_{CC}(\omega',\Delta\omega,\bar{\omega})\rangle\rangle_{ni+j} \equiv \\ &\left[\tilde{\mathbf{v}}_{CC}(2\omega')\boldsymbol{\Delta G}^{<}_{CC}(\bar{\omega}-\omega',\Delta\omega-2\omega') - \boldsymbol{\Delta G}^{<}_{CC}(\bar{\omega}+\omega',\Delta\omega-2\omega')\tilde{\mathbf{v}}_{CC}(2\omega')\right]_{ij}. \end{aligned} \tag{45}$$

Finally inverting $\mathbf{H}(\Delta\omega)$ gives

$$|\boldsymbol{\Delta G}^{<}_{CC}(\Delta\omega,\bar{\omega})\rangle\rangle = \mathbf{G}(\Delta\omega)|B^{(1)}_{CC}(\Delta\omega,\bar{\omega})\rangle\rangle + \frac{1}{\pi}\mathbf{G}(\Delta\omega)\int d\omega'|B_{CC}(\omega',\Delta\omega,\bar{\omega})\rangle\rangle, \tag{46}$$

where

$$\mathbf{G}(\Delta\omega) \equiv \mathbf{H}^{-1}(\Delta\omega). \tag{47}$$

In this notation, a tensor equation appears as a matrix equation and all matrix operations (multiplication, inversion, diagonalization, etc.) can be applied to the rexpressed equation of motion.

The final expression can be formally expanded to n-th order in the perturbation $\tilde{\mathbf{v}}(\Delta\omega)$. In this expansion, we express the TD electronic density in terms of the evolving occupations of the projected device states. The band structure due to the coupling to the electrodes is included directly in the expansion through the energy distribution variable ($\bar{\omega}$) that is also used to calculate the SEs. In the results reported below we study transient current that is appropriately treated at the first order. We also study the effect of initial conditions that include an established non-equilibrium steady state in response to an applied constant bias. We describe next the form of the solution for the steady state case.

2.6. Steady state limit of the perturbative expansion

We now discuss the solution for the steady state, where a constant bias has been turned on for a long time. The Fourier transform of a time independent perturbation $\mathbf{v}(t) = \mathbf{v}_\circ$ (as in eqn. 35) takes the following form in the frequency domain

$$\tilde{\mathbf{v}}(\Delta\omega) = 2\pi\delta(\Delta\omega)\mathbf{v}_\circ. \tag{48}$$

The resulting two-frequency representation of the time-dependent equations of motion is used to derive the time-independent correction to $G^{0,<}$. In this scheme, time independent fields (such as a source-drain bias that has been turned on for a long time) are treated exactly by specializing eqn. 41.

Implementing this scheme for the steady state and without loss of generality, the perturbing potential $\mathbf{v}(t)$ is rewritten as the sum of a time-independent field (of arbitrary strength) $\mathbf{v}_\circ$ and a time dependent perturbing field $\mathbf{v}_{TD}(t)$. Fourier transforming to the frequency domain gives

$$\tilde{\mathbf{v}}(\Delta\omega) = 2\pi\delta(\Delta\omega)\mathbf{v}_\circ + \tilde{\mathbf{v}}_{TD}(\Delta\omega). \tag{49}$$

The resulting two-frequency equation of motion becomes

$$\mathbf{H}_{v_\circ}(\Delta\omega)|\boldsymbol{\Delta G}^{<}_{TD,CC}(\Delta\omega,\bar{\omega})\rangle\rangle = |B^{(1)}_{TD,CC}(\Delta\omega,\bar{\omega})\rangle\rangle + \frac{1}{\pi}\int d\omega'|B_{TD,CC}(\omega',\Delta\omega,\bar{\omega})\rangle\rangle, \tag{50}$$

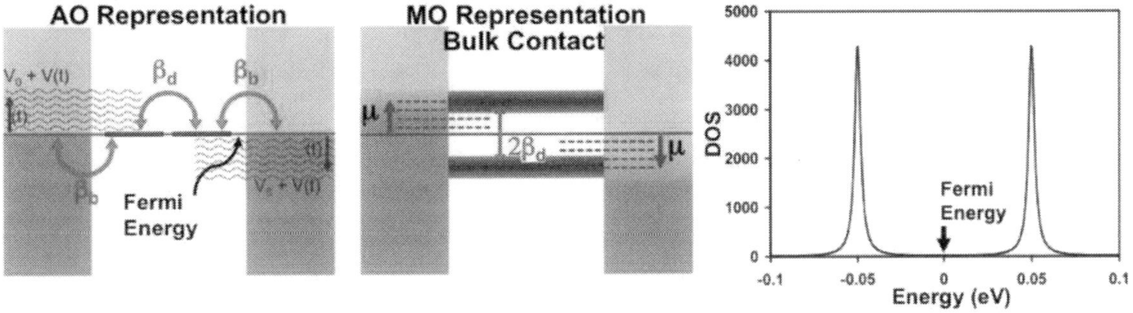

Figure 1. Schematic diagram of the model system. (a) localized/atomic orbital representation, where the strong coupling of the device state to the wires are indicated (b) Diagonalized molecular orbital representation. (c) Broadened electronic density of states.

where the propagating super operator

$$\mathbf{H}_{v_\circ,ni+j,nk+l}(\Delta\omega) \equiv \boldsymbol{\mathcal{H}}_{ijkl}(\Delta\omega) - (v_{\circ,ik}\delta_{lj} - v_{\circ,lj}\delta_{ik}) \tag{51}$$

includes the steady perturbation. The total lesser Green function is now defined in terms of a new time independent initial guess $G^{0,<}_{v_\circ}$ that includes the effects of the time-independent field exactly. In the absence of additional time-dependent perturbations, $G^{0,<}_{v_\circ}$ is the total lesser Green function for the system. The remainder, $\Delta G^{<}_{TD}$, is related to the time dependent perturbations

$$|\boldsymbol{\Delta G}^{<}_{TD,CC}(\Delta\omega,\bar{\omega})\rangle\rangle \equiv |G^{<}_{CC}(\Delta\omega,\bar{\omega})\rangle\rangle - 2\pi\delta(\Delta\omega)|G^{0,<}_{v_\circ,CC}(\bar{\omega})\rangle\rangle, \tag{52}$$

where

$$|G^{0,<}_{v_\circ,CC}(\bar{\omega})\rangle\rangle = |G^{0,<}_{CC}(\bar{\omega})\rangle\rangle + \mathbf{G}_{v_\circ}(0)|B^{(1)}_{v_\circ,CC}(\bar{\omega})\rangle\rangle. \tag{53}$$

Here, $\mathbf{G}_{v_\circ}(\Delta\omega) = \mathbf{H}^{-1}_{v_\circ}(\Delta\omega)$ and

$$|B^{(1)}_{v_\circ,CC}(\bar{\omega})\rangle\rangle_{ni+j} \equiv \left[\mathbf{v}_{\circ,CC}\mathbf{G}^{0,<}_{CC}(\bar{\omega}) - \mathbf{G}^{0,<}_{CC}(\bar{\omega})\mathbf{v}_{\circ,CC}\right]_{ij}. \tag{54}$$

Now, the time dependent "B"-terms are rewritten in terms of the modified initial guess $G^{0,<}_{v_\circ}$ and the time-dependent components of the perturbation ($v_{TD}(\Delta\omega)$):

$$|B^{(1)}_{TD,CC}(\Delta\omega,\bar{\omega})\rangle\rangle_{ni+j} \equiv \left[\tilde{\mathbf{v}}_{TD,CC}(\Delta\omega)\mathbf{G}^{0,<}_{v_\circ,CC}(\bar{\omega}-\Delta\omega/2) - \mathbf{G}^{0,<}_{v_\circ,CC}(\bar{\omega}+\Delta\omega/2)\tilde{\mathbf{v}}_{TD,CC}(\Delta\omega)\right]_{ij} \tag{55}$$

and

$$|B_{TD,CC}(\omega',\Delta\omega,\bar{\omega})\rangle\rangle_{ni+j} \equiv \left[\tilde{\mathbf{v}}_{TD,CC}(2\omega')\boldsymbol{\Delta G}^{<}_{TD,CC}(\Delta\omega-2\omega',\bar{\omega}-\omega') - \boldsymbol{\Delta G}^{<}_{TD,CC}(\bar{\omega}+\omega',\Delta\omega-2\omega')\tilde{\mathbf{v}}_{TD,CC}(2\omega')\right]_{ij}. \tag{56}$$

The last two equations are identical in form to the general equations (44) and (45). In these equations the effect of $\mathbf{v}_\circ$ enters explicitly only into the Hamiltonian super operator, where we replace $\mathbf{H}(\Delta\omega)$ with $\mathbf{H}_{v_\circ}(\Delta\omega)$ and the initial lesser GF ($G^{0,<}$) with $G^{0,<}_{v_\circ}$. We emphasize that the constant perturbation cannot be simply added to the Hamiltonian for calculating the initial lesser Green function $G^{0,<}(\bar{\omega})$ for proper description of existing non-equilibrium conditions. The effect of a source-drain bias on an initially decoherent electronic charge density cannot be modeled by a simple field asymmetry embedded within the Hamiltonian.

Journal of Physics: Conference Series **220** (2010) 012008 doi:10.1088/1742-6596/220/1/012008

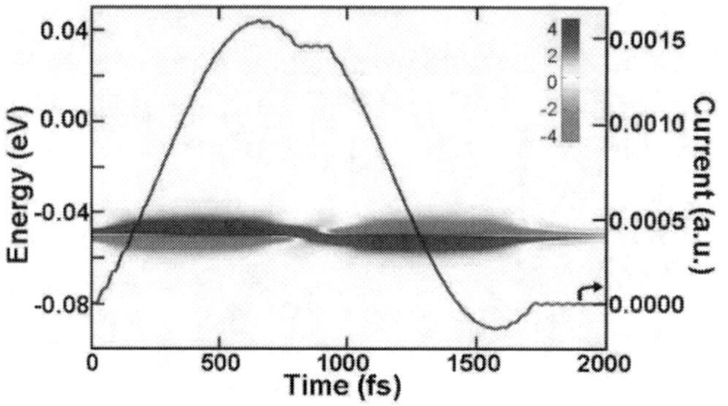

Figure 2. The transient current under the effect of a slowly varying potential. Both the corresponding current density is provided (band structure at the ground state). Note that the right axis corresponds to the current curve while the left axis corresponds to the energy distribution variable

2.6.1. Current evaluation We now describe the evaluation of the current from the time propagated $G^<$ function. The grand canonical expectation value of any dynamical variable (following the KC formalism[104, 106, 108]) is given by

$$\langle \hat{J}(t_1) \rangle = -i \int_{-\infty}^{\infty} d\vec{x}_1 \lim_{x_2 \to x_1} \left[\mathcal{J}(\vec{x}_1) G^<(x_1, x_2) \right] . \tag{57}$$

The expectation value can be expressed as a product of two matrices

$$\langle \mathbf{J}(t_1) \rangle = -iTr \left[\lim_{t_2 \to t_1} \left[\mathcal{O}\mathbf{G}^<(t_1, t_2) \right] \right] = -iTr \left[\lim_{\Delta t \to 0, \bar{t} \to t_1} \left[\mathcal{J}\mathbf{G}^<(\bar{t}, \Delta t) \right] \right] = Tr \left[\lim_{\bar{t} \to t_1} \left[\frac{-i}{2\pi} \int_{-\infty}^{\infty} d\bar{\omega} \mathcal{J}\mathbf{G}^<(\bar{t}, \bar{\omega}) \right] \right] , \tag{58}$$

where $\mathcal{J}_{\mathbf{ij}} = \int_{-\infty}^{\infty} \mathbf{d\tilde{x}} \phi_{\mathbf{i}}(\mathbf{\tilde{x}}) \mathcal{J}(\mathbf{\tilde{x}}) \phi_{\mathbf{j}}(\mathbf{\tilde{x}})$. Let us now consider the specific case of the current density operator:

$$\mathcal{J}(\vec{x}) = -i \left[\vec{\nabla}_x \delta(\vec{x} - \vec{r}) + \delta(\vec{x} - \vec{r}) \vec{\nabla}_x \right] \tag{59}$$

or in an AO basis representation

$$\mathcal{J}_{ji}(\vec{r}) = i \left[\phi_i(\vec{r}) \vec{\nabla} \phi_j(\vec{r}) - \phi_j(\vec{r}) \vec{\nabla} \phi_i(\vec{r}) \right] . \tag{60}$$

The electron current through a given plane is calculated by tracing $[-i\mathcal{J}\mathbf{G}^<]$.

2.6.2. Models We concentrate in this study on the transient current through a one dimensional wire composed of hopping sites, where interactions are included only between neighboring sites. In the considered model, the confined system includes two states which are coupled strongly by β_d to result with a pair of bonding and anti-bonding states separated by $2\beta_d$ ($H_{12} = H_{21} \equiv \beta_d$, where H is the electronic Hamiltonian). This energy level scheme is illustrated in Figure 1, where the atomic orbital representation is provided in part (a) and the corresponding MO picture reflecting the coupling is given in part (b). The two strongly inter-coupled orbitals are interfaced with electrodes to result with a pair of conducting (broadened) bands of states. In

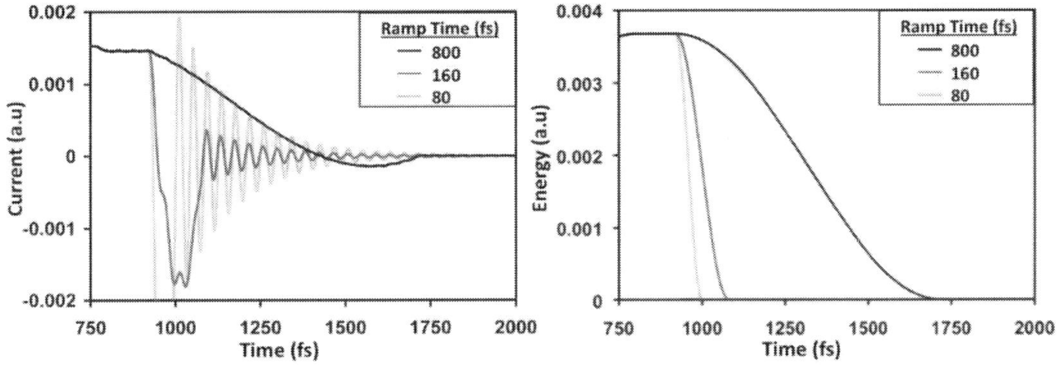

Figure 3. TD currents which differ in the rate of the potential turn off. The 3 TD bias curves are provided to the right.

the AO representation this is achieved by coupling each site of the strongly interacting pair to a different lead. The coupling to the leads is represented by the SEs, $\Sigma = \beta_c^\dagger g_s \beta_c$. In our model, the surface GF (g_s) is an imaginary constant number that implements a wide band approximation for the electronic density of the bulk material projected on the surface site. We use $\beta_d = -0.005(eV), \beta_c = 0.1\beta_d$ where the resulting density of states (DOS) is plotted in part (c).

We note that β and s are the electronic and overlap coupling terms. The diagonal terms of the model Hamiltonian are set to the initial Fermi energy of the system. Accordingly, we express the current operator with the numerical values assigned to the electronic integrals in the Hamiltonian

$$\mathbf{J} = \mathbf{i}\begin{bmatrix} 0 & \beta_d \\ -\beta_d & 0 \end{bmatrix}. \tag{61}$$

The traced quantity $i\mathbf{J}\mathbf{G}^<(\bar{\mathbf{t}}, \bar{\omega})$ provides the TD band structure (energy distibution) of the current operator. In this case it describes the evolving current through the center of the model system. In all calculations reported below in the results section we set the FE to zero ($\mu_0 = 0$). This also defines the on site energies as described in the figure.

We use a sufficiently small value ($\eta_d = 0.002eV$) for the broadening factor added to the imaginary component of the Hamiltonian used to calculate the G^R of the equilibrated system (eq 32). Finally, we note that in all calculations the target bias potential is set to 0.1V unless otherwise stated.

3. Results

We begin by considering the effect of a direct current (DC) potential bias, where we follow the switching temporal effect on the transient current. The current following a (relatively) slow rate for switching off the bias is provided in Figure 2. (Note the current scale is provided to the right). We note that while the switching off time is long, where an oscillatory transient response is almost absent, the transient current dips below zero before the full dissipation of the system to the zero current limit. We then further resolve the time-dependence of the band structure of the current operator. The figure depicts, by a projected color map, the distribution of the current operator ($J(\bar{\omega}, \bar{t})$), where the current at any time can be extracted by integrating over the energy distribution ($\int d\bar{\omega} J(\bar{\omega}, \bar{t})$). The current is associated with contributions from the lower occupied state (ground state), where the upper portion of the band contributes to the positive current. The band structure of the current operator at the switch off is the reflection

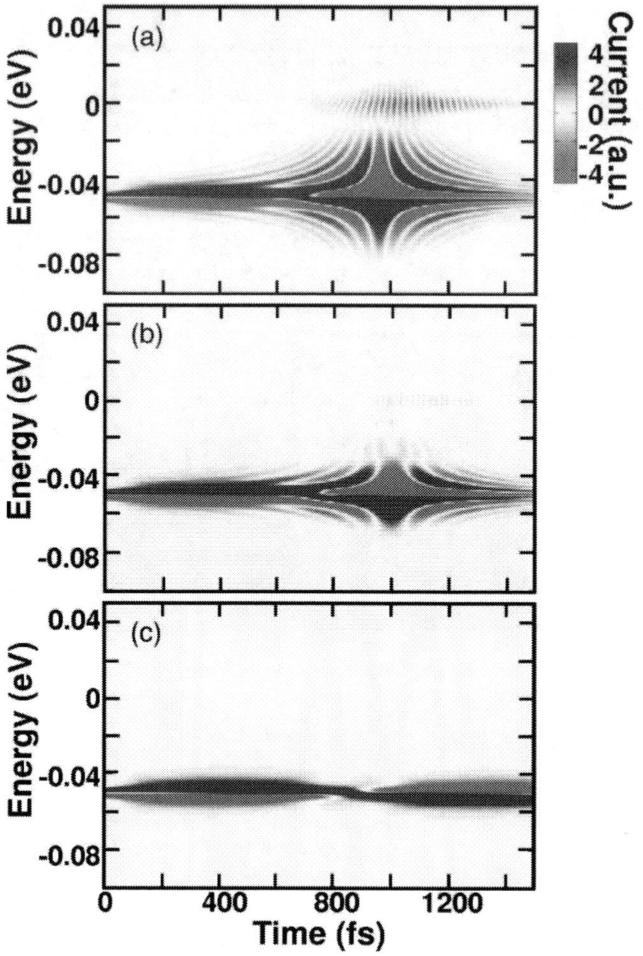

Figure 4. Time-dependent current distributions projected on a color map (I(t) are provided in Figure 3). (a) fastest turn off rate, (b) medium turn off rate, (c) slowest turn off rate.

of the band at the turn on event. At turn off, the positive part of the distribution is at the energetically lower part of the band (further away from the Fermi energy). The band narrows, as the current continues to dissipate.

The current under increased rates of bias switching is considered next. In Figure 3 we provide the current upon varying the rate for switching off the bias. The oscillatory response of the current is shown to increase in amplitude with the rate of switching. It is also evident that the current will oscillate with a larger number of periods when the rate is increased. This is also reflected in the band structure as shown in Figure 4. The oscillatory response is shown to be related to the direct interference of the two states. The quicker rate of the turn off is shown to result in stronger interference due to the two states. The frequency of the oscillation depends on the present energy levels, where the amplitude of the oscillation is determined by the rate of perturbation change.

We now clarify that the temporal features of the current distribution prior to the switching off as indicated in the figure is not a violation of causality. This is related to the nature of the $\bar{t}$ variable, and by taking the full distribution we generate information that is dependent on

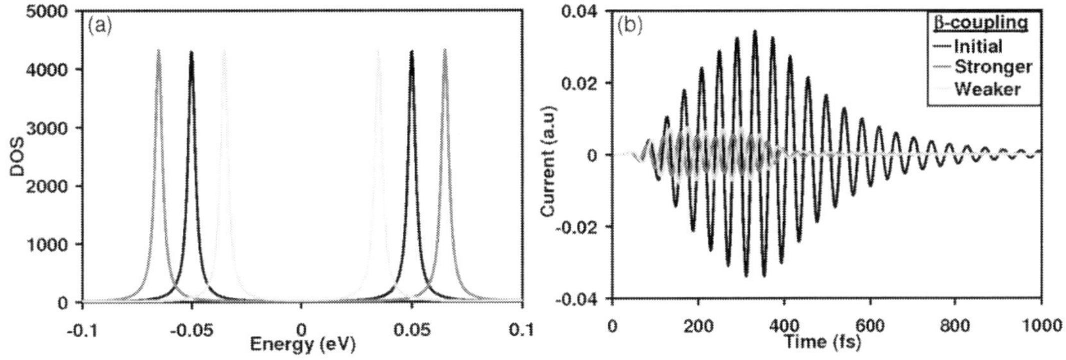

Figure 5. Effect of the coupling strength: (a) The density of states (b) TD current under an AC tuned to the original coupling strength.

the observational time. Indeed, the current operator, which serves as physical probe, illustrates that causality is preserved. This can also be understood as a reflection of the uncertainty principle, where Husimi transformations can be used to reflect the fundamental limitations of measurements. This highlights the importance in the current distributions expressed in Wigner form as further demonstrated below. This representation is used to analyze the underlying quantum effects leading to the oscillatory and directed response of the current.

As apparent by the current distribution due to the quick switching event, the two states can be coupled to interfere and lead to an oscillatory response of the current. Stronger coherence can clearly be achieved by a monochromatic field that is tuned to oscillate at a frequency associated to the energy separating the two energy levels. In Figure 5 we follow the effect of a tuned driving AC bias on the current. The AC amplitude is at 0.01V. We apply an AC pulse tuned exactly to the weakly coupled states and then apply the same AC bias on the two states where their coupling is either enhanced or decreased by 30% relative to the orignal value. The change in the coupling strength, as reflected in Figure 5 (a), either further separates (increased β_d) or diminishes the energy gap between the two states (reduced β_d). In either case, as shown in Figure 5 (b), the responding current oscillations are reduced. The case where the energy levels are further separated leads to a reduced time period of the response (increase of the frequency). This is expected from the nature of the coupling-induced detuning effect. We note that upon the opposite shift in β the period is slightly increasing.

The coherence underlying the response to an AC bias is also reflected in the band structure of the current operator. In Figure 6 the color maps for the current plots of Figure 5 are provided. We follow the effect of varying the coupling strength where the AC remains tuned exactly to the system with the original β coupling parameter. The current band structure of the perfectly tuned case (unshifted β_d) demonstrates the strong amplitude of the AC response by the straight (vertical) bands. These bands are slanted at the transient conditions as shown in Figure 6 (a). For the β detuned cases as shown in Figures 6 (b-c) the slanting is increased and the conductance amplitudes are diminished as expected. The mini-band that appears at a shift below the populated state due to the interference with the excited state depends on the tuning. The detuning diminishes its strength and shifts its location. It is shifted below -0.1eV for the case with the stronger coupling or above that value for the second detuned case. Finally, we also point at the different features of the detuning due to the change in beta as reflected, for example, in the decay regime of the main coherence, see Figures 6 (b-c).

We next consider the effect of a DC bias on a coherent driven system, where the frequency of the applied AC bias is set exactly to the energy difference between the two levels. The effect of

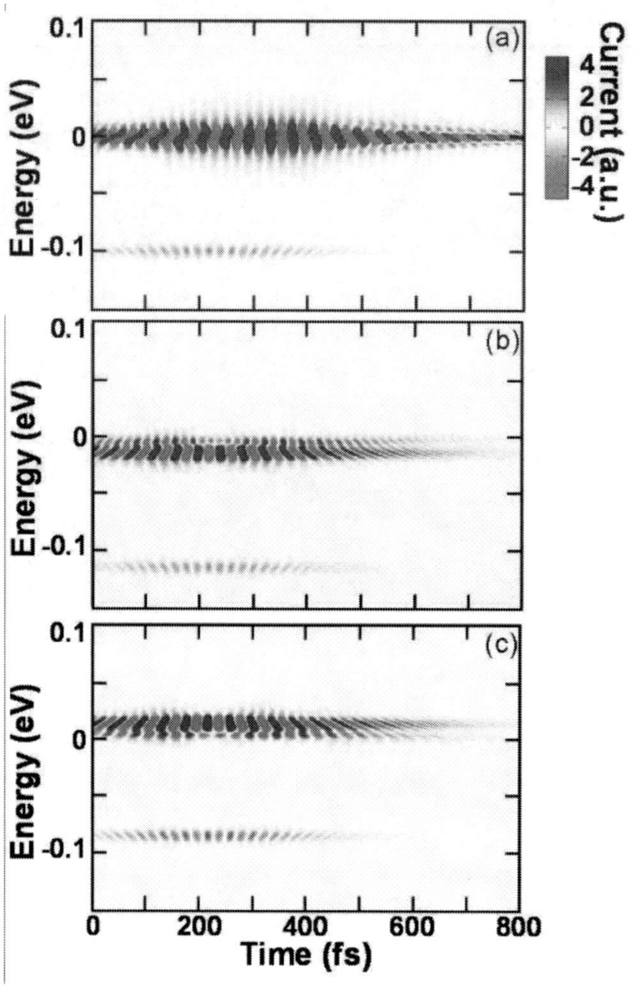

Figure 6. The current band structure corresponding to the AC tuned to the energy difference of the two energy levels under the effect of varying the coupling strength (see Figure 5). (a) initial coupling , (b) stronger coupling, (c) weaker coupling.

the DC bias to detune the response was analyzed before[65]. Here its effect to shift the electronic spectra is demonstrated in Figure 7 (a). The electronic excitation energy is shown to increase due to the DC bias as expected. The applied DC bias also further broadens the spectral peak and diminishes its height. In Figures 7 (b-d) we follow the TD currents with different strengths of an applied DC under the effect of an AC bias that is retuned to the DC affected energy states. The three plots focus each on different time ranges, where Figure 7 (b) provides an overall view, Figure 7 (c) zooms on the initial times where the existing conducting flux is evident. It is also demonstrated that the flux due to the DC bias reduces the amplitude of the oscillatory response, an effect that can be studied only by a TD scheme. Finally Figure 7 (d) shows the tail, where the AC bias is turned off and the system returns to its constant biased state.

The band structure of the corresponding current operator is provided in Figure 8. The effect of the bias to induce constant flux is reflected in the thin band centered about the ground state energy that is enhanced by the increase of the DC bias. As expected the main coherence is

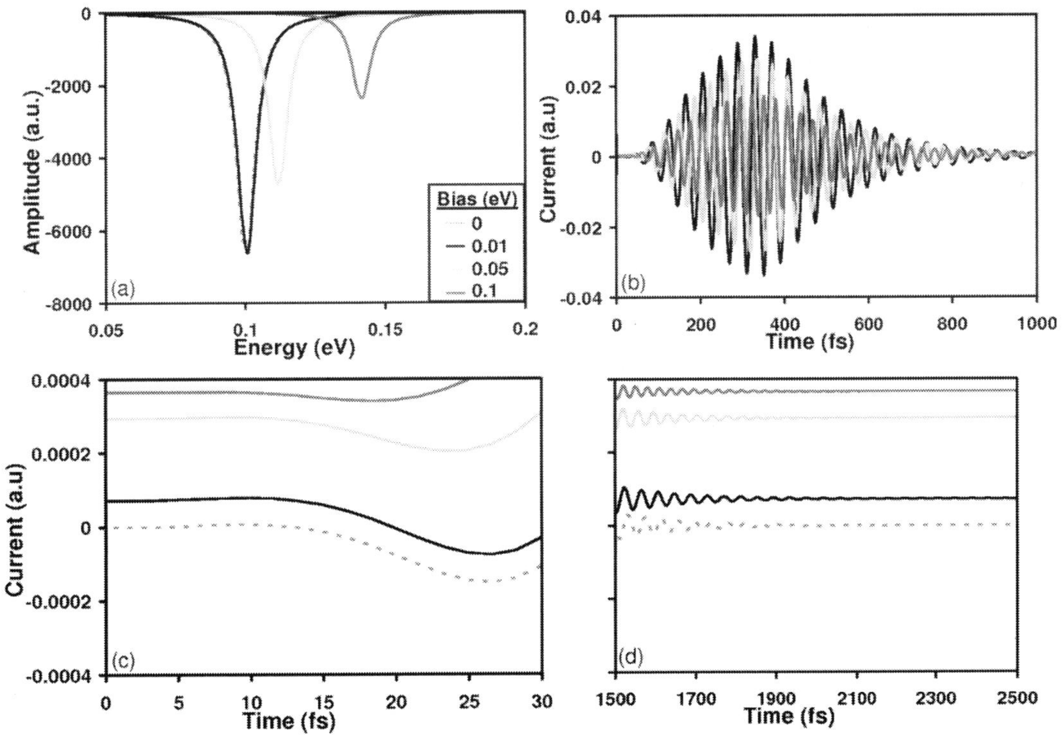

Figure 7. Effects of an existing DC bias on: (a) Electronic spectra of the bias system (b-d) Coherence driven conductance.

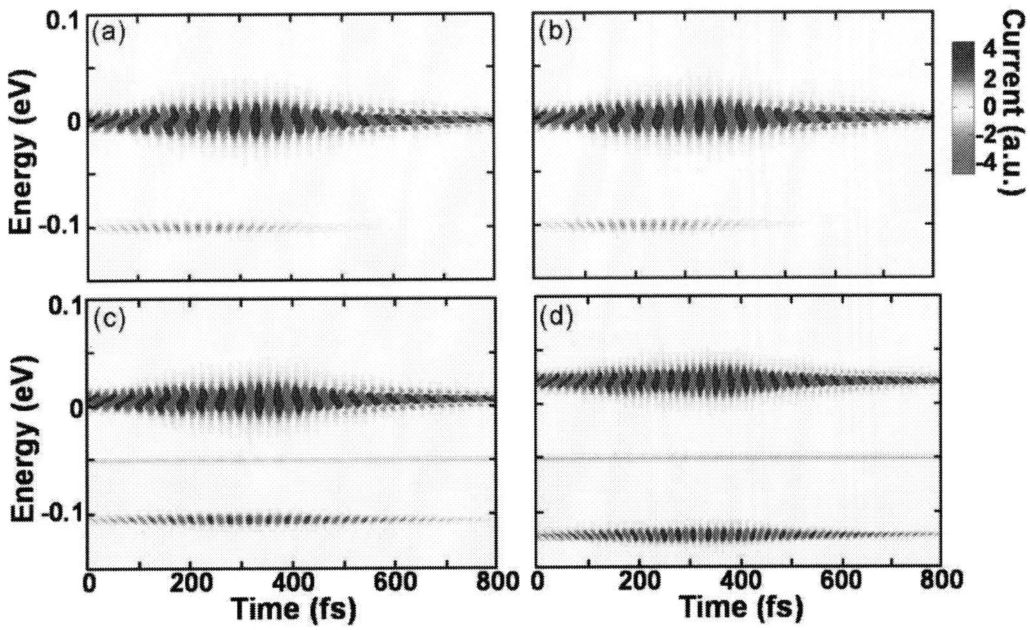

Figure 8. The current distributions of the cases with different applied DC biases. (a) 0 eV, (b) 0.01 eV, (c) 0.05 eV, (d) 0.10 eV.

obtained at the energy exactly between the two interfering states. This main coherence is shown to decrease, in spite of the retuning of the AC bias, in the presence of the DC induced flux. This is a consequence of the DC to reduce the spectral peak that corresponds to the electronic excitation as revealed above. Interestingly, the mini-band of the current operator corresponding to the shift below the ground state features an opposite trend. This mini-band is somewhat enhanced upon the application of the bias. This effect however is still not able to reverse the overall observation of the DC bias reducing the oscillatory response.

4. Concluding Comments

Electron transport through model electronic channels is studied by solving the electronic equations of motion of coupled and bias systems. The electronic equations of motion is represented using Keldysh formalism by Green Functions. Their solution is obtained by expressing the time dependence in the frequency domain, where in order to achieve full time resolution the GFs are defined using two independent time variables. We solve the equations using TD perturbation theory.

The TD approach provides a rigorous treatment of the extended system. The calculations also provide a plethora of information through the resolved band structure of the evolving electronic density. Here, the band structure of the current operator in model electronic channels are solved for. The study considers the transient features of the transport through biased systems. The evolving current operator is used to generate data that highlights the quantum interferences that determine the oscillatory current at transient conditions. The oscillatory component of the current that is noted at transient conditions is associated with interference of the conducting states, where the constant component under steady state conditions when the transient features are dissipated is determined by the broadening extent of the conducting states.

The TD approach is also used to study driven transport. In the considered system, the oscillatory bias perturbation affects the current through the state interferences. The relationships between DC bias and AC-inducing coherences are studied in detail. The calculated current operator is used to reflect the quantum interferences affecting the current under the driving TD perturbations. The ability to analyze in general the possibility for driving the current by applied TD perturbations becomes of increased importance due to experimental related advances.

Acknowledgments

BDD acknowledges University of Michigan and the Petroleum Research Fund of the ACS (through Grant 47118-G6) for financial support. HP would like to acknowledge the University of Michigan Rackham Graduate School Summer Institute for financial support.

References

[1] Bumm L A, Arnold J J, Cygan M T, Dunbar T D, Burgin T P, Jones L I, Allara D L, Tour J M and Weiss P S 1996 *Science* **271** 1705–1707
[2] Reed M A, Zhou C, Muller C J, Burgin T P and Tour J M 1997 *Science* **278** 252–254
[3] Stipe B C, Rezaei M A and Ho W 1998 *Science* **280** 1732–1735
[4] Chen J, Reed M A, Rawlett A M and Tour J M 1999 *Science* **286** 1550–1552
[5] Collier C P, Wong E W, Belohradsky M, Raymo F M, Stoddart J F, Kuekes P J, Williams R S and Heath J R 1999 *Science* **285** 391–394
[6] Park H, Park J, Lim A K L, Anderson E H, Alivisatos A P and McEuen P L 2000 *Nature* **407** 57–60
[7] Collier C P, Mattersteig G, Wong E W, Luo Y, Beverly K, Sampaio J, Raymo F M, Stoddart J F and Heath J R 2000 *Science* **289** 1172–1175
[8] Rueckes T, Kim K, Joselevich E, Tseng G Y, Cheung C L and Lieber C M 2000 *Science* **289** 94–97
[9] Joachim C, Gimzewski J K and Aviram A 2000 *Nature* **408** 541–548
[10] Collins P G, Arnold M S and Avouris P 2001 *Science* **292** 706–709

[11] Holmlin R E, Haag R, Chabinyc M L, Ismagilov R F, Cohen A E, Terfort A, Rampi M A and Whitesides G M 2001 *J. Am. Chem. Soc.* 5075–5085
[12] Cui X D, Primak A, Zarate X, Tomfohr J, Sankey O F, Moore A L, Moore T A, Gust D, Harris G and Lindsay S M 2001 *Science* **294** 571–574
[13] Nitzan A 2001 *Annu. Rev. Phys. Chem.* **52** 681
[14] Tour J M, Rawlett A M, Kozaki M, Yao Y X, Jagessar R C, Dirk S M, Price D W, Reed M A, Zhou C W, Chen J, Wang W Y and Campbell I 2001 *Chem.-Eur. J.* **7** 5118–5134
[15] Reichert J, Ochs R, Beckmann D, Weber H B, Mayor M and v Löhneysen H 2002 *Phys. Rev. Lett.* **88** 176804
[16] Kushmerick J G, Holt D B, Pollack S K, Ratner M A, Yang J C, Schull T L, Naciri J, Moore M H and Shashidhar R 2002 *J. Am. Chem. Soc.* **124** 10654–10655
[17] Mbindyo J K N, Mallouk T E, Mattzela J B, Kratochvilova I, Razavi B, Jackson T N and Mayer T S 2002 *jacs*
[18] Fan F R F, Yang J P, Cai L T, Price D W, Dirk S M, Kosynkin D V, Yao Y X, Rawlett A M, Tour J M and Bard A J 2002 *J. Am. Chem. Soc.* **124** 5550–5560
[19] Park J, Pasupathy A N, Goldsmith J I, Chang C, Yaish Y, Petta J R, Rinkoski M, Sethna J P, Abruna H D, McEuen P L and Ralph D C 2002 *Nature* **417** 722–725
[20] Liang W J, Shores M P, Bockrath M, Long J R and Park H 2002 *Nature* **417** 725–729
[21] Luo Y, Collier C P, Jeppesen J O, Nielsen K A, Delonno E, Ho G, Perkins J, Tseng H R, Yamamoto T, Stoddart J F and Heath J R 2002 *ChemPhysChem* **3** 519
[22] Adams D M, Brus L, Chidsey C E D, Creager S, Creutz C, Kagan C R, Kamat P V, Lieberman M, Lindsay S, Marcus R A, Metzger R M, Michel-Beyerle M E, Miller J R, Newton M D, Rolison D R, Sankey O, Schanze K S, Yardley J and Zhu X Y 2003 *J. Phys. Chem. B* **107** 6668–6697
[23] Metzger R M 2003 *Chem. Rev.* **103** 3803–3834
[24] Nazin G V, Qiu X H and Ho W 2003 *Science* **302** 77–81
[25] Nitzan A and Ratner M A 2003 *Science* **300** 1384–1389
[26] Xu B Q and Tao N 2003 *Science* **301** 1221–1223
[27] Kushmerick J, Lazorcik J, Patterson C, Shashidhar R, Seferos D and Bazan G 2004 *Nano. Lett.* **4** 639–642
[28] Xiao X, Xu B Q and Tao N 2004 *Nano. Lett.* **4** 267–271
[29] Xiao X, Xu B Q and Tao N 2004 *Angew. Chem. Int. Ed.* **45** 6148–6152
[30] Guisinger N, Greene M, Basu R, Baluch A and Hersam M 2004 *Nano. Lett.* **4** 55–59
[31] McCreery R L 2004 *Chem. Mater.* 4477–4496
[32] Fan F R, Yao Y, Cai L, Cheng L, Tour J and Bard A 2004 *Journal of the American Chemical Society* **126** 4035–4042 ISSN 0002-7863
[33] Piva P G, DiLabio G A, Pitters J L, Zikovsky J, Rezeq M, Dogel S, Hofer W A and Wolkow R A 2005 *Nature* **435** 658–661
[34] Xu B Q, Xiao X Y, Yang X, Zang L and Tao N J 2005 *J. Am. Chem. Soc.* **127** 2386–2387
[35] Kitagawa K, Morita T and Kimura S 2005 *J. Phys. Chem. B* **109** 13906–13911
[36] Guo X, Small J P, Klare J E, Wang Y, Purewal M S, Tam I W, Hong B H, Caldwell R, Huang L, O'Brien S, Yan J, Breslow R, Wind S J, Hone J, Kim P and Nuckolls C 2006 *Science* **311** 356–359
[37] Natelson D, Yu L H, Ciszek J W and Keane Z K T J M 2006 *Chem. Phys* **324** 267–275
[38] Keane Z K, Ciszek J W, Tour J M and Natelson D 2006 *Nano. Lett.* **6** 1518–1521
[39] Zhirnov V V and Cavin R K 2006 *Nat. Mater.* **5** 11–12
[40] Tang J, Wang Y, Nuckolls C and Wind S J 2006 *J. Vac. Sci. Technol. B.* **24** 3227–3229
[41] Tang J Y, Wang Y L, Klare J E, Tulevski G S, Wind S J and Nuckolls C 2007 *Angew. Chem. Int. Ed.* **46** 3892–3895
[42] Venkataraman L, Park Y S, Whalley A C, Nuckolls C, Hybertsen M S and Steigerwald M L 2007 *Nano. Lett.* **7** 502–506 ISSN 1530-6984
[43] Chen F, Hihath J, Huang Z, Li X and J T N 2007 *Annu. Rev. Phys. Chem.* **58** 535–564
[44] Chen X, Jeon Y M, Jang J W, Qin L, Huo F, Wei W and Mirkin C A 2008 *J. Am. Chem. Soc.* 8166–8188
[45] Ashwell G J, Wierzchowiec P, Phillips L J, Collins C J, Gigon J, Robinson B J, Finch C M, Grace I R, Lambert C J, Buckle P D, Ford K, Wood B J and Gentle I R 2008 *Phys. Chem. Chem. Phys.* **10** 1859–1866 URL `http://dx.doi.org/10.1039/b719417j`
[46] Kouwenhoven L P, Jauhar S, Orenstein J, McEuen P L, Nagamune Y, Motohisa J and Sakaki H 1994 *Phys. Rev. Lett.* **73** 3443–3446
[47] van der Vaart N C, Godijn S F, Nazarov Y V, Harmans C J P M, Mooij J E, Molenkamp L W and Foxon C T 1995 *Phys. Rev. Lett.* **74** 4702–4705
[48] Keay B J, Zeuner S, Allen S J, Maranowski K D, Gossard A C, Bhattacharya U and Rodwell M J W 1995 *Phys. Rev. Lett.* **75** 4102–4105

[49] Oosterkamp T H, Kouwenhoven L P, Koolen A E A, van der Vaart N C and Harmans C J P M 1997 *Phys. Rev. Lett.* **78** 1536–1539
[50] López R, Aguado R, Platero G and Tejedor C 1998 *Phys. Rev. Lett.* **81** 4688–4691
[51] Brouwer P W 1998 *Phys. Rev. B* **58** R10135–R10138
[52] Oosterkamp T H, Fujisawa T, van der Wiel W G, Ishibashi K, Hijman R V, Tarucha S and Kouwenhoven L P 1998 *Nature* **395** 873–876
[53] Switkes M, Marcus C M, Campman K and Gossard A C 1999 *Science* **283** 1905–1908
[54] Zhou F, Spivak B and Altshuler B 1999 *Phys. Rev. Lett.* **82** 608–611
[55] Nordlander P, Pustilnik M, Meir Y, Wingreen N S and Langreth D C 1999 *Phys. Rev. Lett.* **83** 808–811
[56] Platero G and Aguado R 2004 *Phys. Reports* **395** 1–157
[57] Murphy J E, Beard M C and Nozik A J 2006 *J. Phys. Chem. B* **110** 25455–25461
[58] Zhong Z, Gabor N M, Sharping J E, Gaeta A L and MCeuen P L 2008 *Nature Nanotechnology* **3** 201
[59] Stefanucci G and Almbladh C O 2004 *Phys. Rev. B* **69** 195318–195334
[60] Kurth S, Stefanucci G, Almbladh C O, Rubio A and Gross E K U 2005 *Phys. Rev. B* **72** 035308
[61] Bushong N, Sai N and Di Ventra M 2005 *Nano. Lett.* **5** 2569
[62] Maciejko J, Wang J and Guo H 2006 *Phys. Rev. B* **74** 085324
[63] Hou D, He Y, Liu X, Kang J, Chen J and Han R 2006 *Physica E* **31** 191
[64] Sai N, Bushong N, Hatcher R and Ventra M D 2007 *Phys. Rev. B* **75** 115410 (pages 8)
[65] Prociuk A and Dunietz B D 2008 *Phys. Rev. B* **78** 165112
[66] Stefanucci G, Kurth S, Rubio A and Gross E K U 2008 *Phys. Rev. B* **77** 075339
[67] Kouwenhoven L P, Jauhar S, McCormick K, Dixon D, McEuen P L, Nazarov Y V, van der Vaart N C and Foxon C T 1994 *Phys. Rev. B* **50** 2019–2022
[68] Nowak A M and McCreery R L 2004 *J. Am. Chem. Soc.* **126** 16621–16631
[69] Nowak A M and McCreery R L 2004 *Analytical Chemistry* **76** 1089–1097
[70] Bonifas A P and McCreery R L 2008 *Chemistry of Materials* **20** 3849–3856
[71] Ward D R, Halas N J, Ciszek J W, Tour J M, Wu Y, Nordlander P and Natelson D 2008 *Nano. Lett.* **8** 919–924
[72] Ward D R, Scott G D, Keane Z K, Halas N J and Natelson D 2008 *J. Phys.: Condens. Matter* **20** 374118
[73] Stoof T H and Nazarov Y V 1996 *Phys. Rev. B* **53** 1050–1053
[74] Stafford C A and Wingreen N S 1996 *Phys. Rev. Lett.* **76** 1916–1919
[75] Brune P, Bruder C and Schoeller H 1997 *Phys. Rev. B* **56** 4730–4736
[76] Pedersen M H and Büttiker M 1998 *Phys. Rev. B* **58** 12993–13006
[77] Wang B, Wang J and Guo H 1999 *Phys. Rev. Lett.* **82** 398–401
[78] Moskalets M and Büttiker M 2004 *Phys. Rev. B* **69** 205316
[79] Arrachea L 2005 *Phys. Rev. B* **72** 121306
[80] Wu B H and Cao J C 2006 *Phys. Rev. B* **73** 205318
[81] Agarwal A and Sen D 2007 *Physical Review B (Condensed Matter and Materials Physics)* **76** 235316 (pages 7)
[82] Arrachea L, Naon C and Salvay M 2007 *Phys. Rev. B* **76** 165401
[83] Baer R and Neuhauser D 2003 *Int. J. Quant. Chem.* **91** 524–532
[84] Baer R, Seidman T, Ilani S and Neuhauser D 2004 *J. Chem. Phys.* **120** 3387–3396
[85] Baer R and Neuhauser D 2004 *J. Chem. Phys.* **121** 9803–9807
[86] Burke K, Car R and Gebauer R 2005 *Phys. Rev. Lett.* **94** 146803
[87] Neuhauser D and Baer R 2005 *J. Chem. Phys.* **123** 204105
[88] Qian X, Ju Li J, Lin X and Yip S 2006 *Phys. Rev. B* **73** 035408
[89] Zheng X, Wang F, Yam C Y, Mo Y and Chen G 2007 *Phys. Rev. B* **75** 195127
[90] Gurvitz S A and Prager Y S 1996 *Phys. Rev. B* **53** 15932–15943
[91] Lehmann J, Kohler S, Hänggi P and Nitzan A 2002 *Phys. Rev. Lett.* **88** 228305
[92] Lehmann J, Kohler S, Hanggi P and Nitzan A 2003 *The Journal of Chemical Physics* **118** 3283–3293
[93] Torres L E F F 2005 *Phys. Rev. B* **72** 245339
[94] Harbola U, Esposito M and Mukamel S 2006 *Phys. Rev. B* **74** 235309
[95] Moskalets M and Büttiker M 2002 *Phys. Rev. B* **66** 205320
[96] Moskalets M and Büttiker M 2004 *Phys. Rev. B* **70** 245305
[97] Jauho A P, Wingreen N S and Meir Y 1994 *Phys. Rev. B* **50** 5528–5544
[98] You J Q, Lam C H and Zheng H Z 2000 *Phys. Rev. B* **62** 1978–1983
[99] Zhu Y, Maciejko J, Ji T, Guo H and Wang J 2005 *Phys. Rev. B* **71** 075317 (pages 10)
[100] Arrachea L and Moskalets M 2006 *Phys. Rev. B* **74** 245322 (pages 13)
[101] Arrachea L, Yeyati A L and Martin-Rodero A 2008 *Physical Review B (Condensed Matter and Materials Physics)* **77** 165326 (pages 8)

[102] Baym G 1961 *Physical Review* **124** 287
[103] Baym G 1962 *Physical Review* **127** 1391
[104] Kadanoff L P and Baym G 1962 *Quantum Statistical Mechanics* (New York: Benjamin and Cummings)
[105] Langreth D C 1976 *Linear and Non-linear Electron Transport in Solids* (New-York: Plenum Press)
[106] Bonitz M 1998 *Quantum Kinetic Theory* (Stuttgart: Teubner)
[107] Cini M 1980 *Phys. Rev. B* **22** 5887
[108] Fetter A L and Walecka J D 2003 *Quantum Statistical Mechanics* (Dover)
[109] Prociuk A, Van Kuiken B and Dunietz B D 2006 *J. Chem. Phys.* **125** 204717

Journal of Physics: Conference Series **220** (2010) 012009 doi:10.1088/1742-6596/220/1/012009

Dopants and Roughness induced Resonances in thin Si Nanowire transistors: A self-consistent NEGF-Poisson study

A. Martinez, J. R. Barker, N. Seoane*, A. R. Brown and A. Asenov

Dept. Electronics & Electrical Eng., University of Glasgow,

Glasgow, G12 8LT

United Kingdom

*Dept. Electronics and Computing Science

Univ. Santiago de Compostela, 15782

Spain

antonio@elec.gla.ac.uk

Abstract. Non-Equilibrium Green Function simulations of the effect of discrete ionised dopants and surface roughness in Silicon nanowire transistors show the strong presence of resonances in the transmission coefficients. These resonances or quasi-bound states are the main component in the screening of dopants and play an important role in the current flow. Resonances appear through a self-consistent calculation of the electron density and potential. In this work we study several examples that exhibit different types of resonances. We start with a single impurity case and gradually evolved to a complex case with several random impurities. Interface roughness in a very narrow nanowire could induce resonant cavities as is proven in this paper. The effect of these resonances in the current-voltage characteristic of the transistors is considered in detail.

1. Introduction

Present day bulk Si MOSFET technology is very close to exhausting its possibilities for further device miniaturisation. Silicon nanowire transistors are considered as one of the potential candidates in future Integrated Circuit technologies. The Gate-All-Around (GAA) silicon nanowire transistor architecture exhibits higher mobility and better electrostatic control than the corresponding bulk MOSFET architecture. Two of the most important factors that affect small transistor behaviour, or the electron propagation, are the intrinsic discreteness of the dopant atoms and the roughness of the Si/SiO_2 interface.

Device simulation consists of the self-consistent solution of two equations: firstly, the Poisson equation, which is used to compute the electrostatic potential from the fixed and mobile charge, and through the use of adequate boundary conditions; the other equation, which can be called a transport model, allows for the calculation of the electron and current density. There are different types of transport model, which differ in their degree of sophistication. The drift-diffusion model is the simplest and most widely used in the electronics community. It treats the electron as a charged fluid, which drifts and diffuses. The drift is caused by the local electric field and the diffusion is due to the fact that the electrons have a random kinetic energy of the order of k_BT. This model successfully catches the field-charge interaction and the effect of the temperature. Parameters such as electron mobility and density of states are necessary in this model. The MOSFET transistor is an electrostatic beast, which means that the electric field controls most of the transistor behaviour and induces the current flow. The interplay between the charge and the field, through trying to minimize the electrostatic energy and concentration gradients, determines the device characteristics. Fully functional transistors with body thickness as small as 0.7 nm have been fabricated [1]. Despite the strong confinement induced by the narrow channel, electrons are pushed through the channel by the powerful electric field.

Refined transport models, which treat electrons as Newtonian balls with renormalized masses due to the electron interaction with the perfect crystal, are also deployed. The collision with imperfections and vibrations of the crystal are included through the abrupt change of momentum and energy of the electron after a certain drift time. The drift time is calculated by quantum transition probabilities using the first Born approximation. This model allows for a spectral representation of the density, which at highly inhomogeneous electric fields, deviates from the Maxwellian or Fermi-Dirac distribution. The physical parameters of this model are effective electron masses or energy dispersions, electron-phonon coupling constants, sound velocity, elastic constant, etc., which are more fundamental than the parameters used in drift-diffusion. Average physical parameters such as electric field dependent mobility can be calculated from this model, if low field mobility is known.

Both models have been corrected to accommodate the quantum effects associated with the reduction of the transistor dimensions. The most successful of these corrections is the density gradient. This accounts for a repulsion energy (not associated to electric repulsion) associated with large concentration gradients or a strong localization of electrons, which pushes electrons away from the interfaces and mimics a ground state distribution of electrons in the wells. The origin of this force or potential comes from a striking similarity between the Hamilton-Jacobi equation and the Schrödinger equation for the phase of the wave function. The equation for the phase contains an additional potential, which varies as $\nabla^2\left(\sqrt{n}\right)/\sqrt{n}$, where n is the electron density. This potential is considered as a quantum potential that corrects the classical potential in the transport equation. Using the drift-diffusion model and the density gradient correction, it is possible to simulate the volume inversion i.e. when the electron density becomes larger in the middle of the channel as compared with that at the interface, which is really a property of the ground state. This effect occurs at body thicknesses below 20 nm [2]. Gate capacitances become smaller due to volume inversion and quantum confinement in the triangular well, induced by the band bending at the Si/SiO_2 interface creating the need for a larger gate voltage in order to achieve the same inversion charge.

The intrinsic discreteness of the impurities in the device needs to be considered as the size of the device reduces. Technologically identical devices, which differ in the microscopic spatial configuration of dopants and the number of dopants, present different behaviours due to the non-self-averaging inherent in small structures and the small number of dopants. Capturing of these effects is essential for an accurate description of the current flows in devices.

In this work we present a study of the influences of different types of spatial inhomogeneities such as discrete dopants/surface roughness on the electron transport through a thin Si nanowire transistor [3, 4]. The shape of the self-consistent electrostatic potential around an ionised donor impurity induces resonant states. These states shape the transmission function and are an important element in the propagation of the electron through the nanostructure. Surface roughness can, under certain conditions, also induce resonant levels. Studies of the effect of dopants on the device characteristics in a statistical ensemble of devices, which differ in the particular realisations of disorder, will be presented. In this work we have deployed a full 3D Non-Equilibrium Green Function (NEGF) formalism coupled with the Poisson equation. The NEGF equations are written in the effective mass approximation and use Hartree's approximation for the interacting electron system. The rest of the paper is divided into three parts. Section 2 provides an introduction to the use of NEGF describing the coupling of the NEGF transport equation and the Poisson equation; a brief description of the NEGF-Poisson algorithm is presented. Section 3 describes models of different sources of inhomogeneous potential, such as those created by the roughness of the SiO_2/Si interface and the discreteness of the dopant atoms. We discuss the role of these inhomogeneities in the creation of resonance states that control electron flow. Furthermore, we computed the impact of these resonances on transistor performance. The final section presents the conclusions.

2. Non-Equilibrium Green's Function Approach to Device Simulation

The quantum carrier transport is described using the NEGF approach, which is a generalisation of Landauer's formalism [5, 6, 7, 8] to treat many body systems at room temperature in the context of the one particle Green's function. The Hamiltonian used in the discretisation of the NEGF equations is an effective-mass Hamiltonian that folds the full crystal interaction into the electron effective masses:

$$H = -\frac{\hbar^2}{2}\nabla_i\left(\frac{1}{m}\right)_{ij}\nabla_j + V(\vec{r}) \tag{1}$$

Where $(1/m)_{ij}$ is the inverse of the mass tensor which contains the effective masses and V(r) is the self-consistent electrostatic potential coming from the solutions of Poisson's equations.

The effective masses of the transport valleys are extracted from Tight Binding calculations that capture the dependence of the electron band structure on the nanowire diameter [9]. Due to the small dimensions of the wire cross-section we only include four of the six lowest valleys of the silicon conduction band. The two valleys that have been neglected have transversal masses $0.3m_0$ in the directions perpendicular to the transport direction, resulting in large ground-state energy shifts associated with the transversal confinement. As a result the electron population of these valleys is negligible compared to the other four valleys for the simulated nanowire diameter, temperature and bias conditions. Sources of incoherent scattering such as phonon interaction, and the corresponding self-energies, are not included in our NEGF simulations. We calculate the correlation matrix, $G^<$, using the recursive algorithm described in [10, 11, 12]. From the correlation matrix, the electron and current densities are calculated by the following equations:

$$n(E,x) = iG^<(E,x,x) \tag{2}$$

$$J(E,x) = -i\frac{e\hbar}{2m}(\nabla - \nabla')G^{<}(E,x,x')\Big|_{x=x'} \quad (3)$$

The boundary conditions of the Green's function equations at the contacts, which are given through the contact self-energies, are defined using the algorithm described in [13].

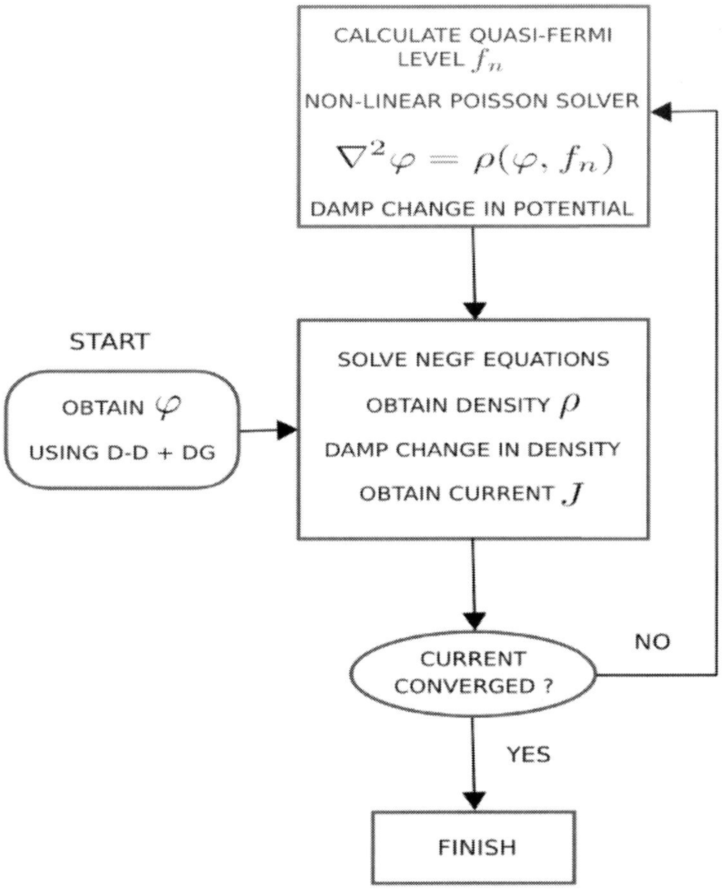

Figure 1. Flow chart of the 3D Poisson-NEGF solution methodology.

Fig. 1 shows the flow chart of our simulator illustrating the computational procedure used to solve the coupled Poisson-NEGF equations. The electrostatic potential and the electron density obtained from a density gradient (DG) solution of the Drift-Diffusion (DD) equations [14] serve as an initial condition for the Poisson-NEGF cycle. The DD solver has Neumann boundary conditions for Poisson's equation in the source and drain instead of the Dirichlet boundary conditions usually used in the DD formalism [15, 16], which matches well with the Green's function boundary conditions. A close initialisation of the potential distribution at the beginning of the Poisson-NEGF loop reduces drastically the number of NEGF iterations.

After the first Poisson-NEGF iteration the change in electron concentration from the initial DG solution to the new NEGF solution is moderated by damping. A gradual change in the electron density prevents oscillatory behaviour of the solution around the impurities leading to divergence. The solution instabilities are associated with the extreme sensitivity of the quantum density to the shape of the attractive potential. This is related to the discrete nature of the quasi-bound states and their energy sensitivity to the shape of the potential.

We have found that solving the non-linear Poisson equation results in a much more stable convergence of the Poisson-NEGF system than if the linear version is used. Once a new electron density is obtained from the NEGF solver, a quasi-Fermi level, f_n, is calculated using the new density and the old potential [17]. This quasi-Fermi level is used to update the electron concentration and the Jacobian when solving the non-linear Poisson equation iteratively (see Fig. 1). Adaptive damping is used after the solution of the Poisson equation to limit the change in potential and to improve convergence. The alternate solutions of Poisson and NEGF are iterated until density and current converge.

3. Modelling of Surface Roughness and Discrete Dopants

The random discrete dopants have been introduced in a 4 nm region of the source/drain (S/D) leads between the channel and the continuously doped S/D regions next to the contacts. Each Si lattice site in these regions is considered and whether this site has a dopant or not is determined using a rejection technique based on the ratio between the nominal doping concentration and the silicon atom density at that point [18]. The resulting total number of dopants in the discrete dopant regions closely follows a Poisson distribution. The charge of each dopant is distributed to the surrounding nodes of the discretisation mesh using the cloud-in-cell technique. The region with continuous doping between the discretely doped regions and the contacts guarantees a homogeneous injection into the source/drain from the reservoirs. The random rough interface between Si and SiO_2 is introduced using the approach described in [19]. The interface is modelled by a randomly generated surface with exponential autocorrelation function and specified rms amplitude and correlation length selected based on results reported in [20]. The correlation length of the order of 2 nm is similar to the one typically used in Monte Carlo simulations used to reproduce the universal mobility curve [21]. The generated 'analogue' random surface is quantised to two levels on the scale of one inter-atomic layer.

The simulated Si NWTs have a 6 nm undoped channel with 2.2x2.2 nm^2 cross-section, 0.8 nm SiO_2 oxide and 10 nm S/D regions doped at 10^{20} cm^{-3}. The transport in the nanowire occurs in the <100> direction. The diameter-dependent effective masses are extracted from sp^3d^5 second-neighbour-basis tight-binding calculations [9]. All the simulations in this work have been done at room temperature.

3.1. Discrete Dopants

In the following subsections we present and discuss the simulation of discrete dopants. Firstly we analyzed the case of a donor in the middle of a doped silicon wire. This example illustrates the relationship between the shape of the potential and the type of resonance/anti-resonance. These resonances will be present in all the cases studied. The second example concentrates on simulations of the effect on the Si gate-all-around nanowire transistor of random discrete donors in the source and drain. This example has practical applications, as random dopant variability is an important issue in the reliability of digital circuits.

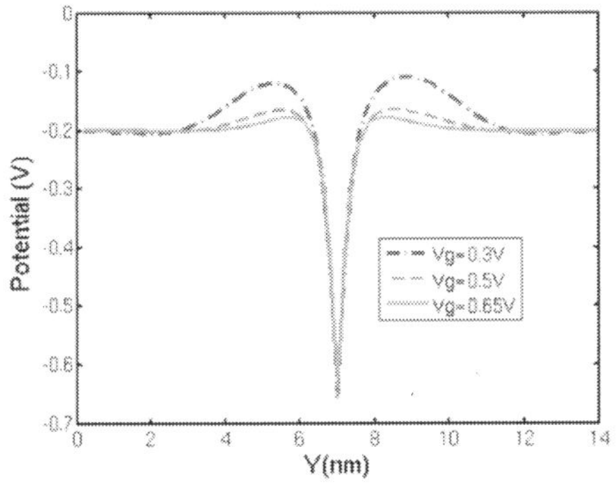

Figure 2. Inverted sombrero type potential with barriers across the middle of the nanowire at indicated gate biases.

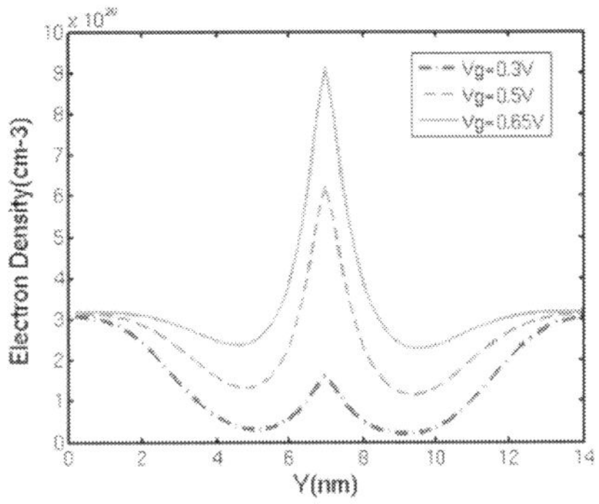

Figure 3. The electron density for the inverted sombrero type potential across the middle of the nanowire at the same gate biases.

3.1.1. *A donor in the middle of the Si wire*.

We introduced a donor like impurity in the middle of a doped nanowire. The impurity is modelled as a charge distributed in the cubic volume of 0.2nm × 0.2nm × 0.2nm which is the range of the dielectric screening length [22]. A tuning of the gate potential can partially control the shape of the impurity potential. This allows us to study the effects of the shape of the impurity potential on the electron transmission. The gate potential introduces a barrier around the impurity that partially repels the electrons from the channel. As the gate potential increases, the electrons are attracted to the channel and their presence screens the potential of the impurity. The density of electrons and its shape around the impurity depends on the self-consistent potential and the nature of the quantum states around it.

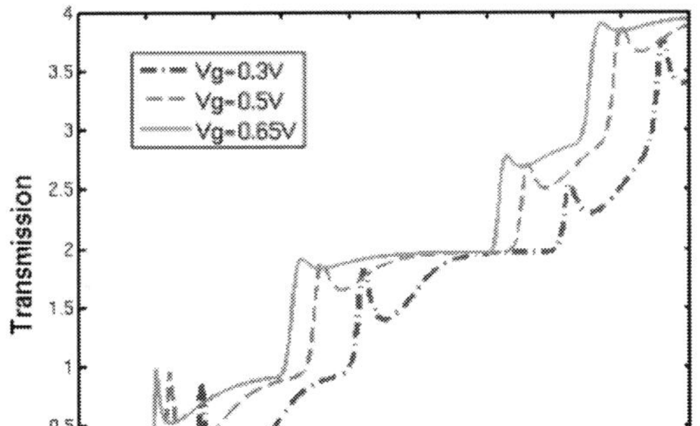

Figure 4. Transmission coefficient from the inverted sombrero type potential along the nanowire from the source to the drain, again at the same gate biases.

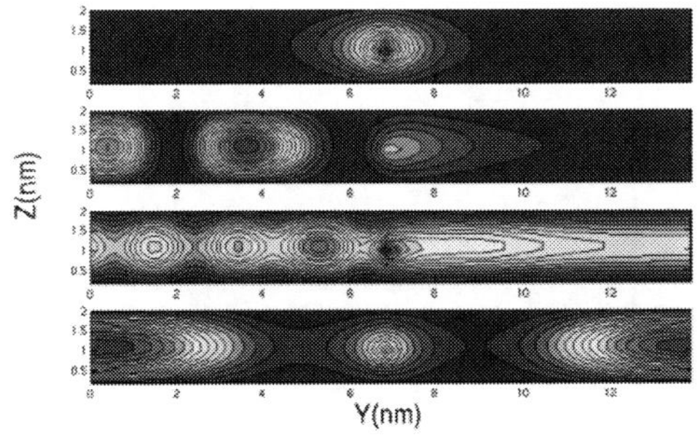

Figure 5. Electron density at different energies along the nanowire. The upper most panel shows the resonance state resulting in a strong electron localization.

That has to account also for the polarization including the image charge induced at the dielectric/semiconductor interface as well as at the dielectric/metal interface. An impurity placed in the middle of the channel will create a deep potential surrounded by potential barriers, the so-called inverted sombrero type potential (also a double Gaussian type potential [23] since it can be constructed from two Gaussians). Fig. 2 shows that this potential initially exhibits wide barriers around the dip at an applied gate bias of 0.3 V. As the potential applied to the metal gate increases, the barriers are lowering and become nearly flat at a gate bias of 0.65 V. Fig. 3 shows the corresponding electron density along the channel of the nanowire at the same gate biases. The corresponding transmission coefficients as a function of the electron energy for this potential are presented in Fig. 4. At Vg=0.3 V, there is a pronounced resonance at an energy of 0.2805 eV

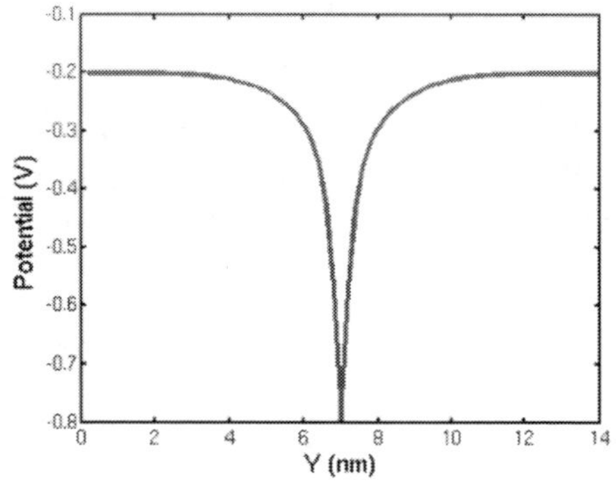

Figure 6. Potential without barrier (A single Gaussian type) across the middle of the nanowire.

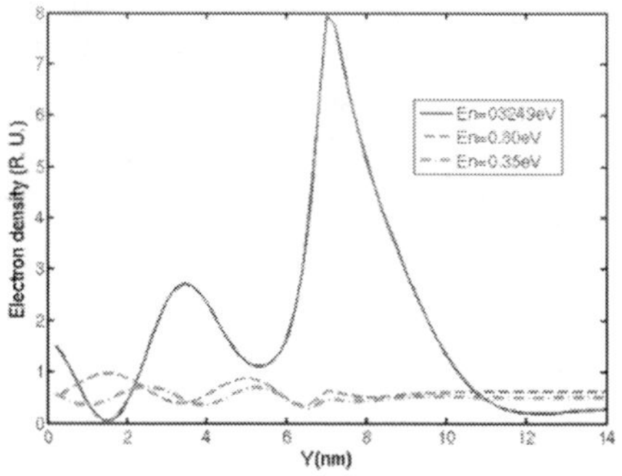

Figure 7. The electron density for the single Gaussian type potential across the middle of the nanowire at the indicated energies.

The transmission at this resonance energy falls to almost zero before the first energy mode of the nanowire starts to appear. The wave function of this resonance mode is strongly localised around the impurity. This can be seen in the upper panels of Fig. 5, which shows the single-electron density at the resonant energy (0.2805eV). The electron wave function is strongly localised at the impurity position with no wave reflection and is fully transmitted. The two middle panels of Fig. 5 show the probability density of electrons with energies 0.35 eV and 0.65 eV corresponding to low and high transmission probability. In contrast to the resonant electron, these electrons have a de-localised wave function.

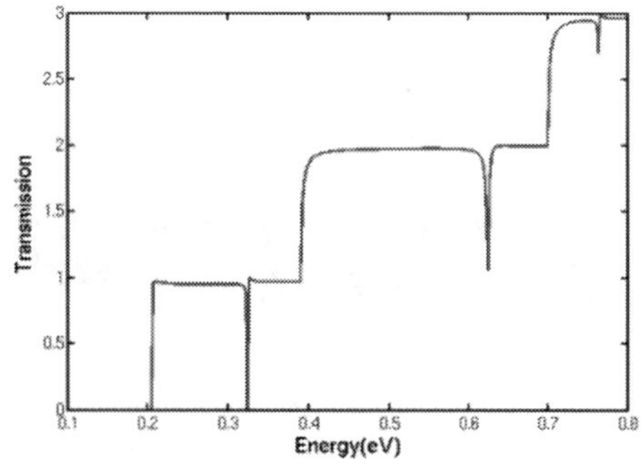

Figure 8. Transmission coefficient from the single Gaussian type potential along the nanowire from the source to the drain, again at an energy of 0.3249 eV. An asymmetric Fano resonance due to a localised level in continuum can be identified.

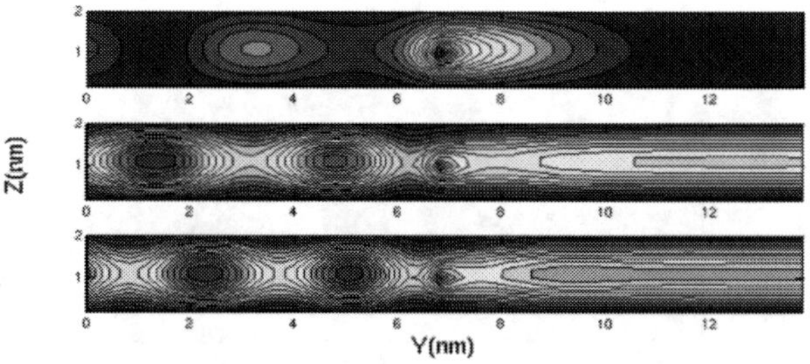

Figure 9. Electron density through the middle of the nanowire. The upper panel shows the resonant state resulting in electron localisation. The middle and lower panels show non-resonant states at which the electron is de-localised.

This de-localised electron wave is partially reflected while the transmission probability increases with the electron energy approaching the next energy mode. The following resonance is less pronounced compared to the first occurrence at the single energy mode leading to much smaller decline in transmission probability.

The situation changes at large gate biases when the shape of the self-consistent potential becomes Coulomb-like (a single Gaussian-type potential) which has no barriers as illustrated in Fig. 6. The electron density will now exhibit a typical asymmetry, as demonstrated in Fig. 7, for indicated energy modes. Fig. 8 shows corresponding transmission with an asymmetric Fano resonance [24]. This resonance appears due to the existence of a localised level in the continuum. Fig. 9 shows the probability density for three different energies. The energy of the electron in the upper panel is close to the Fano resonance and its wave function is strongly localised around the impurity. The Fano resonance state has the largest contribution to the screening of the impurity charge. The electron wave function at Fano resonance will be fully reflected and makes no contribution to the electron transport through the nanowire.

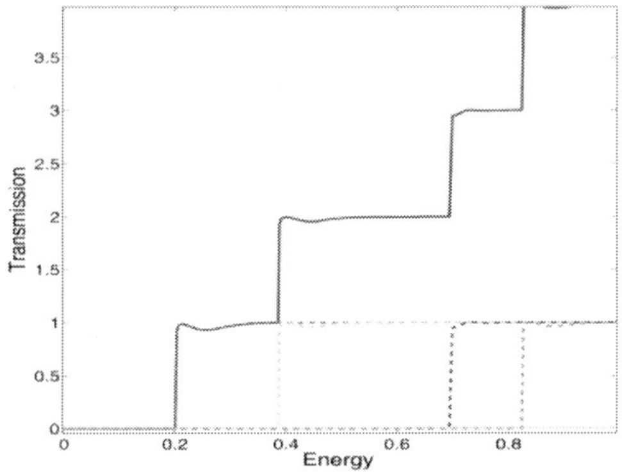

Figure 10. Transmision coefficients for the uncoupled mode case. Full line: the total transmission. Dashed line: transmission of a single mode

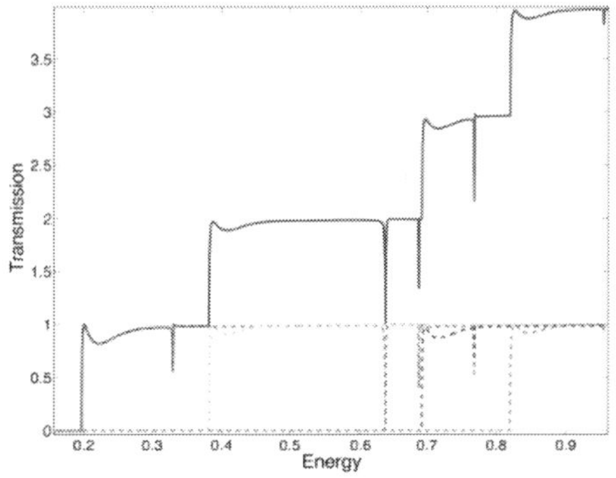

Figure 11. Transmision coefficients for the coupled mode case. Full line: the total transmission. Dashed line: transmission of a single mode

The middle and lower panels of Fig. 9 show non-resonant electron states, which are de-localised thus exhibiting several nodes along the wire. The Fano resonances will always appear at the same amount of energy below the next transversal mode. This observation was already pointed out in [23], which presents indications that the resonances are due to the interaction between the modes. In order to confirm the relation of the Fano resonance with the coupling between the modes with have carried out the simulation in a coupled mode-space approach, which allows the coupling and decoupling of the modes in an undoped wire. The results are presented in Figs. 10 and 11 for the transmission coefficients for the case of the coupled and uncoupled cases respectively. The non-resonant electron states have very small contributions to the screening of the impurity charge but will contribute significantly into the electron transport. Fig. 10 does not exhibit anti-resonances because the modes are not coupled, but in Fig. 11 the anti-resonances are present confirming the hypothesis.

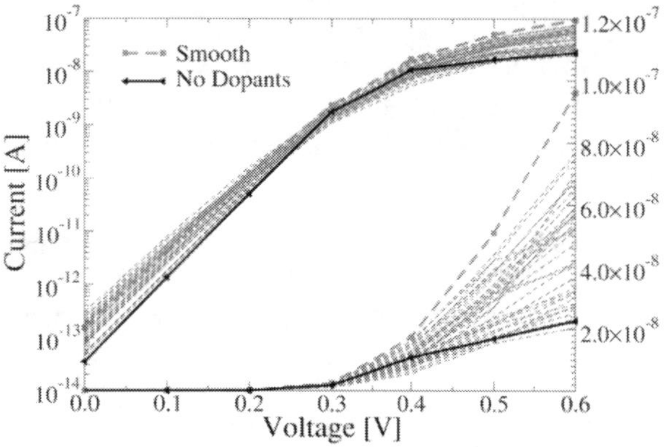

Figure 12 I_D-V_G characteristics of the nanowire transistor with *random dopants* in the source and drain. The full black line is for a device with no dopants in the 'atomistic' regions.

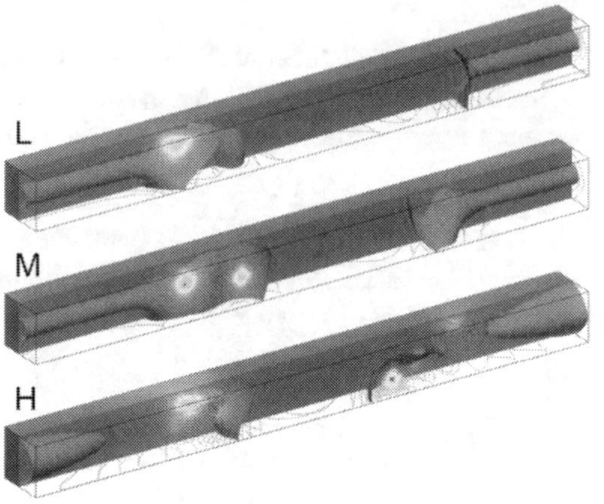

Figure 13. Potential for the *random dopant* devices with lowest, median and highest current at V_G=0.4 V. From top to bottom, the lowest current (L), median current (M) and highest current (H) devices are represented.

3.1.2. *Dopants in the Source and Drain of GAA Si Nanowire Transistor.*

Due to the significant computational burden associated with the 3D NEGF approach, and its slow convergence in the presence of attractive impurity potentials, the statistical simulation study has been restricted to a small statistical sample of 30 randomly generated device configurations. All simulations were carried out at V_D=1 mV in the linear mode of device operation. Fig. 12 shows, on a linear and logarithmic scale, the current-voltage characteristics of the 30 microscopically different NWTs with different random dopant (RD) configurations. The nanowire with continuous doping (labelled as 'smooth' in the figure), and one with no dopants at all in the random dopant regions, are shown for comparison. At V_G<0.3 V, which marks the transition between the sub-threshold and the linear region of the transistor operations, the device configurations with a high concentration of discrete dopants located close to the channel lower the gate barrier potential, leading to a higher current than in the smooth case. At V_G>0.3 V the smooth device always delivers a higher current than in the RD devices. This is mainly associated with the coherent impurity scattering in RD devices, which reduces the current due to partial reflection from the impurity potentials. The degree of backscattering, for a particular RD configuration, depends strongly on the gate voltage. As a consequence, the on-current in 'unlucky' RD devices is reduced to as low as 20% of the current for the smooth device. A particular configuration may have low scattering at low V_G and high scattering at high V_G, or vice versa, relative to other configurations. This sensitivity to the gate voltage is a result of the relative proximity of the discrete dopants to the tail of the channel potential barrier and lead to a crossing of the I_D-V_G curves as can be seen in Fig.12.

There is a variation in the sub-threshold slope of the I_D-V_G curves produced by the different atomistic configurations. Devices with configurations of dopants close to the central axis of the wire, far from the SiO_2 interfaces, will have relatively poor electrostatic control compared with devices with dopants distributed closer to the interfaces and to the channel/source and channel/drain junctions. In addition to the sub-threshold slope variation due purely to electrostatics, which have been observed in drift-diffusion simulations, there is an additional contribution from the varying degrees of source-to-drain tunnelling.

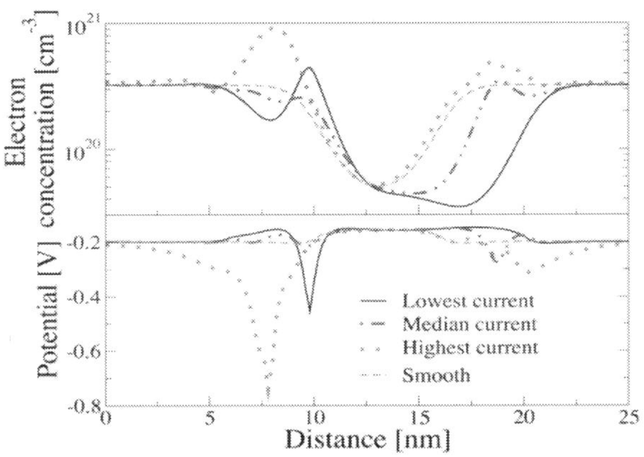

Figure 14. Density and potential along the nanowire at V_G=0.4 V for the smooth and the three *random dopant* devices.

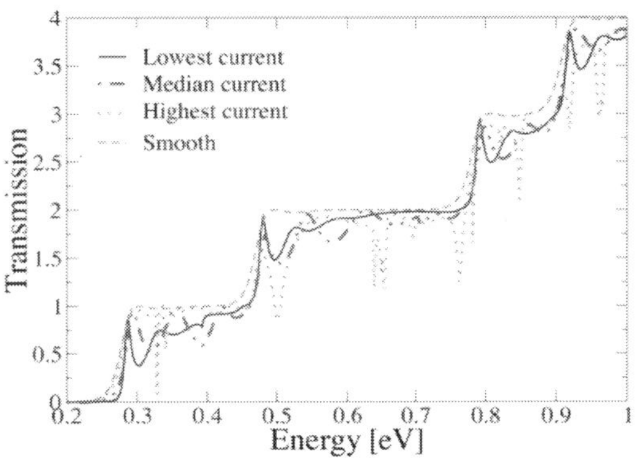

Figure 15. Transmission coefficients at V_G=0.4 V for the smooth and three *random dopant* devices. The early start of the transmission of the highest current device is a consequence of the impurities close to the channel (see Fig. 13).

At V_G=0.4 V, we have investigated the behaviour of the device configurations with: the lowest current (L), the median current (M) and the highest current (H). The corresponding 3D self-consistent electrostatic potential distributions are shown in Fig. 13 (although not all the impurities are evident in the figure). The device with the shortest effective channel length exhibits the largest current. This is clear from Fig. 14, which shows the electron concentration and self-consistent electrostatic potential along the middle of the channel. The device with the largest current has six dopants in the source, producing a region of average dopant concentration much larger than the dopant concentration of the continuously doped smooth device.
The physical effects shaping the impact of the discreteness of the donors on the current-voltage characteristics can be explained qualitatively using the transmission coefficients, which are directly related to the current through the well-known Landauer formula [25],

$$J = \frac{e^2}{h}\int T(\varepsilon)\left(f_S(\varepsilon) - f_D(\varepsilon)\right) d\varepsilon \tag{3}$$

where T is the transmission coefficient and f_S(e)/f_D(e) are the Fermi distributions in the source/drain.

Fig. 15 shows the transmission coefficients for the three RD devices and for the smooth device presented as a reference. For the highest current device the shape of the self-consistent potential around some impurities produces Fano-type resonances as seen in Fig. 14. These result in zero transmission at the resonance energies and can be interpreted as back scattering for the corresponding electrons. The electrons at these resonant energies (or quasi-bound states) are the main source of the screening of the impurity potential. For the other two devices we have transmission resonance peaks due to the inverted sombrero shape of the self-consistent potential of the impurities (see Fig.14). The relative value of the on-current between the L, M and H devices can be understood using the Landauer formula (Eq.3) with the transmission coefficients (Fig. 14) of the L, M and H devices at around 0.3 eV (the Fermi Level in Source is 0.2). Note that the transmission for the smooth device has a stair-type characteristic as expected. Therefore, the irregular shape and dips in the transmission for the devices with discrete impurities can be interpreted as a fraction of carriers being reflected back or back scattered. This feature is not visible at lower gate bias because of the dominant reflection of the carriers by the channel barrier potential

3.2. Surface Roughness

In this section the effect of the interface roughness will be considered. We have assumed rough interface only in the channel and have ignored any roughness between the oxide and the gate material. In order to study the statistical effect associated with the interface roughness the I_D-V_G characteristics of thirty nanowire transistors with different randomly generated interface roughness patterns were simulated and analysed. The collection of individual I_D-V_G characteristics of the devices with different surface roughness configurations is shown in Fig. 16. The devices with a smooth interface, and with the smallest possible cross-section (all the interface roughness steps are in the direction that decreases the body thickness of the channel) are also presented in Fig. 16 for comparison.

The three surface roughness cases selected for analysis produce the lowest, median and the second highest current at all gate biases. The device that produces the highest off-current has a distinct surface roughness pattern and will be analyzed later. At 2.2 nm body thickness even a small intrusion of the interface into the channel is sufficient to raise the ground state of the transverse wave function resulting in the step increases in transmission occurring at higher energies. This effect, together with the roughness-induced scattering, means that the current of the rough devices are generally lower compared to that of the smooth device. It is possible to have a rough device with a higher current than that of the smooth device if there is no overall narrowing of the channel, but the probability for this is very low. The body thickness confinement variation causes a large spread in threshold voltage which (for the 30 devices simulated here) spans 124 mV. In the extreme random roughness case there is a decrease in the on-current by more than 25% compared to that of the smooth device if the threshold voltages are aligned.

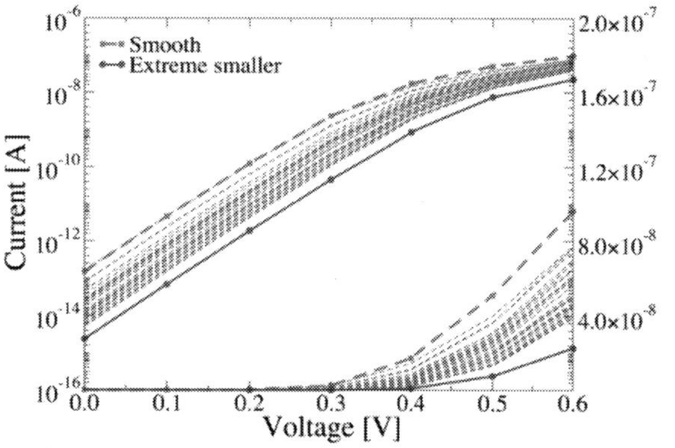

Figure 16. I_D-V_G characteristics of the nanowire transistor with *surface roughness* in the channel. The solid line represents the device with the smallest possible cross-section (2.0×2.0 nm^2) where all the interface roughness steps are in the direction that decreases the body thickness of the channel. The smooth device is also shown for comparison.

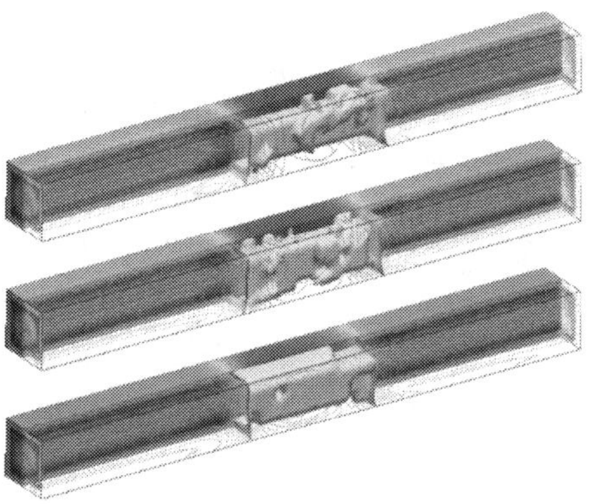

Figure 17. Potential for the low, median and high current (from top to bottom) of the *surface roughness* devices at V_G=0.4 V.

The 3D distribution of the self-consistent electrostatic potential for the three selected devices at high gate (V_G=0.4V) are shown in Fig. 17 with the lowest current device at the top and the highest current device at the bottom. The impact of the reductions in channel body thickness on the potential distribution can be seen in the figure. This together with the corresponding up-shift in the ground-state sub-band resulting from the increased confinement leads to a decrease in the electron current.

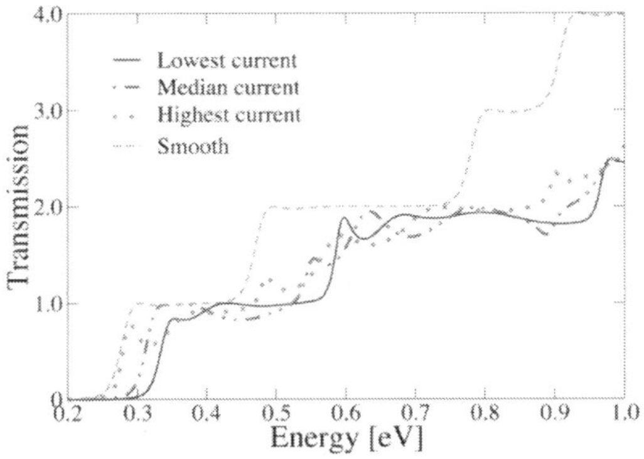

Figure 18. Transmission coefficients at V_G=0.4 V for the smooth and three *surface roughness* devices.

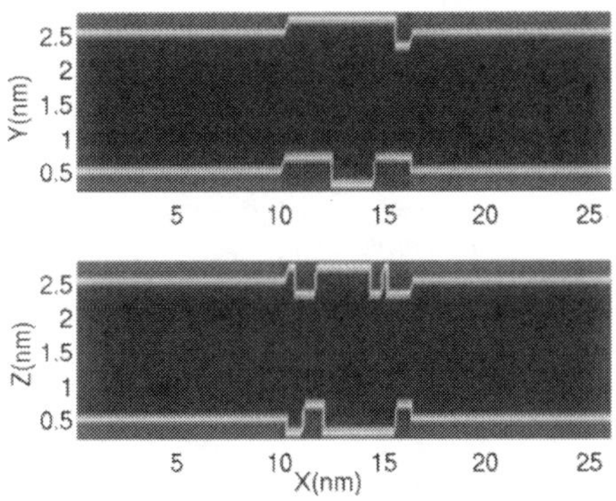

Figure 19. Surface roughness in two perpendicular planes along the channel length for a device in which the surface roughness pattern produces a cavity in the channel.

The transmission coefficients at high gate bias are compared in Fig. 18. It is possible for the surface roughness to result in specific arrangements of body-thickness variations that resemble a resonant structure. The impact of such resonance is evident in the second sub-band of the lowest current device in Fig.18. As mentioned previously, surface roughness can also induce quasi-bound states. To illustrate the point we have selected a device in which the particular surface roughness pattern produces a cavity in the channel. This device has a current slightly larger than the smooth device at low gate bias. Fig. 19 shows the surface roughness in two perpendicular planes along the channel length. There is a well-defined cavity between 10 and 15 nm. Fig. 20 shows the transmission of this device and the smooth device. The transmission coefficient of the rough device shows resonances that are clearly Breit-Wigner-type for the first two sub-bands. The first resonance appears before the onset of the smooth device transmission, implying a large current in this device compared to the smooth one. Another manifestation of the resonances is the two maxima in the middle of the channel in the Local Density of States plotted in Fig. 21. The first sub-band energy is also plotted with a dashed line in the same figure in order to clarify the position of the longitudinal resonant cavity.

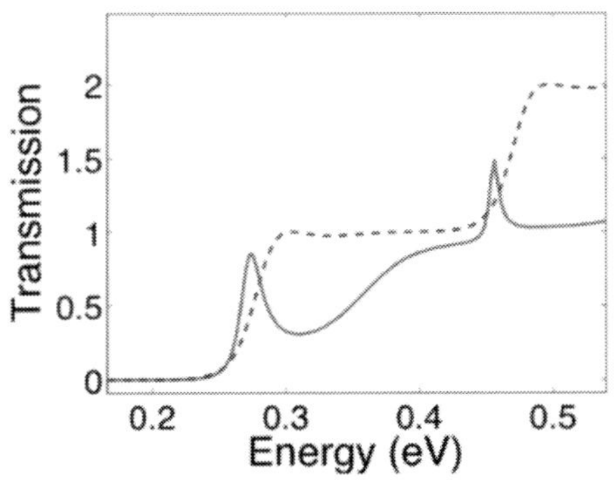

Figure 20. Transmission coefficients for the *surface roughness* device depicted in Fig.18 (full line) and for the smooth device (dashed line).

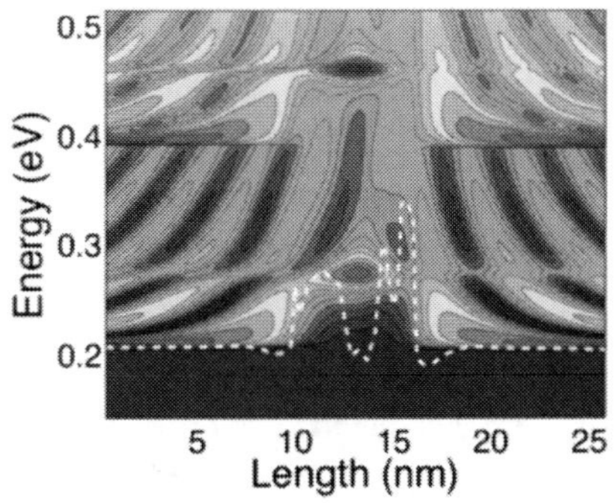

Figure 21. Local density of states as a function of the channel length for the *surface roughness* device represented in Fig.18.

4. Conclusions

A statistical study of variability due to random discrete dopants at the source/drain and interface roughness in a Si nanowire transistor has been carried out using a full 3D NEGF formalism. The role of resonances in electron transport and transistor performance has been widely studied. The relationship between the shape of the potential and the type of resonances has also been investigated in simple and more complex cases. The transmission coefficients have been used to correlate the spatial inhomogeneities of electrostatic potential with device behaviour. In this small transistor we have proven that the self-average of microscopic roughness in the interface and the discreteness of dopants does not occur and a continuum model will fail to describe the relevant physics involved.

References

[1] Uchida, K., Koga, J., Takagi, S.: Experimental Study on Carrier Transport Mechanisms in Double- and Single-Gate Ultrathin-Body MOSFETs — Coulomb Scattering, Volume Inversion, and δTSOI -induced Scattering. In: *IEDM Techn. Dig.* 805–808 2003
[2] Gámiz F and Fischetti M. V. Monte Carlo simulation of double-gate silicon-on-insulator inversion layers: the role of volume inversion *J. Appl. Phys.* **89** 5478 2005
[3] Martinez A, Seoane N Brown A R, Barker J R and Asenov A," 3-D Non-Equilibrium Green's Function Simulation of Non-Perturbative Scattering from Discrete Dopants in the Source and Drain of a Silicon Nanowire Transistor, accepted in *IEEE Trans. Nanotechnol* to be published (2009)
[4] Seoane N, Martinez A, Brown A R, Barker J R and Asenov A, Current Variability in Si Nanowire MOSFETs due to Random Dopants in the Source/Drain Regions: A Fully-3D NEGF Simulation Study, to be published *IEEE-Trans. Electron Dev.* (2009)
[5] Landauer R, "Electrical resistance of disordered one-dimensional lattices", *Phil. Mag.*, **21**, p.863, 1970.
[6] Buttiker M, "Symmetry of electrical conduction", *IBM J. Res. Dev.*, Vol.32, p.317, 1988.
[7] Engquist H L. and Anderson P L, "Definition and measurements of the electrical and thermal resistances", *Phys. Rev. B*, Vol.24, p.1151, 1981.
[8] Imry Y and Landauer R, "Conductance viewed as transmission", *Rev. Mod. Phys*, **71** 306, 1999.
[9] Nehari K, Cavassilas N, Autran J L, Bescond M, Munteanu D and Lannoo M, "Influence of band structure on electron ballistic transport in silicon nanowire MOSFET's: an atomistic study", *Solid-State Electronics*, Vol. 50, p.716, 2006.
[10] Lake, R , Klimeck G Bowen R G and Jovanovic D, "Single and multiband modeling of quantum electron transport through layered semiconductor devices", *J. App. Phys.*, **81**, 7845, 1997.
[11] Svizhenko A, Anantram M P, Govindan T R, Biegel B and Venugopal R, "Two dimensional quantum mechanical modelling of nanotransistors", *J. Apps. Phys.*, **91**, 2343, 2002.
[12] Anantram M P, Svizhenko A and Martinez A, "Erratum: Two-dimensional quantum mechanical modelling of nanotransistors", *J. App. Phys.*, **100**, 119903, 2006.
[13] Venugopal R, Ren Z, Datta S, Jovanovic D and Lundstrom M, "Simulating quantum transport in nanoscale transistors: Real versus mode-space approaches", *J. Appl. Phys.*, **92**, p.3730, 2002.
[14] Asenov A, Slavcheva G, Brown A R, Davies J H and Saini S, "Increase in the Random Dopant Induced Threshold Fluctuations and Lowering in Sub-100 nm MOSFETs Due to Quantum Effects: A 3-D Density-Gradient Simulation Study", *IEEE Trans. Electron Dev.*, **48**, 722-729, 2001.
[15] Brown A R, Martinez A, Bescond M and Asenov A, "Nanowire MOSFET variability: a 3D density gradient versus NEGF approach", *Silicon Nanoelectronics Workshop, 10-11 June, Kyoto, Japan*, 127-128, 2007.
[16] Riddet C, Brown A R, Roy S and Asenov A, "Boundary Conditions for Density Gradient Corrections in 3D Monte Carlo Simulations", *Journal of Computational Electronics*, **7**, 231, 2008.
[17] Anantram M P and Svizhenko A, "Multidimensional modelling of Nanotransistors", *IEEE*

Trans. Electron Devices, **54**, 2100-, 2007.

[18] Frank D J, Taur Y, Ieong M and Wong H S P, "Monte Carlo modeling of threshold variation due to dopant fluctuations", Symposium on VLSI Technology Dig. Techn. Papers, 169-170, 1999.

[19] Martinez A, Bescond M, Barker J R, Svizhenko A, Anantram M P, Millar C and Asenov A, "A self-consistent 3D-Full Real Space NEGF Simulator for studying Nonperturbative effect in Nano-MOSFETs", *IEEE Trans. Electron Devices*, **54,** 2213-2222, 2007.

[20] Goodnick A, Ferry D K, Wilmsen C W, Lilienthal A, D. Fathy and Krivanek O L, "Surface roughness at the Si (100)-SiO_2 interface", *Phys. Rev. B., Condens. Matter* **32** 8171-8186, 1985.

[21] Takagi S, Toriumi A, Iwase M and Tango H, "On the universality of inversion layer mobility in Si MOSFET's: Part I-effects of substrate impurity concentration", *IEEE Trans. Electron Devices*, **41**, 2357-236, 1994.

[22] Lannoo M and Bourgoin J, "Point defects in semiconductors", Springer Series in Solid State Sciences, vol. 35, Springer-Verlag, Heilderberg, 1983.

[23] Bardarson J H, Magnusdottir I, Gudmundsdottir G, Tang C S, Manolescu A and Gudmundsson G, "Coherent electronic transport in a multimode quantum channel with Gaussian-type scatterers", Phys. Rev. B. **70** 245308 2004

[24] Vargiamidis V and Polatoglou H M, "Temperature dependence of the Fano effect in quantum wires with short- and finite-range impurities", Phys. Rev B **75** 153308 2007

[25] Datta S, "Quantum Transport, Atom to transistor", *Cambridge University Press*, 2005.

Journal of Physics: Conference Series **220** (2010) 012010 | doi:10.1088/1742-6596/220/1/012010

Ab initio vibrations in nonequilibrium nanowires

A P Jauho[1,2], M Engelund[1], T Markussen[1], and M Brandbyge[1]

[1] Dept. of Micro og Nanotechnology, DTU Nanotech, Technical University of Denmark, Building 345east, 2800 Kongens Lyngby, Denmark

[2] Aalto University, Department of Applied Physics, P.O. Box 11100, FI-00076 AALTO, Finland

E-mail: Antti-Pekka.Jauho@nanotech.dtu.dk

E-mail: Mads.Engelund@nanotech.dtu.dk

E-mail: Troels.Markussen@nanotech.dtu.dk

E-mail: Mads.Brandbyge@nanotech.dtu.dk

Abstract. We review recent results on electronic and thermal transport in two different quasi one-dimensional systems: Silicon nanowires (SiNW) and atomic gold chains. For SiNW's we compute the ballistic electronic and thermal transport properties on equal footing, allowing us to make quantitative predictions for the thermoelectric properties, while for the atomic gold chains we evaluate microscopically the damping of the vibrations, due to the coupling of the chain atoms to the modes in the bulk contacts. Both approaches are based on the combination of density-functional theory, and nonequilibrium Green's functions.

1. Introduction

Modeling charge transport in nanostructures under nonequilibrium conditions via *ab initio* methods is nowadays a well-established computational approach [1]. Electronic transport, however is intimately coupled to the motion of the ions comprising the nanostructure: the charges may excite ionic motion which leads to local heating and energy relaxation. On the other hand, the ionic motion itself is an important channel for heat transport between the electrodes. The vibrations within the nanostructure couple to the motion of the atoms in the bulk contacts, thus becoming damped or, equivalently, they obtain a finite life-time. The inelastic transport properties of nanostructures is a very active research field, where progress occurs on many fronts. In this paper we review two particular subfields in this general area. (i) We compute the ballistic electronic and thermal transport properties of Si-nanowires, paying specific attention to intentional nanostructuring. The main goal is to examine whether the thermoelectric properties can be enhanced via nanoengineering. Treating electronic and vibrational properties on equal footing (based on nonequilibrium Green's functions) is a key element in our work. Presently it is not computationally feasible to consider inelastic scattering in these systems which typically consist of thousands of atoms. The important issue is thus to find a good compromise between microscopic accuracy, and the macroscopic properties of real, nanowire based devices. (ii) In smaller systems, such as the experimentally studied atomic gold wires, a fully microscopic approach has become tractable. The vibrational motion excited in the atomic chains due to electron transport is damped because the localized vibrations couple to the vibrations in the electrodes surrounding the gold chain. It is commonplace to describe this damping by phenomenological life-times; here, however, we describe a recently developed *ab initio* method to actually compute these life-times. Experimentally, a very rich behavior as a function of the elongation of the wire is observed, and the extreme sensitivity to stress, as revealed by our calculations, may form an important ingredient in explaining this behavior.

2. Ab initio modeling of electronic and thermal transport in Silicon nanowires (SiNW)

2.1. Background remarks

Silicon nanowires (SiNW) are presently intensively investigated because of there many potential applications in future nanoelectronics (for a review, see Rurali [2]). Recent ground-breaking experiments have indicated yet another application: *rough* SiNWs can be efficient thermoelectric materials although bulk silicon is not [3, 4]: they conduct charge well but have a low heat conductivity. A measure for the performance is given by the figure of merit $ZT = G_e S^2 T/\kappa$, where G_e, S, and T, are the electrical conductance, Seebeck coefficient, and (average) temperature, respectively. The heat conductivity has both electronic and phononic contributions, $\kappa = \kappa_e + \kappa_{ph}$. Materials with $ZT \sim 1$ are regarded as good thermoelectrics, but $ZT > 3$ is required to compete with conventional refrigerators or generators [5]. Recent theoretical works predict $ZT > 3$ in ultra-thin SiNWs [6, 7, 8]. The high performance SiNWs in Ref. [3] were deliberately produced with a very rough surface, and the high ZT is attributed to increased phonon-surface scattering which decreases the phonon heat conductivity, while the electrons are less affected by the surface roughness. The extraordinary low thermal conductivity measured in rough SiNWs is supported by recent calculations [9, 10]. Surface disorder will, however, begin to affect the electronic conductance significantly in very thin wires [11] and thereby reduce ZT [7, 8].

For defect-free wires, the room temperature phononic conductance scales with the cross-sectional area [12]. Contrary, for ultra-thin wires with diameters $D \leq 5\,\text{nm}$, the electronic conductance is not proportional to the area: it is quantized and given by the number of states at the band edges. Decreasing the diameter to this range thus decreases κ_{ph} while keeping G_e almost constant. An optimally designed thermoelectric material would scatter phonons but leave the electronic conductance unaffected, even for the smallest wires.

2.2. Optimizing the nanowire design

Instead of introducing surface disorder, which affects the electronic conductance in the thin wires, we have recently addressed the question whether other surface designs could lead to improved thermoelectric performance [13]. In a similar spirit, Lee *et al.* [14] proposed already earlier nanoporous Si as an efficient thermoelectric matrial. Blase and Fernández-Serra [15] have shown that covalent functionalization of SiNW surfaces with alkyl molecules leaves the electronic conductance unchanged, because the molecular states are well separated in energy from the nanowire bandedges, see Fig. 1. If the alkyl molecules scatter the phonons, such functionalized SiNWs would be promising candidates for thermoelectric applications. Experimental alkyl functionalization of SiNWs was reported in Ref. [16].

Another possible surface design are branched SiNWs, so-called nanotrees. Nanotrees have been synthesized in III-VI semiconductors [17, 18] and in silicon [19, 20]. The stability and electronic structure of silicon nanotrees have recently been addressed theoretically [21, 22]. The thinner branches will have a larger band gap than the main wire ("trunk"), see Fig. 1. Close to the band edges, the electronic scattering is therefore weak.

The presence of an alkyl molecule or a nanowire branch leads both to a reduction, $\Delta\kappa$, of the thermal conductance and a reduction, ΔG, of the electronic conductance. As we show in Ref. [13], at room temperature (RT) the ratio $\Delta\kappa/\Delta G > 50$ for the alkyl functionalized SiNWs, and $\Delta\kappa/\Delta G > 20$ for a nanotree. By engineering the SiNW surfaces it is thus possible to reduce the

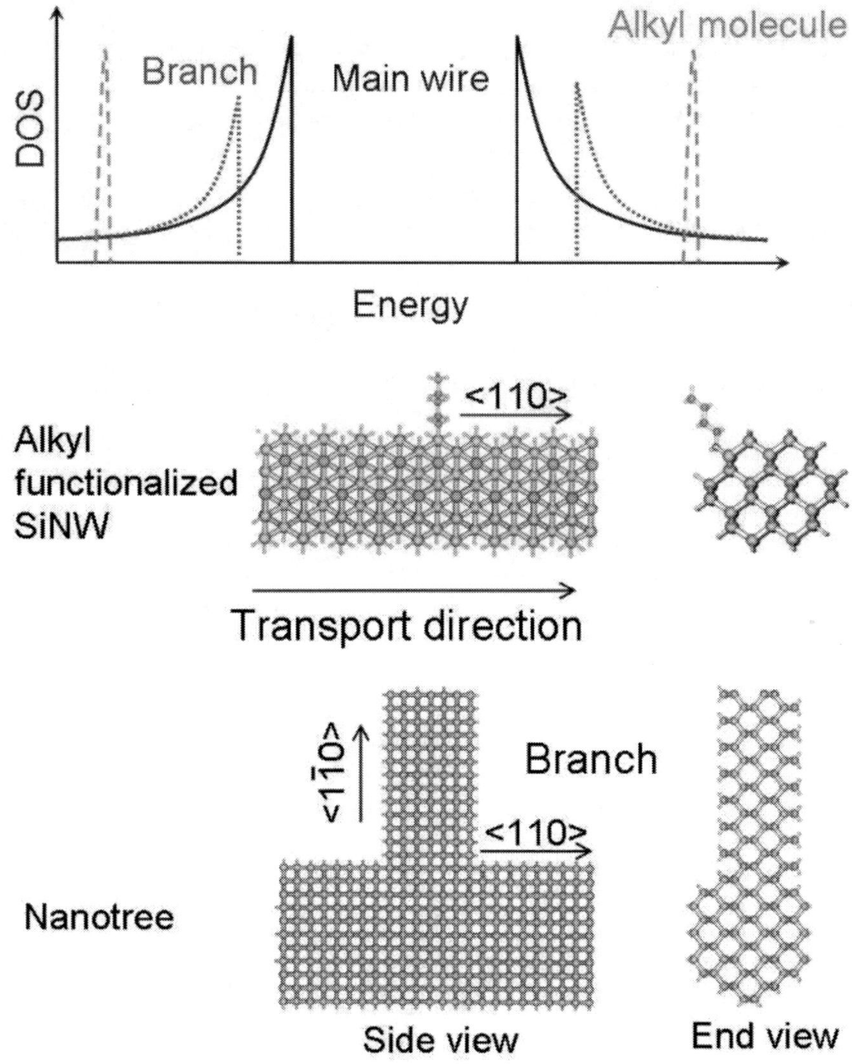

Figure 1. (Top) Sketch of electronic density of states in the alkyl functionalized SiNW (middle) and in the nanotree (bottom). The trunk in the nanotree is broader than the branch and thus has a smaller bandgap. For both structures electrons and holes close to the band edges are weakly scattered while phonons are strongly scattered.

phonon conductance while keeping the electronic conductance almost unaffected. Such surface decorated SiNWs would be promising candidates for nanoscale thermoelectric applications.

2.3. Systems

We consider two specific systems shown in Fig. 1. The first is an alkyl functionalized SiNW with a wire diameter of 12 Å, and with the wire oriented along the $\langle 110 \rangle$ direction. The alkyl (C_nH_{2n+1}) is attached to the H-passivated nanowire replacing a H atom. The second system is a nanotree, where a small diameter (12 Å) branch is attached to a larger diameter (20 Å) trunk. The trunk is oriented in the $\langle 110 \rangle$ direction while the branch is oriented along the $\langle 1\bar{1}0 \rangle$ direction, and is thus perpendicular to the trunk. The length of the branch, L_B, is varied.

2.4. Methods

The electronic Hamiltonian, **H**, and overlap matrix, **S**, of the alkyl functionalized SiNWs are obtained from local orbital DFT calculations [23]. The calculations are performed on super-cells containing 7 wire unit cells, as indicated in Fig. 1.

For the nanotrees, we use a tight-binding (TB) model since these systems contain > 1100 atoms, too many for our DFT implementation. The electronic TB Hamiltonian describing the nanotree is calculated using a 10 band $sp^3d^5s^*$ nearest-neighbor orthogonal TB parametrization [24, 25]. We recently applied the same TB model to study thermoelectric properties of surface disordered SiNWs [8]. The same TB parameters were recently also applied to study SiNW band structures [26] and surface roughness [11, 27].

The phononic system is described using the Tersoff empirical potential (TEP) model [28, 29] for both the nanotree and the functionalized SiNW. For pristine wires, we have recently shown that the TEP model agrees well with more elaborate DFT calculations [12]. We limit our description to the harmonic approximation, thus neglecting phonon-phonon scattering. The harmonic approximation is always valid at low temperatures. In bulk Si, the room temperature anharmonic phonon-phonon relaxation length at the highest frequencies is $\lambda_a(\omega_{max}) \sim 20\,\text{nm}$ and increases as $\lambda_a \propto \omega^{-2}$ at lower frequencies [30]. Experimental studies of silicon films [31] showed that the effective mean free path of the dominant phonons at room temperature is ~300 nm. For relatively short wires with lengths $L \leq 100\,\text{nm}$ the anharmonic effects thus seem to be of limited importance, and the harmonic approximation is expected to be good.

We calculate the electronic conductance from the electronic transmission function, $\mathcal{T}_e(\varepsilon)$ following the standard non-equilibrium Green's function (NEGF)/Landauer setup, where the scattering region (i.e. the regions shown in Fig. 1) is coupled to semi-infinite, perfect wires [32]. Explicitly, the transmission function is given by

$$\mathcal{T}_e(\varepsilon) = \text{Tr}[G^r(\varepsilon)\Gamma_L(\varepsilon)G^a(\varepsilon)\Gamma_R(\varepsilon)], \tag{1}$$

where the retarded and advanced Green functions $G^{r,a}$ can be calculated recursively, and the coupling matrices $\Gamma_{L,R}$ are evaluated with a DFT calculation. A discussion of the validity of this expression can be found, e.g., in Ref.[32]. The electronic quantities in the ZT formula can then be written as [33, 34, 35] $G_e = e^2 L_0$, $S = L_1(\mu)/[eT\,L_0(\mu)]$ and $\kappa_e = [L_2(\mu) - (L_1(\mu))^2/L_0(\mu)]/T$ where $L_m(\mu)$ is given by

$$L_m(\mu) = \frac{2}{h}\int_{-\infty}^{\infty} \mathrm{d}\varepsilon\, \mathcal{T}_e(\varepsilon)(\varepsilon-\mu)^m \left(-\frac{\partial f(\varepsilon,\mu)}{\partial\varepsilon}\right). \tag{2}$$

The phonon transmission function, $\mathcal{T}_{ph}(\omega)$, at frequency ω is calculated in a similar way as the electronic transmission with the substitutions $\mathbf{H} \to \mathbf{K}$ (**K** is the force constant matrix) and $\varepsilon\mathbf{S} \to \omega^2\mathbf{M}$, where **M** is a diagonal matrix with the atomic masses [36, 37, 38]. The phonon thermal conductance is

$$\kappa_{ph}(T) = \frac{\hbar^2}{2\pi k_B T^2}\int_0^{\infty} \mathrm{d}\omega\, \omega^2\, \mathcal{T}_{ph}(\omega)\, \frac{e^{\hbar\omega/k_BT}}{(e^{\hbar\omega/k_BT}-1)^2}. \tag{3}$$

2.5. Charge transport

Figures 2 (a) and (b) show the calculated hole and electron transmissions for a pentyl (C_5H_{11}) functionalized SiNW. Notably, the transmission is nearly perfect close to the bandedges, in agreement with the findings of Ref. [15]. Figures 2 (c) and (d) show the projected density of states (PDOS) on the wire and on the pentyl. The high transmission regions in panel (a) and (b) are seen to correspond with regions of vanishing PDOS on the pentyl molecule. Likewise,

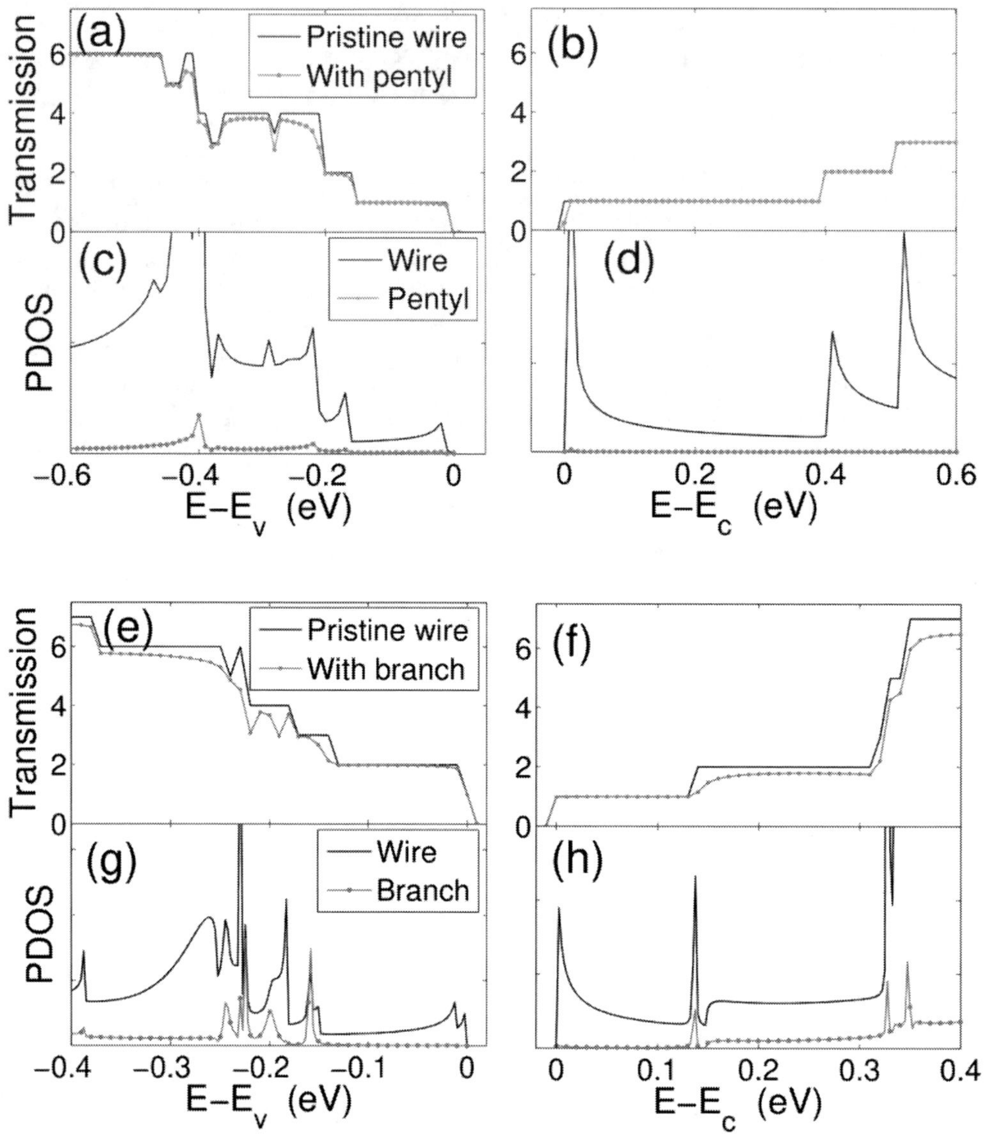

Figure 2. Hole and electron transmissions for pentyl functionalized, $D = 12\,\text{Å}$ SiNW (a)-(b) and for a nanotree with $D = 20\,\text{Å}$ trunk and $D = 12\,\text{Å}$ branch (e)-(f). Panels (c)-(d) and (g)-(h) show the PDOS on the wire and on the pentyl and branch, respectively. The energy scales are relative to the valence band edge, E_v for the holes (left column) and relative to the conduction band edge, E_c for electrons (right column).The DFT band gap of the $D = 12\,\text{Å}$ wire is 1.65 eV, while the TB band gap of the $D = 20\,\text{Å}$ wire is 1.77 eV (from Ref.[13]).

at energies in the valence band where scattering is observed, there is a relatively large PDOS at the pentyl.

Figures 2 (e) and (f) show the transmission through the nanotree. The branch length is 15.4 Å. Again, the transmission close to the band edges is nearly perfect. Figures 2 (g) and

(h) show the PDOS on the main wire and on the branch. We again observe a correspondence between perfect transmission and low PDOS on the branch.

The almost perfect transmissions close to the band edges can be qualitatively understood from the schematic drawing in Fig. 1 (top). The HOMO and LUMO level of the pentyl are located deep inside the bands [15] and the molecular states are thus not accessible for electrons or holes close to the band edges. For the nanotree, the branch has a smaller diameter and thus a larger bandgap. Electrons or holes in the trunk, with energies close to the band edges, are not energetically allowed in the branch and therefore do not 'see' the branch. In addition to the energy considerations, the spatial distribution of the Bloch state also plays a role: the first valence and conduction band Bloch states of the main wire have more weight in the center of the wire than at the edge [8].

2.6. *Thermal transport*

Figure 3 shows the temperature dependence of the thermal conductance ratios κ/κ_0, where κ_0 is the pristine wire thermal conductance, which in the low energy limit equals the universal thermal conductance quantum, $\kappa_Q(T) = 4(\pi^2 k_B^2 T/3h)$ [39]. Panel (a) shows the ratios for wires with alkyles, C_nH_{2n+1}, with different lengths, $n = 3, 5, 7$. The thermal conductance at RT is reduced by $\sim$ 10%, and the overall behavior does not depend on the alkyl length. The inset shows the phonon transmission at low phonon energies. Note the resonant dips in the transmission, where exactly one channel is closed yielding a transmission of three. These dips are associated with an increased local phonon density of states at the alkyl molecule at the resonant energies, corresponding to a localized vibrational mode. Such Fano-like resonant scattering is well-known from electron transport [40]. A phonon eigenchannel analysis [41] shows that the transmission dips are due to a complete blocking of the rotational mode in the wire. The corresponding localized alkyl phonon mode is a vibration in the plane perpendicular to the wire axis.

Panel (b) in Fig. 3 shows the thermal conductance ratio for nanotrees with different branch length, L_B. There is only a weak dependence on L_B at low temperatures, and at RT the four curves basically coincide showing a thermal conductance reduction of 17% of the nanotree compared to the pristine wire. Again we observe resonant transmission dips for the nanotrees. Two channels - the rotational and one flexural mode - close completely at the resonance due to two quasi-localized vibrational modes in the branch. These phonon backscattering resonances are responsible for the dip in the κ/κ_0 ratio around $T = 10\,\mathrm{K}$. Notice that all the conductance ratios approach unity in the low temperature limit. This is because the four acoustic modes transmit perfectly in the limit $\omega \rightarrow 0$ [39].

We may vary the thermoelectric figure of merit, ZT, by varying the chemical potential. Typically ZT displays a maximum for μ close to the band edge [8, 6]. Figure 4 shows the maximum ZT values for the pentyl functionalized SiNW (squares), the nanotree (circles), and surface disordered SiNWs (triangles), where disorder is modeled by introducing surface silicon vacancies. The diameter of the surface disordered wire is $D = 20\,\text{Å}$ and it is oriented along the $\langle 110 \rangle$ direction. The calculational details are given in Ref. [8]. The curves show ZT as a function of the number (N) of pentyl molecules/nanotree branches/silicon vacancies.

Figure 4 shows that increasing the number of scattering centers, i.e. the number of pentyl molecules or nanotree branches increases the ZT for both hole transport (a) and electron transport (b). In the case of holes in the pentyl functionalized SiNWs, the ZT reaches an almost constant level of $ZT = 0.7$ at $N = 40$, but in all other cases ZT increases throughout the range. Increasing the density of molecules/nanotree branches or increasing the length of the wire will thus increase the thermoelectric performance. The surface disordered wires (triangles) show an increasing ZT vs N but at values significantly lower than the two other surface modified wires.

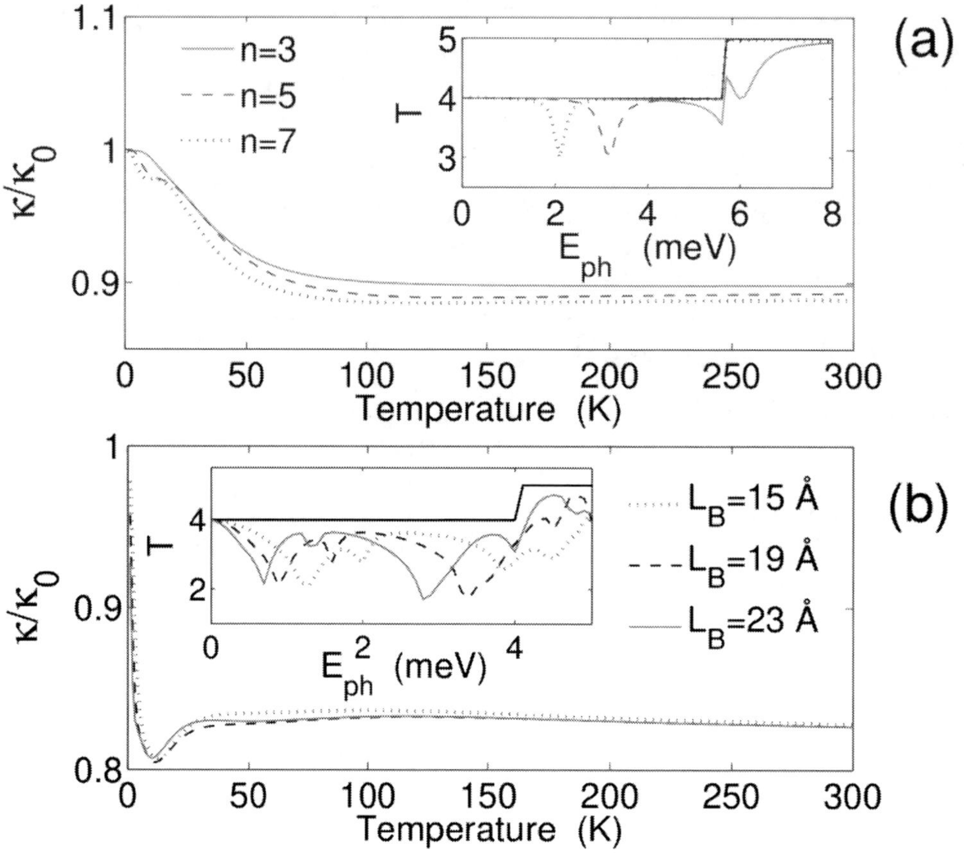

Figure 3. Thermal conductance ratio κ/κ_0 for (a) pentyl functionalized SiNWs with different alkyl lengths, and (b) for nanotrees with different branch lengths, L_B. The nanotree trunks are again $D = 20$ Å with 12 Å diameter branches. The insets show the phonon transmission function at low energies. Fano-like resonant scattering is observed in both systems (from Ref.[13]).

2.7. Discussion

A number of idealizations have been made in our calculations, and we next assess their significance. The structures we have considered represent plausible choices, dictated by computational limitations, but do not necessarily match quantitatively real structures. Thus, for example, a surface decorated SiNW will also be rough, and one should consider the combined effect of all scattering mechanisms. We have not carried out optimizations neither with respect to the attached molecules nor with respect to the geometry of the nanotrees. Electron-phonon and phonon-phonon scattering will affect both the electronic and thermal conductances and the obtained ZT values [7]. We do not expect to reach quantitative agreement with experiment but believe to have identified important trends: In SiNW based thermoelectrics, surface decorations in terms of added molecules or nanowire branches seem to be a better approach than surface disorder in the ultra-thin limit.

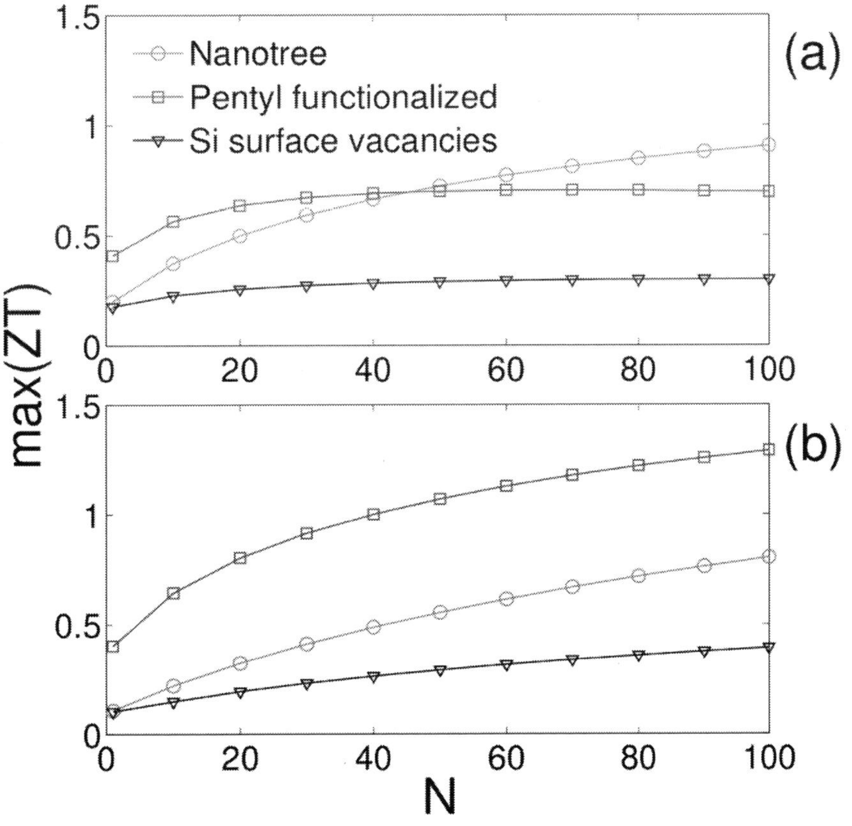

Figure 4. Thermoelectric figure of merit, ZT, for p-type (a) and n-type (b) wires. N is the number of pentyl molecules (squares), nanotree branches (circles) and silicon surface vacancies (triangles) in the wire (from Ref.[13]).

3. Microscopic theory for damping of vibrations in atomic gold wires

3.1. Introductory remarks

As pointed out in the beginning of this review, a particularly important issue in molecular electronics concerns the energy exchange between the charge carriers and the molecular contact. Thus, the local Joule heating resulting from the current passing through the contact, and its implications to the structural stability of such contacts are presently under intense investigation [42, 43, 44, 45, 46]. Experimentally, local heating in molecular conductors in the presence of the current has been inferred using two-level fluctuations [47] and Raman spectroscopy [48].

Mono-atomic chains of metal atoms [49] are among the simplest possible atomic-scale conductors. The atomic gold chain is probably the best studied atomic-sized conductor, and a great deal of detailed information is available from experiments and related theoretical studies; see Ref.[50] for an extensive list of references.

In the case of a gold chain Agraït *et al* [51] reported well-defined inelastic signals in the current-voltage characteristics. These signals were seen as a sharp 1% drop of the conductance at the on-set of back-scattering due to vibrational excitation when the voltage equals the vibrational energy. Especially for the longer chains (6-7 atoms), the vibrational signal due to the Alternating Bond-Length (ABL) mode [52, 53], dominates. This resembles the situation

Journal of Physics: Conference Series **220** (2010) 012010 doi:10.1088/1742-6596/220/1/012010

of an infinite chain with a half-filled electronic band where only the zone-boundary phonon can back-scatter electrons [54] due to momentum conservation.

The inelastic signal gives a direct insight into how the frequency of the ABL-mode depends on the strain of the atomic chain. This frequency can also be used to infer the bond strength. The signature of heating of the vibrational mode is the non-zero slope of the conductance versus voltage beyond the on-set of excitation: with no heating the curve would be flat. Fits to the experiment on gold chains using a simple model [55] suggest that the damping of the excitation can be significant. However, the experiments in general show a variety of behaviors and it is not easy to infer the extent of localization of the ABL vibration or its damping in these systems[1].

In order to address the steady-state effective temperature of the biased atomic gold chain theoretically, it is necessary to consider the various damping mechanisms affecting the localized vibrations, such as their coupling to the vibrations in the contact, or to the phonons in the surrounding bulk reservoirs. Here we outline how to calculate the vibrational modes in atomic gold chains and their coupling and the resulting damping due to the phonon system in the leads; a full account is available in our recent paper [56]. We work within the harmonic approximation and employ first principles density functional theory (DFT) for the atomic chain and the contacts [57] while a potential model is used for the force constants of the leads [58].

Here we focus on chains between two (100)-surfaces. We consider chain-lengths of 3-7 atoms and study the behavior of their vibrations and damping when the chains are stretched. TEM micrographs indicate that the chains are suspended between pyramids, so in our calculations we add the smallest possible fcc-stacked pyramid to link the chain to the given surfaces.

3.2. Method

In the microscopic modeling of the dynamics of the atoms forming the junction one must distinguish between different kinds of atoms: those belonging to the chain, those belonging to the pyramid joining the chain to the substrate, and substrate atoms. The technical issues of how these different types of atoms are treated are discussed in Ref.[56]. A quantity of central importance to all analysis is the mass-scaled dynamical matrix, $\mathbf{K}$,

$$\mathbf{K}_{ij} = \frac{\hbar^2}{\sqrt{m_i m_j}} \frac{\partial^2 E}{\partial u_i \partial u_j} \quad , \tag{4}$$

where E is the total energy of the system, u_i is the coordinate corresponding to the i'th degree of translational freedom for the atoms of the system. m_i is the mass of the atom that the i'th degree freedom belongs to. $\mathbf{K}$ governs the the evolution of the vibrational system within the harmonic approximation. The Newton equation of motion reads

$$\mathbf{K} u_\lambda = \epsilon_\lambda^2 u_\lambda \quad , \tag{5}$$

where λ denotes a mode of oscillation in the system and ϵ_λ is the corresponding quantization energy. The evaluation of $\mathbf{K}$ proceeds as follows. Finite difference DFT calculations were used for the nanostructure, while for the surfaces we used an empirical model due to Tréglia and Desjonqures [58]. The position of the interface between the region treated by DFT and the region treated by the empirical model is a parameter that can be varied; here one must find a value for which the physical results converge keeping the computational labor manageable.

3.3. Green's functions projected on the atomic chain

All properties of interest in the present context can be derived from the Green's function $\mathbf{D}$,

$$[(\epsilon + i\eta)^2 \mathbf{I} - \mathbf{K}]\mathbf{D}(\epsilon) = \mathbf{I} \equiv \mathbf{MD} \quad , \tag{6}$$

[1] N. Agraït, private communication.

where $\eta = 0^+$ and we defined the inverse of the Green's function by $\mathbf{M} = \mathbf{D}^{-1}$. Specifically, we shall need the Green's function $\mathbf{D}_{DD}$ projected onto the region close to the atomic chain. Our procedure is based on a method due to Mingo *et al* [59] which has previously been tested in an investigation of finite Si nanowires between Si surfaces. We find [56]

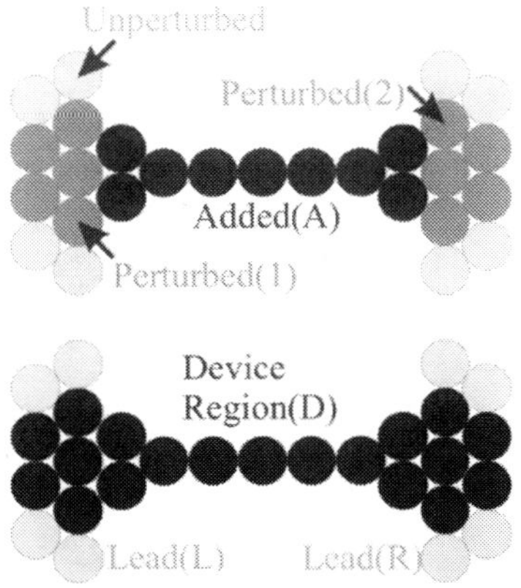

Figure 5. [Top] The forces between surface atoms within next-nearest neighbor distance(4.08Å) of the added atoms are perturbed by the presence of the added atoms. [Bottom] The device region is where the coupling between the atoms is different from the values for the two unperturbed surfaces. The coupling between the device region and the leads is considered to be unperturbed.

$$\begin{aligned} \mathbf{D}_{DD} &= [\mathbf{M}_{DD} - \mathbf{M}_{D\alpha}(\mathbf{M}_{\alpha\alpha})^{-1}\mathbf{M}_{\alpha D}]^{-1} \\ &= [\mathbf{M}_{DD} - \mathbf{\Pi}_{DD}]^{-1} \quad , \end{aligned} \tag{7}$$

where the index $\alpha = (L, R)$, i.e., the left and right unperturbed surface, while $D = \{1, A, 2\}$ labels the regions in left contact, the atomic chain, and the right contact, that are different from the bulk (see Fig. 5). This expression defines the self-energy $\mathbf{\Pi}_{DD} = \mathbf{M}_{D\alpha}(\mathbf{M}_{\alpha\alpha})^{-1}\mathbf{M}_{\alpha D}$ which has the matrix structure

$$\mathbf{\Pi}_{DD} = \begin{pmatrix} \mathbf{M}_{1L}(\mathbf{M}_{LL})^{-1}\mathbf{M}_{L1} & \mathbf{0} & \mathbf{0} \\ \mathbf{0} & \mathbf{0} & \mathbf{0} \\ \mathbf{0} & \mathbf{0} & \mathbf{M}_{2R}(\mathbf{M}_{RR})^{-1}\mathbf{M}_{R2} \end{pmatrix}. \tag{8}$$

This object can be evaluated as follows. First, in the limit of large regions 1 and 2, the coupling elements $\mathbf{M}_{L1}$ and $\mathbf{M}_{R2}$ must approach those of the unperturbed surface, $\mathbf{M}^S_{L1}$ and $\mathbf{M}^S_{R2}$, respectively. In what follows, we shall make the approximation that the regions 1 and 2 are chosen so, that this condition is satisfied sufficiently accurately. Second, we note that the matrix $\mathbf{M}_{\alpha\alpha}$ is *indistinguishable* from the matrix $\mathbf{M}^S_{\alpha\alpha}$, as long as the involved atoms are outside the perturbed regions 1 or 2. Therefore, we can write

$$\begin{aligned} \mathbf{M}_{1L}(\mathbf{M}_{LL})^{-1}\mathbf{M}_{L1} &\simeq \mathbf{M}^S_{1L}(\mathbf{M}^S_{LL})^{-1}\mathbf{M}^S_{L1} \equiv \mathbf{\Pi}^S_{11} \\ \mathbf{M}_{2R}(\mathbf{M}_{RR})^{-1}\mathbf{M}_{R2} &\simeq \mathbf{M}^S_{2R}(\mathbf{M}^S_{RR})^{-1}\mathbf{M}^S_{R2} \equiv \mathbf{\Pi}^S_{22}, \end{aligned} \tag{9}$$

where the accuracy increases with increasing size of regions 1 and 2. On the other hand, using the definition of the self-energy, we have

$$\begin{aligned} \mathbf{\Pi}^S_{11} &= \mathbf{M}^S_{11} - (\mathbf{D}^S_{11})^{-1} \\ \mathbf{\Pi}^S_{22} &= \mathbf{M}^S_{22} - (\mathbf{D}^S_{22})^{-1}, \end{aligned} \tag{10}$$

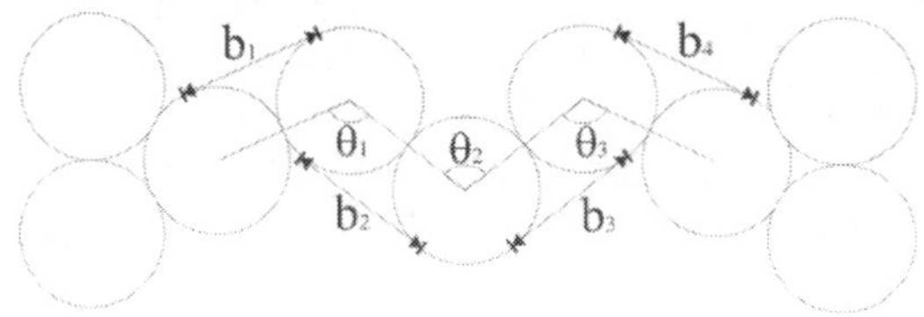

Figure 6. Distances and angles used to define the average bond length, $B = \langle b_j \rangle$, and the average bond angle, $T = \langle \theta_j \rangle$, respectively

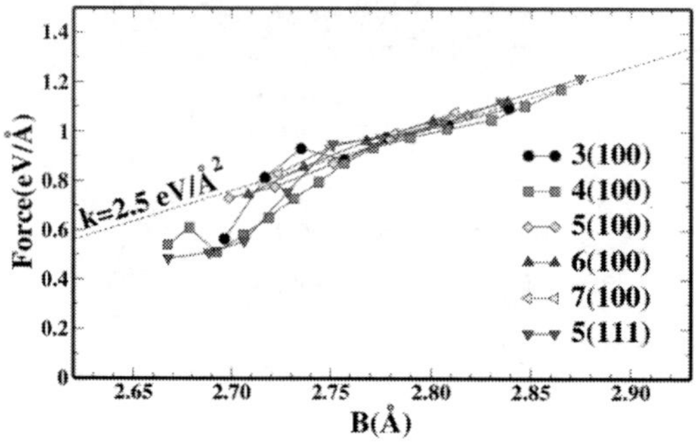

Figure 7. Force as function of average bond length B.

where $\mathbf{D}^S_{ii}, i = 1, 2$ is the projection of the *unperturbed* Green's functions onto the atoms in regions 1,2, respectively. This object is evaluated by exploiting the periodicity in the ideal surface plane. The Fourier transform of $\mathbf{M}^S$ in the parallel directions has a tridiagonal block structure and we can solve for its inverse very effectively using recursive techniques (see e.g. Sancho *et al* [60]). An analysis of the convergence properties of this procedure is given in Ref.[56].

To sum up, the calculation is preformed in the following steps: (i) Start with perfect leads and specify the device in between them. (ii) The atoms in the leads where $\mathbf{K}$ is perturbed by the presence of the device are identified. (iii) The unperturbed surface Green's function $\mathbf{D}^S$ is found via k-point sampling and then used to construct the self-energy, Eqs.(9–10). (iv) The perturbed Green's function is then found using this self-energy via Eqs.(7–8). With the self-energy at hand, we can define the life-times of the various modes in a standard fashion (see Ref. [56] for details), and we now proceed to present some of our results.

3.4. Geometrical Structure, the Dynamical Matrix, and Q-factors

Here we report results for the geometrical structure of the chains as a function of the chain elongation. For each geometry (identified by the number of atoms in the chain, the surface orientation and the type of contact to substrate) a range of calculations were performed with the two surfaces at different separations, with the separations incremented in equally spaced steps. To be able to compare chains of different lengths and between different surfaces we define the average bond length, $B = \langle b_j \rangle$, as the average length between neighboring atoms within the chain, where j runs over the number of bonds in the chain (see Fig. 6). B is useful because it is closely related to the experimentally measurable force [50] on the chain. The close relationship between B and the force is demonstrated in Fig. 7. We note that the force vs. B curves to a good approximation follows a straight line with a slope of $k = 2.5$ eV/Å^2, which can be interpreted as the spring constant of the bonds in the chain.

As the systems are stretched it is mostly the bonds in the chain that are enlongated. At low B Fig. 8 shows that the longer chains adopt a zig-zag confirmation at low average bond length.

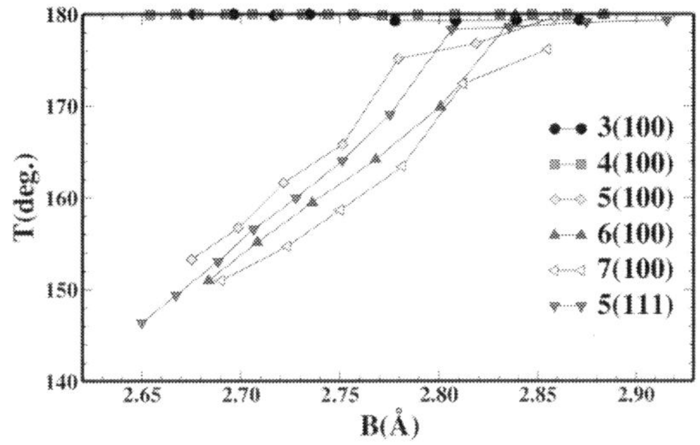

Figure 8. Average bond angle $T = \langle\theta_j\rangle$ as a function of the average bond length. The long chains adopt a zig-zag structure ($T < 180°$) at low B while the short chains remain linear.

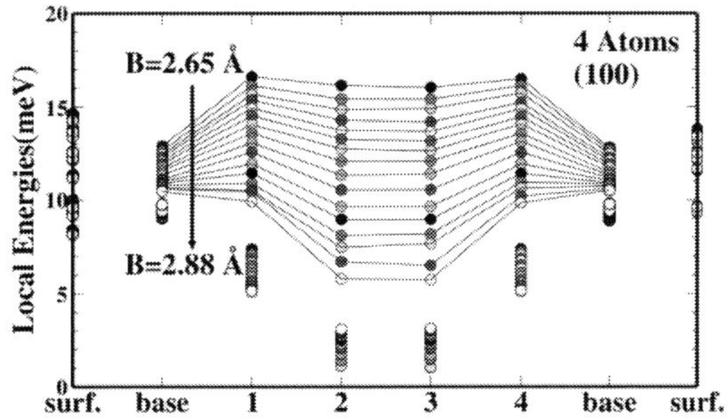

Figure 9. Local energies of a 4 atom chain between two (100)-surfaces at different strains. The largest eigenvalues are connected by a line to guide the eye.

The 3- and 4-atom chains, however, remain linear within the investigated range. Furthermore, the longer chains have a similar variation in the average bond angle. Our results justify the earlier theoretical studies by Frederiksen *et al* [53] and Sánchez-Portal *et al* [61].

We next investigate the energies that are related to different types of movement by analyzing the eigenmodes and eigenvalues of selected blocks of **K**. In particular, we can consider the local motion of individual atoms or groups of atoms, freezing all other degrees of freedom, by picking the corresponding parts of **K**. The square root of the positive eigenvalues of the reduced matrix, which we call local energies, give the approximate energy of a solution to the full **K** that has a large overlap with the corresponding eigenmode, if the coupling to the rest of the dynamical matrix is low. The negative eigenvalues of a block are ignored since they correspond to motion that is only stabilized by degrees of freedom outside the block.

The behavior of the dynamical matrix in terms of local energies is illustrated in Fig. 9. In the chain the local energies are quickly reduced with increased strain while the dynamical matrix of the surfaces is hardly affected at all. The middle bonds in the chain are the ones that are strained and weakened most when the surfaces are moved apart. It is also where the chain is expected to break [62].

Fig. 10 shows how motion parallel to the chain occurs at higher energies than perpendicular motion, and that the LO type motion of the ABL modes has the highest energy. The local energies of the ABL/LO type motion move past the local energies of the pyramid as the strain is increased. In this way the ABL/LO modes can in some sense act as a probe of the contacts.

Figure Fig. 11 shows the projected DOS for a chain with 4 atoms at an intermediate strain. Notice the large variation in the width of the peaks. The peaks with a small width correspond

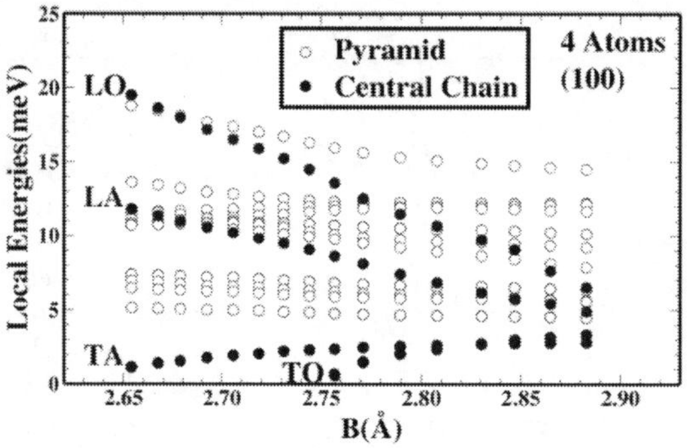

Figure 10. Local energies for selected blocks of **K** as a function of the average bond length in the 4 atom chain. Since the central chain in this case consists of 2 atoms we can classify the eigenvectors as LO: longitudinal optical, LA: longitudinal acoustic, TO: transverse optical (doubly degenerate) or TA: transverse acoustic (doubly degenerate).

to modes that have the largest amplitude in the chain, while the peaks with a large width correspond to modes with large amplitude on the interface.

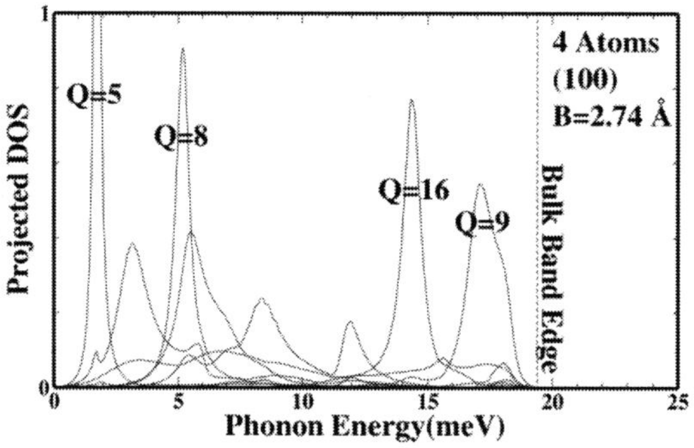

Figure 11. Projected DOS onto a representative selection of the vibrational modes of the device region.

In Fig. 12 we present the Q-factor, spatial localization and peak energy of all modes for chains with 3-7 atoms between (100) surfaces. The ABL modes are of special interest. These modes have been identified by previous theoretical and experimental studies as the primary scatterers of electrons[51, 52, 55, 63, 64, 65, 66]. The ABL modes are easily identified in Fig. 12 since they have the highest energy of the modes that are spatially localized to the central chain (black or dark gray on the figure). Modes corresponding to transverse motion of the central chain are also clearly visible. These modes are energetically and spatially localized, but are of limited interest because of a low electron-phonon coupling.

Certain ABL modes are very long-lived. At low strains, ABL modes lie outside the bulk (and surface) band and have, in the harmonic approximation, an infinite Q-factor. In reality the Q-factor will be limited by electron-phonon and anharmonic interactions. At higher strain the ABL modes move inside the bulk band and one observes a great variation in the corresponding Q-factors. When the peak energy lies inside the bulk band it will mostly be the structure of the connection between the bulk crystal and the chain that determines the width of the peak.

The long chains tend to have longer lived ABL/LO modes due to the larger ratio between the size of the chain and the size of its boundary. The 7-atom chain is especially interesting since it

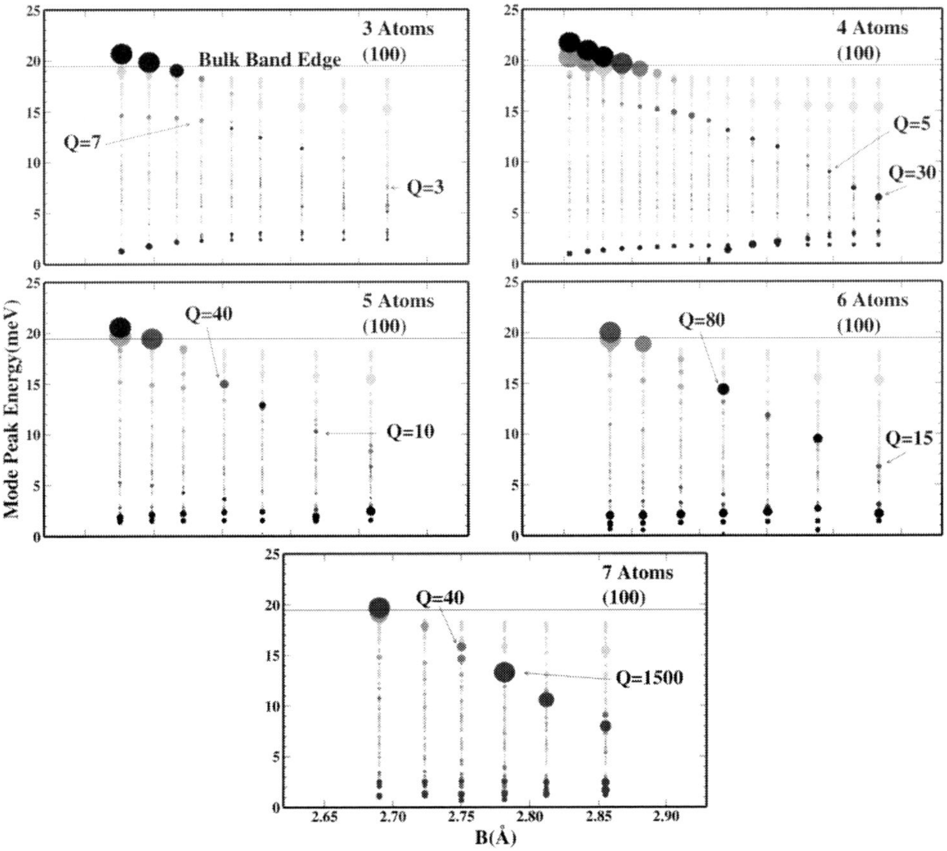

Figure 12. The vibrational modes for chains with 3-7 atoms between two 100-surfaces. The center of the disks are positioned at the peak of the projection of vibrational DOS on the mode in question. The area of a disk is proportional to the Q_λ, but is limited to what corresponds to a Q-factor of 250. The gray level is a measure of the localization of the mode, black being most localized.

has an ABL/LO type mode with a damping of 5 meV at one strain, while at another strain the ABL/LO mode has a damping of 300 meV, i.e., more than one order-of-magnitude variation in the harmonic damping of the primary scatterer of electrons due to only a 0.03 Å change in the average bond length!

The largest damping of an ABL-mode for these systems is $\gamma_\lambda \approx 1$ meV, which is still significantly lower than the ≈ 20 meV band width. This can be attributed to fact, noted above, that there always exists a large mismatch in local energies moving from the central part of the chain to the rest of the system (see Fig. 9).

Previous studies by Frederiksen *et al* [53] obtained a rough estimate for the variation of the non-electronic(harmonic and anharmonic) damping of 5-50 μeV for the longer chains by fitting the experimental IETS signals of Agraït *et al* [54] to a model calculation. The estimated peak energies lie well within the bulk band for all the recorded signals. The reason the non-electronic damping rate can be extracted is because the excitation of vibrations and damping of vibrations through electron-hole creation are both proportional to the strength of the electron-phonon coupling. This means that the step in the experimental conductance, when the bias reaches the phonon energy, can be used to estimate strength of the electron-phonon interaction and thereby the electron-hole pair damping. The slope in the conductance beyond this step can then be used to extract the total damping. By subtracting the electron-hole pair damping from the total damping we get an estimate of the sum of harmonic and anharmonic contributions to the damping.

The estimate in Ref. [53] agrees well with our lowest damping of 5 μeV. The highest damping we have found was ≈ 400 μeV found for the 6 atom chain which is an order of magnitude larger than the upper limit of Ref. [53]. We believe that this discrepancy can be largely attributed to the difficulty in extracting the necessary parameters from experiments when the harmonic damping is large. Furthermore, for the 6-7 atom chains we observe that the high damping occurs at low strain, where the electron-phonon coupling is weak [54].

3.5. Conclusion and Discussion

Our most important finding is that the values of the harmonic damping for the ABL modes can vary by over an order of magnitude with strain. Even with small variations in the strain, the harmonic damping can exhibit this strong variation. This extreme sensitivity may explain the large variations seen experimentally in different chains.

The range of the harmonic damping also depends strongly on the number of atoms in the chain since we see a clear increase in localization going from a 6- to a 7-atom chain. The chain with 7 atoms really stands out, since it, in addition to having very localized modes in general, it also has the greatest variation in harmonic damping. This strong variation in the harmonic damping of the ABL/LO-modes suggests that accurate atomistic calculations of the vibrational structure is necessary to predict the inelastic signal.

The techniques described here can be combined with electronic transport calculations to predict the inelastic signal in the *IV* characteristic of a system. This will be done in future work, where we will also eliminate the use of the empirical model for the leads and use ab-initio parameters for the entire system.

4. Summary

We have reviewed two applications where Density Functional Theory and Nonequilibrium Green's Functions are combined to analyze the electronic and thermal transport properties of nanoscale systems of current interest. We believe that there are many other systems where similar techniques can be successfully used: just one example is a recent study [67] where the current induced edge reconstruction of graphene nanoribbons is analyzed using the presented theoretical framework.

Acknowledgments

We thank the Danish Center for Scientific Computing (DCSC) and Direktør Henriksens Fond for providing computer resources. The authors would like to thank Thomas Frederiksen for helpful discussions and Nicolas Agraït showing his unpublished experimental results. TM acknowledges the Denmark-America foundation for financial support. APJ is grateful to the FiDiPro program of the Finnish Academy.

References

[1] M. Brandbyge *et al.*, Physical Review B **65**, 165401 (2002).
[2] R. Rurali, Reviews of Modern Physics **82**, 427 (2010).
[3] A. Hochbaum *et al.*, Nature (London) **451**, 163 (2008).
[4] A. I. Boukai *et al.*, Nature (London) **451**, 168 (2008).
[5] A. Marjumdar, Science **303**, 777 (2004).
[6] T. T. M. Vo *et al.*, Nano Lett. **8**, 1111 (2008).
[7] E. Ramayya *et al.*, Nano 08, 8th IEEE Conference on Nanotechnology, 339–342 (2008).
[8] T. Markussen, A.-P. Jauho, and M. Brandbyge, Phys. Rev. B **79**, 035415 (2009).

[9] P. Martin *et al.*, Phys. Rev. Lett. **102**, 125503 (2009).
[10] D. Donadio and G. Galli, Phys. Rev. Lett. **102**, 195901 (2009).
[11] M. P. Persson *et al.*, Nano Lett. **8**, 41464150 (2008).
[12] T. Markussen, A.-P. Jauho, and M. Brandbyge, Nano Lett. **8**, 3771 (2008).
[13] T. Markussen, A.-P. Jauho, and M. Brandbyge, Phys. Rev. Lett. **103**, 055502 (2009).
[14] J.-H. Lee, G. A. Galli, and J. C. Grossman, Nano Letters **8**, 3750 (2008).
[15] X. Blase and M.-V. Fernández-Serra, Phys. Rev. Lett. **100**, 046802 (2008).
[16] H. Haick *et al.*, J. Am. Chem. Soc. **128**, 8990 (2006).
[17] K. A. Dick *et al.*, J. Cryst. Growth **272**, 131 (2004a).
[18] K. A. Dick *et al.*, Nat. Mater. **3**, 380 (2004b).
[19] L. Fonseca, O. Resto, and F. Sola, Appl. Phys. Lett. **87**, 113111 (2005).
[20] G. S. Doerk *et al.*, J. Mat. Chem. **18**, 5376 (2008).
[21] M. Menon *et al.*, J. Comput. Theor. Nanosci. **4**, 252 (2007).
[22] P. V. Avramov *et al.*, Nano Lett. **7**, 2063 (2007).
[23] J. M. Soler *et al.*, J. Phys.: Condens. Matter **14**, 2745 (2002).
[24] T. B. Boykin, G. Klimeck, and F. Oyafuso, Phys. Rev. B **69**, 115201 (2004).
[25] Y. Zheng *et al.*, IEEE Trans. Electron Devices **52**, 1097 (2005).
[26] Y. M. Niquet *et al.*, Phys. Rev. B **73**, 165319 (2006).
[27] M. Luisier, A. Schenk, and W. Fichtner, Appl. Phys. Lett. **90**, 102103 (2007).
[28] J. Tersoff, Phys. Rev. B **38**, 9902 (1988).
[29] J. Tersoff, Phys. Rev. B **39**, 5566 (1989).
[30] N. Mingo and L. Yang, Phys. Rev. B **68**, 245406 (2003).
[31] Y. S. Ju and K. E. Goodson, Appl. Phys. Lett. **74**, 3005 (1999).
[32] H. Haug and A.-P. Jauho, *Quantum Kinetics in Transport and Optics of Semiconductors*, vol. 123 of *Springer Solid State Series* (Springer, 2008), 2nd ed.
[33] U. Sivan and Y. Imry, Phys. Rev. B **33**, 551 (1986).
[34] K. Esfarjani, M. Zebarjadi, and Y. Kawazoe, Phys. Rev. B **73**, 085406 (2006).
[35] A. M. Lunde and K. Flensberg, J. Phys.: Condens. Matter **17** (2005).
[36] T. Yamamoto and K. Watanabe, Phys. Rev. Lett. **96**, 255503 (2006).
[37] N. Mingo, Phys. Rev. B **74**, 125402 (2006).
[38] J.-S. Wang, J. Wang, and N. Zeng, Phys. Rev. B **74**, 033408 (2006).
[39] K. Schwab *et al.*, Nature (London) **404**, 974 (2000).
[40] J. U. Nockel and A. D. Stone, Phys. Rev. B **50**, 17415 (1994).
[41] M. Paulsson and M. Brandbyge, Phys. Rev. B **76**, 115117 (2007).
[42] G. Schulze *et al.*, Physical Review Letters **100**, 136801 (2008).
[43] Y. Teramae *et al.* Applied Physics Letters **93**, 083121 (2008).
[44] Z. F. Huang *et al.* Nature Nanotechnology **2**, 698 (2007).
[45] M. Galperin *et al.* Physical Review B **75**, 155312 (2007).
[46] D. A. Ryndyk *et al.* Physical Review B **78**, 085409 (2008).
[47] M. Tsutsui *et al.* Nano Letters **8**, 3293 (2008a).
[48] Z. Ioffe *et al.* Nature Nanotechnology **3**, 727 (2008).
[49] G. Rubio-Bollinger *et al.* Phys. Rev. Lett. **87**, 026101 (2001).
[50] N. Agraït, A. L. Yeyati, and J. M. van Ruitenbeek, Physics Reports **377**, 81 (2003).
[51] N. Agraït *et al.* Physical Review Letters **88**, 216803 (2002b).
[52] T. Frederiksen *et al.* Physical Review Letters **93**, 256601 (2004), 0031-9007.
[53] T. Frederiksen *et al.* Physical Review B **75**, 205413 (2007a), 1098-0121.
[54] N. Agraït *et al.* Chemical Physics **281**, 231 (2002a).
[55] M. Paulsson, T. Frederiksen, and M. Brandbyge, Physical Review B **72**, 201101(R) (2005).
[56] M. Engelund, M. Brandbyge, and A. P. Jauho, Phys. Rev. B **80**, 045427 (2009).
[57] J. M. Soler *et al.* Journal of Physics: Condensed Matter **14**, 2745 (2002).
[58] G. Tréglia and M. C. Desjonquéres, Journal de Physique **46**, 987 (1985).
[59] N. Mingo *et al.* Physical Review B **77**, 033418 (2008).
[60] M. P. L. Sancho, J. M. L. Sancho, and J. Rubio, Journal of Physics F-Metal Physics **14**, 1205 (1984).
[61] D. Sánchez-Portal *et al.* Phys. Rev. Lett. **83**, 3884 (1999).
[62] P. Velez, S. A. Dassie, and E. P. M. Leiva, Chemical Physics Letters **460**, 261 (2008).
[63] T. Frederiksen *et al.* Physical Review B **75**, 235441 (2007b), 1098-0121.
[64] J. K. Viljas *et al.* Physical Review B **72**, 245415 (2005).
[65] E. Hobi, A. Fazzio, and A. J. R. da Silva, Physical Review Letters **100**, 056104 (2008).
[66] J. Hihath *et al.* Nano Letters **8**, 1673 (2008).

[67] M. Engelund *et al.* Phys. Rev. Lett. **104**, 036807 (2010).

Progress in Nonequilibrium Green's Functions IV	IOP Publishing
Journal of Physics: Conference Series **220** (2010) 012011	doi:10.1088/1742-6596/220/1/012011

Unified theory of double-time quantum transport and quantum diffusion

P. Kleinert

Paul-Drude-Intitut für Festkörperelektronik, Hausvogteiplatz 5-7, 10117 Berlin, Germany

E-mail: kl@pdi-berlin.de

Abstract. Based on the nonequilibrium Green function technique, a unified theory of quantum transport and quantum diffusion is developed from quantum-kinetic equations of the full carrier distribution function $f^<(\boldsymbol{k}, \boldsymbol{\kappa}|T, t)$. The rigorous two-time approach is equally applicable to quantum transport as well as quantum diffusion that proceeds either via extended states or by hopping between localized carriers. The approach focuses on the entanglement between the microscopic and macroscopic time regimes and on the related non-Markovian evolution in both time channels. For a one-dimensional superlattice model subject to quantizing dc and ac electric fields, both phononless transport and diffusion mechanisms are identified that have a strict double-time origin. The experimental verification of this intricate quantum effect seems to be feasible by studying quantum-box superlattices at low temperatures.

1. Introduction

Diffusion implying spreading of particles, of probabilities, or of information is a paradigm, which is widely used in physical, biological, geological, and social sciences. The widespread application of the diffusion equation is caused by fundamental physical and mathematical findings dating back to the early studies by Fourier and Laplace about two hundred years ago [1]. At that time, the physical approach focused on the heat propagation through a solid. Fourier derived a parabolic partial differential equation in time and three spatial coordinates, which became later known as the diffusion equation. In a completely other field, namely the theory of probability, the notion of diffusion played also an important role. Laplace proved the central limit theorem, which states that the sum of independent and identically distributed random variables asymptotically approaches the normal distribution, which is likewise a solution of the diffusion equation. In 1905, both facets of diffusion, namely its physical and probabilistic aspects, were put together by Einstein in his famous molecular kinetic theory. By combining the microscopic, stochastic diffusion of particles with the phenomenological description of diffusion on the macroscale, he was able to determine Avogadro's number from the phenomenological diffusion parameter. This ingenious achievement provided further strong evidence for the molecular structure of matter.

Another related topic appeared about four decades later in solid state physics, namely hot electron transport that is theoretically accounted for by the kinetic Boltzmann equation [2]. An extension of this semiclassical theory was soon anticipated, when Martin and Schwinger communicated in their pioneering work the theory of nonequilibrium quantum statistics [3]. Especially on the basis of the Keldysh nonequilibrium Green function approach, the theory of high-field quantum transport has been developed to a high level of sophistication and has been reviewed not only in many papers but also in textbooks. Compared with these activities, there are only few contributions to the related theory of quantum diffusion in solids. One reason for this imbalance could be the famous Einstein relation, which is a consequence of the steady-state fluctuation theorem (for a review, see Ref. [4]) that relates the diffusion coefficient to the mobility. Consequently, near thermal equilibrium, an independent measurement of the diffusion coefficient does not add any information that is not already available from the mobility. However, under nonequilibrium conditions, quantum diffusion unfolds its own independence that is mirrored in a respective general quantum-mechanical description. Indeed, there are research fields, in which quantum diffusion is a central notion. We mention the study of atomic migration in crystalline solids, when tunneling becomes essential. Examples encompass tunneling in quantum crystals such as the diffusion of He^3 in solid He^4, hydrogen subbarrier motion in insulators and metals, the tunneling motion of atoms in insulating glasses, and the kinetics of heavy electrons.

To the best of the authors knowledge, there seems to be no systematic study of quantum diffusion of carriers under nonequilibrium conditions in semiconductor structures. This disparity is strange in view of many applications that found the notion of quantum diffusion, for instance, in the theory of weak localization, in the field of spintronics, or in device simulations based on drift-diffusion equations. It is therefore the aim of this paper to reduce the gap between the compelling achievements in the field of nonlinear quantum transport on the one hand and quantum diffusion of carriers in semiconductors on the other hand. The unified approach covers nonequilibrium quantum transport and quantum diffusion on the same footing and can profit thereby from well established methods developed in the nonequilibrium quantum transport theory.

The treatment is based on the above mentioned Einstein methodology to mutually interchange the descriptions via physical and stochastic diffusion. Accordingly, the analysis starts from the Fokker-Planck equation for the conditional probability P to find a particle at the space-time point $\boldsymbol{r}$, t, provided it occupied the lattice site $\boldsymbol{r}_0$ at an earlier time $t = 0$. From the Laplace transformed equation for the probability of propagation along the z axis

$$sP(\boldsymbol{r}-\boldsymbol{r}_0 \mid s) = \delta(\boldsymbol{r}-\boldsymbol{r}_0) + v_z(s)\frac{\partial}{\partial z}P(\boldsymbol{r}-\boldsymbol{r}_0 \mid s) + D_{zz}(s)\frac{\partial^2}{\partial z^2}P(\boldsymbol{r}-\boldsymbol{r}_0 \mid s), \tag{1}$$

the transport coefficients namely the drift velocity $v_z(s)$ and the diffusion coefficient $D_{zz}(s)$ can be expressed by the moments of the probability propagator P, which is therefore the main quantity of the approach. The pure stochastic Kolmogorov forward Eq. (1) is transferred to a physical picture for quantum transport by a reinterpretation of the conditional probability P in terms of a vacuum expectation value [5]

$$P(\boldsymbol{r}_1,\boldsymbol{r}_2;\boldsymbol{r}_3,\boldsymbol{r}_4|s) = \frac{1}{Z}\int_0^\infty dt e^{-st}\mathrm{Tr}_{\mathrm{ph}}\Big\{\langle 0|\psi(\boldsymbol{r}_2)e^{iHt/\hbar}\psi^\dagger(\boldsymbol{r}_4)\psi(\boldsymbol{r}_3)e^{-iHt/\hbar}\psi^\dagger(\boldsymbol{r}_1)|0\rangle\Big\}, \tag{2}$$

in which Fermionic creation $\psi^\dagger(\boldsymbol{r})$ and annihilation $\psi(\boldsymbol{r})$ operators appear for carriers at the lattice site $\boldsymbol{r}$. An average over the heat-bath Hamiltonian H_{ph} is carried out. Z denotes the partition function and H the total Hamiltonian including the contribution of the electric field. After this physical reinterpretation of the stochastic approach, a more general two-particle correlation function P appears that satisfies a Bethe-Salpeter equation, from which all needed information about quantum transport as well as quantum diffusion are obtained within a unified approach. The most familiar result, which can be derived by this straightforward exact procedure, is the quantum-kinetic Boltzmann equation with a field-dependent scattering term that gives rise to intra-collisional field effects [6, 7]. It is worth mentioning that this unified point of view elucidates in turn general features of the well established transport theory. The main conclusion of this type refers to the principal equivalence of representations that are designed to account for transport via localized and extended states. This equivalence of both different descriptions holds also for the quantum diffusion of carriers in semiconductors. The ability to cover quantum transport and quantum diffusion both via extended and localized states by using principally equivalent pictures is a general character of the unified theory of transport coefficients [8].

Extensions of the unified approach sketched above are possible in many directions. A quantum extension, which we will treat in the next Section, and which goes completely beyond the semiclassical approximation, takes into account the double-time character of the diffusion propagator. It is shown that interesting quantum effects are due to this generic two-time dependence of nonequilibrium one-particle Green functions.

2. Double-time approach

The consideration of transport coefficients that account for double-time effects could likewise start from a Bethe-Salpeter equation for the general four-point probability propagator P, which depends on four time variables. However, a simplified derivation is possible and profits from a procedure that is in a sense opposite to the one-time approach because it starts from double-time Green functions with the related distribution functions and provides expressions for transport coefficients by generalizing the one-time results. From the analysis of the unified one-time approach, it becomes evident that in general the current density is expressed by the time derivative of the dipole operator. In order to calculate this observable within a quantum-statistical framework, exactly two independent two-point nonequilibrium Green functions are needed. From a physical point of view, most attractive are the statistical and spectral components $G^{\gtrless}$ of the Green matrix, which count loosely speaking the available states as well as their occupancy. For simplicity, let us treat a non-degenerate electron gas in a semiconductor structure by the Green functions $G^{\gtrless}(\boldsymbol{k},\boldsymbol{\kappa}|T,t)$, which satisfy coupled Dyson equations in the momentum representation [9]

$$\left[i\hbar\frac{\partial}{\partial t}-\varepsilon(\boldsymbol{k})+ie\boldsymbol{E}(t)\nabla_{\boldsymbol{k}}\right]G^{\gtrless}(\boldsymbol{k}t\mid\boldsymbol{k}'t')=\pm\hbar\int d\boldsymbol{k}_1\Big\{\int_{t'}^{t}dt_1\Sigma^{\gtrless}(\boldsymbol{k}t\mid\boldsymbol{k}_1t_1)G^{\gtrless}(\boldsymbol{k}_1t_1\mid\boldsymbol{k}'t')$$
$$+\int_{-\infty}^{t'}dt_1\Sigma^{\gtrless}(\boldsymbol{k}t\mid\boldsymbol{k}_1t_1)G^{\lessgtr}(\boldsymbol{k}_1t_1\mid\boldsymbol{k}'t')-\int_{-\infty}^{t}dt_1\Sigma^{\lessgtr}(\boldsymbol{k}t\mid\boldsymbol{k}_1t_1)G^{\gtrless}(\boldsymbol{k}_1t_1\mid\boldsymbol{k}'t')\Big\},\tag{3}$$

in which the self-energies $\Sigma^{\gtrless}$ encompass elastic and inelastic scattering on impurities and phonons. To illustrate the general approach, we treat the self-consistent Born approximation with a spatially nonlocal, but temporally local coupling. The kinetic energy of carriers is denoted by $\varepsilon(\boldsymbol{k})$, and the applied electric field $\boldsymbol{E}(t)$ is expressed by the vector potential $\boldsymbol{A}(t)$ via the equation $d\boldsymbol{A}(t)/dt=e\boldsymbol{E}(t)/\hbar$. To make the approach more transparent, center of mass and relative coordinates are introduced according to the shift of variables $\boldsymbol{k}\to\boldsymbol{k}+\boldsymbol{\kappa}/2$, $\boldsymbol{k}'\to\boldsymbol{k}-\boldsymbol{\kappa}/2$ and $t\to T-t/2$, $t'\to T+t/2$. In the new representation, the $\boldsymbol{\kappa}$ and t dependence do not appear, when the model system is homogeneous with respect to space and time, respectively. The derivation proceeds by separating out a phase factor and by introducing the nonequilibrium carrier distribution function $f^{<}(\boldsymbol{k},\boldsymbol{\kappa}|T,t)$ from the equation

$$\overline{G}^{<}(\boldsymbol{k},\boldsymbol{\kappa}|T,t)=G^{<}(\boldsymbol{k},\boldsymbol{\kappa}+\boldsymbol{A}(T,t)|T,t)=if^{<}(\boldsymbol{k}\pm\boldsymbol{A}(T,t)/2,\boldsymbol{\kappa}|T,t)\tag{4}$$

$$\times \exp\left\{ \frac{i}{2\hbar} \int_{-t/2}^{t/2} d\tau \left[\varepsilon \left(\boldsymbol{k} - \frac{\boldsymbol{\kappa}}{2} + \boldsymbol{A}(T+\tau) - \frac{1}{2}\left[\boldsymbol{A}(T+\frac{t}{2}) + \boldsymbol{A}(T-\frac{t}{2}) \right] \right) \right.\right.$$
$$\left.\left. + \varepsilon \left(\boldsymbol{k} + \frac{\boldsymbol{\kappa}}{2} + \boldsymbol{A}(T+\tau) - \frac{1}{2}\left[\boldsymbol{A}(T+\frac{t}{2}) + \boldsymbol{A}(T-\frac{t}{2}) \right] \right) \right] \right\},$$

valid for $t \gtrless 0$, where $\boldsymbol{A}(T,t) = \boldsymbol{A}(T-t/2) - \boldsymbol{A}(T+t/2)$. Equating the time derivative of the dipole operator, the current density is expressed by the Keldysh Green Function $G^<$, which can be replaced by the carrier distribution function $f^<$. The result is given by

$$\boldsymbol{j}(T) = ien \sum_{\boldsymbol{k}} \nabla_{\boldsymbol{\kappa}} \frac{\partial}{\partial T} f^<(\boldsymbol{k}, \boldsymbol{\kappa}|T, t=0) \Big|_{\boldsymbol{\kappa}=\boldsymbol{0}}, \tag{5}$$

where n denotes the carrier density. What is surprising and unexpected in Eq. (5) is the appearance of the second wavevector $\boldsymbol{\kappa} = \boldsymbol{k} - \boldsymbol{k}'$, which refers to a possible inhomogeneity of the system. Such a dependence usually occurs in the treatment of diffusion, when the spreading of "excess carriers" is monitored. Indeed, the diffusion of an initial δ-like inhomogeneity in the carrier density requires for its theoretical description the quantity $\nabla_{\boldsymbol{\kappa}} f^<(\boldsymbol{k}, \boldsymbol{\kappa}|T,t)|_{\boldsymbol{\kappa}=\boldsymbol{0}}$ and not only the particular distribution function $f^<(\boldsymbol{k}, \boldsymbol{\kappa}=\boldsymbol{0}|T,t)$ that refers to a homogeneous system. Within the unified approach, a similar dependence on both quasi-momenta $\boldsymbol{k}$ and $\boldsymbol{\kappa}$ appears also for the field-mediated drift velocity or current density. This more general definition of the current density is necessary to simultaneously account for transport via localized and extended electronic states.

Exploiting the quantum-kinetic equation for the quantity $f^<(\boldsymbol{k}, \boldsymbol{\kappa}|T,t)$, which is derived from the Dyson Eq. (3) for $G^<$ by taking into account Eq. (4), another equivalent expression for the current density is obtained

$$\boldsymbol{j}(T) = en \sum_{\boldsymbol{k}} \boldsymbol{v}_{\text{eff}}(\boldsymbol{k}) f^<(\boldsymbol{k}|T,0), \tag{6}$$

in which an effective drift velocity appears that depends on both time scales as well as both wave vectors $\boldsymbol{k}$ and $\boldsymbol{\kappa}$

$$\boldsymbol{v}_{\text{eff}}(\boldsymbol{k}) = \boldsymbol{v}(\boldsymbol{k}) + i \int_{-\infty}^{\infty} \mathrm{d}t_1 \frac{f^<(\boldsymbol{k}|T,-t_1)}{f^<(\boldsymbol{k}|T,0)} \sum_{\boldsymbol{k}'} \boldsymbol{W}_1(\boldsymbol{k}, \boldsymbol{k}'|T, t_1). \tag{7}$$

The vectorial scattering contribution $\boldsymbol{W}_1(\boldsymbol{k}, \boldsymbol{k}'|T, t_1)$ is given by $\nabla_{\boldsymbol{\kappa}} W(\boldsymbol{k}, \boldsymbol{k}', \boldsymbol{\kappa}|T,t)|_{\boldsymbol{\kappa}=\boldsymbol{0}}$, with W denoting the electric field dependent scattering probability [10]. The second contribution on the right-hand side of Eq. (7) vanishes, whenever the interaction Hamiltonian commutes with the dipole operator. Fortunately, this happens, for instance, for the Fröhlich electron-phonon coupling and the Coulomb interaction. However, this condition is not always satisfied. A famous counterexample is the transport of small polarons, where the $\boldsymbol{W}_1$ contribution to the current even dominates. Again, it is noted that in general not only the distribution functions $f^{\gtrless}(\boldsymbol{k}, \boldsymbol{\kappa}=\boldsymbol{0}|T,t)$ enter the approach, but in addition also the $\boldsymbol{\kappa}$ gradient of the full function $f^{\gtrless}(\boldsymbol{k}, \boldsymbol{\kappa}|T,t)$ at $\boldsymbol{\kappa}=\boldsymbol{0}$. The $\boldsymbol{\kappa}$ dependence reappears in Eq. (7) via the vector field $\boldsymbol{W}_1$. Without the second term on the right-hand side of Eq. (7), there is no current, when the states do not exhibit any dispersion ($\boldsymbol{v}(\boldsymbol{k}) = \nabla_{\boldsymbol{k}}\varepsilon(\boldsymbol{k})/\hbar = \boldsymbol{0}$). Consequently, the theory cannot cope with hopping transport in this case. It is just the $\boldsymbol{W}_1$ contribution in Eq. (7), which is needed for the exact derivation of an equivalent representation that is designed for the treatment of hopping [8].

A second feature of Eqs. (6) and (7) consists in its double-time nature. Even the distribution function $f^<(\boldsymbol{k}|T,0)$, which enters Eq. (6), depends effectively on both time scales T and t. This fact is due to the non-Markovian time behavior of the quantum-kinetic equations [10], which are obtained from the Dyson Eq. (3). The mixing of the "macroscopic" (T) and "microscopic" (t) time channels is a generic property of nonequilibrium dynamics. In the conventional understanding, there is only a one-dimensional time flow from the past to the future. However, the more fundamental quantum-statistical point of view reveals an entangled time hierarchy that constitutes in the background, mediated by a second time variable, the evolution of real macroscopic objects. The entanglement of these histories is avoided by adopting the generalized Kadanoff-Baym ansatz [11, 12, 13], by which the dependence on the time parameter t is effectively disregarded [9]. To give a specific result for the quantum-kinetic equation, let us treat a non-degenerate electron gas in the self-consistent Born approximation. Assuming that the slow time scale is inert against fast microscopic fluctuations ($T+\Delta t\to T$), the double-time "Boltzmann" equation has the form

$$\left[\frac{\partial}{\partial T}+\frac{i}{\hbar}\left(\varepsilon(\boldsymbol{k}+\frac{\boldsymbol{\kappa}}{2})-\varepsilon(\boldsymbol{k}-\frac{\boldsymbol{\kappa}}{2})\right)-\frac{\partial \boldsymbol{A}(T,t)}{\partial t}\nabla_{\boldsymbol{k}}+\frac{\partial \boldsymbol{A}(T,t)}{\partial T}\nabla_{\boldsymbol{\kappa}}\right]f^<(\boldsymbol{k},\boldsymbol{\kappa}|T,t)$$
$$=\sum_{\boldsymbol{k}_1}\int_{-\infty}^{\infty}dt_1 f^<(\boldsymbol{k}_1,\boldsymbol{\kappa}|T,t-t_1)W(\boldsymbol{k}_1,\boldsymbol{k},\boldsymbol{\kappa}|T,t,t_1), \tag{8}$$

where the scattering probability W includes scattering-in and scattering-out contributions. This equation, which reduces to the well-known results in the one-time approximation [6], reminds on the gauge-invariant quantum-kinetic equation, in which memory effects due to collisions are taken into account [14, 15].

The diffusion coefficient is obtained by a similar double-time approach, which is also based on the method of moments applied to the Fokker-Planck equation, by which both transport coefficients are defined. By collecting both the regular and the irregular part of the diffusion tensor $\widehat{D}$, we obtain

$$\widehat{D}(T) = \sum_{\boldsymbol{k}}\left\{\boldsymbol{v}(\boldsymbol{k})\otimes\boldsymbol{\varphi}(\boldsymbol{k}|T,0)+i\int_{-\infty}^{\infty}\mathrm{d}t_1\boldsymbol{\varphi}(\boldsymbol{k}|T,-t_1)\otimes\sum_{\boldsymbol{k}'}\boldsymbol{W}_1(\boldsymbol{k},\boldsymbol{k}'|0,t_1)\right\}$$
$$-\frac{1}{2}\sum_{\boldsymbol{k}}\int_{-\infty}^{\infty}\mathrm{d}t_1 f^<(\boldsymbol{k}|T,-t_1)\sum_{\boldsymbol{k}'}\widehat{W}_2(\boldsymbol{k},\boldsymbol{k}'|0,t_1), \tag{9}$$

which has a similar structure as Eq. (6). There is an additional contribution to $\widehat{D}$, which is due to the second derivative with respect to $\boldsymbol{\kappa}$: $\widehat{W}_2=\nabla^2_{\kappa_i\kappa_j}W|_{\boldsymbol{\kappa}=\boldsymbol{0}}$. The first term on the right-hand side of Eq. (9) mainly differs from the result for the current density in Eqs. (6) and (7) by the appearance of the vector field $\boldsymbol{\varphi}(\boldsymbol{k}|T,t)$, which takes over the role played by the distribution function $f^<$ in Eq. (6). This quantity satisfies the sum rule

$$\sum_{\boldsymbol{k}}\boldsymbol{\varphi}(\boldsymbol{k}|T,t)=\boldsymbol{0}, \tag{10}$$

and is related to the gradient of the distribution function $\boldsymbol{g}(\boldsymbol{k}|T,t)=i\nabla_{\boldsymbol{\kappa}}f^<(\boldsymbol{k},\boldsymbol{\kappa}|T,t)|_{\boldsymbol{\kappa}=\boldsymbol{0}}$ via the equation

$$\boldsymbol{\varphi}(\boldsymbol{k}|s,t)=\boldsymbol{g}(\boldsymbol{k}|s,t)-\frac{1}{s}f^<(\boldsymbol{k}|s,t)\sum_{\boldsymbol{k}'}\boldsymbol{v}_{\text{eff}}(\boldsymbol{k}'|s,t)f^<(\boldsymbol{k}'|s,t). \tag{11}$$

The distinct character of quantum diffusion under nonequilibrium conditions is reflected by this quantity that satisfies its own kinetic equation. For the sake of comparison, let us treat the same model that leads to the kinetic Eq. (8). The corresponding double-time "quantum Boltzmann" equation for the vector field φ has the form

$$\left[\frac{\partial}{\partial T} - \frac{\partial \boldsymbol{A}(T,t)}{\partial t}\nabla_{\boldsymbol{k}}\right]\varphi(\boldsymbol{k}|T,t) = \sum_{\boldsymbol{k}_1}\int_{-\infty}^{\infty} dt_1 \varphi(\boldsymbol{k}_1|T,t-t_1)W(\boldsymbol{k}_1,\boldsymbol{k}|T,t,t_1)$$
$$+\boldsymbol{v}(\boldsymbol{k})f^<(\boldsymbol{k}|T,t) - \sum_{\boldsymbol{k}_1}\boldsymbol{v}(\boldsymbol{k}_1)f^<(\boldsymbol{k}_1|T,t) \tag{12}$$
$$+i\sum_{\boldsymbol{k}_1}\int_{-\infty}^{\infty} dt_1 f^<(\boldsymbol{k}_1|T,t-t_1)\left[\boldsymbol{W}_1(\boldsymbol{k}_1,\boldsymbol{k}|T,t,t_1) - \sum_{\boldsymbol{k}_2}\boldsymbol{W}_1(\boldsymbol{k}_1,\boldsymbol{k}_2|T,t,t_1)\right].$$

Again, it is concluded that there is no closed equation for the one-time function $\varphi(\boldsymbol{k}|T,t=0)$, which underlines the strict double-time character of the approach. In addition, Eqs. (9) to (12) are sufficiently general so that they are likewise applicable to quantum diffusion via extended and localized electronic states. Furthermore, all remarks concerning the entangled time evolution as well as the $\boldsymbol{\kappa}$ dependence that apply to the current density are also valid for Eqs. (9) to (12), which describe double-time quantum diffusion of carriers in semiconductors.

As mentioned, in the hopping regime and in one-time approximation, both the drift velocity and the diffusion coefficient vanish in the absence of any inelastic scattering. However, taking into account the double-time character of the quantum statistics, a phononless transport mechanism can be identified in the regime of Wannier-Stark localization. This is shown in the next Section.

3. Application: phononless transport

The double-time dependence of kinetic coefficients has to be taken into account, for instance, in ultra-short-time laser spectroscopy of semiconducting specimens [16, 17], in the consideration of aging phenomena as well as for the treatment of the nonexponential relaxation in glassy systems that break time translation symmetry [18, 19, 20].

In this Section, a robust high-field, steady-state transport mechanism in a semiconductor superlattice is studied that strictly originates from the double-time dependence of the correlation functions. The model is a one-dimensional, weakly-coupled semiconductor superlattice with the dispersion relation

$$\varepsilon(k) = \frac{\Delta}{2}(1-\cos(kd)), \tag{13}$$

where Δ denotes the miniband width and d the lattice constant of the superlattice. Along the superlattice axis, sufficiently strong dc and ac electric fields are applied

$$E(t) = E_{dc} + E_{ac}\cos(\omega_{ac}t), \tag{14}$$

so that Wannier-Stark localization occurs due to Bloch oscillations ($\Omega_{\text{dc,ac}}\tau \gg 1$, with τ being an effective scattering time and $\Omega_{\text{dc,ac}} = eE_{\text{dc,ac}}d/\hbar$ are the Bloch frequencies). Both the periodicity in the time domain and the Stark-ladder structure of the energy spectrum are accounted for by the discrete Fourier representation

$$f^{\gtrless}(\boldsymbol{k}|T,t) = \sum_{l,m=-\infty}^{\infty} f_l^{\gtrless}(m,t)e^{ilkd+im\omega_{\text{ac}}T}, \tag{15}$$

which is used to solve the quantum-kinetic equations for $f^<$ by taking into account elastic scattering on white-noise potential fluctuations. According to Eq. (5), the current density is calculated from the equation

$$j = en\frac{\Delta d}{2\hbar}\frac{1}{2i}\left[f^<_{l=-1}(m=0,t=0) - f^<_{l=1}(m=0,t=0)\right]. \tag{16}$$

Under the condition of strong quantizing electric fields, Wannier-Stark localization takes place, and the quantum-kinetic equation for the Fourier coefficients $f_l^<(T,0)$ reduces to the form

$$\left\{\frac{\partial}{\partial T} + il\Omega_{dc} + il\Omega_{ac}\cos(\omega_{ac}T)\right\} f_l^<(T,0) \tag{17}$$
$$= U\sum_{k,q} e^{-ilkd}\int_0^\infty \mathrm{d}t_1\left[\Phi_{0t_1}(k,q) - \Phi^*_{0t_1}(k,q)\right]$$
$$\times\left\{f_0^<(T-\frac{t_1}{2},-t_1)f_0^>(T-\frac{t_1}{2},t_1) - f_0^<(T-\frac{t_1}{2},t_1)f_0^>(T-\frac{t_1}{2},-t_1)\right\},$$

where U denotes the strength of the short-range impurity potential. The phase factor $\Phi_{t,t_1}(k,q)$, which enters this equation, is given by

$$\Phi_{t,t_1}(k,q) = \exp\left\{\frac{i}{\hbar}\int_{t_1}^{t}\mathrm{d}\tau\left[\varepsilon\left(k+q+A(\tau+T-\frac{t}{2}-t_1)-A(T-\frac{t}{2})\right)\right.\right.$$
$$\left.\left.-\varepsilon\left(k+A(\tau+T-\frac{t}{2}-t_1)-A(T-\frac{t}{2})\right)\right]\right\}. \tag{18}$$

From Eq. (17), it is found that no current can flow through the superlattice, when the double-time dependence is discarded $[f_0^{\gtrless}(T,t) \to f_0^{\gtrless}(T)]$. This observation confirms the fact that under Wannier-Stark localization there is no current without any inelastic scattering. However, within the two-time approach, there exists a phononless current contribution even in the Wannier-Stark regime. An analytical solution is found by a Fourier transformation with respect to the additional time dependence and by adopting the ansatz

$$f^<_{l=0}(m=0,\omega) = f^>_{l=0}(m=0,\omega)f(\omega), \tag{19}$$

in which the frequency dependent function $f(\omega)$ accounts for the time variable t, which appears only in the double-time approach, when time translation invariance is broken. Using the formal solution of the kinetic Eq. (17), the current density of the one-dimensional superlattice is given by

$$j = env_d,\quad v_d = \frac{\pi U\Delta^2 d}{8\hbar^2}\frac{J_\nu(\nu')J_{-\nu}(\nu')}{\omega_{ac}^2\sin(\pi\nu)}\sum_k\frac{J_k^2(\nu')}{k+\nu} \tag{20}$$
$$\times\int_{-\infty}^{\infty}\frac{\mathrm{d}\omega}{2\pi}f_0^>(0,\omega)f_0^>(0,\omega+\omega_{ac}(k+\nu))\left[f(\omega+\omega_{ac}(k+\nu))-f(\omega)\right],$$

with the frequency parameters $\nu = \Omega_{\mathrm{dc}}/\omega_{\mathrm{ac}}$ and $\nu' = \Omega_{\mathrm{ac}}/\omega_{\mathrm{ac}}$. J_k denotes theBessel function. From Eq. (20), it is concluded that the phononless current disappears in the strict one-time approach ($f(\omega)$ =const.), which is an approximation completely in the spirit of the Kadanoff-Baym ansatz. The specific phononless current contribution only appears in the Wannier-Stark regime, when the combined double-time history is taken into account in the equation of motion. Consequently, this steady-state current is a signature of the double-time character of quantum statistics. The result for the current density in Eq. (20) that is valid under quantizing electric fields has a familiar structure consisting of the combined density of states multiplied by the difference of occupation numbers for the final and initial states. It should be noted, however, that the ω dependence in this equation has its strict origin on fast quantum effects on the microscopic time scale, which are completely disregarded in the one-time approximation. An experimental study of this intricate phononless quantum transport should be possible by studying a strongly biased semiconductor superlattice, which is subject to a strong perpendicular magnetic field that quantizes the in-plane carrier motion. To effectively suppress inelastic scattering due to phonons, the lattice temperature of this quantum-box superlattice should be extremely low.

The treatment of the diffusion coefficient proceeds in a similar manner. Starting from Eq. (9), we obtain for the one-dimensional model

$$D = \sum_k \frac{1}{\hbar}\frac{\mathrm{d}\varepsilon(k)}{\mathrm{d}k}\frac{\omega_{ac}}{2\pi}\int_0^{2\pi/\omega_{ac}} \mathrm{d}T\varphi(k|T,t=0), \tag{21}$$

which is complemented by the quantum-kinetic equation for the Fourier coefficients $\varphi_l(T,t)$ given by

$$\begin{aligned}&\left\{\frac{\partial}{\partial T} + il\Omega_{dc} + il\Omega_{ac}\cos(\omega_{ac}T)\cos\left(\frac{\omega_{ac}t}{2}\right)\right\}\varphi_l(T,t) \\ &= i\frac{\Delta d}{4\hbar}\left\{f^<_{l+1}(T,t) - f^<_{l-1}(T,t) - (f^<_1(T,t) - f^<_{-1}(T,t))\right\} \\ &+\frac{d}{2\pi}\int_0^{2\pi/d}\mathrm{d}ke^{-ilkd}\sum_{k_1}\int_{-\infty}^{\infty}\mathrm{d}t_1\varphi(k_1|T,t-t_1)W(k_1,k|T,t,t_1).\end{aligned} \tag{22}$$

The formal solution of this equation is used to derive an analytic expression for the diffusion coefficient in Eq. (21). For a weakly coupled superlattice, the result has the Tien-Gordon [21] form

$$\begin{aligned}D(\Omega_{\rm dc},\Omega_{\rm ac}) &= \sum_{k=-\infty}^{\infty} J_k\left(\frac{\Omega_{\rm ac}}{\Omega_{\rm dc}}\right)\widetilde{D}(\Omega_{\rm dc}+k\omega_{\rm ac},\Omega_{\rm ac}) \\ \widetilde{D}(\Omega_{\rm dc},\Omega_{\rm ac}) &= \frac{d}{2}v_d(\Omega_{\rm dc},\Omega_{\rm ac})\coth\left(\frac{\hbar\Omega_{\rm dc}}{\Delta}\right),\end{aligned} \tag{23}$$

in which, however, the drift velocity v_d depends both on the Bloch frequency of the dc and ac electric field. This suggestive result applies to the Wannier-Stark hopping regime under the condition that carriers thermalize more quickly in a given well than they need to escape by tunneling. The double-time quantum character of this high-field phononless diffusion mechanism has the same origin as the phononless current treated above. It appears only in an approach that goes beyond the Kadanoff-Baym ansatz. The simple model of a one-dimensional superlattice has been adopted here to derive analytical results that describe phononless quantum transport and quantum diffusion on an equal footing. The double-time effect itself seems to be sufficiently robust that it should survive in more realistic calculations.

Further progress in the development of a unified approach of carrier drift and diffusion in semiconductors is expected from the treatment of collective diffusion in the presence of many-body interaction as well as from a study of quantum transport and quantum diffusion in finite semiconductor devices. Another promising direction of research includes anomalous diffusion in the unified quantum picture of carrier drift and diffusion.

4. Conclusion

The Fokker-Planck equation, which has many applications both in physics and mathematics, serves as a suitable basis for the development of a unified theory of carrier drift and diffusion. By interpreting the main quantity of this equation, namely the probability propagator, in terms of an averaged vacuum expectation value, a quantum theory of transport coefficients is straightforwardly derived. The general results that refer to carrier drift and diffusion in semiconductors are equally applicable to the treatment of localized and extended states. By an exact transformation, the hopping transport picture is converted into a description that is more designed for band transport problems and vice versa. Based on the theory of nonequilibrium Green functions, the unified approach is extended to account for the double-time dependence of the one-particle propagators. Within the self-consistent Born approximation with respect to scattering, double-time kinetic equations are obtained that have the structure of a non-Markovian Boltzmann equation, where the scattering rates depend on the electric field. As an application of the unified quantum approach, a phononless current contribution is identified that has its origin in the double-time dependence of Green functions.

References

[1] Narasimhan T N 2009 *Physics Today* **July** 48
[2] Davydov B I 1937 *Zh. exper. teor. Fiz.* **7** 1069
[3] Martin P and Schwinger J 1959 *Phys. Rev.* **115** 1342
[4] Evans D J and Searles D J 2002 *Adv. Phys.* **51** 1529
[5] Kudinov E K and Firsov Y A 1966 *Sov. Phys. JETP* **22** 603
[6] Haug H and Jauho A P 1996 *Quantum Kinetics in Transport and Optics of Semiconductors* (*Springer Series in Solid-State Sciences* vol 123) (Berlin, Heidelberg, New York: Springer Verlag)
[7] Bryksin V V and Kleinert P 2003 *J. Phys.: Condens. Matter* **15** 1415
[8] Kleinert P and Bryksin V V 2003 *Phys. Letters A* **317** 315
[9] Bryksin V V and Kleinert P 1999 *J. Phys. A: Math. Gen.* **32** 2731
[10] Kleinert P 2009 *Phys. Rev. E* **79** 051107
[11] Lipavský P, Špička V and Velický B 1986 *Phys. Rev. B* **34** 6933
[12] Špička V and Lipavský P 1995 *Phys. Rev. B* **52** 14615
[13] Velický B, Kalvová A and Špička V 2008 *Phys. Rev. B* **77** 041201
[14] Kremp D, Bornath T, Bonitz M and Schlanges M 1999 *Phys. Rev. E* **60** 4725
[15] Bonitz M, Bornath T, Kremp D, Schlanges M and Kraeft W D 1999 *Contrib. Plasma Phys.* **39** 329
[16] Bonitz E M 2000 *Progress in Nonequilibrium Green's Functions* (Singapore: World Scientific)
[17] Eckstein M and Kollar M 2008 *Phys. Rev. B* **78** 205119
[18] Rubí E M and Pérez-Vicente C 1997 *Complex Behavior of Glassy Sytems* (Berlin: Springer Verlag)
[19] Jaroszyński J and Popović D 2006 *Phys. Rev. Lett.* **96** 037403
[20] Jaroszyński J and Popović D 2007 *Phys. Rev. Lett.* **99** 046405
[21] Tien P R and Gordon J P 1963 *Phys. Rev.* **129** 647

Journal of Physics: Conference Series **220** (2010) 012012 doi:10.1088/1742-6596/220/1/012012

Time-dependent quantum transport with superconducting leads

Gianluca Stefanucci, Enrico Perfetto and Michele Cini

Dipartimento di Fisica, Università di Roma Tor Vergata, Via della Ricerca Scientifica 1, I-00133 Rome, Italy

E-mail: gianluca.stefanucci@roma2.infn.it

Abstract. We present results on the microscopic dynamics of electrons in nanoscale systems coupled to superconducting leads. By solving the time-dependent Bogoliubov-deGennes equations for the Nambu-Gorkov Keldysh Green's function we are able to calculate the current and the charge density when the system is perturbed by bias voltages. For scattering of electrons across a normal-superconducting surface we provide a time-dependent picture of the Andreev reflections. When the nanoscale system is contacted to two DC biased superconducting leads the amplitude of the current oscillations at even multiple of the bias can be extracted by Fourier transforming the time-dependent results. In the transient regime the dwelling time is inversely proportional to the bias in agreement with the occurrence of multiple Andreev reflections.

1. Introduction

The progressive miniaturization of electronic devices has led to the emerging field of nanoelectronics. Due to the reduced dimensionality, heat losses in integrated circuits of semiconductors or normal conductors will be the main reason of instabilities like, e.g., bond-breakings, electromigrations, etc. [1, 2, 3, 4]. Heat losses are considerably reduced in superconducting devices. Thus, superconducting nanoelectronics offers a promising and timely alternative to the semiconductor nanotechnology [5, 6, 7]. A large theoretical and experimental effort is currently devoted to, e.g., the realization of quantum bits (QUBIT) using atomic-size quantum point contacts or quantum dots coupled to superconducting leads. The state of a QUBIT can then be manipulated using electromagnetic pulses of the duration of few nano-seconds or even faster. The description of the ultrafast electron dynamics in such *Josephson nano-junctions* is not only of considerable fundamental interest but also of importance for their potential applications in future electronics.

The study of the electronic response to time-dependent ac field or train pulses in nanodevices chemically attached to superconducting leads has so far remained largely unexplored. There are several difficulties related to the construction of a feasible scheme already at a mean-field level. The system is an open system, the electronic energy scales are 2-3 orders of magnitude larger than a typical superconducting gap, the problem is intrinsically time-dependent even for DC biases (Josephson oscillations), the scattering states extends from $z = -\infty$ to $z = +\infty$, with z the longitudinal direction, and the possible formation of Andreev bound states (ABS) give rise to persistent oscillations in the density and current. The time-evolution of localized wave-packets scattering across a superconducting-normal (SN) interface was explored long ago [8, 9]. More recently the analysis has been extended to scattering states [10, 11, 12] in standard quantum transport geometries. However, there has been no attempt to calculate the time-dependent response of a superconductor-normal-superconductor (SNS) junction.

In this work we propose and illustrate a general scheme suited to study time-dependent (TD) quantum transport in SNS junctions. The propagation scheme is tested in popular model Hamiltonians that can describe quantum point contacts, atomic contacts, atomic chains, etc. We provide the first ever calculation of the time-dependent transient response of a Josephson nano-junction and illustrate the role played by the multiple Andreev reflections in the on-set of the Josephson regime. We also show the Andreev-reflection space-time pattern in long atomic chains sandwiched between a normal and a superconducting lead.

The plan of the paper is as follows: In the next Section we describe the theoretical framework based on the equation of motion for the Nambu-Keldysh Green's function in the BCS approximation. The Green's function projected on the region of interest obeys a Dyson-like equation with an embedding self-energy which accounts for the openness of the system. In Section 3 we briefly illustrate the main ideas of the propagation algorithm while in Section 4 we present the numerical results. Conclusions and outlooks are summarized in Section 5.

2. Equation of motion for the Nambu-Keldysh Green's function in the BCS approximation

We consider electrons interacting through an attractive Gorkov point-contact interaction $W_g(\mathbf{r})$. Without loss of generality we assume that the system is in equilibrium at negative times, $t < 0$. We are interested in studying the dynamics of the system if an external time-dependent perturbation is switched on at $t = 0$ [13, 14]. The system is described by a Hamiltonian $\hat{H}(t) = \hat{K}(t) + \hat{H}_g$ which is the sum of a quadratic term

$$\hat{K}(t) = \sum_{\sigma} \int \mathrm{d}\mathbf{r}\mathrm{d}\mathbf{r}' \; \psi_{\sigma}^{\dagger}(\mathbf{r}) \, \langle \mathbf{r} | \boldsymbol{K}_{\sigma}(t) | \mathbf{r}' \rangle \, \psi_{\sigma}(\mathbf{r}'), \qquad \sigma = \uparrow, \downarrow \tag{1}$$

with $\boldsymbol{K}_{\sigma}(t)$ a hermitean matrix, and of the following interaction operator

$$\hat{H}_g = \int \mathrm{d}\mathbf{r} \; W_g(\mathbf{r}) \, \psi_{\uparrow}^{\dagger}(\mathbf{r}) \psi_{\downarrow}^{\dagger}(\mathbf{r}) \psi_{\downarrow}(\mathbf{r}) \psi_{\uparrow}(\mathbf{r}). \tag{2}$$

The BCS approximation consists in discarding the correlations between the motion of two electrons added to the system. In normal systems the BCS approximation is the same as the Hartree-Fock approximation. In superconducting systems we further have to account for anomalous diagrams describing the amplitude of finding the system in the same state after the removal or the addition of two electrons. The normal and anomalous Keldysh-Green's functions are defined according to

$$\langle \mathbf{r} | \boldsymbol{G}_{\sigma}(z; z') | \mathbf{r}' \rangle = \frac{1}{i} \langle T_{\mathrm{K}} \left\{ \psi_{\sigma}(\mathbf{r}, z) \psi_{\sigma}^{\dagger}(\mathbf{r}', z') \right\} \rangle, \qquad \sigma = \uparrow, \downarrow \tag{3}$$

$$\langle \mathbf{r} | \boldsymbol{\mathcal{F}}(z; z') | \mathbf{r}' \rangle = \frac{1}{i} \langle T_{\mathrm{K}} \left\{ \psi_{\downarrow}(\mathbf{r}, z) \psi_{\uparrow}(\mathbf{r}', z') \right\} \rangle, \tag{4}$$

$$\langle \mathbf{r} | \overline{\boldsymbol{\mathcal{F}}}(z; z') | \mathbf{r}' \rangle = -\frac{1}{i} \langle T_{\mathrm{K}} \left\{ \psi_{\uparrow}^{\dagger}(\mathbf{r}', z') \psi_{\downarrow}^{\dagger}(\mathbf{r}, z) \right\} \rangle, \tag{5}$$

where z, z' run on the Keldysh contour γ which is an oriented contour composed by an upper branch going from 0 to ∞, a lower branch going from ∞ to 0 and a purely imaginary (thermal) segment going from 0 to $-i\beta$ (here β is the inverse temperature). The operator T_{K} in the above definitions is the contour ordering operator and move operators with later contour variable to the left (an extra minus sign has to be included for odd permutations). We have introduced the short hand notation

$$\langle T_{\rm K}\{\ldots\}\rangle = \frac{{\rm Tr}\left[e^{\beta\mu\hat{N}}T_{\rm K}\left\{e^{-i\int_\gamma {\rm d}\bar{z}\,\hat{H}(\bar{z})}\ldots\right\}\right]}{{\rm Tr}\left[e^{\beta\mu\hat{N}}T_{\rm K}\left\{e^{-i\int_\gamma {\rm d}\bar{z}\,\hat{H}(\bar{z})}\right\}\right]}, \tag{6}$$

for denoting averages. In Eq. (6) the quantity μ is the chemical potential and $\hat{N}$ is the total number of particle operator. For z on the upper/lower branch of γ the Hamiltonian $\hat{H}(z)$ is equal to $\hat{H}(t)$, with t the distance of z from the origin. For z on the thermal segment $\hat{H}(z) = \hat{H}(0)$. Introducing the Nambu-Gorkov Keldysh-Green's function

$$\underline{\boldsymbol{G}}(z;z') = \begin{bmatrix} \boldsymbol{G}_\uparrow(z;z') & -\boldsymbol{\mathcal{F}}^T(z';z) \\ \overline{\boldsymbol{\mathcal{F}}}(z;z') & -\boldsymbol{G}_\downarrow^T(z';z) \end{bmatrix}, \tag{7}$$

where the superscript T denotes the transpose of the indicated quantity, the equation of motion in the BCS approximation can be cast in a rather elegant form

$$\left\{ i\frac{\overrightarrow{\rm d}}{{\rm d}z}\underline{\boldsymbol{1}} - \underline{\boldsymbol{K}}(z) \right\} \underline{\boldsymbol{G}}(z;z') = \delta(z-z'). \tag{8}$$

In Eqs. (8) the effective BCS Hamiltonian reads

$$\underline{\boldsymbol{K}}(z) = \begin{bmatrix} \boldsymbol{K}_\uparrow(z) + \boldsymbol{\Sigma}_\uparrow(z) & \boldsymbol{\Delta}(z) \\ \overline{\boldsymbol{\Delta}}(z) & -\boldsymbol{K}_\downarrow^T(z) - \boldsymbol{\Sigma}_\downarrow^T(z) \end{bmatrix}, \tag{9}$$

where $\boldsymbol{\Sigma}_{\uparrow/\downarrow}(z)$ is the Hartree-Fock self-energy (which in this case is the same as the Hartree self-energy since the Gorkov interaction is point-like) while

$$\langle \mathbf{r}|\boldsymbol{\Delta}(z)|\mathbf{r}'\rangle = i\delta(\mathbf{r}-\mathbf{r}')W_g(\mathbf{r})\,\langle \mathbf{r}|\boldsymbol{\mathcal{F}}(z^+;z)|\mathbf{r}\rangle \tag{10}$$

is the so called pairing potential or order parameter. From the definition of the normal and anomalous Green's function the equation of motion has to be solved with the boundary conditions

$$\underline{\boldsymbol{G}}(0_-;z') = -e^{\beta\mu\underline{\boldsymbol{\sigma}}}\,\underline{\boldsymbol{G}}(-i\beta;z'), \qquad \underline{\boldsymbol{G}}(z;0_-) = -\underline{\boldsymbol{G}}(z;-i\beta)\,e^{-\beta\mu\underline{\boldsymbol{\sigma}}}, \tag{11}$$

where $\underline{\boldsymbol{\sigma}} = \begin{bmatrix} \boldsymbol{1} & 0 \\ 0 & -\boldsymbol{1} \end{bmatrix}$ is the third Pauli matrix.

Let us now specialize to the transport setup. The one-particle Hilbert space can be partitioned in a L/R electrode and a central region C, see Fig. 1. For simplicity we here also consider the case $\boldsymbol{K}_\uparrow(t) = \boldsymbol{K}_\downarrow(t) = \boldsymbol{\mathcal{E}}(t) + \boldsymbol{\mathcal{V}}$. The matrix $\boldsymbol{\mathcal{E}}(t) = \boldsymbol{\mathcal{E}} + \boldsymbol{\mathcal{U}}(t)$ descibes the normal part of the isolated L, R and C regions and is composed of an equilibrium time-independent matrix $\boldsymbol{\mathcal{E}}$ and of a time-dependent local perturbation $\boldsymbol{\mathcal{U}}(t)$ which represents the bias (in the leads) and/or a gate voltage (in the central region), $\boldsymbol{\mathcal{U}}_{ij}(t) = \delta_{ij}U_i(t)$. The matrix $\boldsymbol{\mathcal{V}}$ accounts for the hopping between different regions.

Introducing the uncontacted Green's function $\underline{\boldsymbol{g}}$ which obeys the equation of motion (8) with $\boldsymbol{\mathcal{V}} = 0$ and the same boundary conditions as $\underline{\boldsymbol{G}}$ the equation of motion for $\underline{\boldsymbol{G}}$ projected onto region CC takes the form

$$\left\{ i\frac{\overrightarrow{\rm d}}{{\rm d}z}\underline{\boldsymbol{1}} - \underline{\boldsymbol{K}}_{CC}(z) \right\} \underline{\boldsymbol{G}}(z;z') = \delta(z-z') + \int {\rm d}\bar{z}\,\underline{\boldsymbol{\Sigma}}(z;\bar{z})\,\underline{\boldsymbol{G}}_{CC}(\bar{z};z'), \tag{12}$$

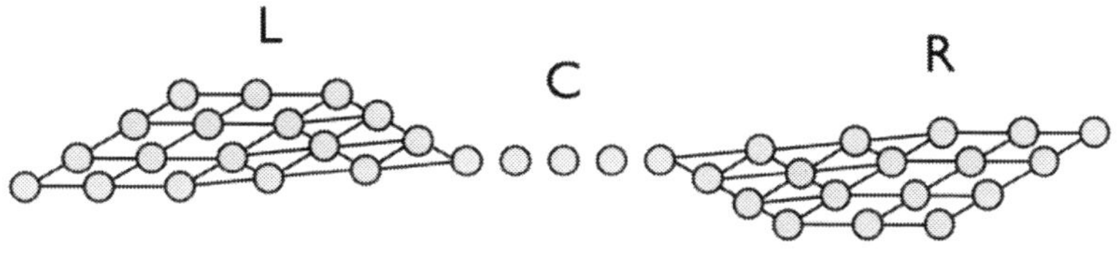

Figure 1: Schematic of the transport set-up

where the embedding self-energy is defined according to

$$\underline{\mathbf{\Sigma}}(\bar{z};\bar{z}') = \sum_{\alpha=L,R} \underline{\mathbf{\Sigma}}_\alpha(\bar{z};\bar{z}') = \sum_{\alpha=L,R} \underline{\boldsymbol{\mathcal{V}}}_{C\alpha}\, \underline{\boldsymbol{g}}_{\alpha\alpha}(\bar{z};\bar{z}')\, \underline{\boldsymbol{\mathcal{V}}}_{\alpha C}. \tag{13}$$

We here wish to emphasize that the equation of motion is defined on the generalized Keldysh contour. Converting the above equation in equations for real times we obtain a set of coupled equations known as Kadanoff-Baym equations [15, 16, 17, 18] recently implemented to study correlated quantum transport [19, 20]. Below we briefly describe how to solve Eq. (12) with a pure embedding self-energy.

3. Algorithm for the time-propagation

Physically observable quantities like the current and the density can be extracted from the lesser component of the Green function. From Eq. (8) one can show that $\underline{\boldsymbol{G}}^{\lessgtr}(t;t') = i\sum_q f^{\lessgtr}(E_q)|\Psi_q(t)\rangle\langle\Psi_q(t')|$ where $f^<(\omega) = 1/[1+\exp(\beta\omega)]$ is the Fermi function and $f^>(\omega) = f^<(\omega) - 1$. The states $|\Psi_q(0)\rangle$ are eigenstates of $\underline{\boldsymbol{K}} - \mu\underline{\boldsymbol{\sigma}}$ with eigenvalue E_q and evolve in time according to the Schrödinger equation $i\frac{\mathrm{d}}{\mathrm{d}t}|\Psi_q(t)\rangle = \underline{\boldsymbol{K}}(t)|\Psi_q(t)\rangle$, which, in this case, is equivalent to the time-dependent Bogoliubov-deGennes (BdG) equation. For the calculation of the initial states we used the generalized waveguide approach of Ref. [21] properly extended to superconducting leads. We consider homogeneous time-dependent perturbations in electrode $\alpha = L, R$, i.e., $\boldsymbol{K}_{\uparrow/\downarrow,\alpha\alpha}(t) = \boldsymbol{K}_{\uparrow/\downarrow,\alpha\alpha}(0) + U_\alpha(t)\mathbf{1}_\alpha$ (where $\mathbf{1}_\alpha$ is the identity matrix in region α). Assuming that the constriction does not alter the equilibrium of the macroscopically large superconducting reservoirs we have

$$\underline{\boldsymbol{K}}_{\alpha\alpha}(t) = e^{-i\mu_\alpha(t)\underline{\boldsymbol{\sigma}}_\alpha}\left(\underline{\boldsymbol{K}}_{\alpha\alpha}(0) + U_\alpha(t)\underline{\boldsymbol{\sigma}}_\alpha\right)e^{i\mu_\alpha(t)\underline{\boldsymbol{\sigma}}_\alpha}, \tag{14}$$

with $\underline{\boldsymbol{\sigma}}_\alpha$ the third Pauli matrix in electrode α and $\mu_\alpha(t) = \mu t + \int_0^t \mathrm{d}\bar{t}\, U_\alpha(\bar{t})$. In writing Eqs. (14) we have made use of the fact that a bulk superconductor perturbed by a homogeneous time-dependent potential has a pairing potential with a time-dependent phase which guarantess the gauge invariance of the equations. For region C there are no restrictions on the time-dependence of the perturbation. The BdG equation for every single one-particle state $|\Psi\rangle$ is numerically solved using the Crank-Nicholson formula

$$\left(\underline{\mathbf{1}} + i\delta\underline{\boldsymbol{K}}^{(m)}\right)|\Psi^{(m+1)}\rangle = \left(\underline{\mathbf{1}} - i\delta\underline{\boldsymbol{K}}^{(m)}\right)|\Psi^{(m)}\rangle, \tag{15}$$

where $|\Psi^{(m)}\rangle = |\Psi(t_m)\rangle$, $\underline{\boldsymbol{K}}^{(m)} = \frac{1}{2}\left[\underline{\boldsymbol{K}}(t_{m+1}) + \underline{\boldsymbol{K}}(t_m)\right]$, and $t_m = 2\delta m$. Eq. (15) is of no practical use since the matrices representing the operators are infinite-dimensional. It is possible, however, to extract from Eq. (15) an equation for the projection of $|\Psi\rangle$ on region C only [22, 23]

$$\left(\underline{\mathbf{1}}_C + i\delta\underline{\boldsymbol{K}}_{\text{eff}}^{(m)}\right)|\Psi_C^{(m+1)}\rangle = \left(\underline{\mathbf{1}}_C - i\delta\underline{\boldsymbol{K}}_{\text{eff}}^{(m)}\right)|\Psi_C^{(m)}\rangle$$

$$
\begin{aligned}
&- i\delta \sum_{\alpha=L,R} \underline{z}_\alpha^{(m)} \underline{K}_{C\alpha}\, \underline{g}_{\alpha\alpha}^{m} \left(\underline{1}_\alpha + \underline{g}_{\alpha\alpha}\right) |\Psi_\alpha^{(0)}\rangle \\
&- \delta^2 \sum_{\alpha=L,R} \sum_{j=0}^{m-1} \underline{z}_\alpha^{(m)} \left(\underline{Q}_\alpha^{(j+1)} + \underline{Q}_\alpha^{(j)}\right) \underline{\bar{z}}_\alpha^{(m-1-j)} \\
&\times \left(|\Psi_C^{(m-j)}\rangle + |\Psi_C^{(m-1-j)}\rangle\right). \qquad (16)
\end{aligned}
$$

The effective Hamiltonian $\underline{K}_{\rm eff}^{(m)} = \underline{K}_{CC}^{(m)} - i\delta \sum_{\alpha=L,R} \underline{z}_\alpha^{(m)} \underline{Q}_\alpha^{(0)} \underline{\bar{z}}_\alpha^{(m)}$ has the size of region C and is expressed in terms of the phase factors $\underline{z}_\alpha^{(m)} = \frac{1}{2}\left(e^{-i\mu_\alpha(t_{m+1})\underline{\sigma}_C} + e^{-i\mu_\alpha(t_m)\underline{\sigma}_C}\right)$, and embedding matrices $\underline{Q}_\alpha^{(m)} = \frac{1}{2}\underline{K}_{C\alpha}\, \underline{g}_{\alpha\alpha}^{m} \left(\underline{1}_\alpha + \underline{g}_{\alpha\alpha}\right) \underline{K}_{\alpha C}$ with propagator $\underline{g}_{\alpha\alpha} = (\underline{1}_\alpha - i\delta \underline{K}_{\alpha\alpha})/(\underline{1}_\alpha + i\delta \underline{K}_{\alpha\alpha})$. More details on the practical implementation of the numerical algorithm will be given in a future publication.

4. Numerical results

In this Section we present results on NNS and SNS junctions obtained using the time-evolution algorithm described in the previous Section. In a typical superconducting nano-junction set-up the band-width of the leads, W, is the largest energy scale. If the leads are connected to a central region with hopping integrals $V_{\rm hop} \ll W$ the time-dependent density and current in region C depend only on the ratio $\Gamma = 8V_{\rm hop}^2/W$; the quantity Γ provides the broadening of the energy levels of region C. The geometrical structure of the leads is, therefore, of no relevance and in the calculations below we model the superconducting leads as one-dimensional semi-infinite tight-binding chains with nearest neighbor hopping $V_L = V_R = V$; the band-width is then $W = 4|V|$. To illustrate the fundametal physical phenomena occuring in a nano-scale Josephson junction we here consider one-dimensional central regions with nearest neighbor hopping V_C. Numbering the site of region C from left to right as $1, 2, \ldots, N_C$ we further consider that the hopping between L and C occur only via the end-site of L and site 1 of C while the hopping between R and C via the end-site of R and site N_C of C. The corresponding hopping parameters are denoted by V_{LC} and V_{RC}. We wish to observe that these kind of model junctions can describe several physically relevant situations, ranging from quantum-point contacts to quantum dot arrays or atomic chains. The pairing potential is taken to be constant in lead $\alpha = L, R$ and equal to Δ_α. In the central region we set the pairing potential to 0. This means that the pairing potential in the central region is not calculated self-consistently but rather constitues an input of the problem. All calculations presented below have been performed with the chemical potential $\mu = 0$ and the temperature $T = 1/\beta = 0$.

The first system we study is an NNS junction with $V = 50$, $V_{LC} = V_{RC} = 7$ and $N_C = 21$ sites with nearest neighbor hopping $V_C = 7.5$. The pairing potential is chosen as $\Delta_L = 0$ and $\Delta_R = 1$. With these parameters the broadening $\Gamma \sim 2$ and the superconducting gap is therefore $2\Delta_R \sim \Gamma$: this is the so called intermediate coupling regime. The generic right-going scattering state can be written as a superposition of a spin-up electron ($c_{kL\uparrow}^\dagger + R_+ c_{-kL\uparrow}^\dagger$) and a spin-down hole $c_{-kL\downarrow}$ wave-functions in lead L and as a superposition of the two independent superconducting quasi-particles, or bogolons, $\alpha_{kR}^\dagger = u_k c_{kR\uparrow}^\dagger - v_k c_{-kR\downarrow}$ and $\beta_{kR}^\dagger = u_k c_{-kR\downarrow}^\dagger + v_k c_{kR\uparrow}$ wave-functions in lead R. The reflection and transmission coefficients as a function of energy are displayed in Fig. 2. In the energy window considered the probability amplitude R_- for an electron to be reflected as a hole is rather small except in the vicinity of the middle of the gap. A perfect Andreev reflection (AR) occurs at zero energy, i.e., the spin-up electron is entirely reflected as a spin-down hole. In perfect Andreev conditions the incident $k\uparrow$ electron is not reflected, so it goes across the boundary and must be found in lead R either as such or as a $(k\uparrow, -k\downarrow)$ Copper pair arising from the incoming electron bound to the one that left the spin-down hole

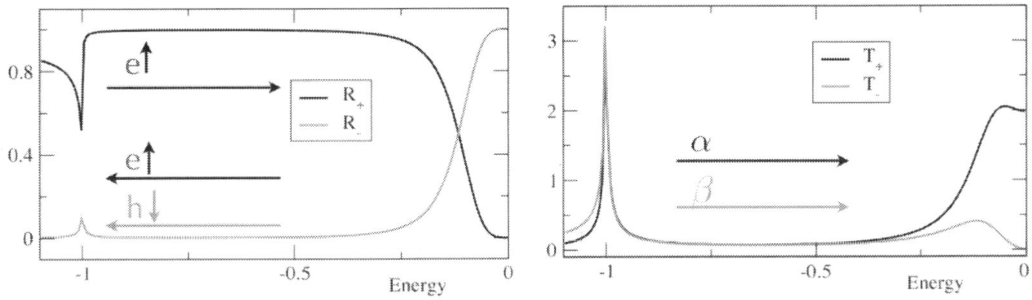

Figure 2: Reflection and transmission amplitudes for a spin-up electron incident from lead L as a function of its energy. Left panel: Reflection coefficient for spin-up electron and spin-down hole in lead L. Right panel: Transmission coefficients for the bogolon wave-functions α and β.

in the L lead. By creating a β bogolon in lead R one would reach a many-body state where the incident electron is destroyed. This cannot happen since that electron is transmitted with probability 1. Infact such an amplitude may be seen to vanish in the right panel of Fig. 2. The α bogolon amplitude instead is allowed and it corresponds to detecting a Cooper pair. There is considerable structure also at energy $E = -1$, at the bottom of the gap. The AR probability is of the order 1/2 (left panel). The right panel shows that one has essentially the same chance of observing an α or β bogolon. Both events can be understood as incomplete Andreev scatterings.

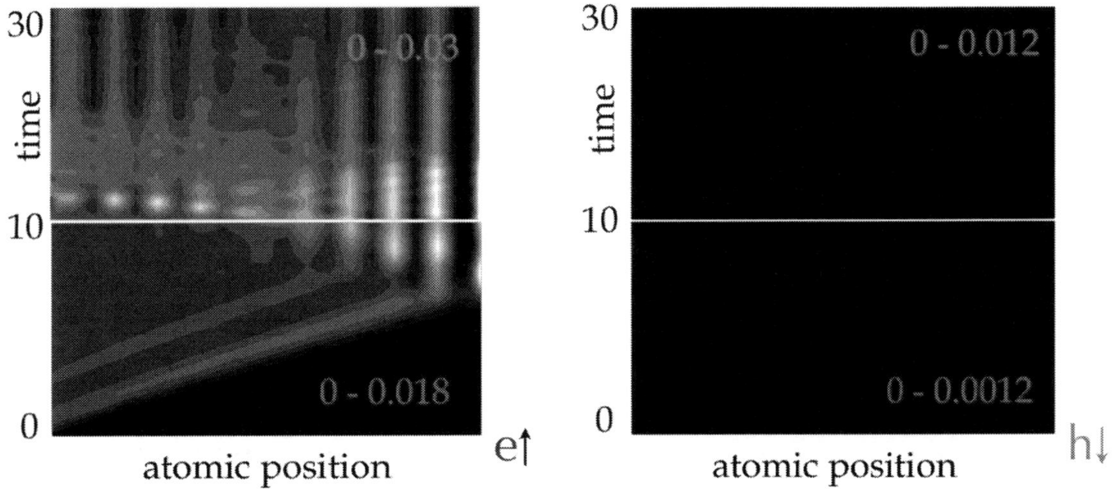

Figure 3: Contour plot of the spin-up electron (left panel) and spin-down hole (right panel) density change along the $N_C = 21$ sites of the central chain as a function of time. The applied bias is $U_L = -0.5$. The numbers in each contour plot represent the minimum (black) and maximum (white) value of the density.

To provide a time-dependent picture of the AR we consider the system in which the normal lead L is initially uncontacted and filled with electrons up to the chemical potential $\mu = 0$. At time $t = 0$ we switch on the contact V_{LC} as well as a DC bias U_L and propagate in time *only* the spin-up electron states in lead L with energy in the range $(-1, 0)$. In Fig. 3 we display the contour plot of the change in the spin-up electron density and spin-down hole density in the

$N_C = 21$ sites of the chain for bias $U_L = -0.5$. We observe that the change in density must be entirely attributed to the propagation of spin-up electrons from lead L. As expected from Fig. 2 AR's are strongly suppressed for $U_L = -0.5$ and the electrons are entirely reflected as electrons. Correspondingly, no change in the hole-density is detected, see right panel of Fig. 3. In the long-time limit the density profile remains constant and the period of its corrugation agrees with the period of the Friedel oscillations.

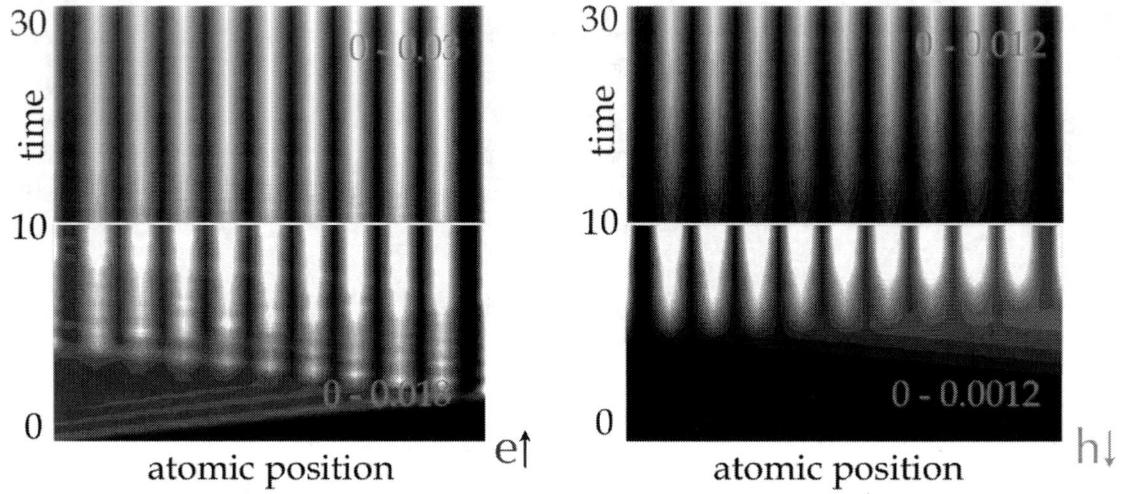

Figure 4: Same as in Fig. 3 but with bias $U_L = 0$.

The situation is completely different if $U_L = 0$. In this case electrons at the Fermi energy undergo a perfect AR and we should be able to appreciate a change in the spin-down hole density. Fig. 4 shows that this is indeed the case. We notice that the electron density front hits the superconducting lead earlier than in Fig. 3 since the Fermi velocity is larger. No change in the hole density is visible before the electrons have time to scatter. From that time on, however, we can clearly observe a hole density front propagating backward. Due to charge conservation a carrier with charge two times larger than the electron charge, i.e., a Copper pair, is formed in lead R and moves right. Fig. 4 thus provides a time-dependent picture of the normal-current supercurrent conversion at the NS inteface.

We next consider an SNS junction with $V = 100$, $V_{CL} = V_{CR} = -7$, $\Delta_L = \Delta_R = 1$ and $N_C = 2$ sites connected with a hopping integral $V_C = -1.5$. The system admits two couple of degenerate ABS at energy $\sim \pm 0.89$. In Fig. 5, left panel, we show the time-dependent change of the density in site 1 and 2 when a DC bias U_L is switched on in the left lead. After a transient the density oscillates in time and the period of the oscillations is $T = 2\pi/(2U_L)$, as expected. The transient time is longer the smaller is the bias. This is caused by the occurrence of multiple AR at the junction. An electron with energy U_L with respect to the top of the valence band has to undergo Δ/U_L reflections before can be transmitted. To the best of our knowledge Fig. 5 contains the first explict calculation of the time-dependent response of a SNS junction to a constant DC bias. The transient regime, the on-set of the Josephson oscillations and the Josephson regime are all clearly visible.

An extra merit of performing real-time evolutions is the possibility of calculating the frequency components of the Josephson current by simply Fourier transforming the time-dependent curves in the Josephson regime. At present the calculation of the Fourier components of the current

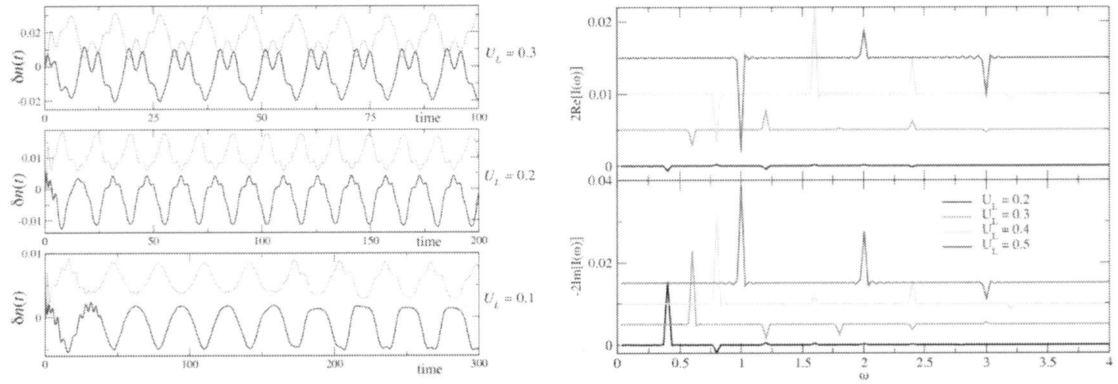

Figure 5: Left panel: Time dependent density change $\delta n(t) \equiv n(t) - n(0)$ in site 1 [black (thick)] and 2 [red (thin)] when a DC bias $U_L = 0.1, 0.2, 0.3$ is applied to the left superconducting lead. The other parameters are given in the main text. The onset of the Josephson regime is clearly visible. Right panel: Discrete Fourier transform of the time-dependent current flowing between site 1 and 2. The curves corresponding to bias $U_L = 0.n$ are shifted upward by $0.005(n-2)$ for clarity.

is performed by combining scattering formalisms with Flocquet theory and can be shown to be equivalent to the solution of a tridiagonal matrix system[24, 25, 26]. In Fig. 5, right panel, we report the real and imaginary part of the discrete Fourier transform of the current flowing between site 1 and 2. The curves exhibit peaks at frequency $2nU_L$ as expected. We observe that the method here used can also be employed to calculate the frequency spectrum of the current when the system is exposed to an AC bias. Such study will permit us to obtain the intensity of the Shapiro spikes in nanoscale transport experiment.

5. Conclusions and outlooks

We reported preliminary results of a novel time-propagation scheme applied to NNS and SNS nano-junctions. The entire approach is based on the BCS approximation to deal with superconducting correlations in the leads. However, it can be combined with the superconducting extension of density functional theory (DFT) [27] and of TDFFT [28] to go beyond a mean field description and include exchange-correlation effects. We have been able to provide a time-dependent picture of a non-interacting electron gas scattering across a normal-superconducting interface and to observe the occurrence of AR's. We also calculated the transient response of a SNS junction to a sudden switch on of the bias. The transient regime is dominated by multiple AR's and is longer the smaller is the bias. After the transient regime a Josephson regime sets in and both current and density oscillate in time. The frequency spectrum of the current can be computed by a simple discrete Fourier transform and reveals the presence of sharp peaks at even multiple of the bias. Our results lend themselves to experimental tests and may be of relevance for technological applications.

References

[1] R. S. Sorbello, Solid State Phys. **51**, 159 (1997).
[2] T. N. Todorov, J. Hoekstra, and A. P. Sutton, Phys. Rev. Lett. **86**, 3606 (2001).
[3] M. Di Ventra, S. T. Pantelides, N. D. Lang, Phys. Rev. Lett. **88**, 046801 (2002).
[4] C. Verdozzi, G. Stefanucci and C.-O. Almbladh, Phys. Rev. Lett. **97**, 046603 (2006).
[5] C. J. Lambert and R. Raimondi, J. Phys.: Condens. Matter **10**, 901 (1998).
[6] Y. Makhlin, G. Schön and A. Shnirman, Rev. Mod. Phys. **73**, 357 (2001).

[7] J.-P. Cleuziou, W. Wernsdorfer, V. Bouchiat, T. Ondarchuhu and M. Monthioux, Nature Nanotech. **1**, 251 (2006).
[8] R. Kümmel, Z. Phys. **218**, 472 (1969).
[9] H. De Raedt, K. Michielsen and T. M. Klapwijk, Phys. Rev. B **50**, 631 (1994)
[10] A. Jacobs and R. Kümmel, Phys. Rev. B **64**, 104515 (2001).
[11] Y. Xing, Q.-f. Sun and J. Wang, Phys. Rev. B **75**, 125308 (2007).
[12] E. Perfetto, G. Stefanucci and M. Cini, Phys. Rev. B **80**, 205408 (2009).
[13] M. Cini, Phys. Rev. B **22**, 5887 (1980).
[14] G. Stefanucci, and C.-O. Almbladh, Phys. Rev. B **69**, 195318 (2004).
[15] L. P. Kadanoff and G. Baym, *Quantum Statistical Mechanics* (Benjamin, New York, 1962).
[16] P. Danielewicz, Ann. Phys. (N.Y.) **152**, 239 (1984).
[17] N.-H. Kwong and M. Bonitz, Phys. Rev. Lett. **84** 1768 (2000).
[18] N. E. Dahlen and R. van Leeuwen, Phys. Rev. Lett., **98** 153004 (2007).
[19] P. Myöhänen, A. Stan, G. Stefanucci and R. van Leeuwen, Europhys. Lett. **84**, 67001 (2008).
[20] P. Myöhänen, A. Stan, G. Stefanucci and R. van Leeuwen, Phys. Rev. B **80**, 115107 (2009).
[21] G. Stefanucci, E. Perfetto, S. Bellucci and M. Cini, Phys. Rev. B **79**, 073406 (2009).
[22] S. Kurth, G. Stefanucci, C.-O. Almbladh, A. Rubio and E. K. U. Gross, Phys. Rev. B **72**, 035308 (2005).
[23] G. Stefanucci, E. Perfetto and M. Cini, Phys. Rev. B **77**, 075339 (2008).
[24] Q.-f. Sun, H. Guo and J. Wang, Phys. Rev. B **65**, 075315 (2002).
[25] D. F. Martinez, J. Phys. A **36**, 9827 (2003).
[26] G. Stefanucci, S. Kurth, A. Rubio and E. K. U. Gross, Phys. Rev. B **77**, 075339 (2008).
[27] L. N. Oliveira, E. K. U. Gross and W. Kohn, Phys. Rev. Lett. **60**, 2430 (1988).
[28] O.-J. Wacker, R. Kümmel and E. K. U. Gross, Phys. Rev. Lett. **73**, 2914 (1994).

Journal of Physics: Conference Series **220** (2010) 012013 doi:10.1088/1742-6596/220/1/012013

Quantum breathing mode of interacting particles in harmonic traps

Sebastian Bauch, David Hochstuhl, Karsten Balzer, and Michael Bonitz

Institut für Theoretische Physik und Astrophysik, Christian-Albrechts-Universität Kiel, Leibnizstrasse 15, 24098 Kiel, Germany

E-mail: bauch@theo-physik.uni-kiel.de

Abstract. The breathing mode — the uniform radial expansion and contraction of a system of interacting particles — is analyzed. Extending our previous work [Bauch et al 2009 *Phys. Rev. B.* **80** 054515] we present a systematic analysis of the breathing mode for fermions with an inverse power law interaction potential $w(r) \sim r^{-d}$ with $d = 1, 2, 3$ in the whole range of coupling parameters. The results thus cover the range from the ideal "gas" to the Wigner crystal-like state. In addition to exact results for two particles obtained from a solution of the time-dependent Schrödinger equation we present results for $N = 4, 6$ from multiconfiguration time-dependent Hartree-Fock simulations.

1. Introduction

Quantum systems in traps and their time-dependent properties are of growing interest in many fields, including correlated electrons in metal clusters [1] or quantum dots [2, 3, 4] and ultracold Bose and Fermi gases in traps or optical lattices, for recent overviews see e.g. [5, 6]. Particular attention has recently been devoted to Bose-Einstein condensation in low dimensions [7] and to the analysis of nonideality (interaction) effects [8, 9, 10, 11], including superfluidity and crystallization, see [12].

In the case of strong coupling, the collective static and dynamic properties of these systems are largely determined by their collective modes [8, 9, 6]. It was shown in a recent paper [13] that, among them, the monopole or breathing mode (BM) which is easily excited experimentally [9] carries particularly valuable information on the system dimensionality d, the spin statistics of the particles and on the form and relative strength λ_d of their pair interaction given by Eqs. (4, 7-9).

In [13], the authors concentrated on the case of Coulomb interaction. Here, this analysis is extended to a more general interaction potential of inverse power form, $w(r) \sim r^{-d}$. In particular, new numerical results are presented for the cases $d = 2$ and $d = 3$. Moreover, the previous exact results were limited to two particles whereas for $N = 3$ and $N = 4$ time-dependent Hartree-Fock calculations were performed. Here we report results for $N \leq 6$ from Multiconfiguration Time-dependent Hartree-Fock calculations (MCTDHF) which include correlation corrections to the mean-field case.

The paper is organized as follows: first we introduce the system under consideration and our approach to model the basic properties of interacting trapped particles at arbitrary coupling by means of solving the two-particle time-dependent Schrödinger equation in one spatial dimension. Following the discussion of the excitation and (theoretical) measurement of the collective modes in such a quantum system we switch to the case of $1/r^2$ interaction. Here, analytical results for the breathing mode offer profound benchmarks for our approach. We then investigate the fundamental cases of Coulomb and dipole interacting particles within a broad range of coupling strengths in detail and apply MCTDHF methods for more than two particles. We close with a conclusion and an outlook for further calculations, addressing the open questions arising from our work.

2. Model and solution procedure

Let us consider a system of N interacting particles in a harmonic trapping potential. The time-dependent Schrödinger equation (TDSE) for such a system reads

$$i\hbar\frac{\partial}{\partial t}\Psi(\boldsymbol{r}_1,...,\boldsymbol{r}_N) = \hat{H}\Psi(\boldsymbol{r}_1,...,\boldsymbol{r}_N,t), \tag{1}$$

with the N-particle Hamilton operator, given by

$$\hat{H} = \sum_{i=1}^{N}\hat{h}_i + \sum_{i\neq j}^{N} w(|\boldsymbol{r}_i - \boldsymbol{r}_j|)\,. \tag{2}$$

The single-particle hamiltonian, $\hat{h}_i$, is defined as the sum of the kinetic energy and the external single particle potential,

$$\hat{h}_i = \hat{t}_i + v_i(\boldsymbol{r}_i)\,. \tag{3}$$

The interaction is given by a binary interaction potential, depending on the distance of particle i and j,

$$w(|\boldsymbol{r}_i - \boldsymbol{r}_j|) = \frac{a_d}{|\boldsymbol{r}_i - \boldsymbol{r}_j|^d}\,, \tag{4}$$

with a d-dependent interaction parameter a_d (e.g. $a_1 = e^2/4\pi\epsilon_0$). Finally, one arrives at

$$i\hbar\frac{\partial}{\partial t}\Psi(\boldsymbol{r}_1,\ldots,\boldsymbol{r}_N;t) = \left(\frac{1}{2m}\sum_i \hat{P}_i^2 + \frac{1}{2}m\Omega^2\sum_i \boldsymbol{r}_i^2 + a_d\sum_{i<j}\frac{1}{|\boldsymbol{r}_i - \boldsymbol{r}_j|^d}\right)\Psi(\boldsymbol{r}_1,\ldots,\boldsymbol{r}_N;t)\,. \tag{5}$$

We will concentrate on identical particles of equal masses, charges, dipole moments and spin projections in this work.

We now introduce scaled quantities, $\tilde{\boldsymbol{r}}_i = \sqrt{\frac{m\Omega}{\hbar}}\boldsymbol{r}_i$, $\tilde{\boldsymbol{P}}_i = \sqrt{\frac{1}{m\hbar\Omega}}\boldsymbol{P}_i$ and $\tilde{t} = t/\Omega$, which correspond to a rescaling of lengths, times and energies ($\tilde{E} = E/\hbar\Omega$). With the spatial coordinate representation of $\tilde{p}_i = -i\frac{\partial}{\partial \tilde{\boldsymbol{r}}_i}$ and omitting the tilde symbol in the following, we arrive at the dimensionless form of Eq. (5):

$$i\frac{\partial}{\partial t}\Psi(\boldsymbol{r}_1,\ldots,\boldsymbol{r}_N;t) = \left(-\frac{1}{2}\sum_i\frac{\partial^2}{\partial \boldsymbol{r}_i^2} + \sum_i \boldsymbol{r}_i^2 + \underbrace{\frac{m a_d}{\hbar^2}l_0^{2-d}}_{\lambda_d}\sum_{i<j}\frac{1}{|\boldsymbol{r}_i - \boldsymbol{r}_j|}\right)\Psi(\boldsymbol{r}_1,\ldots,\boldsymbol{r}_N;t)\,, \tag{6}$$

where λ_d is the interaction parameter, which contains all physical constants occurring in Eq. (5). For the cases considered in this work, we obtain:

(i) Coulomb interaction:

$$\lambda_1 = \frac{e^2}{4\pi\epsilon_0}\frac{m}{\hbar^2}\cdot l_0 \tag{7}$$

(ii) inverse square interaction:

$$\lambda_2 = a_2\frac{m}{\hbar^2} \tag{8}$$

(iii) and dipole interaction:

$$\lambda_3 = a_3\frac{m}{\hbar^2}\cdot\frac{1}{l_0} \tag{9}$$

with the oscillator length $l_0 = \sqrt{\hbar/m\Omega}$. The parameter λ_d describes the coupling of the system, i. e., the relative interaction strength between the particles, which can be tuned in experiments, e. g. by changing the frequency of the external trapping potential. We note here, that for $d = 2$ λ_2 is independent of l_0 and thus independent of Ω. Therefore, other physical parameters need to be changed (e.g. charge, mass, ...) in order to vary the interaction strength.

2.1. Breaking down to a two-particle problem

It has been shown for classical systems, corresponding to the case of $\lambda \to \infty$, that a purely radial and uniform breathing mode exists only in the following cases[14]:

(i) for clusters with special symmetries for any interaction

(ii) for every N if the interaction potential has the form $w(r) \propto 1/r^n$ or $w(r) \propto \log r$

Furtheron, it has been proven, that the classical breathing mode is *N-independent* (which is also true for the non-interacting quantum gas with $\lambda = 0$), which lets us conclude, that most of the properties should be present in the fundamental two-body problem. However, the independence of the breathing frequency on the particle number N is still an open question in the case of interacting quantum particles at intermediate coupling strengths which is addressed in the following section.

The TDSE for two interacting particles in a one-dimensional harmonic trap follows directly from Eq. (6) and reads as

$$i\frac{\partial}{\partial t}\Psi(x_1,x_2,t) = \left[-\frac{1}{2}\left(\frac{\partial^2}{\partial x_1^2}+\frac{\partial^2}{\partial x_2^2}\right)+\frac{r_1^2+r_2^2}{2}+\frac{\lambda_d}{(x_1-x_2)^d}\right]\Psi(x_1,x_2,t)\,. \tag{10}$$

To avoid divergencies in this low-dimensional system, the interaction potential needs to be regularized. We follow the "standard" procedure by using the soft-core potential

$$w(x_1,x_2) = +\frac{\lambda_d}{[(x_1-x_2)^2+\kappa^2]^{d/2}}\,, \tag{11}$$

with a small cut-off parameter κ. A brief remark on the influence of this parameter on our results is given below in section 3.1 in the discussion of the $1/r^2$ potential. A more detailed discussion, also addressing the influence on different states (antisymmetric/symmetric), is to be found in [13]. We may conclude here, that our results do not significantly depend on the special choices of κ, and are thus relevant for the limiting case of $\kappa \to 0$.

Obviously, the two-particle problem can be simplified to a (coupled system) of two ODEs, by transformation to (rescaled) center of mass and relative coordinates and utilizing a product ansatz $\Psi(R,r;t) = \phi(R/\sqrt{2};t)\cdot\varphi(\sqrt{2}r;t)$ for the wave function. In one spatial dimension, one finally obtains

$$i\partial_t\phi(R,t) = \left(-\frac{1}{2}\frac{\partial^2}{\partial R^2}+\frac{R^2}{2}\right)\phi(R,t) \tag{12}$$

for the scaled center of mass problem and

$$i\partial_t \varphi(r,t) = \left(-\frac{\partial^2}{2\partial r^2} + \frac{1}{2}r^2 + \frac{\lambda_d}{2^{d/2} r^d} \right) \varphi(r,t) \tag{13}$$

for the TDSE of the scaled relative part of the wave function. This already indicates the occurrence of two types of motions — one class of (interaction-dependent) oscillations governed by the TDSE of the relative problem, Eq. (13), in the following indicated with ω_r and one for the (interaction-independent) center of mass problem, labeled by ω_R. However, the general version of the splitting involves a time-dependent constant which couples both modes, and it is not clear from the beginning, if this mode coupling introduces additional spectral features. For an illustration of the time-dependent density, see the supplementary video to ref.[13]. Therefore, the full two-particle TDSE, Eq. (10), is solved numerically by using a grid-based Crank-Nicolson approach in combination with operator-splitting for the time evolution.

2.2. Initial state and excitation of the breathing mode

We initially prepare the system in its anti-symmetric ground sate, i. e. the energetically lowest state with a wave function obeyeing $\Psi(x_1, x_2, t=0) = -\Psi(x_2, x_1, t=0)$. This corresponds to a system of two spin-polarized fermions. However, for 1D systems with our type of interaction (singular at $x_1 = x_2$ besides the smoothing by κ), the Bose-Fermi-mapping is applicable and our results hold also for bosonic systems [15]. The symmetry of the wave function is, of course, preserved during (real) time propagation of the TDSE. For different types of interactions and interaction parameters λ, this initial state is obtained by propagating the full two-particle TDSE in imaginary time.

To excite the radial mode, we switch off the external trap. Then, the interaction and the quantum mechanical dispersion (especially in the case of small λ) drive the particles out of their equilibrium position during TDSE propagation in real time. After a short period of time (typically $t_{\text{off}} = 0.1$) the trap is restored and the breathing mode is excited. This type of excitation is completely radial, and therefore no dipole moment is induced (which would lead to a dipole or sloshing/Kohn mode of frequency Ω in the observable $\langle x \rangle(t)$, cf. Appendix). The special choice of t_{off} and the explicit switching procedure (e.g. smooth or instantaneously) play no significant role and do no affect the resulting breathing oscillations, cf. [13] for a detailed investigation including also other types of (time-dependent) excitations.

2.3. Observing the breathing mode

The breathing mode manifests in a characteristic oscillation of an adequate observable, which has to be sensitive on monopole oscillations. Among others, one possible quantity is the expectation value of x^2, which corresponds in our case to the expectation value of the potential energy. This quantity measures the width $\left(\langle x^2 \rangle - \langle x \rangle\right)(t)$ of the density in the trap, since the linear term $\langle x \rangle$ is absent owing to the vanishing dipole moment. An alternative approach is the calculation of $\langle |x| \rangle$ used as an observable in experiments on classical systems in [16]. However, the results for the breathing frequencies are the same [13]. We will only consider the case of $\langle x^2 \rangle(t)$ in the present work.

To extract the breathing frequencies ω_{r} and ω_{R}, we used a fitting procedure, which involves two independent frequencies:

$$f(t, \omega_{\text{r}}, \omega_{\text{R}}) = a \sin[\omega_R(t - t_0)] + b \sin[\omega_r(t - t_0')] + c\,. \tag{14}$$

The frequency ω_{R} related to the center of mass wave function has been fixed to the known exact analytical value of $\omega_{\text{R}} = 2$ which agrees perfectly with our simulation data. Thus, the existence of both modes results in a clearly visible beating in the time series of the potential energy in the simulation.

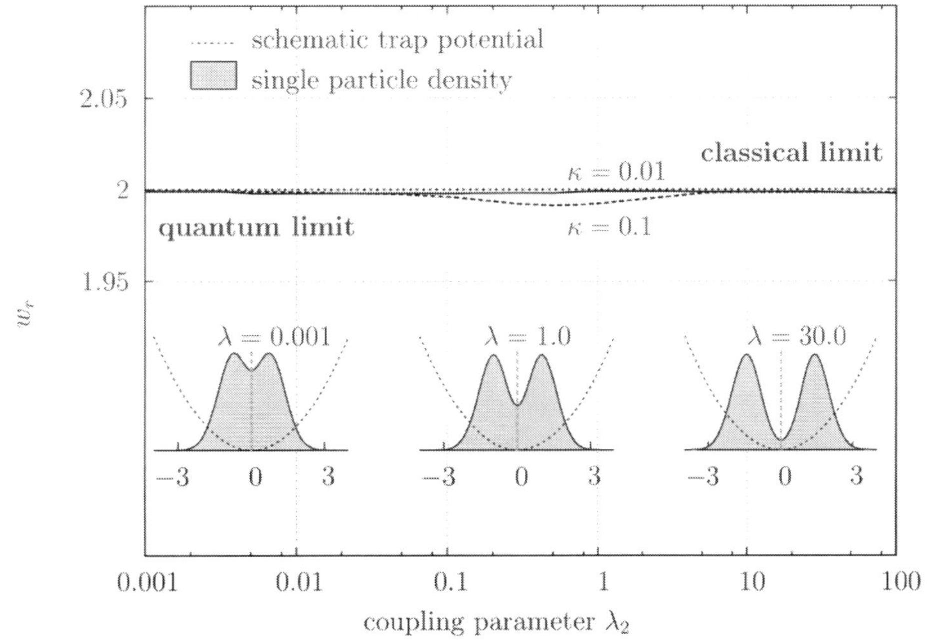

Figure 1. Breathing mode of two interacting particles in a harmonic trap for $1/r^2$ interaction. Shown is the mode ω_r associated with the relative wave function for two different values of the interaction smoothing parameter κ, cf. Eq. (11). The insets show the single-particle density (grey marked area) in the trap (schematically drawn by dashed lines), prior to excitation of the breathing mode for three different coupling regimes. The breathing mode is excited by turning off the trap for a duration of $t_{\text{off}} = 0.1$ and measuring the oscillation of $\langle x^2 \rangle(t)$. The mode associated with the center-of-mass problem, $\omega_R = 2$, is not shown.

3. Results for two particles

3.1. $1/r^2$ *interaction*

The breathing modes in a many-particle system interacting via a $1/r^2$ potential are well known, both classically and in quantum mechanics. The corresponding frequency (of the relative mode) ω_r is independent of the coupling parameter λ and attains the universal value of $\omega_r = 2$ [17] which coincides with the pure quantum mode ω_R, being also independent of λ. Furtheron, the full N-particle spectrum in a one-dimensional trap has been calculated analytically [19]. Therefore, this special type of particle-particle interaction is a useful and powerful test for the methods, for the preparation of the initial state via imaginary time propagation (accuracy) as well as for the time-propagation yielding the collective modes. The results obtained provide a powerful benchmark which then can be generalized to other types of interaction ($d \neq 2$).

The results from our TDSE calculations are shown in Figure 1 for two values of the (necessary) regularization parameter κ. According to [17], the exact solution *without* κ is λ-independent. However, due to the potential smoothing for $\kappa = 0.1$ we find a slight modification at intermediate coupling (smaller than one percent) which drops if one decreases κ further. That lets us conclude, that calculations with antisymmetric (spin-polarized) wave functions give the correct value for the breathing frequency — also in the case of other types of interactions, cf. [13] for a discussion concerning symmetric initial states. Figure 1 also gives the calculated single-particle densities before excitation of the breathing modes, obtained by imaginary time propagation for three chosen values of λ. One clearly sees the transition from a delocalized quantum state ($\lambda = 0.001$) with large overlap of both density peaks to a state with well-separated densities, corresponding

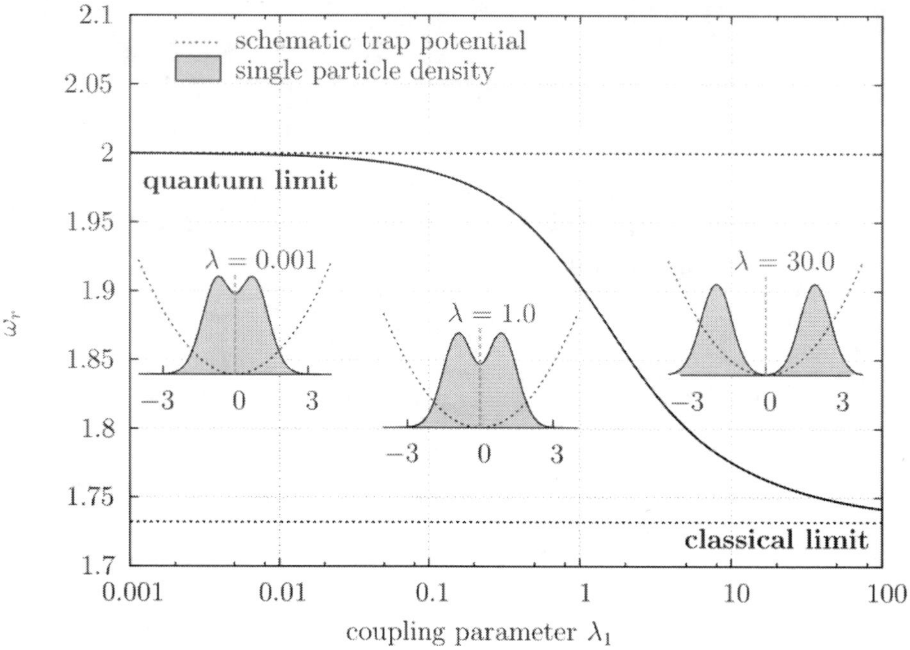

Figure 2. Breathing mode of two Coulomb interacting particles in a harmonic trap. All parameters are as in Fig. 1.

to a classical-like behavior ($\lambda = 30$).

3.2. Coulomb interaction

The second case is the Coulomb type interaction, $d = 2$. Figure 2 shows the result for $\omega_r(\lambda)$ for two particles in a harmonic trap. Both known limits, the quantum limit $\omega_r = 2$ of the non-interacting quantum gas and the classical case, $\omega = \sqrt{3}$ are indicated by dotted lines. With increasing λ, the mode ω_r decreases towards the classical regime. In between both limits, a smooth monotonic transition from the non-interacting to the strongly-coupled case is to be found. In analogy to the previous case, the single-particle densities in the trap are shown for the different interaction regimes, showing the same trend as in the $1/r^2$ case: for the ideal system, the double-peak structure in the single-particle density of two spin-polarized fermions in a harmonic trap is retrieved, whereas in the limit of large λ, the density separates into two distinct peaks being of perfect Gaussian shape exhibiting no significant overlap.

3.3. Dipole interaction

The case of dipole interaction, $d = 3$, is important for the investigation of dipolar quantum gases in traps and/or electron-hole bilayers [12, 18]. The classical value is known to be $\omega_r = \sqrt{5}$ whereas the quantum limit for $\lambda = 0$ is again $\omega_r = 2$. The result for two particles in a harmonic trap is given in Figure 3. One again recognizes the smooth, monotonic transition from the non-interacting to the classical regime. In contrast to Coulomb, the breathing mode ω_r increases with stronger interaction. For $\lambda = 100$, the classical case is not yet reached, in contrast to the case of Coulomb interaction. The explanation is to be found in the densities. In the case of dipole interaction, the density peaks have a non-vanishing overlap also in the case of large λ (case $\lambda = 30$ shown), which is not the case for Coulomb interaction, cf. Fig. 2 $\lambda = 30.0$. Therefore, the classical-like regime is reached only for larger λ. We note, that calculations for larger values

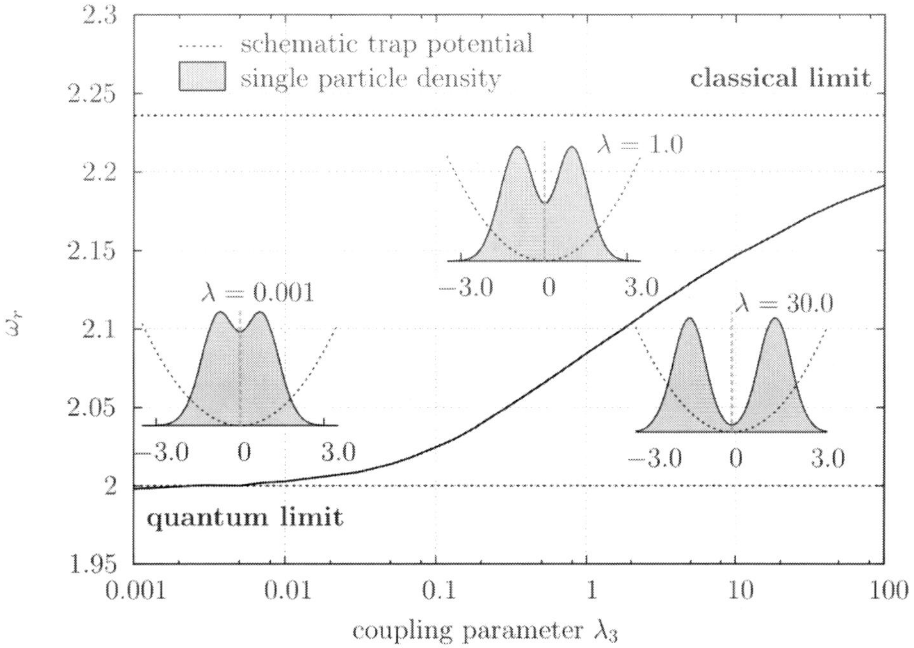

Figure 3. Breathing mode of two dipole interacting particles in a harmonic trap. All parameters are as in Fig. 1.

of λ become very demanding, because the potential gets very steep and the necessary numerical grid to represent the wave function (and the potentials) needs to be very fine.

Figure 4 shows a comparison of the breathing behavior for Coulomb and dipole systems, where the appropriate parameter for comparison is the oscillator length l_0, cf. Eqs. (7) and (9). In this representation, the differences of Coulomb and dipole interacting particles become apparent: whereas for Coulomb interaction the high density regime (small l_0) corresponds to small λ (ideal quantum gas) and the low density regime (large l_0) behaves classically, the opposite is true for dipole interaction. Here, the ideal quantum regime is observed for low density whereas the case of strong coupling corresponding to well-separated, classically behaving particles, occurs at high densities.

4. Larger particle numbers

To investigate the behavior of the breathing mode related to the relative problem for more than two particles, we performed time-dependent multi-configuration Hartree-Fock simulations, where the inclusion of correlation effects can be tuned via a parameter M determining the involved Slater determinants to represent the N-particle wave function. For a detailed description of the method and references for further reading, see [20] and [21] in the present volume. We concentrate on the fundamental case of Coulomb interaction ($d = 1$) in the following. Figure 5 shows the result for the relative breathing frequency, ω_r for up to $N = 6$ particles obtained via the multi-configuration Ansatz. The correlation parameter M has been adjusted to $M = N + 2$ to account for correlation beyond the mean-field Hartree-Fock level. The

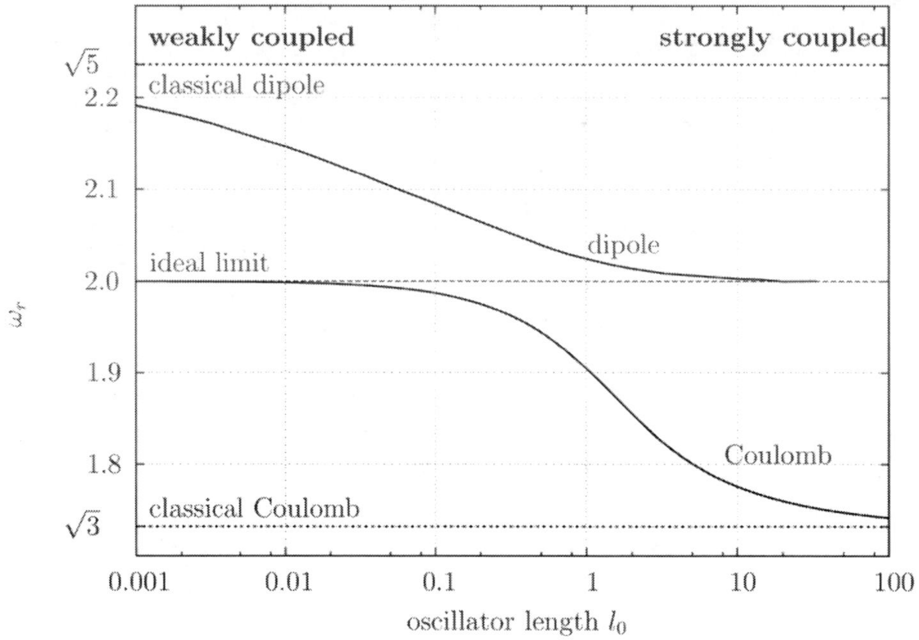

Figure 4. Breathing mode of two interacting (dipole/Coulomb interaction) particles in a harmonic trap plotted vs. characteristic length scale l_0. The regimes of strong and weak coupling are indicated.

comparison with the TDSE result for two particles (bold black line) shows perfect agreement with the MCTDHF calculations for two particles. However, for larger particle numbers, additional Slater determinants are needed, thus the degree of included correlations differs for different particle numbers. The inspection of the λ_1 dependence of ω_r reveals a N-independent shape, a smooth transition from the weakly interacting quantum gas to the strongly coupled classical case. However, in the transition region of intermediate coupling a weak dependence on N can be spotted, but this might be due to additional correlation effects beyond the $M = N + 2$ approximation.

5. Conclusion and outlook

In summary, we have computed the breathing modes of interacting one-dimensional quantum systems for different types of interaction. In all cases, two different modes occur, one center-of-mass related breathing motion independent of the coupling of the particles, and one relative mode. The latter one can be identified as the classical-alike mode, which asymptotically attains in all investigated cases the classical value for the breathing frequency. Furtheron the dependence on the particle number has been investigated via the powerful, in general exact, multi-configurational Hartree Fock method revealing a similar structure of the dependence on the coupling of the breathing mode associated with the relative oscillation. However, the question whether the breathing mode is independent of the particle number for arbitrary coupling, as it is in the two limits, remains unanswered and is postponed to a forthcoming work.

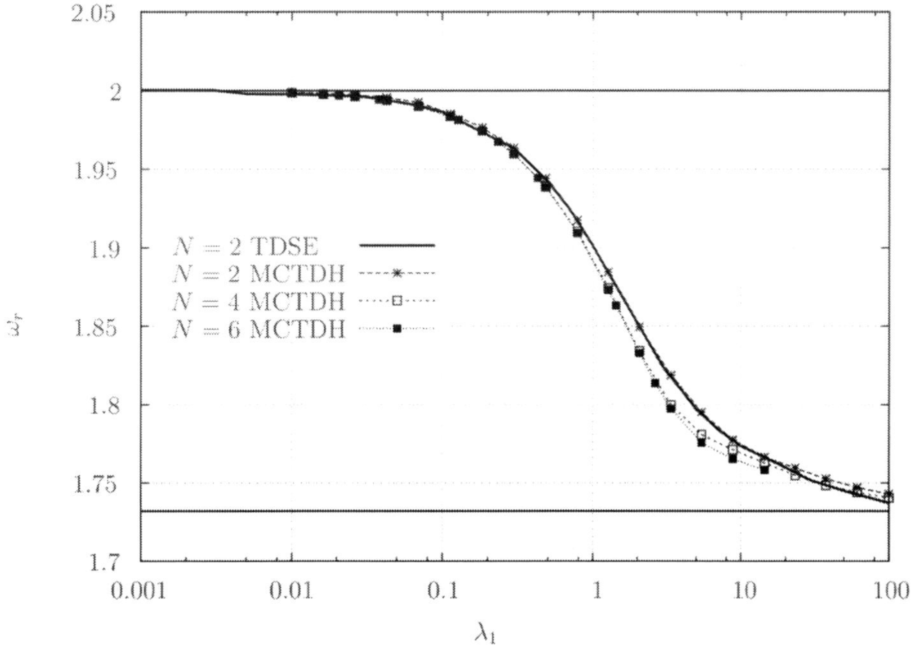

Figure 5. Dependence of the breathing mode associated with the relative motion, ω_r, on the particle number N for $N = 2, 4, 6$ Coulomb interacting particles in a harmonic trap. The results are obtained via the direct solution of the TDSE and via the multi-configuration Hartree Fock ansatz. The correlation tuning parameter M in the MCTDHF method has been set to $M = N + 2$.

Appendix A. Derivation of (ideal) breathing frequency ω_{R}

We consider the center of mass problem, cf. Eq. (12), in one spatial dimension, which is nothing else than a simple harmonic oscillator problem ($R \equiv x$):

$$i\partial_t \phi(x,t) = \left(-\frac{1}{2}\frac{\partial^2}{\partial x^2} + \frac{x^2}{2}\right)\phi(x,t) \ . \tag{A.1}$$

Any initial excitation, e. g. created by switching off the trap for a small period of time, can be written as a superposition of eigenstates ϕ_m,

$$\phi(x,t=0) = \sum_m c_m \phi_m(x) \ , \tag{A.2}$$

where

$$\left(-\frac{1}{2}\frac{\partial^2}{\partial x^2} + x^2\right)\phi_m = \left(m + \frac{1}{2}\right)\phi_m \ . \tag{A.3}$$

The (time-dependent) solution of Eq. (A.1) is then simply given by

$$\phi(x,t) = \sum_m c_m \phi_m(x) e^{-iE_m t} \ , \tag{A.4}$$

where $E_m = m + 1/2$ denotes the eigenvalues associated with the eigenfunction ϕ_m. The measurable quantities, giving the collective modes of the system, are physical observables of the

Journal of Physics: Conference Series **220** (2010) 012013	doi:10.1088/1742-6596/220/1/012013

form $\langle x^k\rangle(t)$ for $k = 0, 1, \ldots$ Their time dependence follows as

$$\langle x^k\rangle(t) = \sum_{ij} c_i^* c_j e^{-i(E_j-E_i)t}(x^k)_{ij}\,, \tag{A.5}$$

with matrix elements $(x^k)_{ij} = \int \mathrm{d}x \phi_i^*(x) x^k \phi_j(x)$. Therefore, $\langle x^k\rangle(t)$ may oscillate with frequencies $\omega^{(k)} = |E_j - E_i|, i \neq j$, depending on the initial state c_i and the corresponding matrix elements $(x^k)_{ij}$.

We now consider only the relevant cases for k:

(i) $k = 1$ (*dipole moment*): the frequency for dipole oscillations is given by $\omega^{(1)} = 1$, because $(x)_{ij} \propto \delta_{ij} \pm 1$. This leads to the well-known Kohn (sloshing mode)

(ii) $k = 2$ (*potential energy*): $\langle x^2\rangle(t)$ may oscillate with $\omega^{(2)} = 2$, since $(x^2)_{ij} \neq 0$ only for $i = j$ and $i = j \pm 2$.

For our type of radial excitation, no dipole moment is induced. Hence, only the second case is relevant, and the observable oscillation of the center of mass part of the two-particle wave function attains the λ-independent value of $\omega^{(2)} = \omega_R = 2$. As the derivation of the center-of-mass wave function is N-independent, also the oscillation of the non-interacting, ideal quantum gas behaves uniformly, in analogy to the case of the strongly coupled, classical case.

References

[1] Baletto F and Ferrando R 2005 *Rev. Mod. Phys.* **77** 371
[2] Filinov A V, Bonitz M and Lozovik Yu E 2001 *Phys. Rev. Lett.* **86** 3851
[3] Filinov A V, Lozovik Yu E and Bonitz M 2000 *Phys. Stat. Sol. (b)* **221** 231
[4] Reimann S M and Manninen M 2002 *Rev. Mod. Phys.* **74** 1283
[5] Bloch I 2005 *Nature Phys.* **1** 23
[6] Giorgini S, Pitaevskii L P and Stringari S 2008 *Rev. Mod. Phys.* **80** 1215
[7] Görlitz A, Vogels J M, Leanhardt A E, Raman C, Gustavson T L, Abo-Shaeer J R, Chikkatur A P,Gupta S, Inouye S, Rosenband T and Ketterle W 2001 *Phys. Rev. Lett.* **87** 130402
[8] Menotti C and Stringari S 2002 *Phys. Rev. A* **66** 043610
[9] Moritz H, Stöferle Th, Köhl M and Esslinger T 2003 *Phys. Rev. Lett.* **91** 250402
[10] Pedri P, De Palo S, Orignac E, Citro R and Chiofalo M K 2008 *Phys. Rev. A* **77** 015601
[11] Bloch I, Dalibard J, Zwerger W 2008 *Rev. Mod. Phys.* **80** 885
[12] Filinov A, Böning J, Bonitz M and Lozovik Yu E 2008 *Phys. Rev. B* **77** 214527
[13] Bauch S, Balzer K, Henning C and Bonitz M 2009 *Phys. Rev. B* **80** 054515
[14] Henning C, Fujioka K, Ludwig P, Piel A, Melzer A and Bonitz M 2008 *Phys. Rev. Lett.* **101** 045002
[15] Girardeau M 1960 *J. Math. Phys.* **1** 516
[16] Melzer A, Klindworth M and Piel A 2001 *Phys. Rev. Lett.* **87** 115002
[17] Geller M R and Vignale G 1996 *Phys. Rev. B* **53** 6979
[18] Filinov A, Ludwig P, Bonitz M, and Lozovik Yu E 2009 *J. Phys. A* **42** 214016
[19] Calogero F 1971 *J. Math. Phys.* **12** 419
[20] Bonitz M, Hochstuhl D, Bauch S, and Balzer K 2010 *Contrib. Plasma Phys.* **80** 54–59
[21] Hochstuhl D, Bauch S and Bonitz M 2009 Proceedings of conference PNGF IV this issue

Journal of Physics: Conference Series **220** (2010) 012014 doi:10.1088/1742-6596/220/1/012014

Transport through a vibrating quantum dot: Polaronic effects

T Koch†, J Loos‡, A Alvermann†, A R Bishop§, and H Fehske†

† Institute of Physics, Ernst-Moritz-Arndt University Greifswald, 17487 Greifswald, Germany
‡ Institute of Physics, Academy of Sciences of the Czech Republic, 16200 Prague, Czech Republic
§ Theory, Simulation and Computation Directorate, Los Alamos National Laboratory, Los Alamos, New Mexico 87545, USA

E-mail: holger.fehske@physik.uni-greifswald.de

Abstract. We present a Green's function based treatment of the effects of electron-phonon coupling on transport through a molecular quantum dot in the quantum limit. Thereby we combine an incomplete variational Lang-Firsov approach with a perturbative calculation of the electron-phonon self energy in the framework of generalised Matsubara Green functions and a Landauer-type transport description. Calculating the ground-state energy, the dot single-particle spectral function and the linear conductance at finite carrier density, we study the low-temperature transport properties of the vibrating quantum dot sandwiched between metallic leads in the whole electron-phonon coupling strength regime. We discuss corrections to the concept of an anti-adiabatic dot polaron and show how a deformable quantum dot can act as a molecular switch.

1. Introduction

Recent progress in nanotechnology allows for the fabrication of electronic devices with organic molecules as the active elements, which may constitute an alternative to conventional semiconductor technology in the search for further miniaturisation. The basic example for such a device is a single organic molecule contacted with metallic leads. The molecule can be described as a quantum dot, i.e. as a system of finite size that is coupled to macroscopic charge reservoirs. The electronic transport properties of such molecular electronic components depend on the geometry of the molecule and the properties of the molecule-lead contact, as well as on the average charge of the molecule [1]. Since the molecule is so small, quantisation of energy levels becomes important. Furthermore, the molecule is susceptible to specific structural changes in the presence of charge carriers. The energy of these deformations or vibrations can be comparable to the kinetic energy of passing particles, whose mobility may therefore be substantially affected. This process seems to play a fundamental role in the observed non-linear behaviour of basic molecular devices, including negative differential resistance and hysteresis [2]. With strong coupling to vibrational degrees of freedom, current rectification and fast current switching may be realised [3].

2. Model

To investigate transport in a molecular quantum dot, we consider the model Hamiltonian

$$H = \sum_{k,a} E_k c_{ka}^\dagger c_{ka} + \Delta d^\dagger d - g\omega_0 (b^\dagger + b) d^\dagger d + \omega_0 b^\dagger b - \frac{t_d}{\sqrt{N}} \sum_{k,a} \left(d^\dagger c_{ka} + c_{ka}^\dagger d \right) . \quad (1)$$

Here $c_{ka}^\dagger$ (c_{ka}) are creation (destruction) operators of non-interacting electrons with energy E_k ($k = 1, \dots, N$) in the left and right lead ($a = l, r$). We assume that the leads are semi-infinite one-dimensional chains with a semi-elliptical density of states $\varrho(\xi) = \frac{1}{N}\sum_k \delta(\xi - E_k) = (2/\pi W^2)\sqrt{W^2 - \xi^2}\,\Theta(W^2 - \xi^2)$, where W is the half bandwidth. The quantum dot is modelled by a single level Δ with fermionic operators $d^{(\dagger)}$. Working with spinless fermions, we take a large local Coulomb repulsion at the quantum dot for granted. To describe the deformation of the molecule, an electron at the dot couples via a Holstein-like term ($\propto g$) to a local phonon mode $b^{(\dagger)}$ of energy ω_0. The last term in (1) allows for the dot-lead particle transfer ($\propto t_d$).

In analogy to Holstein's small polaron theory [4], we expect – for sufficiently large electron-phonon (EP) coupling g and frequency ω_0 – the formation of a polaron-like state at the dot, corresponding to an electron with an accompanying phonon cloud. To account for this effect, a generalised Lang-Firsov-transformation $\widetilde{H} = UHU^\dagger$ with $U = \exp\{\widetilde{g}(b^\dagger - b) d^\dagger d\}$ and $\widetilde{g} = \gamma g$ is applied, where $\gamma \in [0, 1]$ is a variational parameter. The transformed Hamiltonian reads

$$\widetilde{H} = \sum_{k,a} E_k c_{ka}^\dagger c_{ka} + \widetilde{\Delta} d^\dagger d - C_d d^\dagger d + \omega_0 b^\dagger b - \sum_{k,a} \left(C_t d^\dagger c_{ka} + C_t^\dagger c_{ka}^\dagger d \right) , \quad (2)$$

with the renormalised dot-level $\widetilde{\Delta} = \Delta - \varepsilon_p \gamma(2 - \gamma)$ and the polaron binding energy $\varepsilon_p = g^2 \omega_0$. As γ grows from zero to one, our approach interpolates between a weak coupling ansatz ($\gamma = 0$, $\widetilde{H} = H$), and the complete Lang-Firsov-transformation ($\gamma = 1$) restricted to large phonon frequencies and strong EP coupling, when the direct EP coupling via $C_d = g\omega_0(1 - \gamma)(b^\dagger + b)$ is replaced by a phonon-affected dot-lead transfer term $C_t = (t_d/\sqrt{N})\exp\{-\widetilde{g}(b^\dagger - b)\}$. In this way, we are able to describe the system for a large range of parameter values. Note that the effects of a finite Coulomb interaction might be included by considering instead of the Hamiltonian (1) Hubbard/Anderson-Holstein-type models [5, 6]. Here the influence of the phonons is mainly to suppress the repulsion between the electrons at the molecular orbitals [7]. When the energy scales set by the Coulomb and EP interactions become comparable bipolaronic states may form at the quantum dot [5].

3. Theoretical approach

Our main interest lies in the single particle spectrum at the dot and the linear conductance in the case of vanishing voltage bias between the leads. Both of these quantities can be obtained from the polaronic spectral function $A_{dd}(\omega) = \lim_{\delta \to 0^+} [\mathrm{Im}\, G_{dd}^R(\omega + \mathrm{i}\delta)]$, where the retarded Green function G_{dd}^R corresponds to the (polaronic) operators $d^{(\dagger)}$ in the transformed Hamiltonian in equation (2). We base our calculations on the equations of motion of generalised Matsubara Green functions in equilibrium [8, 9]

$$G_{dd}(\tau_1, \tau_2; \{V\}) = -\frac{1}{\langle S \rangle} \langle \mathcal{T}_\tau d(\tau_1) d^\dagger(\tau_2) S \rangle . \quad (3)$$

The mean value and the time dependences in (3) are determined by $\widetilde{H} - \mu \hat{N}$, where μ is the equilibrium chemical potential of the system and $\hat{N}$ denotes the particle number operator. The S-matrix

$$S = \mathcal{T}_\tau \exp\left\{ -\int_0^\beta \mathrm{d}\tau\, V_t(\tau) C_t(\tau) + \bar{V}_t(\tau) C_t^\dagger(\tau) + V_d(\tau) C_d(\tau) \right\} \quad (4)$$

describes the coupling to the components of a fictitious external potential $\{V\}$.

3.1. Dot spectral function

Starting from the equations of motion for the dot Green function G_{dd} and the lead-dot transfer Green function $G_{cd;ka}$, we express the occuring interaction terms by functional derivatives with respect to $\{V\}$ [8], e.g.,

$$\frac{\langle T_\tau C_t(\tau_1) c_{ka}(\tau_1) d^\dagger(\tau_2) S[V]\rangle}{\langle S[V]\rangle} = -\bar{C}_t(\tau_1;\{V\})\, G_{cd;ka}(\tau_1,\tau_2;\{V\}) + \frac{\delta G_{cd;ka}(\tau_1,\tau_2;\{V\})}{\delta V_t(\tau_1)}, \quad (5)$$

with the interaction coefficients $\bar{C}_t^{(\dagger)}(\tau;\{V\}) \equiv \langle T_\tau C_t^{(\dagger)}(\tau) S\rangle/\langle S\rangle$. The resulting coupled equations lead to a functional differential equation for the polaronic self energy $\Sigma_{dd}(\tau_1,\tau_2;\{V\}) = G_{dd}^{(0)-1}(\tau_1,\tau_2) - G_{dd}^{-1}(\tau_1,\tau_2;\{V\})$ (see equation (17) in [9]), which can be evaluated in a two-step process. First, we neglect terms with functional derivatives of Σ_{dd}. We then insert the result, $\Sigma_{dd}^{(1)}$, into these derivatives and only keep terms up to second order in $\bar{C}_t^{(\dagger)}$. In this way, we find

$$\begin{aligned}\Sigma_{dd}^{(2)}(\tau_1,\tau_2;\{V\}) &= -\bar{C}_d(\tau_1;\{V\})\delta[\tau_1-\tau_2] + \sum_{k,a} \bar{C}_t(\tau_1;\{V\}) G_{cc;\,ka}^{(0)}(\tau_1,\tau_2)\bar{C}_t^\dagger(\tau_2;\{V\}) \quad (6)\\ &+\sum_{k,a} G_{cc;\,ka}^{(0)}(\tau_1,\tau_2)\left[\frac{1}{\langle S\rangle}\langle T_\tau C_t(\tau_1)C_t^\dagger(\tau_2)S\rangle - \bar{C}_t(\tau_1;\{V\})\bar{C}_t^\dagger(\tau_2;\{V\})\right]\\ &+G_{dd}^{(1)}(\tau_1,\tau_2;\{V\})\left[\langle T_\tau C_d(\tau_1)C_d(\tau_2)S\rangle - \bar{C}_d(\tau_1;\{V\})\bar{C}_d(\tau_2;\{V\})\right].\end{aligned}$$

Here $G_{dd}^{(1)}$ denotes the Green function determined by the first order self energy and $G_{cc;\,ka}^{(0)}$ is the free Green function of the lead states. We then let $\{V\}\to 0$ and calculate the correlation functions of the interaction coefficients assuming an independent Einstein oscillator. After Fourier transformation and summation over bosonic Matsubara frequencies, the retarded self energy follows in the low-temperature approximation $\beta\omega_0 \gg 1$ as [9]

$$\begin{aligned}\Sigma_{dd}^{(2)}(\omega+\mathrm{i}\delta) &= 2\,t_d^2\,e^{-\tilde{g}^2}\int_{-W}^{W}\mathrm{d}\xi\,\varrho(\xi)\,\frac{1}{\omega+\mathrm{i}\delta-(\xi-\mu)} \quad (7)\\ &+2\,t_d^2\,e^{-\tilde{g}^2}\sum_{s\geq 1}\frac{(\tilde{g}^2)^s}{s!}\int_{-W}^{W}\mathrm{d}\xi\,\varrho(\xi)\left(\frac{n_F(\xi-\mu)}{\omega+\mathrm{i}\delta-(\xi-\mu)+s\omega_0}+\frac{1-n_F(\xi-\mu)}{\omega+\mathrm{i}\delta-(\xi-\mu)-s\omega_0}\right)\\ &+[(1-\gamma)g\omega_0]^2\int_{-\infty}^{+\infty}\mathrm{d}\omega' A_{dd}^{(1)}(\omega')\left(\frac{n_F(\omega')}{\omega+\mathrm{i}\delta-\omega'+\omega_0}+\frac{1-n_F(\omega')}{\omega+\mathrm{i}\delta-\omega'-\omega_0}\right).\end{aligned}$$

Within our iterative scheme, the corresponding spectral function $A_{dd}^{(2)} \equiv A_{dd}$ is evaluated in a two step process using

$$A_{dd}^{(n)}(\omega) = -\frac{1}{\pi}\lim_{\delta\to 0^+}\frac{\mathrm{Im}\,\Sigma_{dd}^{(n)}(\omega+\mathrm{i}\delta)}{\left[\omega+\mu-\tilde{\Delta}-\mathrm{Re}\,\Sigma_{dd}^{(n)}(\omega+\mathrm{i}\delta)\right]^2+\left[\mathrm{Im}\,\Sigma_{dd}^{(n)}(\omega+\mathrm{i}\delta)\right]^2}, \quad (8)$$

where the first order self energy $\Sigma_{dd}^{(1)}$ is given by the first two terms in (7). The self energy $\Sigma_{dd}^{(2)} \equiv \Sigma_{dd}$ accounts for multi-phonon processes as well as finite particle densities. As $g\to 0$, our model (1) reduces to the Fano-Anderson Hamiltonian for a rigid impurity in a 1D lattice. Then only the first term in (7) remains and gives the exact self energy. For finite EP coupling, the spectrum contains multiple phononic side bands. If the condition $\omega_0 < W - |\mu|$ is fulfilled, these bands overlap and $\mathrm{Im}\,\Sigma_{dd} \neq 0$ along the whole ω-axis. Otherwise, a localised polaron-like state may exist in the intervals where $\mathrm{Im}\,\Sigma_{dd} = 0$.

For the numerical evaluation of the spectral function we keep δ in (8) as a small positive parameter ($\delta \lesssim 5 \cdot 10^{-3}$). This avoids the problematic evaluation of principal value integrals in $\mathrm{Re}\,\Sigma_{dd}$ for $\delta \to 0$. The resulting spectral function fulfils the sum rule $\int A_{dd}(\omega)\mathrm{d}\omega = 1$, which is preserved in our approximations.

3.2. Ground state energy

The variational parameter γ is determined by minimisation of the ground state energy $E = \langle \widetilde{H} - \mu\hat{N}\rangle$ with respect to γ. We identify the statistical averages in E with expressions like equation (5), letting $\tau_1 \to \tau_2^-$ and $\{V\} \to 0$. As in the evaluation of $\Sigma_{dd}^{(1)}$, we neglect the functional derivatives in equation (5), so that $\langle C_t d^\dagger c_{ka}\rangle \approx -\langle C_t\rangle G_{cd;ka}(\tau_1,\tau_2)\big|_{\tau_1\to\tau_2^-}$. Applying the same approximation to the equations of motion of the Green functions, we determine their Fourier transforms in the complex plane to first order as

$$G_{cc;ka}(z) \approx G^{(0)}_{cc;ka}(z) + \langle C_t^\dagger\rangle\langle C_t\rangle\big[G^{(0)}_{cc;ka}(z)\big]^2 G_{dd}(z)\,, \tag{9}$$

$$G_{cd;ka}(z) \approx -\langle C_t^\dagger\rangle G^{(0)}_{cc;ka}(z) G_{dd}(z)\,. \tag{10}$$

For the variation of γ we can omit the constant lead energy related to $G^{(0)}_{cc;ka}$. In equilibrium, $\langle b^\dagger b\rangle$ and $\langle C_d\rangle$ vanish as $T \to 0$, and we obtain

$$\begin{aligned} E \;=\; & -2\,t_d^2\,e^{-\widetilde{g}^2}\int_{-W}^{\mu}\mathrm{d}\xi\,\varrho(\xi)\int_0^{+\infty}\mathrm{d}\omega'\,A_{dd}(\omega')\left[\frac{2}{\omega'-(\xi-\mu)}+\frac{\xi-\mu}{(\omega'-(\xi-\mu))^2}\right] \\ & +2\,t_d^2\,e^{-\widetilde{g}^2}\int_{\mu}^{W}\mathrm{d}\xi\,\varrho(\xi)\int_{-\infty}^{0}\mathrm{d}\omega'\,A_{dd}(\omega')\left[\frac{2}{\omega'-(\xi-\mu)}+\frac{\xi-\mu}{(\omega'-(\xi-\mu))^2}\right] \\ & +(\widetilde{\Delta}-\mu)\int_{-\infty}^{0}\mathrm{d}\omega' A_{dd}(\omega')\,. \end{aligned} \tag{11}$$

Via A_{dd}, $\widetilde{\Delta}$ and $\widetilde{g}$, the energy E, given by equation (11), depends on γ and can be used to determine the extremal variational parameter γ_{min}.

3.3. Conductance

We finally calculate the linear conductance L starting from the Meir-Wingreen formula [10], which expresses the current J through the dot in terms of the retarded non-equilibrium Green function G^R_{dd} of the electronic operators d in (1). For equal coupling to the left and right lead, and finite voltage bias $\Phi = -(\mu_l - \mu_r)/\mathrm{e}$, with e being the elementary (positive) charge, the current takes the simple form

$$J = -\mathrm{e}\,t_d^2\int_{-W}^{W}\mathrm{d}\xi\,\varrho(\xi)\,[n_F(\xi-\mu_l)-n_F(\xi-\mu_r)]\,\mathrm{Im}\,G^R_{dd}(\xi)\,. \tag{12}$$

Here $n_F(\xi) = (e^{\beta\xi}+1)^{-1}$ is the Fermi distribution function of the isolated leads in thermal equilibrium, at respective chemical potential $\mu_{l/r}$. For small voltage bias, i.e. $\mu_{l,r} = \mu \pm \delta\mu/2$, we can express the current as $J = -L\delta\mu/\mathrm{e}$. Then, the linear conductance $L = \lim_{\delta\mu\to 0}\{-\mathrm{e}J/\delta\mu\}$ results from (12) as

$$L = \mathrm{e}^2\pi\,t_d^2\int_{-W}^{W}\mathrm{d}\xi\,\varrho(\xi)\,[-n_F'(\xi-\mu)]\,A^{\mathrm{e}}_{dd}(\xi-\mu)\,, \tag{13}$$

where the electronic spectral function $A^{\mathrm{e}}_{dd}(\omega) = -\lim_{\delta\to 0^+}[\mathrm{Im}\,G^R_{dd}(\omega+\mathrm{i}\delta)]/\pi$ is now calculated in equilibrium. Based on the factorisation of the statistical averages of phonon and polaron

Journal of Physics: Conference Series **220** (2010) 012014 doi:10.1088/1742-6596/220/1/012014

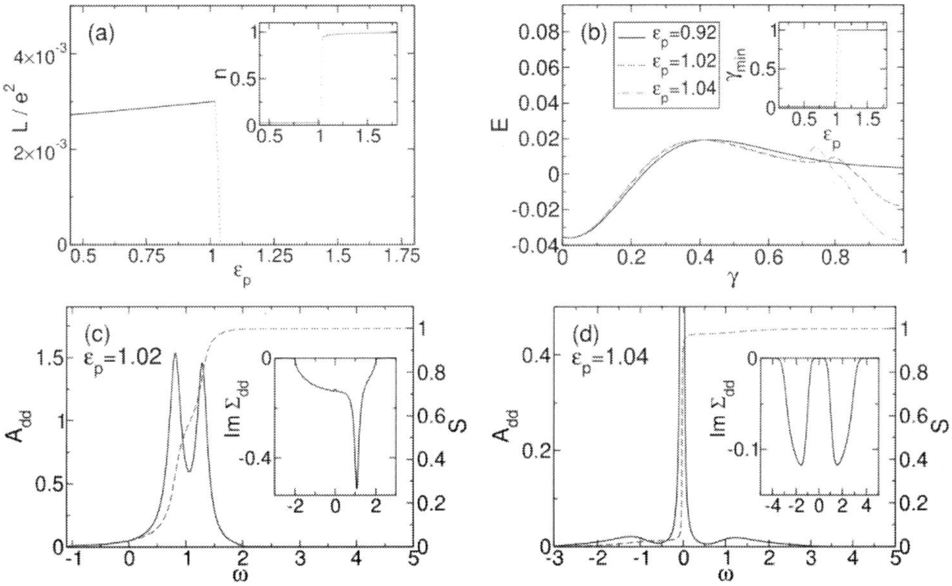

Figure 1. For model parameters $\omega_0 = 0.05$, $\Delta = 1$, $t_d = 0.25$, $\mu = 0$. Panel (a): Linear conductance L and occupation of the dot n (inset) as functions of ε_p. At $\varepsilon_p^c = 1.03$ an abrupt transition takes place. Panel (b): Dot energy E as a function of γ for several EP coupling strengths. At ε_p^c, a second minimum at $\gamma = 1$ becomes the global minimum, causing a jump in the extremal parameter γ_{min} (inset). Panels (c) and (d): Spectral function $A_{dd}(\omega)$, integrated spectral weight $S(\omega) = \int_{-\infty}^{\omega} \mathrm{d}\omega' A_{dd}(\omega')$ and imaginary part of the self energy $\Sigma_{dd}(\omega)$ for ε_p in the vicinity of the transition, showing the sudden formation of a long-living polaron-like state.

variables, a relation between A_{dd}^e and the polaronic A_{dd} has been derived in equation (40) of [11]:

$$A_{dd}^e(\omega) = e^{-\tilde{g}^2} \sum_{s \geq 0} \frac{(\tilde{g}^2)^s}{s!} \left[A_{dd}(\omega - s\omega_0)\Theta(\omega - s\omega_0) + A_{dd}(\omega + s\omega_0)\Theta(-\omega - s\omega_0) \right] . \tag{14}$$

Then, from (13) and (14), the linear conductance in the low-temperature approximation is

$$L = \mathrm{e}^2 \pi \, t_d^2 \, e^{-\tilde{g}^2} \varrho(\mu) A_{dd}(0) . \tag{15}$$

As usual, it depends on the accessibility of dot states at the Fermi level. In addition the conductance exhibits a Lang-Firsov renormalisation $\tilde{t}_d = t_d e^{-\tilde{g}^2/2}$ of the dot-lead transfer integral, in accordance with the results for the Holstein model.

4. Numerical results and discussion

With our choice of a semi-elliptical lead density of states, the non-interacting system ($\epsilon_p = 0$) is translational invariant if the dot-lead transfer t_d is equal to the value $t = W/2$ for electron hopping along the leads. In the following, we fix $t = 1$ and set $t_d = 0.25$, representing the weak dot-lead coupling case.

4.1. Adiabatic regime

Let us first consider the adiabatic regime ($\omega_0 = 0.05 < t_d$), where the phononic time-scale is much slower than the electronic time-scale and the deformation of the dot adjusts quasi-statically

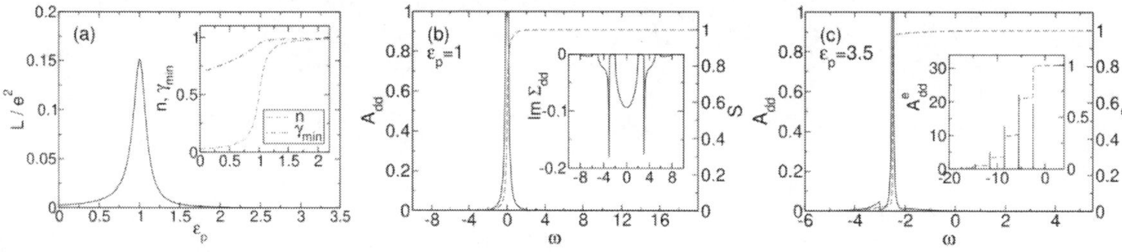

Figure 2. For model parameters $\omega_0 = 3$, $\Delta = 1$, $t_d = 0.25$, $\mu = 0$. Panel (a): Because E has a single minimum at γ_{min} for all ε_p, a smooth crossover with $\gamma_{min} \geq 0.7$ occurs (inset). The linear conductance reaches a maximum at $\varepsilon_p \approx \Delta$, when the peak in A_{dd} is shifted to the Fermi level (c.f. panel (b)). Panel (c): For large ε_p, a bound polaron-like state may exist in the intervals where $\mathrm{Im}\,\Sigma_{dd} = 0$. Then the electronic spectrum exhibits a multi-peak structure.

to the average electronic occupation. We set $\mu = 0$ and $\Delta = 1$, so that without EP coupling ($\varepsilon_p = 0$) the dot simply acts as an impurity leading to scattering. For these parameters the conductance $L = 3 \cdot 10^{-3}$ is small, i.e. far below its value $L \approx 0.16$ for $\Delta = 0$ (see figure 1a). As ε_p increases, L and the dot occupation $n = \langle d^\dagger d \rangle$ grow only marginally. At a critical EP coupling $\varepsilon_p^c = 1.03$ a sudden transition to $n \approx 1$ takes place, resulting in a drop of L by eight orders of magnitude. This behaviour can be traced back to a jump in γ_{min}, as figure 1b shows. For small ε_p, $E(\gamma)$ has a single minimum at $\gamma_{min} = 0.02 \ll 1$. Hence there is almost no renormalisation of Δ or t_d and L is nearly independent of ε_p. However, the third term in (7) contributes and the single peak in $A_{dd}^{(1)}$ is added to $\mathrm{Im}\,\Sigma_{dd}$, resulting in two maxima of equal spectral weight around $\omega = \Delta$ in A_{dd} (see figure 1c). At $\varepsilon_p = \varepsilon_p^c$, when the gain in potential energy overcompensates the loss in kinetic energy, a second minimum of $E(\gamma)$ appears and becomes the global minimum. As a result γ_{min} jumps to unity, which corresponds to a complete Lang-Firsov transformation. The dot level ($\widetilde{\Delta} = \Delta - \epsilon_p = -0.03$) is shifted below the Fermi level and the dot-lead transfer is reduced by a factor of $\exp\{-\widetilde{g}^2\}$ with $\widetilde{g}^2 \geq 20$, so that L vanishes. Because $\omega_0 \ll W$, the spectrum consists of overlapping phononic bands around a pronounced central peak, which signals the formation of a long-living polaron-like state at the quantum dot. We know from variational approaches to the polaron problem, that the jump in γ_{min} may be an artefact of our variational ansatz. In the Holstein model with EP interaction at every lattice site, the formation of a heavy polaron can occur as a sharp but always continuous crossover for small phonon frequencies. In contrast, a true phase transition from zero to finite dot occupation n is found for a single electron at the quantum dot with $\Delta > 0$ [12]. This phase transition becomes more pronounced as ω_0 gets smaller. In [5], similar behaviour is found in a generalised Holstein-Hubbard model with site-dependent potentials and EP couplings. Therefore, our results do not contradict the overall physical picture, and the variational Lang-Firsov transformation simulates the rapid adiabatic transition by a discontinuous change in γ_{min}.

4.2. Anti-adiabatic regime

In the anti-adiabatic regime with $\omega_0 = 3 \gg t_d$, the dot deformation adjusts instantaneously to the presence of an electron. As figure 2a shows, our approach is better suited to this regime: For $\mu = 0$ and $\Delta = 1$, $E(\gamma)$ has a single minimum $\gamma_{min} \geq 0.7$ for all EP coupling strengths and we find a smooth crossover to $n \approx 1$. At the quantum dot, both the renormalisation of t_d and Δ affect the transport. Since the phonon number $g^2 = \varepsilon_p/\omega_0$ is small in the anti-adiabatic case, the shift of the dot level is the predominant effect. With increasing ε_p, the repulsive potential $\widetilde{\Delta}$ is continuously lowered and the conductance L grows, reaching almost the maximum of the

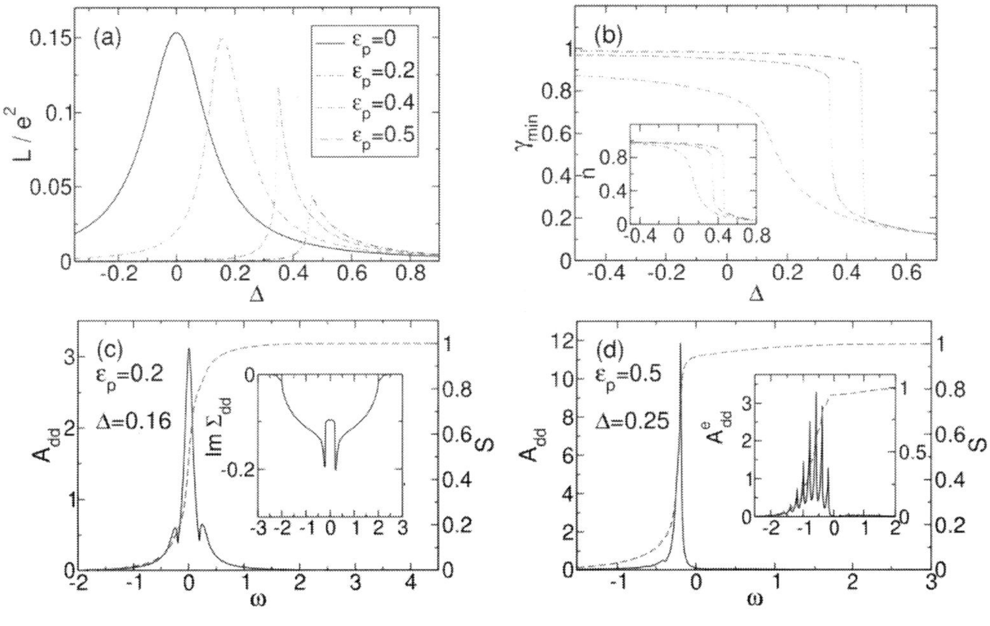

Figure 3. For model parameters $\omega_0 = 0.2$, $t_d = 0.25$, $\mu = 0$. Panels (a) and (b): L and γ_{min} as a function of the quantum dot level. For small $\varepsilon_p = 0.2$, we find a smooth crossover. As ε_p grows, a jump-like transition occurs and the dot acts as a molecular switch. Panel (c): Spectral function at the point of maximum conductance in the smooth crossover. Panel (d): Due to overlapping phononic bands, no bound state exists even for large ε_p.

$\Delta = 0$ case when $\varepsilon_p \approx \Delta$. Here the peak in A_{dd} is shifted to the Fermi level (see figure 2b) and we can speak of phonon-assisted transport. Because $\omega_0 > W$, the self energy features few non-overlapping phonon bands (see inset of figure 2b). For large ε_p, a bound polaron-like state forms when $\widetilde{\Delta}$ is located in the intervals with $\operatorname{Im}\Sigma_{dd} = 0$, as figure 2c shows for $\varepsilon_p = 3.5$ and $\widetilde{\Delta} \approx -2.5$. The electronic spectrum then exhibits the typical structure of multiple Poisson-weighted peaks (inset). Note that the delta peaks have finite width due to our numerical parameter $\delta > 0$.

The results show that, as for the Holstein model, the complete Lang-Firsov transformation is restricted to large phonon frequency and strong electron-phonon coupling. Away from this limit, γ_{min} significantly differs from one, and our approach allows for important corrections in the regimes of weak coupling and moderate to small phonon frequencies.

4.3. Intermediate phonon frequency

Next we investigate the intermediate regime, where phononic and electronic energies become comparable ($\omega_0 = 0.2$). We consider an experimentally relevant situation by keeping the EP coupling fixed and varying the dot level (gate voltage). In the case of a rigid quantum dot ($\varepsilon_p = 0$), scattering off the dot potential reduces the conductance. Therefore, in figure 3a, L exhibits a maximum at $\Delta = 0$ and decreases symmetrically as $|\Delta|$ grows. For finite EP coupling $\varepsilon_p = 0.2$, this maximum is shifted to $\Delta > 0$, where the repulsive dot potential is compensated by the EP interaction. The spectral function then shows a peak at the Fermi level and two phonon satellites (see figure 3c). As we see from figure 3b, the optimal variational parameter is a continuous function of Δ taking values from 0.1 to 0.9. Therefore, the effective renormalisation of $\widetilde{t}_d$ and $\widetilde{\Delta}$ depends, via γ_{min}, on the dot level Δ itself. In contrast to the result for a complete Lang-Firsov transformation with fixed $\gamma = 1$ (cf. figure 5 of [13]), the conductance maximum

Journal of Physics: Conference Series **220** (2010) 012014 doi:10.1088/1742-6596/220/1/012014

is shifted by less than ε_p, and L decreases asymmetrically away from this point. In accordance with Galperin *et al* [13] and Mitra *et al* [14], we find no phonon side peak in $L(\Delta)$. If ε_p is set to larger values ($\varepsilon_p = 0.4$ and 0.5), a transition in the dot occupation appears which is again related to a jump in γ_{min}. La Magna and Deretzis [2] found a similar behaviour: for intermediate phonon frequencies and strong EP coupling their variational ansatz showed bistabilities causing a sudden occupation of the quantum dot (cf. figs. 2c and 2d in [2]). As the dot level falls below a critical value, the charging of the quantum dot is accompanied by a drop in the conductance. In this way the quantum dot acts as a simple molecular switch. Now the spectrum shows a pronounced maximum at $\tilde{\Delta}$, which is below the Fermi level. Note that, due to overlapping phonon bands, no bound state exists even for large ε_p (figure 3d).

5. Conclusion

In the present contribution we studied transport through a deformable quantum dot, realised e.g. by an organic molecule in contact to leads. Our treatment is based on Green's functions equations, with the dot Green's function as the central quantity of our calculation. EP interaction significantly affects the shape of the associated dot spectral function. Since at low temperature the conductance of the quantum dot is determined by the spectral weight close to the Fermi energy, it changes accordingly. The basic effect, with dramatic consequences for the conductance, is the formation of a "localised" polaron-like dot state. Similar to the Holstein polaron, this effect can be captured in a Lang-Firsov approach. The virtue of our analytical approach lies in the variational determination of the Lang-Firsov parameter. This allows us to account for the basic polaronic effects both in the adiabatic and anti-adiabatic regimes, which are distinguished by their different influence on the spectral function and, accordingly, on the conducteance (briefly summarised in figure 4). Particularly for comparable phonon and electron time-scales we find interesting physical behaviour, which motivates further investigation of the quantum dot system, e.g. with respect to applications as a current switch.

The limitations of our study suggest two directions for improvements. First, the inclusion of Coulomb interaction at the dot. Many interesting effects we present here arise from the competition between electron-phonon interaction and a repulsive dot potential. Coulomb interaction gives rise to an effective repulsive dot potential, whose strength depends on the dot population. This may allow for even stronger non-linear behaviour, as then the transition

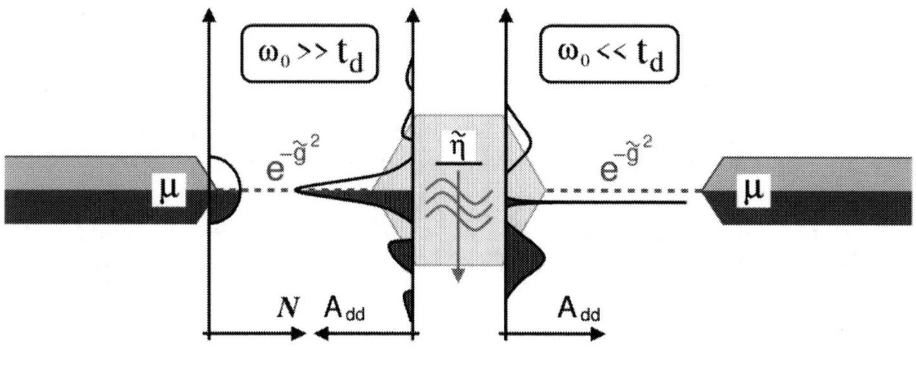

Figure 4. Sketch of the basic polaronic effects on transport through a vibrating quantum dot. In addition to a renormalisation of the dot-lead transfer integral, in the anti-adiabatic regime, a continuous lowering of the dot potential towards the Fermi level enhances the transport. In the adiabatic regime, the sudden formation of a quasi-localised polaron state causes a drop of the conductance.

between an effectively attractive or repulsive dot depends on the dot population through both interaction mechanisms. Second, all our approximations neglect correlations between phononic and polaronic degrees of freedom (note however that the Lang-Firsov transformation introduces strong correlations between the phononic and electronic degrees of freedom). In this way, the variational Lang-Firsov transformation can be used to map the entire problem approximately onto a purely electronic problem, as in reference [2]. This reference reports results similar to ours, e.g. for the dot population, but also for artificial bistabilities corresponding to different local minima of the groundstate energy, which are a common feature of these variational approaches. As a first step towards the reintroduction of electron-phonon correlations our treatment includes the weak-coupling contribution to the self-energy, which partly accounts for retardation of electron-phonon interaction at finite phonon frequencies. A detailed comparison of our results to those presented in reference [2] has to be given elsewhere, and the improvement of our treatment along the lines indicated is the subject of future work.

Acknowledgements. We are grateful to one referee for bringing reference [2] to our attention. H.F. acknowledges the hospitality at the Institute of Physics, Czech Academy of Sciences, Prague.

6. References

[1] M. Galperin, M. A. Ratner, and A. Nitzan. Molecular transport junctions: vibrational effects. *J. Phys. Condens. Matter*, 19:103201, 2007.

[2] A. La Magna, I. Deretzis. Phonon driven nonlinear behavior in molecular devices. *Phys. Rev. Lett.*, 99:136404, 2007.

[3] A. M. Bratkovsky. Current rectification, switching, polarons, and defects in molecular electronic devices. *Springer Series in Materials Science*, 103:623, 2007.

[4] T. Holstein. Studies of polaron motion. Part I. The molecular-crystal model. *Ann. Phys. (N.Y.)*, 8:325, 1959.

[5] H. Fehske, G. Wellein, J. Loos, and A. R. Bishop. Localized polarons and doorway vibrons in finite quantum structures. *Phys. Rev. B*, 77:085117, 2008.

[6] J. Mravlje and A. Ramšak. Electron transport through molecules in the Kondo regime: the role of molecular vibrations. arXiv:0912.3536.

[7] H. Fehske, D. Ihle, J. Loos, U. Trapper, and H. Büttner. Polaron formation and hopping conductivity in the Holstein-Hubbard model. *Z. Phys. B*, 94:91, 1994.

[8] L. P. Kadanoff and G. Baym. *Quantum Statistical Mechanics*. Benjamin/Cumming Publishing Company, Reading, Massachusetts, 1962.

[9] J. Loos, T. Koch, A. Alvermann, A. R. Bishop, and H. Fehske. Phonon affected transport through molecular quantum dots. *J. Phys. Condens. Matter*, 21:395601, 2009.

[10] Y. Meir and N. S. Wingreen. Landauer formula for the current through an interacting electron region. *Phys. Rev. Lett.*, 16:2512, 1992.

[11] J. Loos, M. Hohenadler, and H. Fehske. Spectral functions of the spinless Holstein model. *J. Phys. Condens. Matter*, 18:2453, 2006.

[12] A. Alvermann and H. Fehske. Chebyshev approach to quantum systems coupled to a bath. *Phys. Rev. B*, 77:045125, 2008.

[13] M. Galperin, A. Nitzan, and M. A. Ratner. Resonant inelastic tunneling in molecular junctions. *Phys. Rev. B*, 73:045314, 2006.

[14] A. Mitra, I. Aleiner, and A. J. Mills. Phonon effects in molecular transistor: Quantal and classical treatment. *Phys. Rev. B*, 69:245302, 2004.

Journal of Physics: Conference Series **220** (2010) 012015 doi:10.1088/1742-6596/220/1/012015

Open Quantum Dots in Graphene: Scaling Relativistic Pointer States

D K Ferry[1]**, L Huang, R Yang, Y-C Lai, R Akis**
School of Electrical, Computer, and Energy Engineering and Center for Solid State Electronics Research, Arizona State University, Tempe, AZ 85287-5706 USA

E-mail: ferry@asu.edu

Abstract. Open quantum dots provide a window into the connection between quantum and classical physics, particularly through the decoherence theory, in which an important set of quantum states are not "washed out" through interaction with the environment—the pointer states provide connection to trapped classical orbits which remain stable in the dots. Graphene is a recently discovered material with highly unusual properties. This single layer, one atom thick, sheet of carbon has a unique bandstructure, governed by the Dirac equation, in which charge carriers imitate relativistic particles with zero rest mass. Here, an atomic orbital-based recursive Green's function method is used for studying the quantum transport. We study quantum fluctuations in graphene and bilayer graphene quantum dots with this recursive Green's function method. Finally, we examine the scaling of the domiant fluctuation frequency with dot size.

1. Introduction

The manner in which the quantum states of a system evolve into their classical counterparts has been of interest since the advent of quantum theory [1]. While dynamical systems may be purely chaotic or regular, the most ubiquitous in nature are those whose phase space is instead *mixed* [2]. The study of quantum behavior in open systems is important, also, to the *decoherence* theory [3]. This latter theory connects the properties of open quantum systems to the existence of a set of *pointer states*, which remain robust in the presence of the environmental coupling, eventually correlating with regular classical orbits. The evidence for such pointer states has been provided in the studies of single open quantum dots [4,5]. And, it has clearly been demonstrated that these quantum states are highly correlated with the classical states that arise in the mixed phase space and satisfy the Kolmogorov-Arnold-Moser behavior [6] expected in such systems [7,8]. It has also been shown that these pointer states exist as families, which include offspring, that lead to periodicities in a magnetic field, and that new states can exist in arrays of quantum dots [9].

Graphene has been the subject of a considerable level of interest in the past few years [10]. This interest seems to originate from the astonishing difference between graphene and other well-known two-dimensional semiconductor systems. This single layer, one atom thick, sheet of carbon has atoms which are arranged into a honeycomb lattice, which produces a unique bandstructure where

[1] To whom any correspondence should be addressed.

Journal of Physics: Conference Series **220** (2010) 012015 doi:10.1088/1742-6596/220/1/012015

the bands and electron transport are mainly derived by the surface normal p_z orbitals. Near the K point, the band structure exhibits a linear energy-momentum relationship, and is effectively described by the Dirac equation, in which charge carriers imitate relativistic particles with zero rest mass. This strictly two-dimensional material exhibits exceptionally high crystalline and electronic quality, and despite its short history, has already revealed an abundance of new physics and potential applications, as well as a very high mobility at room temperature [11,12]. Quantum dots have been described for graphene [13,14], and single-electron charging has been observed in small graphene dots [15,16]. Studies of these dots have shown chaotic behavior [16], but behavior similar to normal semiconductor dots has been found [17]. Importantly, scarring behavior and the tendency to support pointer states has been found in graphene dots which would support chaotic and mixed behavior [18]. Importantly, conduction fluctuations have been suggested [18,19], in keeping with recent experiments in bilayer graphene quantum dots [20].

In this paper, we discuss the computation of conductance through open quantum dots in single-layer and bilayer graphene, using the recursive Green's function formalism. In particular, we discuss the scaling of magnetic field and energy with dot size for the normalized behavior of conductance fluctuations which arise from phase-space tunneling to the pointer states [5]. In the next section, we discuss the properties of graphene and its band structure which are important for these calculations. We then turn to a description of the recursive Green's function technique and its formulation in an atomic basis. Finally, we discuss the observed conductance fluctuations, their dominant characteristics and periods and how these scale with quantum dot size.

2. Graphene

Graphene is a single layer of carbon atoms arranged in a hexagonal honeycomb structure. The proper unit cell contains two atoms per unit cell, which are denoted as atom A and atom B, as shown with two different colors in Fig. 1(a). This leads to an equivalent Brillouin zone with two minima at K and K', as shown in Fig. 1(b). These latter two points are known as the Dirac points for reasons that become clear below. The C-C nearest neighbor distance is a = 0.142 nm, but the length of the primitive lattice vector shown in the figure is 0.246 nm. Carbon atoms possess four valence electrons. Three of these (one s orbital and two p orbital electrons) form tight bonds, known as σ bonds, with three neighboring atoms in the graphene plane. The fourth one is the p_z state, directed normal to the lattice. The σ bonds form a deep valence band which will not play a role in the conductivity. The p_z orbitals form the π band and this constitutes the band of interest. The nature of graphene is such that the most common method of computing the band structure is simple tight-binding, in which both the nearest neighbor interaction and the second-neighbor interactions are important. The former is typically identified by a general matrix element γ_0, while the latter is denoted by the parameter γ_0'. Each atom has 3 nearest neighbors of the opposite type and 6 second neighbors of the same type, and the tight-binding formulation leads to [21]

$$\begin{aligned} E(\mathbf{k}) &= E_0 \pm \gamma_0\sqrt{3+f(\mathbf{k})} - \gamma_0' f(\mathbf{k}) \\ f(\mathbf{k}) &= 2\cos\left(\sqrt{3}k_y a\right) + 4\cos\left(\frac{\sqrt{3}}{2}k_y a\right)\cos\left(\frac{3}{2}k_x a\right) \end{aligned} \quad . \tag{1}$$

This band structure is shown in Fig. 2. At the K and K' points the bands touch. Expanding around these points, for small wave vectors away from these points, we find that the energies are approximately [10]

$$E = E_0 + 3\gamma_0' \pm \frac{3\gamma_0 a}{2}k + O(k^2) \quad . \tag{2}$$

Journal of Physics: Conference Series **220** (2010) 012015 doi:10.1088/1742-6596/220/1/012015

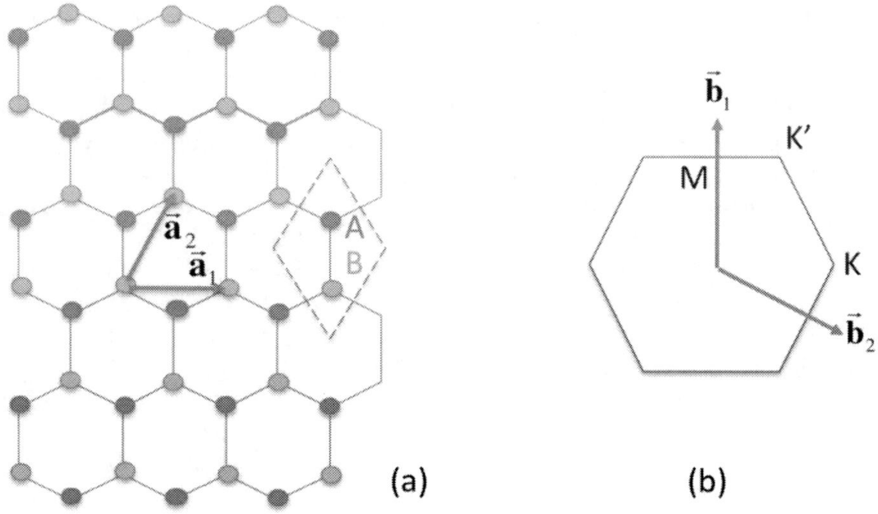

Fig. 1. (a) The hexagonal atomic structure of graphene. The unit cell (dashed line) contains two atoms, denoted as atom A and atom B. Here, the unit vectors are $\mathbf{a}_1 = a(\sqrt{3},0)$, $\mathbf{a}_2 = a(\frac{\sqrt{3}}{2},\frac{3}{2})$, where $a = 0.142$ nm. (b) The brillouin zone of graphene contains two valleys at K and K'. The reciprocal lattice vectors are $\mathbf{b}_1 = \frac{2\pi}{a}\left(\frac{1}{3},-\frac{1}{\sqrt{3}}\right)$, $\mathbf{b}_2 = \frac{2\pi}{a}\left(0,\frac{2}{3}\right)$.

Without loss of generality, we can set the sum of the first two terms to zero, and then we note that the bands are linear bands, with the upper sign for electrons and the lower sign for holes. Moreover, these bands are chiral, in that the positive slope band has positive helicity and the negative slope band has negative helicity. The helicity arises from the pseudo-spin describing the two atomic contributions to the wave function. For our purpose here, we note that these linear bands are Dirac-like in which we can write the energy as

$$E = \pm\hbar v_F k \quad , \tag{3}$$

for which

$$v_F = \frac{3\gamma_0 a}{2\hbar} \tag{4}$$

is the effective "speed of light." One can use angle-resolved photo-emission data to check the energy value at M and K points [22], and this gives a value of γ_0 as 2.52±0.09 eV, and a Fermi velocity of 8.15±0.28 × 10^5 m/s. The second-neighbor interaction breaks the electron-hole symmetry around the Dirac point and we find a value of 0.083 eV for this parameter. Usually, however, one retains only the first-neighbor interaction and often takes γ_0 as 3 eV for convenience, which leads to a Fermi velocity of 9.7 × 10^5 m/s.

An important peculiarity of graphene is the Dirac band structure, which corresponds to particles with zero rest mass. That is, the effective mass of the particles at the Dirac point is exactly zero. Away from the Dirac point, the effective mass may be readily obtained by utilizing the equivalence of crystal momentum and quasi-particle momentum, as

$$m^* = \frac{\hbar k_F}{v_F} = \frac{\hbar}{v_F}\sqrt{\pi n} \quad , \tag{5}$$

a result that has been confirmed by cyclotron resonance studies [23]. In an equivalent fashion, the density of states, with both spin and valley degeneracy of 2, is given by

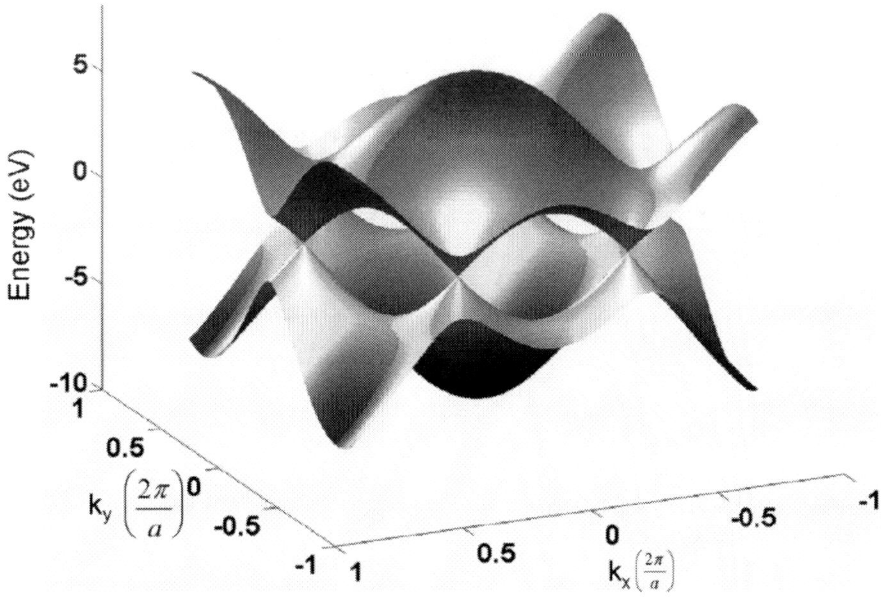

Fig. 2. The bandstructure of graphene [γ_0 = 2.52 eV, γ_0' = 0.083 eV, E_0 = -0.25 eV].

$$\rho(E) = \frac{2A}{\pi(\hbar v_F)^2} E \quad . \tag{6}$$

In intrinsic graphene, the Fermi level lies exactly at the Dirac point. The conductance does not go to zero at this point, but takes a value $\sim 4e^2/\pi h$ [24], which arises from quantum considerations of the zero bandgap. It has been reported that disorder in graphene can create electron-hole puddles, which cause an increase of the conductance at the Dirac point, as well as locally shift the Fermi energy from the Dirac point [25,26]. These puddles were proposed theoretically and have been observed experimentally [27,28]. The origin of the disorder remains an open question. In general, transport at room temperature is dominated by impurity scattering, where the impurities typically reside in the oxide upon which the graphene is placed. However, recent studies, in which a high-dielectric constant solvent has been placed around the graphene, has shown mobilities well above 40,000 cm^2/Vs [11], a value thought to be the intrinsic phonon limited mobility [12].

In bilayer graphene, one has the problem of arranging the two lattices with respect to each other. In general, these two layers are oriented in such a manner that the atoms of one sublattice in layer 1 are directly below the atoms of a sublattice in layer 2. The second set of atoms in layer 1 sit below the voids in the hexagons of layer 2. Still there arise two methods of stacking. In the Bernal stacking, the A atoms of layer 2 sit over the B atoms of layer 1 [29,30,31], but it has also been suggested that the A atoms of layer 2 sit over the A atoms of layer 1 [10,32]. Obviously these two differ only in the notation assigned to the atoms of layer 2. The major interaction, for band structure purposes, is between the two atoms sitting directly over one another, and this is denoted γ_1. Cyclotron resonance studies have given a value close to 0.43 eV for the electrons and 0.52 eV for the holes [33] for this parameter, but a value of 0.39 also has been suggested [34]. Other interactions that

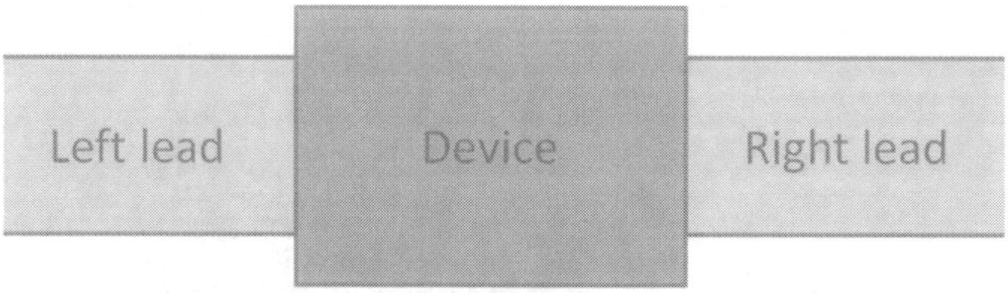

Fig. 3 A conceptual visualization of the separation of the entire system into a pair of leads plus the actual device.

have been considered are between the A atom of layer 1 and the next neighbor in layer 2 (γ_3) and between the two non-aligned atoms (γ_4).

3. The Recursive Green's Function Method

The Green's function has been applied to the study of small devices for a great many years. The most general formulation splits the overall system into three parts: a left contact, a right contact, and the active device under study, as can be seen in Fig. 3. Then, the total system Hamiltonian can be written in reduced block form as

$$H = \begin{bmatrix} H_L & H_{LD} & 0 \\ H_{DL} & H_D & H_{DR} \\ 0 & H_{RD} & H_R \end{bmatrix} . \quad (7)$$

The blocks labelled $H_{L(R)D}$ and their adjoints represent the coupling between the two contacts and the device itself. One example of this type of calculation is to study the structure of the oxide between a crystalline Si layer and a poly-Si layer [35]. It is usual to develop a coupling in which the contacts are reflected into the device Hamiltonian as self-energies, and conductance then developed from the reduced Hamiltonian for the device alone, as described by Fisher and Lee [36]. A modern approach to this is termed the *contact block reduction method* [37]. Nevertheless, this often still results is a Hamiltonian whose size is too large to effectively diagonalize. For this reason, the recursive Green's function (RGF) was developed [38,39]. This is easily generalized to include the effects of a magnetic field [40].

Before proceeding, however, it is worth making a comment. Many people claim to study transport with the non-equilibrium Green's functions (NEGF), but limit themselves to the ballistic case. When this is done, the lesser-than and greater-than correlation functions reduce to the normal retarded and advanced functions plus some self-energies which arise from interactions with the contacts [41,42], and the two Green's functions have their equilibrium form. The calculations here incorporate the same approach. While the NEGF are not specifically mentioned, any extension to non-ballistic transport would require the use of these two extra correlation functions, which can also be treated by recursive techniques.

3.1 The Slice Hamiltonian

The first step in applying the RGF method is to discretize the entire space of the nanostructure that is of interest. In the case of graphene, the atomic structure provides this discretization, and this is known as an atomic basis (as opposed to a lattice basis used in the case of known energy bands). Here, we assume a form for the e.g. left contact Green's function and then use this to connect to the first slice of the device via a coupling interaction term, which is part of H_{LD}. It is only part of this term, as this latter term couples the left contact to the *entire* device, not just to the first slice. Hence, one part of the interation is to build up this overall interaction and the modification of the Green's function for the left contact, which we will denote as $G(0,0)$. To implement this, everything begins with the basic slice. In Fig. 4, we illustrate a small graphene quantum dot connected to two narrow ribbon leads. In this figure, we denote the basic slice for the case of zig-zag boundaries. The case for arm-chair boundaries can be done as easily [43,44], but we will stick to the former situation in this discussion. Each slice contains an equal number of A and B atoms. The graphene lattice itself is defined by three nearest-neighbor vectors, which are defined by

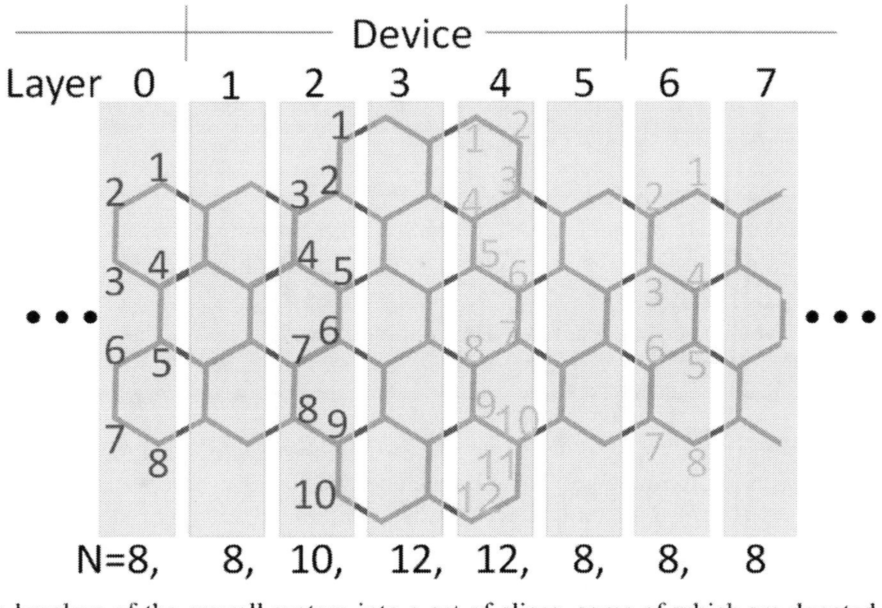

Fig. 4 The breakup of the overall system into a set of slices, some of which are denoted as contact slices.

$$\mathbf{d}_1 = a(0,-1) \quad , \quad \mathbf{d}_2 = \frac{a}{2}(\sqrt{3},1) \quad , \quad \mathbf{d}_3 = \frac{a}{2}(-\sqrt{3},1) \tag{8}$$

in the (x,y) plane, where a is the nearest neighbor distance of 0.142 nm. Let us consider slice 2, shown in Fig. 4, for example. The slice Hamiltonian for this slice is zero with the exception of the first off-diagonal terms. These terms have the nearest-neighbor coupling in these first off diagonal terms. If we have periodic lattices, then the description of the near-neighbor interactions includes the phase shifts that give rise to the Bloch sums. Here, however, we do not have this periodicity in either direction. Hence, the coupling terms in the interaction are merely the value of the overlap integral for wave functions on two adjacent atoms with the Hamiltonian. This is given the value

$$t_{i,i\pm1} = \gamma_0 \ . \tag{9}$$

When we consider bilayer graphene, we double the number of atoms in each slice, and pursue the more complicated inter-slice couplings that add off diagonal terms to the larger Hamiltonian for the slice.

The Hamiltonian which couples the slices is somewhat more complicated. Normally, on a rectilinear grid, this coupling Hamiltonian would be diagonal. Here, however, it is more difficult due to the positions of the various atoms in the slice. Let us consider, for example, the coupling between slices 1 and 2 (the atoms designations are the same as for slice 0). Note from Fig. 4 that atom 1, in slice 1, is coupled to atom 3, in slice 2. Hence,

$$\left(H_{12}\right)_{1,3} = \gamma_0 \ . \tag{10}$$

By the same token, atom 4, in slice 1, is coupled to atom 4, in slice 2. This gives a diagonal term of the form

$$\left(H_{12}\right)_{4,4} = \gamma_0 \ . \tag{11}$$

It is important to note that this coupling Hamiltonian is not required to be Hermitian, as it is basically a self-energy term (only the total Hamiltonian is required to be Hermitian). One may be concerned about the lack of the phase shifts here, but we remind ourselves that the coupling is a simple two center integral. The basic structure of the interslice coupling Hamiltonians defines the crystal topology and this, together with the fact that all of the overlap integrals between nearest neighbors are equal, sets the nature of the atomic lattice, ensuring the hexagonal structure. Another important point in the interslice coupling term is that this Hamiltonian matrix is not square. Rather, we note that slice 1 and slice 2 have different numbers of atoms, so that the coupling Hamiltonian will have different numbers of rows and columns.

3.2 Adding a Magnetic Field

When an external magnetic field is applied, the hopping terms γ_0 discussed above acquire an additional phase due to the Peierls' phase factor. This additional phase factor is given by

$$\vartheta_{\mathbf{r},\mathbf{r}+\mathbf{d}} = 2\pi\phi_{\mathbf{r},\mathbf{r}+\mathbf{d}} / \phi_0 \ , \tag{12}$$

where $\phi_0 = h/e$ is the flux quantum. The coupled flux is given by

$$\phi_{\mathbf{r},\mathbf{r}+\mathbf{d}} = \int_{\mathbf{r}}^{\mathbf{r}+\mathbf{d}} \mathbf{A}\cdot d\mathbf{r} \ . \tag{13}$$

In this expression, $\mathbf{d}$ is any of the three nearest-neighbor vectors that appear in (9), and the full x and y shift from one atom to the next must be included. Typically, however, one uses the Landau gauge in which $\mathbf{A} = (-By,0)$, so that only the x shift is incorporated, although the y position directly affects the amount of that phase shift. Hence, by the time one has moved across the slice, the entire flux through the slice has been incorporated. Generally, however, one evaluates the y value at points midway between the two atoms being coupled.

An important point with the magnetic field in structures in which the slices may have different extent in the transverse direction is the position of $y = 0$. This needs to be at the center of the slice, so that there is no discontinuity in the vector potential as one moves from one slice to the next.

3.3 The Recursive Procedure

The recursion begins by establishing a self-energy correction which acts upon the first slice, but is based upon the left contact and its connection to the first slice. That is, we define the left self-energy by the action

$$\Sigma_L = H_{10}G_{00}H_{01} \ , \quad \Gamma_L = i\left(\Sigma_L - \Sigma_L^+\right) \ , \tag{14}$$

where

$$G_{00} = \left(EI - H_0 + i\eta\right)^{-1} \ , \tag{15}$$

and a small damping factor has been added for convergence purposes. Here, I is a unit matrix and E is the Fermi energy. We then recompute the Green's function for the 0th slice to be

$$G(0,0) = \left(EI - H_0 + \Sigma_L + i\eta\right)^{-1} \ . \tag{16}$$

We now can propagate across the structure using the recursion for $j > 1$ as

$$G(j,j) = \left(EI - H_j - H_{j,j-1}G(j-1,j-1)H_{j-1,j}\right)^{-1} \quad . \tag{17}$$

But, we will also need another Green's function in order to compute the transmission, and this is found from the auxiliary functions

$$\begin{aligned} G(0,j) &= G(0,j-1)H_{j-1,j}G(j,j) \\ G(j,0) &= G(j,j)H_{j,j-1}G(j-1,0) \end{aligned} \quad . \tag{18}$$

At the right contact, we have to connect to the boundary. Let us assume that the last device slice has $s = L$, then the right contact Green's functions become

$$\begin{aligned} G(L+1,L+1) &= \left(EI - H_R + \Sigma_R - H_{L+1,L}G(L,L)H_{L,L+1}\right)^{-1} \\ G(0,L+1) &= G(0,L)H_{L,L+1}G(L+1,L+1) \\ G(L+1,0) &= G(L+1,L+1)H_{L+1,L}G(L,0) \end{aligned} \quad . \tag{19}$$

Here, the right self-energy has been computed in exactly the same manner as (14), but with the right contact Hamiltonian. The transmission through the total system is now found from (one doesn't worry about the velocities since they are all equal to the Fermi velocity)

$$T = Tr\left\{\Gamma_L G(0,L+1)\Gamma_R G^+(L+1,0)\right\} \quad . \tag{20}$$

In many cases, one also wants to know the reflection coefficient, in order to make sure that flux is conserved in the system. In this case, one needs to also compute the interaction of the entire system on the Green's function for the left contact. Hence, we need an additional recursion

$$G(0,0) = G(0,0) + G(0,j-1)H_{j-1,j}G(j,0) \quad . \tag{21}$$

The reflection coefficient is then essentially the trace of the squared magnitude of $G(0,0)$.

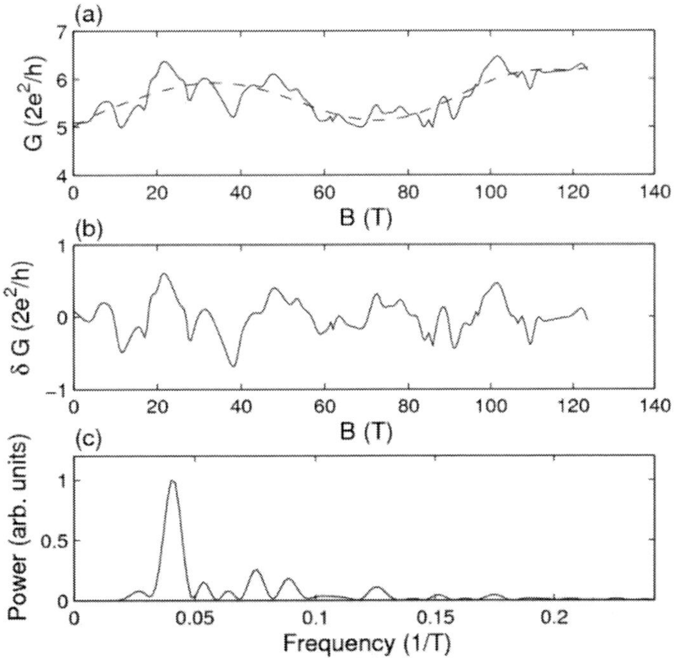

Fig. 5 (a) Conductance through a 20 nm square graphene quantum dot with 4 nm leads. (b) The conductance fluctuations after the background (slowly varying conductance shown as the dashed line in (a)) is removed. (c) Fourier spectrum of the conductance fluctuations showing that a single dominant frequency governs the fluctuations.

The charge density may be found at each atomic site by utilizing the slice Green's function $G(j,j)$. In much the same way that the density of states is found, we find the probability density, at the Fermi energy, at site i in slice j to be

$$\rho_{ij} = -\frac{1}{\pi}\mathrm{Im}\left\{\left[G(j,j)\right]_{ii}\right\} . \tag{22}$$

To find the total density, we need to integrate over all energies up to the Fermi energy (measured from zero for the electrons). This is usually a integral sensitive to energy gridding, and a contour along a complex energy is often used to ease this problem.

3.4 An Example

As an example, we compute the conductance through a 20 × 24 nm square graphene, single layer quantum dot, in which leads of 4 nm width are applied. In Fig. 5, we plot the total conductance, the fluctuations after the background has been subtracted and the Fourier spectra of these fluctuations. Here, the so-called background that is subtracted is found by smoothing the overall curve. The tunnelling to the pointer states gives rise to Fano resonances, and these can be smoothed away to yield the background conductance. In experiment, the procedure is usually to raise the temperature to a few degrees, where the fluctuations are heavily damped and this curve used for the subtraction. The spirit is the same, as the smoothing gives an effective temperature effect. For reasons that will become clear in the next section, the magnetic field is rather high, but it is clear that a single narrow band of frequencies dominate the fluctuations. This is in keeping with the results found on normal semiconductor quantum dots in which the set of pointer states, and their periodic reappearance at particular magnetic fields corresponds to the existence of quantum Darwinism in this system [9].

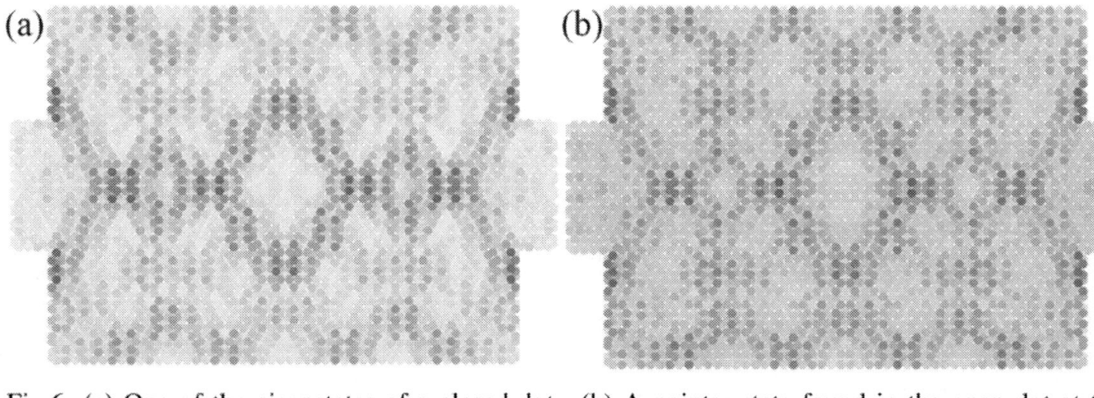

Fig.6 (a) One of the eigenstates of a closed dot. (b) A pointer state found in the open dot at the same energy as that of the state in (a).

Another interesting aspect is the comparison between the eigenstates of the closed system, which can be computed by diagonalizing the full Green's function determined from the Hamiltonian (7) and the pointer states which remain in the open system, which are computed from the density at each atomic site found from the Green's function. In Fig. 6, we show results for a rectangular shaped dot, and compare one eigenstate found in the closed dot with a pointer state found in the open dot. The agreement between these two results clearly shows the connection between the pointer states and the eigenstates, and that the latter are states which have not undergone decoherence.

4. Scaling in Quantum Dots

In this section, we would like to examine how the dominant frequency of the fluctuation, as seen in Fig. 5(c), varies with dot size. In general, these fluctuations, due to phase space tunnelling to the pointer states, only occur in the low magnetic field regime [45]. Once the cyclotron radius fits within the dot, then edge states begin to form, and the transport is dominated by the quantum Hall effect. Consequently, we need to scale the magnetic field with the dot size. In general, we can develop the scaling factor for the maximum of the magnetic field by setting the dot size D equal to the cyclotron radius at the Fermi energy, as

$$D = l_B^2 k_F \ , \tag{23}$$

where

$$l_B = \sqrt{\frac{\hbar}{eB}} \tag{24}$$

is the magnetic length. The Fermi wavelength is found from the Dirac band energy $E = \hbar v_F k_F$, with the Fermi velocity being given by (4). Then, we find that the scaling is given by

$$D = \frac{2\hbar(E/\gamma_0)}{3eaB} \ , \tag{25}$$

or the maximum magnetic field is given by

$$B_{\max} = \frac{2\hbar(E/\gamma_0)}{3eaD} \ . \tag{25a}$$

We follow this scaling in the range of magnetic field that is studied in the set of dots that we examine.

Here, we will examine single layers of graphene, using $\gamma_0 = 3$, as well as bilayers of graphene. For the bilayers, we assume the Bernal stacking with parameters $\gamma_0 = 3.16$, $\gamma_1 = 0.39$ and $\gamma_3 = 0.315$ [34]. Here, γ_1 represents the coupling between the two aligned A and B' atoms of the first and second layers. The other constant represents the coupling between the nonaligned B (in the center of the hexagon) and A' atoms of the second layer. In Fig. 7, we show a sketch of the 20 nm dot and its leads. For dot sizes W = 20-50 nm, we will use a 4 nm lead width and an energy of $E/\gamma_0 = 0.8$. For larger dots, the lead opening is scaled up as $4(W/50)$ nm and the energy is reduced as $E/\gamma_0 = 0.8(50/W)$. The fluctuations for the smallest dot, for graphene, is shown in Fig. 5(a). The fluctuations for the bilayer dot are quite similar, and we compare these in Fig. 8. Our interest is in the dominant frequency of the fluctuations, and we summarize the results for a series of dots in Fig. 9.

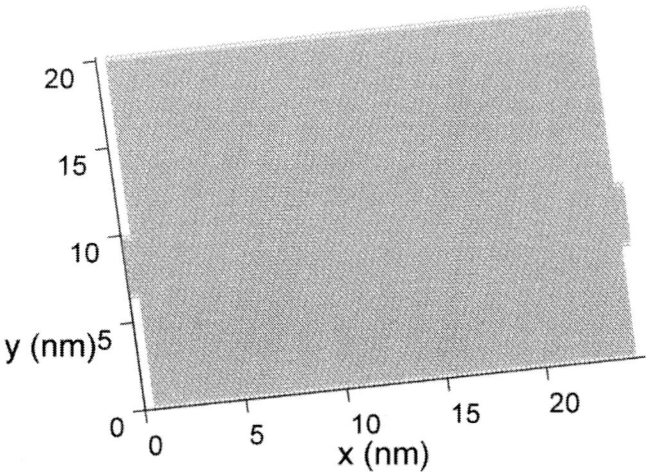

Fig. 7 A schematic drawing of the bilayer graphene quantum dot, showing the connection of the two leads.

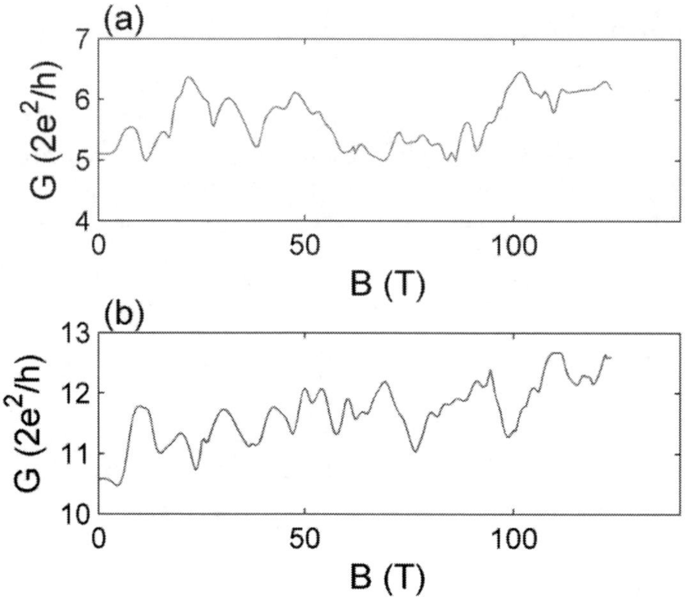

Fig. 8 Fluctuations observed in a single graphene layer dot (a) and in a bilayer dot (b).

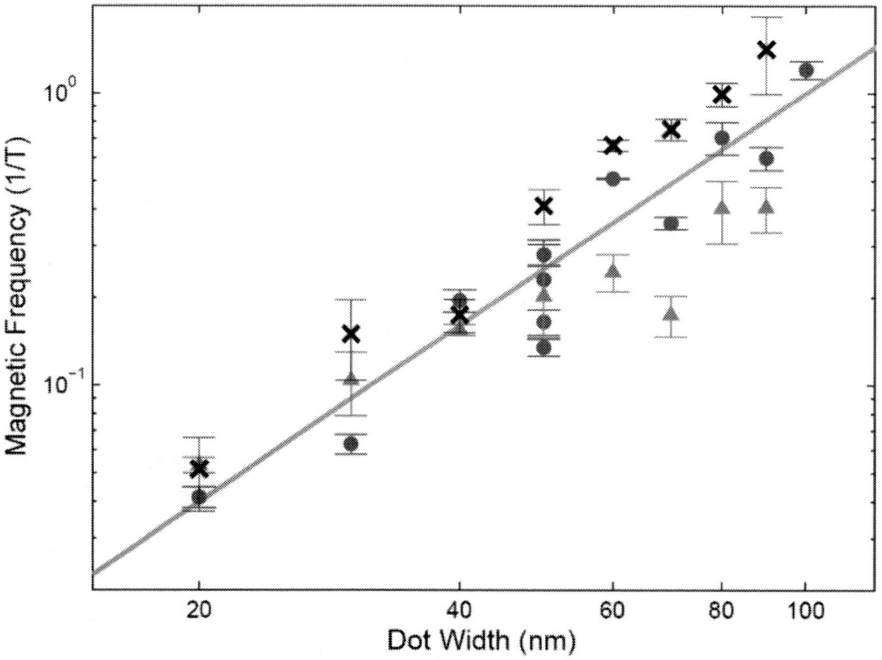

Fig. 9 The variation of the dominant magnetic frequency with the dot size. The monolayer graphene is plotted as solid circles while the bilayer graphene is plotted with the triangles. Bilayer graphene with a factor of 2 lower energy is plotted as the crosses. It is not clear whether or not a real difference exists between the various results.

There is a clear change in behavior as the size of the dot increases beyond ca. 50 nm. For smaller dot sizes, the magnetic frequency is largely the same in both the monolayer and bilayer dots. However, for the larger dots, the two frequencies clearly diverge. Recent experiments in a much larger quantum dot in bilayer graphene suggests a magnetic frequency of about 9 T^{-1} at a dot size of 0.5 μm [20]. Scaling of the single layer graphene data in Fig. 9 to this size would suggest a magnetic frequency 2-3 times as large. There is insufficient data in the figure to estimate whether the bilayer data would extrapolate to a value near that observed, but it is certain that the bilayer would likely have a much lower magnetic frequency. We can only hope that, as we examine larger dots, we can extend these curves to a size that matches that used in the experiment.

It is also worth noting that there is no obvious power law relationship between the dot size and the magnetic frequency in Fig. 9, although the solid line represents a scaling with the area of the dot. In previous studies of dot size in semiconductor dots, experiments on dots in the range of 0.7-1.2 μm on a side showed a scaling with the dot edge size [46,47]. Simulations, however, exhibited multiple frequencies in these large dots and a transition from behavior in which the magnetic frequency scaled with the dot area in smaller dots [48], typically with an edge less than 300 nm in extent. If we take all the curves together in Fig. 9, then it appears that scaling with the dot area seems to be appropriate for these small dots, as was found in the semiconductor dots. One point worth noting is that if we extend the solid curve in Fig. 9 to a dot size commensurate with the recent experiments [20], then the expected frequency is approximately 9 T^{-1}, which is the value found in the experiments.

5. Conclusions

Here, we have examined the nature of fluctuations in quantum dots at low magnetic fields, in which transport is thought to exhibit phase space tunneling to pointer states within the dots. These fluctuations are seen in graphene, with its "relativistic" massless Dirac band structure, just as in semiconductor quantum dots, with similar characteristic features at similar sizes. The use of the recursive Green's function method, with an atomic basis set by the individual atoms of graphene, captures the important aspects of the transport and fluctuations, and remains an important tool for studies of transport in a wide range of mesoscopic structures.

The authors would like to thank S. M. Goodnick, R. S. Shishir, and I. Knezevic for many helpful discussions.

---

[1] Wheeler J A, Zurek W H, *Quantum Theory and Measurement* (Princeton University Press, Princeton, N. J., 1983).
[2] Ketzmerick R, *Phys. Rev. B* **54**, 10841 (1996).
[3] Zurek W H, *Rev. Mod. Phys.* **75**, 715 (2003).
[4] Ferry D K, Akis R, Bird J P, *Phys. Rev. Lett.* **93**, 026803 (2004).
[5] Ferry D K, Akis R, Bird J P, *J. Phys. Cond. Matt.* **17**, S1017 (2005).
[6] Broer H W, *Bull. Amer. Math. Soc.* **41**, 507 (2004).
[7] de Moura A P S, Lai L-C, Akis R, Bird J P, Ferry D K, *Phys. Rev. Lett.* **88**, 236804 (2002).
[8] Brunner R, Meisels R, Kuchar F, Akis R, Ferry D K, Bird J P, *Phys. Rev. Lett.* **98**, 204101 (2007).
[9] Brunner R, Akis R, Ferry D K, Kuchar F, Meisels R, *Phys. Rev. Lett.* **101**, 024102 (2008).
[10] Castro Neto A H, Guinea F, Peres N, Novoselov K S, Geim A, *Rev. Mod. Phys.* **81**, 109 (2009).
[11] Chen F, Xia J, Ferry D K, Tao N J, *Nano Lett.* **9**, 2571 (2009).
[12] Shishir R S, Ferry D K, *J. Phys. Cond. Matt.* **21**, 232204 (2009).
[13] Silvestrov P G, Efetov K B, *Phys. Rev. Lett.* 98, 016802 (2007).
[14] Pereira J M, Jr., Vasilopoulos P, Peeters F M, *Nano Lett.* 7, 946 (2007).

[15] Stampfer C, Güttinger J, Molitor F, Graf D, Ihn T, Ensslin K, *Appl. Phys. Lett.* **92**, 012102 (2008).
[16] Ponomarenko L A, Schedin F, Katsnelson M I, Yang R, Hill E W, Novoselov K S, Geim, A K, *Science* **320**, 356 (2008).
[17] Libisch F, Stampfer C, Burgdörfer J, *Phys. Rev. B* **79**, 115423 (2009).
[18] Huang L, Lai Y-C, Ferry D K, Goodnick S M, Akis R, *Phys. Rev. Lett* **103**, 054101 (2009).
[19] Huang L, Lai Y-C, Ferry D K, Akis R, Goodnick S M, *J. Phys. Cond. Matt.* **21**, 344203 (2009).
[20] Ujiie Y, Morimoto T, Aoki N, Ferry D K, Bird J P, Ochiai Y, *J. Phys. Cond. Matt., in press.*
[21] Wallace P R, *Phys. Rev.* **71**, 622 (1947).
[22] Bostwick A, Ohta T, Seyller T, Horn K, Rotenberg E, *Nature Phys.* **3**, 37 (2006).
[23] Novoselov K S, Geim A, Morozov S V, Jiang D, Katsnelson M I, Grigorieva I V, Dubonos S V, Firsov A A, *Nature* **438**, 197 (2005); Bostwick A, Ohta T, Seyller T, Horn K, Rotenberg E, *Nature Phys.* **3**, 37 (2006).
[24] Fradkin E, *Phys. Rev. B* **33**, 3257 (1986); Ludwig A, Fisher M, Shankar R, Grinstein G, *Phys. Rev. B* **50**, 7526 (1994); Aleiner I L, Efetov K B, *Phys. Rev. Lett.* 97, 236801 (2006); Atland A, *Phys. Rev. Lett.* **97**, 236802 (2006); Peres N M, Guinea F, Castro Neto A H, *Phys. Rev. B* **73**, 125411 (2006); Fritz L, Schmalian J, Müller M, Sachdev S, *Phys. Rev. B* **78**, 085416 (2008); Kashuba A, *Phys. Rev. B* **78**, 085415 (2008); Bardarson J H, Tworzydło J, Brouwer P W, Beenakker C W J, *Phys. Rev. Lett.* **99**, 106801 (2007).
[25] Rossi E, Das Sarma S, *Phys. Rev. Lett.* **101**, 166803 (2008).
[26] Katsnelson M I, Novoselov K S, Geim A K, *Nature Phys.* **2**, 620 (2006).
[27] Hwang E H, Adam S, Das Sarma S, *Phys. Rev. Lett.* **98**, 186806 (2007).
[28] Martin J, Akerman N, Ulbricht G, Lohmann T, Smet J H, von Klitzing K, Yacoby A, *Nature Phys.* **4**, 144 (2008).
[29] McCann E, Falko V I, *Phys. Rev. Lett.* **96**, 086805 (2006).
[30] Castro E V, Novoselov K S, Morozov S V, Peres N M R, Lopes dos Santos J M B, Nilsson J, Guinea F, Geim A K, Castro Neto A H, *Phys. Rev. Lett.* 99, 216802 (2007).
[31] Park C-H, Giustino F, Cohen M L, Louie S G, Nano Lett. 8, 4229 (2008).
[32] Li Z Q, Henriksen E A, Jiang Z, Hao Z, Martin M C, Kim P, Stormer H L, and Basov D N, *Phys. Rev. Lett.* 102, 037403 (2009).
[33] Henriksen E A, Jiang Z, Tung L-C, Schwartz M E, Takita M, Wang Y-J, Kim P, Stormer H L, *Phys. Rev. Lett.* **100**, 087403 (2008).
[34] Koshino M, Ando T, *Phys. Rev. B* **73**, 245403 (2006).
[35] Demkov A A, Zhang X, Drabold D A, *Phys. Rev. B* **64**, 125306 (2001).
[36] Fisher D S, Lee P A, *Phys. Rev. B* **23**, 6851 (1981).
[37] Mamaluy D, Vasileska D, Sabathil M, Zibold T, Vogl P, *Phys. Rev. B* **71**, 245321 (2005).
[38] Lee P A, Fisher D S, *Phys. Rev. Lett.* **47**, 882 (1981).
[39] Thouless D J, Kirkpatrick S, *J. Phys. C* **14**, 235 (1981).
[40] Ando T, *Phys. Rev. B* **44**, 8017 (1991).
[41] Meir Y, Wingreen, N S, *Phys. Rev. Lett.* **68**, 2512 (1992).
[42] Ferry D K, Goodnick, S M, Bird, J P, *Transport in Nanostructures*, 2nd Ed. (Cambridge Univ. Press, 2009) Sec. 9.4.4.
[43] Wakabayashi K, Fujita M, Ajiki H, Sigrist M, *Phys. Rev. B* **59**, 8271 (1999).
[44] Li T C, Lu S-P, *Phys. Rev. B* **77**, 085408 (2008).
[45] Akis R, Bird J P, Vasileska D, Ferry D K, de Moura A P S, Lai Y-C, in *Electron Transport in Quantum Dots*, Ed. by Bird J P (Kluwer, Boston, 2003) 209-276.

[46] Bird J P, Ishibashi, K, Aoyagi Y, Sugano T, Akis R, Ferry D K, Pivin D P Jr., Connolly K M, Taylor R P, Newbury R, Olatona D M, Micolich A, Wirtz R, Ochiai Y, Okubo Y, *Chaos Sol. & Fractals* **8**, 1299 (1997).

[47] Ferry D K, Bird J P, Akis R, Pivin D P, Jr., Connolly K M, Ishibashi K, Aoyagi, Y, Sugano T, Ochiai Y, *Jpn. J. Appl. Phys.* **36**, 3944 (1997).

[48] Holmbert N, Akis R, Pivin D P Jr., Bird J P, Ferry, D K, *Semicond. Sci. Technol.* **13**, A21 (1999).

Journal of Physics: Conference Series **220** (2010) 012016 doi:10.1088/1742-6596/220/1/012016

Artificial damping in the Kadanoff-Baym dynamics of small Hubbard chains

M. Puig von Friesen, C. Verdozzi and C.-O. Almbladh

Mathematical Physics and European Theoretical Spectroscopy Facility (ETSF), Lund University, 22100 Lund, Sweden

E-mail: marc@teorfys.lu.se, cv@teorfys.lu.se, coa@teorfys.lu.se

Abstract. We perform a comparative study of exact and approximate time-evolved densities in small Hubbard chains. The approximate densities are obtained via many-body perturbation theory (Hartree-Fock, 2nd Born, GW and T-matrix approximations) within the framework of the time-dependent Kadanoff-Baym equations. Benchmarking approximate results against exact ones allows us to address two rather fundamental issues in the non equilibrium dynamics of strongly correlated systems. I) A characterisation of the performance of several standard MBAs in the non-equilibrium regime. Having a definite notion of how good a specific MBA can be is highly relevant to its application to cases (typically, infinite systems) where exact solutions are not available. Our results show that the T-matrix approximation is overall superior to the other MBAs, at all electron densities. II) A scrutiny of the whole idea of Many Body Perturbation Theory in the Kadanoff-Baym sense, when applied to finite systems. The surprising outcome of our study is that during the time evolution, the KBE develop an unphysical steady state solution. This is a genuinely novel feature of the time-dependent KBE, i.e. is not inherited from possible limitations/approximations in the calculation of the initial state. Our extensive numerical characterisation gives robust evidence that the problem occurs in general, whenever MBPT is applied to finite systems, and approximate self energies based upon infinite partial summations are used. We also offer some more conceptual and general consideration on the dependence of this behaviour on the number of particles and system size. This is followed by our conclusions and glimpses of future work.

1. Introduction

The Kadanoff-Baym equations (KBE) [1, 2] are one of the main frameworks to describe quantum systems out of equilibrium [3, 5]. Recently, the KBE have received renewed attention due to the increased capability of solving them numerically [6, 7, 8, 9, 10, 11, 12, 13, 14, 15, 16]. A main advantage of the KBE is their direct connection to Many Body Perturbation Theory (MBPT): One can build, in a systematic way, approximations for the one-particle Greens function, G, the key quantity in the KBE. The range of validity of the Many Body Approximations (MBAs) can be studied by comparing the approximate results with exact numerical ones for finite strongly correlated systems. Naturally, in finite closed systems one cannot address some important issues such as the onset of the steady state regime, or the role of dissipation, as they for example occur in quantum transport geometries [14]. This approach has, however, a second fundamental advantage, i.e. it unravels the role of finite size effects in the KBE.

In addition, the study of small clusters with strong interactions in a time dependent (TD) environment is interesting of its own: in recent years, the topic has grown in importance, boosted by various technological advances such as coupled quantum dots [17] in nanoscale devices and the emerging field of atomtronics [18] in optical lattices. In reality, nanosystems often contain hundreds or thousands of atoms; such sizes are very difficult to treat, especially if we wish to describe their dynamics and excitations. Typically, one then represents the real system by a Hubbard-like model with much fewer degrees of freedom (core electrons eliminated by pseudopotential techniques, perhaps the host material in a quantum dot system is taken into account via effective-mass approximations, etc.) which account only for the active orbitals.

Small clusters in non equilibrium attached to leads have been studied in the stationary limit [12] as well in real-time [14]. In this paper we do not consider the role of reservoirs, and focus on the real-time evolution of small clusters (more specifically, open-ended Hubbard chains), through full CI calculations as well as within MBAs by propagation of the KBE. We consider the Hartree-Fock approximation (HFA), the 2nd Born approximation (BA), the GW-approximation (GWA) [19] and the T-matrix approximation (TMA) [20, 21] all of which, a part from HFA, have self energies which are non local in space and time. These four approximations are all conserving [1], which is a central requirement of the KBE approach. We make a spin-dependent treatment of the GWA [12, 22], which among other things mitigates the problem of self screening as discussed in [23, 24]. Another common approach to microscopic TD phenomena is Time Dependent Density Functional Theory (TDDFT) [25]. A well established connection exists between TDDFT and MBPT on the Keldysh contour [26]. Here, to highlight this link, we will obtain MBPT-based exchange-correlation (xc) potentials via TD reverse engineering [27], using the TD densities from the KBE.

The plan of this paper is as follows: In Section 2, we describe the model systems considered; this is followed by a survey of the KBE (Section 3), and then by a brief description of the MBAs we used (Section 4). The TDDFT approach is summarised in Section 5; in Sections 6 and 7, we present ground state and TD results, respectively; Section 8 describes in more detail specific features of the TD densities; in Section 9, we briefly address the significance of long-range interactions, i.e. beyond the onsite term U. Finally, our conclusions and outlook are presented in Section 10.

2. Model system

The systems we investigate are open-ended Hubbard chains, described by the following Hamiltonian:

$$H=\sum_{R\sigma}\epsilon_R\hat{n}_{R\sigma}-V\sum_{(RR')\sigma}a^{\dagger}_{R\sigma}a_{R'\sigma}+U\sum_{R}\hat{n}_{R\uparrow}\hat{n}_{R\downarrow}+\sum_{R\,\sigma}w_R\left(t\right)\hat{n}_{R\sigma} \tag{1}$$

Here $\hat{n}_{R\sigma}=a^{\dagger}_{R\sigma}a_{R\sigma}$, $\sigma=\uparrow,\downarrow$, $\langle RR'\rangle$ denotes nearest neighbour sites, ϵ_R are the onsite energies. The $V=1$ is the hopping parameter, U is the interaction energy and $w_R\left(t\right)$ is a local external field which can be of any shape in time t and space. U and $w_R\left(t\right)$ are given in units of V. We consider chains of lengths $L=2,4,6$ and $N_e=2,4,6$ number of electrons (1/6 and half filling); we take spin-up and -down electrons equal in number, $N_\uparrow=N_\downarrow$; this will hold under time evolution since H has no spin-flip terms. In the rest of the paper, the filling will be assigned by $n=N_\uparrow/L$. Our exact solution are obtained via short iterative Lanczos propagation.

3. Kadanoff-Baym equations

To describe the system out of equilibrium, we determine the one-particle Greens function $G\left(t_1,t_2\right)$ by solving its equation of motion [1, 26]

$$\left(i\partial_{t_1}-h\left(t_1\right)\right)G\left(t_1,t_2\right)=\delta\left(t_1,t_2\right)+\int_{\gamma}\Sigma\left(t_1,t\right)G\left(t,t_2\right)dt \tag{2}$$

Journal of Physics: Conference Series **220** (2010) 012016 doi:10.1088/1742-6596/220/1/012016

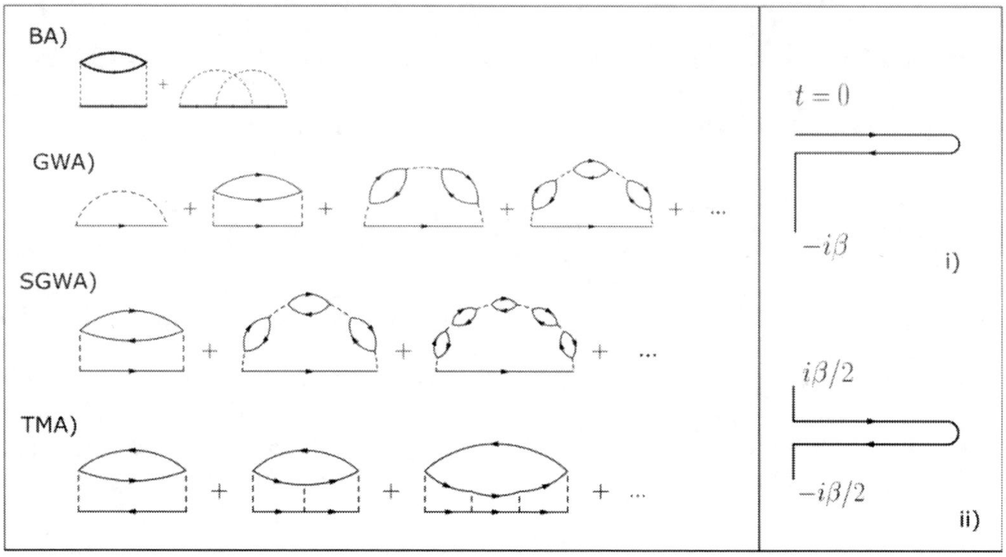

Figure 1. Contours and approximation schemes.

and the adjoint one corresponding to t_2. Here h is the single particle Hamiltonian, γ represents the Keldysh contour [2] and Σ is the self energy, whose functional dependence on G is established by the approximation used. Instead of using the more conventional contour in Fig. (1i), we use the equivalent contour in Fig. (1ii) [5] which is numerically more favourable and has an analytical limit when $T \to 0$. On this contour the G and Σ decay exponentially at $t_1, t_2 \to \pm i\beta/2$ and therefore we only need dense meshpoints close to zero. This is in contrast to the traditional contour Fig. (1ii) where one needs many meshpoints close to zero as well as close to $-i\beta$ due to the periodicity properties of G and Σ [4].

The KBE specialised to the lesser component and t_1 for this contour become

$$
\begin{aligned}
(i\partial_{t_1} - h(t_1))\, G^<(t_1,t_2) = &\int_0^{t_1} d\bar{t}\, \left[\Sigma^R(t_1,\bar{t})\, G^<(\bar{t},t_2) + \Sigma^<(t_1,\bar{t})\, G^A(\bar{t},t_2)\right] \\
&+ \frac{1}{i}\int_0^{\beta/2} d\bar{\tau}\, \left[\Sigma^<(t_1,-i\bar{\tau})\, G^>(-i\bar{\tau},t_2) + \Sigma^>(t_1,i\bar{\tau})\, G^<(i\bar{\tau},t_2)\right]. \qquad (3)
\end{aligned}
$$

Here the superscripts $<, >, R, A$ have their usual meaning [1, 2]. In this paper we will work at zero temperature, $T \to 0$. We obtain the correlated ground state by solving, by simple iteration, the Dyson's equation

$$G = G_0 + G_0\Sigma[G]G$$

self-consistently, with $(\epsilon - h)G_0 = 1$. We then use this as the initial state [3] for the time evolution. Eq. (3) is solved by propagation using a predictor-corrector method similar to that presented in [28]. The conservation laws as well as the time reversal symmetry are obeyed up to arbitrary accuracy. We use a meromorphic representation for the spectral functions as the system is finite [29]. As an example, the Green's function can be written as $G_{RR'}(\epsilon) = \sum_j A^j_{RR'}(\epsilon - a_j)^{-1}$, where $A^j_{RR'}$ are residue matrices and a_j are pole positions in the single particle orbital representation. This representation is very convenient as convolutions become simple matrix products [30] and are obtained analytically: Given two functions

$$B_{RR'}(\epsilon) = \sum_j \frac{B^j_{RR'}}{\epsilon - b_j}\ ,\ D_{RR'}(\epsilon) = \sum_k \frac{D^k_{RR'}}{\epsilon - d_k} \tag{4}$$

then, for the convolution $C_{RR'}(\epsilon)$, we get

$$C_{RR'}(\epsilon) = \int B_{RR'}(\epsilon')\, D_{RR'}(\epsilon - \epsilon')\, \frac{d\epsilon'}{2\pi i} = \sum_{j,k} B^j_{R'R} D^k_{R'R} \frac{1}{\epsilon - b_j - d_k} \tag{5}$$

The number of poles increases rapidly at every iteration. To keep this under control in a numerically manageable way, we merge small and close poles according to some importance threshold criteria, by replacing two poles by an effective one. The residue matrix of the effective pole is the sum of the two old residue matrices and the pole position is the center of gravity of the poles being replaced. Here the "mass" of a pole is the trace of the corresponding residue matrix.

4. Many-body approximations (MBAs)

In a one-band Hubbard model, the interaction can be treated either spin-dependent, $U\sum_R n_{R\uparrow}n_{R\downarrow}$ or spin-independent, $\frac{1}{2}U\sum_{R\sigma\sigma'} a^\dagger_{R\sigma} a^\dagger_{R\sigma'} a_{R\sigma'} a_{R\sigma}$. These treatments are evidently equivalent in any order by order expansion such as the HFA or BA (Fig. (1)) as the diagrams suffering from self interaction are exactly compensated by corresponding exchange diagrams. In approximations based upon partial summations, however, this equivalence is in general lost. To illustrate this point we consider the GWA, $\Sigma(12) = G(12)\,W(12)$ both spin-independent (GWA), which gives $W = U + UPW$ and spin-dependent (SGWA), which results in $W = UPU + (UP)^2 W$, where $P(12) = G(12)\,G(21)$ in both treatments. In this paper, the TMA, $\Sigma(12) = U^2 G(21)\,T(12)$, is treated only spin-dependent and $T = \phi - \phi U T$ where $\phi(12) = G(12)\,G(12)$.

The computational cost increases considerably with increasing complexity of the self energy, along the sequence HFA, BA, GWA and TMA. The main difference is that in the GWA and TMA, in addition to propagating G, we also need, in parallel, to solve the integral equations for W or T. As an example, the equation for $T^<$ is

$$\begin{aligned} T^<(t_1,t_2) = \Phi^<(t_1,t_2) - U\int_0^{t_1} d\bar{t}\left[\Phi^R(t_1,\bar{t})\,T^<(\bar{t},t_2) + \Phi^<(t_1,\bar{t})\,T^A(\bar{t},t_2)\right] \\ -\frac{U}{i}\int_0^{\beta/2} d\bar{\tau}\left[\Phi^<(t_1,-i\bar{\tau})\,T^>(-i\bar{\tau},t_2) + \Phi^>(t_1,i\bar{\tau})\,T^<(i\bar{\tau},t_2)\right], \end{aligned} \tag{6}$$

which is solved iteratively. The solution of the coupled integral equations (3, 6) for the G and the W or T is done self-consistently at every time step.

5. TDDFT from MBPT and the effective potential

Our KBE results for the TD densities also provide insight for TDDFT approaches to strongly correlated systems (e.g., to the Hubbard model). Given a specific MBA, from the resulting TD density we obtain the corresponding effective potential $v_{eff} = v_H + w + v_{xc}$, where v_H is the Hartree potential and v_{xc} the exchange correlation potential. In practice, this is done via a numerical reverse engineering procedure [27], by minimising $|n(t) - n_{KS}(t)|$ at each time-step, where $n_{KS} = \sum_\nu^{occ} |\psi_\nu^{KS}|^2$ is the Kohn-Sham density, found by solving $i\dot{\psi}_\nu^{KS} = \left(\hat{t} + v_{eff}\right)\psi_\nu^{KS}$ and with the kinetic energy given by $\hat{t} = -V\sum_{\langle RR'\rangle\sigma} a^\dagger_{R\sigma} a_{R'\sigma}$.

6. The initial, ground state spectral function

We start by solving the Dyson's equation self-consistently to obtain the initial G for propagating the KBE.

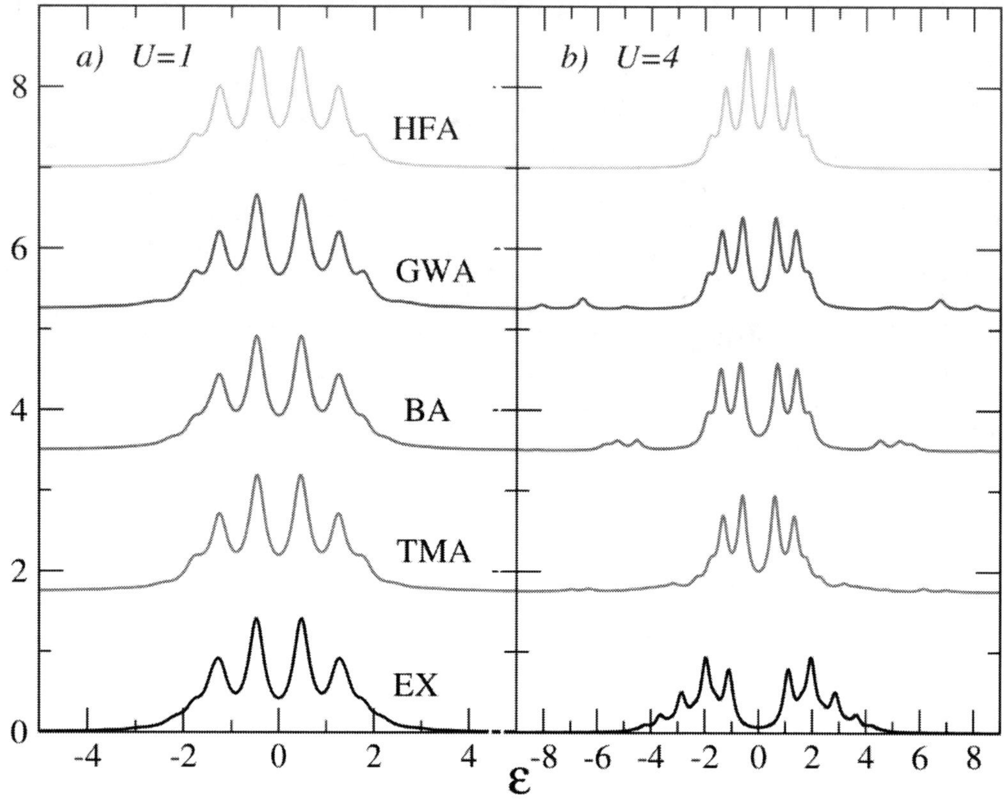

Figure 2. Ground state spectral functions for L=6 at half filling and different interaction strengths U. The curves correspond to exact (black), TMA (red), BA (green), GWA (blue) and HFA (orange). The curves are shifted for clearer comparison and we have used a Lorentzian broadening $\Gamma = 0.2$.

In Figs. (2, 3), we present results for the hole and electron contributions to the spectral density for different particle concentrations and interaction strengths. The performance of MBAs generally deteriorates with increasing U. The cases we display are the same to be considered for the TD results. Already the ground state results reveal some basic and generic features of the different many-body approximations we employ to evolve our clusters in time. Such features, discussed in Figs. (2, 3), are common to other clusters with different size, fillings and interactions strengths, whose results we do not show here.

The case of half filling is shown in Fig. 2. We note that all curves exhibit electron-hole symmetry, and that, on increasing U, a correlation gap opens. This feature is reproduced at different extent by the approximate schemes. The BA gives the best description, yet it largely underestimates the gap. We note that our non-magnetic HFA solutions reproduce, albeit shifted in energy, the non-interacting spectral density. At large U, the exact solution displays satellite structures which mimic the upper and lower Hubbard bands in extended systems. Such behaviour is however incorrectly described by all the approximations, which (especially in the case of the BA and the GWA) also introduce spurious satellites further away from the band.

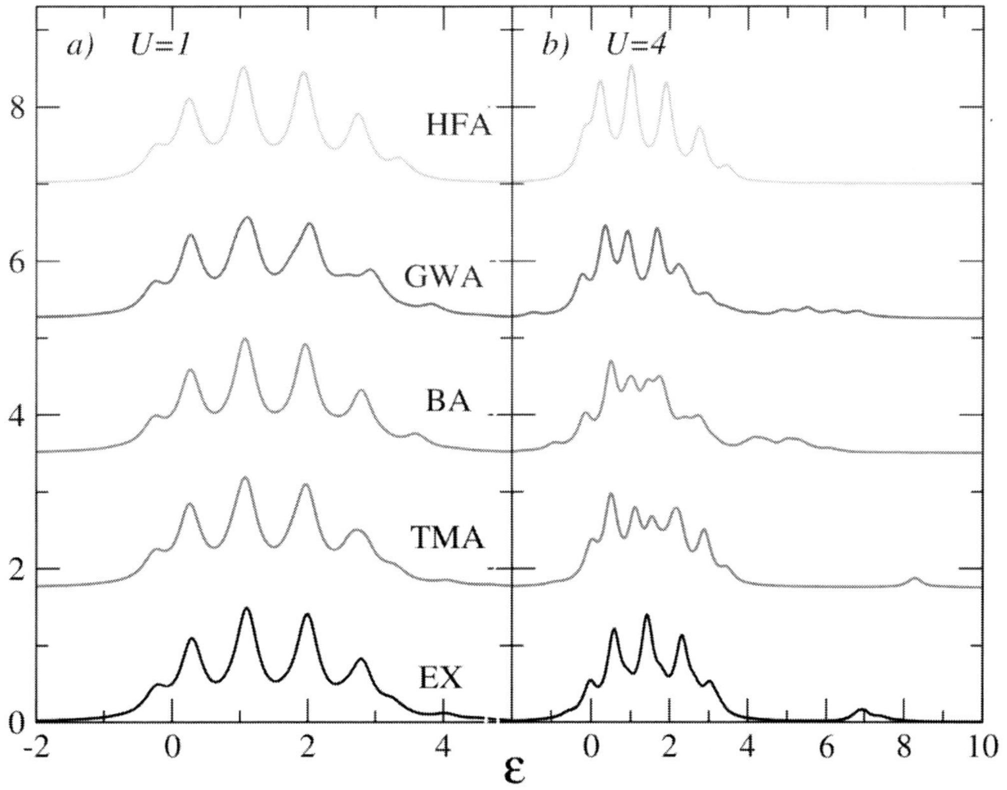

Figure 3. Ground state spectral functions for L=6 at $n = 1/6$ and different interaction strengths U. The curves correspond to exact (black), TMA (red), BA (green), GWA (blue) and HFA (orange). The curves are shifted for clearer comparison and we have used a Lorentzian broadening $\Gamma = 0.2$.

Results for the low density regime are presented in Fig. (3). For U=1, the main effect of the interaction is an increased asymmetry in the band region; the overall agreement between exact and approximate results is rather good, especially for the BA and the TMA (both the BA and the GWA slightly overestimate the shoulder on the high end of the band region). For U=4, the most notable feature in the exact solution is a satellite feature at high energy (about 6.5 at the bottom of panel 3b; in an extended system, for U=4 this would be a two-electron anti-bound state and it would lie outside the band continuum). Such satellite is well reproduced by the TMA (although its distance from the band region is overestimated). In contrast, the satellite is smeared out in the BA and GWA, and obviously absent in the HFA (in the HFA, there are no correlation effects). In the band region, for U=4, there is a moderate agreement between exact and approximate results.

We also compared GWA against SGWA results: with the latter we noted a slight improvement, which is expected, as the SGWA includes fewer faulty diagrams; yet, the SGWA is still worse than the BA or the TMA. Since the improvement is marginal, results are not shown here (however, some SGWA results will presented for the TD densities below). It is worth noting that, using a MBA expansion in terms of non-magnetic propagators, the SGWA has a magnetic instability as a function of U. In the dimer, where the poles of $W[G_0]$ are $\epsilon = \pm\sqrt{4V^2 \pm 2VU}$, this occurs for $U \geq 2V$ (such symmetry breaking is absent in the dimer exact solution).

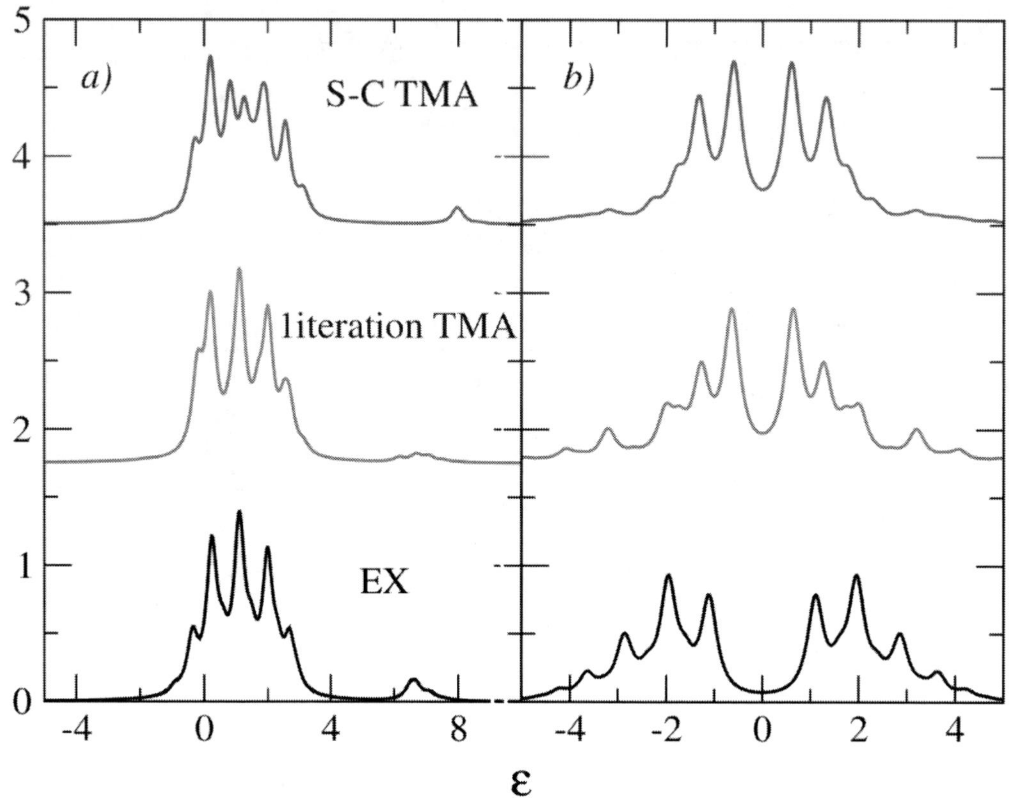

Figure 4. First iteration and self-consistent TMA spectral function versus exact one. L=6 and in a): $n = 1/6$ and in b): $n = 1/2$. The curves correspond to exact (black), first iteration TMA (green) and self-consistent TMA (blue). The curves are shifted for clearer comparison and we have used a Lorentzian broadening $\Gamma = 0.2$.

We note that the spectral functions are better at first iteration, see Fig. (4); this corresponds to the known fact that self-consistent partial diagrammatic summations often deteriorate the spectral properties compared to non self-consistent ones [31, 32]. Self consistency is required for dynamically conserving approximations, and for total energy calculations; but different partial summation criteria should be specifically adopted for spectral densities [31]. For example, in some cases, vertex corrections are required to remove artefacts introduced by self consistency [30].

We conclude this section by remarking that all MBAs based on partial summations involve infinitely many diagrams, and this results in infinitely many, but discrete, number of poles in the ground state spectral functions [33]. This is in contrast to the exact solution which, as a consequence of the finiteness of phase space of the system, has only a finite number of poles.

7. Time dependence

For $t < 0$, the system is in its ground (initial) state, as discussed in the previous section. At $t > 0$, an external field is applied. In our calculations, we have employed different kinds of external fields, but in this paper we present results only for the form $w_R(t) = w_0 \delta_{R,1} \Theta(t)$, i.e. we let the (stepwise) perturbation to act only on the leftmost, $R = 1$, site. The time is given in units of the inverse hopping parameter $(1/V)$.

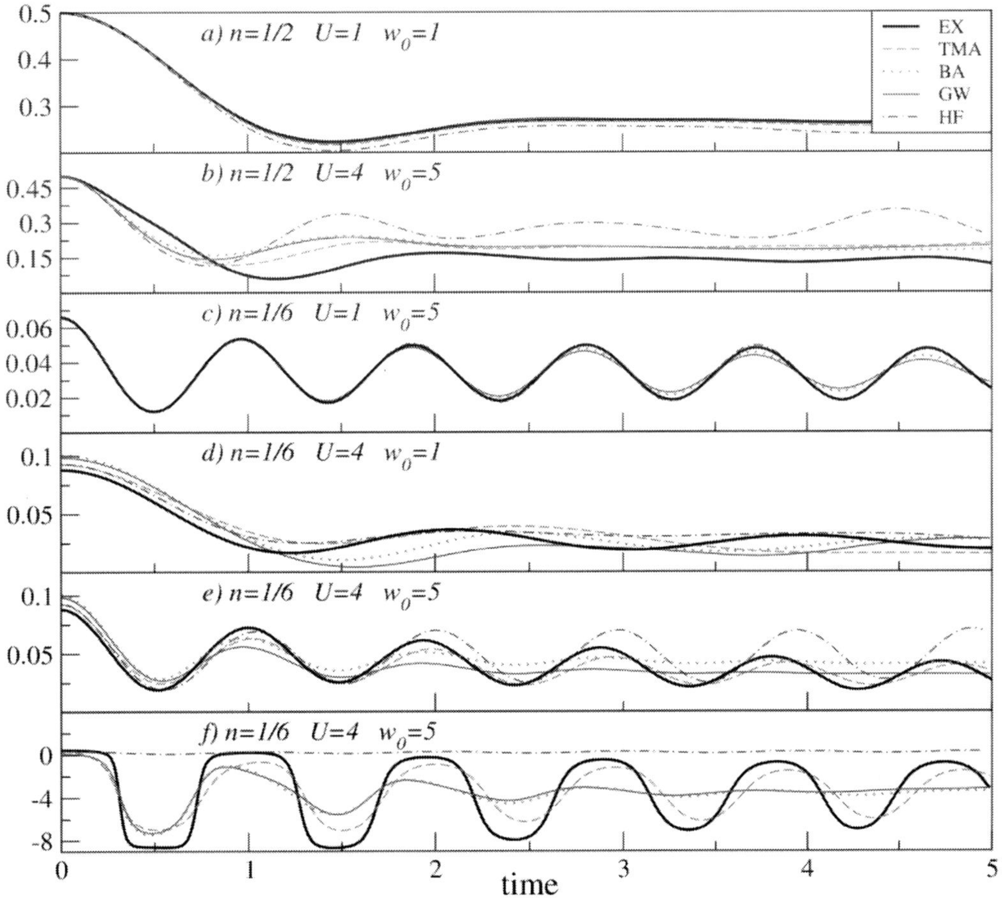

Figure 5. Time-dependent densities and v_{eff} on site 1 for $L = 6$. The curves correspond to exact (thick solid black), TMA (dashed red), BA (dotted green), GWA (thin solid blue) and HFA (brown dashed dot)

In Fig. (5), panels a-e), we show time-dependent densities corresponding to the initial states shown in Fig. (3). In panel f), we present results for v_{eff} in a selected case. All curves represent the dynamics on site 1. In the panels a) and b), $n = 1/2$ while in panels c) to f), $n = 1/6$. In panel a) we show the simplest case ($U = 1, w_0 = 1$) for which all MBAs give a good description of the density. In panel b), the interaction as well as the external field are increased and we see clearly that the HFA performance is poor while the other MBAs are very similar and closer to the exact density. In panel c) we see that the HFA and TMA are on top of the exact density while the BA and GWA start to deviate after some time (with BA being the slowest). This implies that even if the ground states are not too different, the BA and GWA fail to describe the non-linear response. In panel d) we see that none of the MBAs give a satisfactory description. We attribute this to the fact that the ground state spectral function is not well described in the band region, which is important when the external field is weak. In panel e), however, the TMA is seen to perform very well (contrary to the other MBAs which either lack or completely misplaces the satellite) as the influence of the satellite becomes essential. It is also clearly seen that similarly as in panel c) the BA and GWA fail in the non-linear regime. In panel f) we show the corresponding v_{eff} for the different MBAs. The comparison of exact versus MBPT-based xc-potentials for TDDFT shows trends similar to those for the densities: in particular, the v_{eff}

of the TMA is superior to those of the other MBAs.

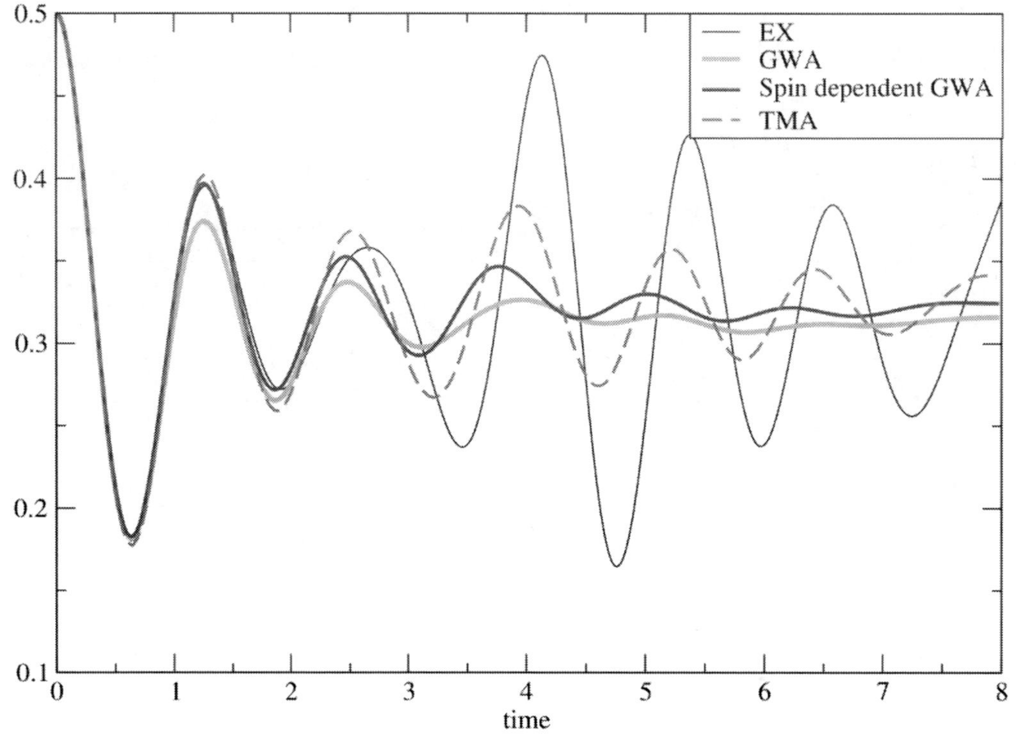

Figure 6. Time dependent densities for $L = 4$, $U = 1$, $w_0 = 5$ and $n = 1/2$. The curves correspond to exact (black), GWA (thick orange) and SGWA (thick blue)

In Fig. (6) we compare GWA, SGWA and TMA. Because of the magnetic instability in the SGWA, we can compare only for weak interaction, $U < 2$. We see that the SGWA is slightly better than its spin-independent counterpart but inferior to the TMA.

As general comment to our findings, the GWA is known to provide a good screening of the Coulomb interaction, while the TMA gives a better description if the interaction is short ranged, especially in the low density regime. The overall good performance (both for the ground state as well as for the time evolution) of the TMA for the short-ranged Hubbard Hamiltonian is in agreement with previous studies of ground state properties of clusters [30, 34]. In other words, the performance of the different approximations in the ground state has non negligible relevance to the TD behaviour of the system. In particular, an external perturbation redistributes the electrons also in the unoccupied energy levels, and two basic spectral features, the band-gap and the satellite, play a key role in the TD evolution.

8. Damping

When we propagate our system under the action of a strong external field we reach an artificial steady state, see Fig. (7). (By a steady state we mean a state where the single-particle properties are independent of time, and where the one-particle Green's function depends only on the time difference). Hints of this behaviour are already present in Fig. (5), but here we wish discuss this feature in greater detail.

The damping is not a numerical artefact, Fig. (8). Particle and energy conservation are strictly obeyed within our numerical accuracy. When we reverse the direction of time in the

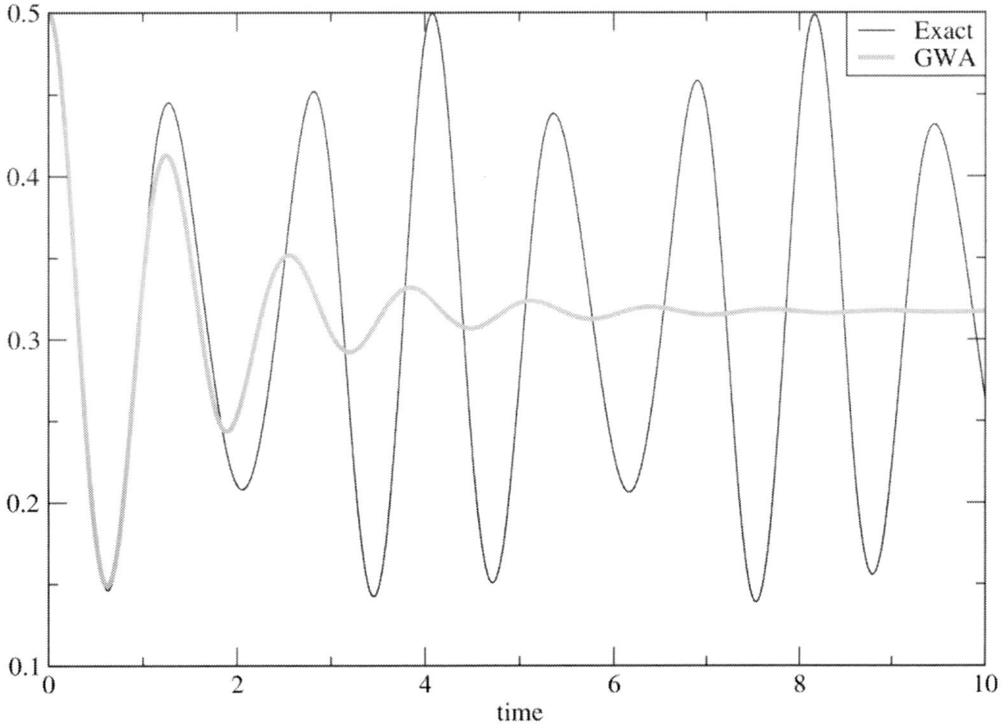

Figure 7. Densities for $L = 2, n = 1/2, U = 1, w_0 = 5$, exact (thick orange) and GWA (black). Damping of GWA density versus exact solution.

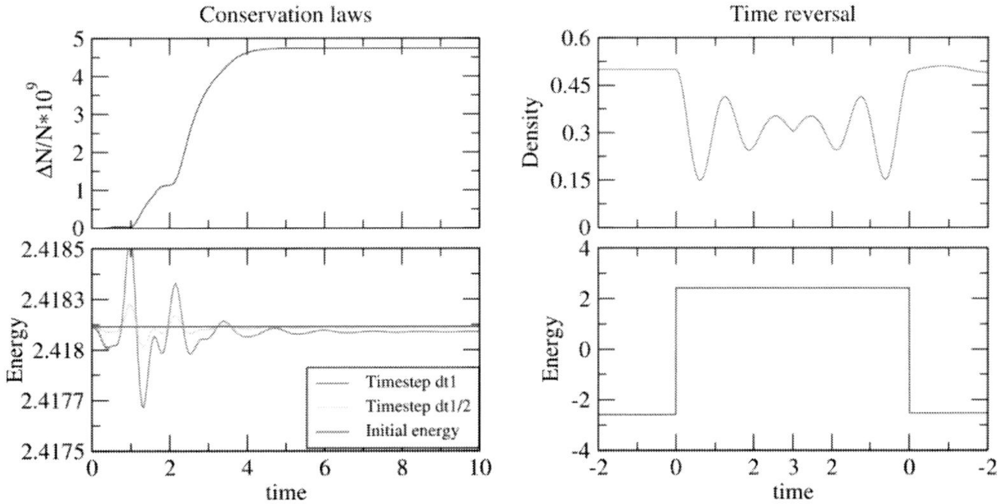

Figure 8. Conservation laws (left) and time reversal (right) for $L = 2, n = 1/2, U = 1, w_0 = 5$.

propagation, the system goes back to the initial state and remains there. The damping rate increases with the strength of the external field and is absent in the regime where linear response applies. In this limit the dynamics is described by the Bethe-Salpeter equation, with a kernel

$\delta\Sigma/\delta G$. The latter would have a discrete spectrum in our MBAs, and so would the resulting density response. This leads to a non-damped dynamics.

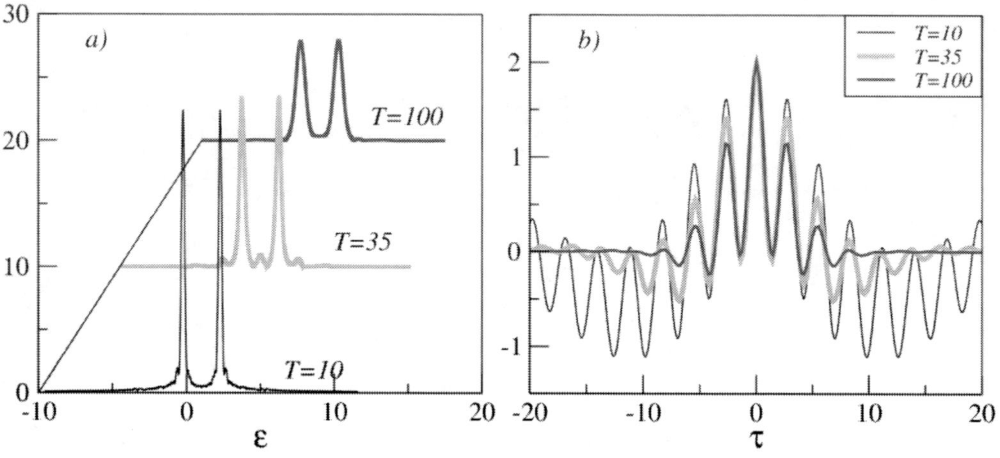

Figure 9. Time dependent spectral functions for $L = 2, n = 1/2, U = 1, w_0 = 2$, a): in energy space and b): in time space. The different curves correspond to $T = 10$ (black), $T = 35$ (thick orange) and $T = 100$ (thick blue)

To study the non-linear response behaviour, a convenient quantity is the instantaneous spectral function

$$A(T,\omega) = -Tr\,Im \int_{-2T}^{2T} e^{i\omega\tau} \left[G^> - G^<\right] \left(T + \frac{\tau}{2}, T - \frac{\tau}{2}\right) d\tau, \tag{7}$$

where $T = (t_1 + t_2)/2$ and $\tau = (t_1 - t_2)$, and its counterpart in time space. At the steady state, the spectral function gets broadened in energy space, Fig. (9a), and damped in time space, Fig. (9b). Note that, for our dimer, the exact instantaneous spectral function would continue to oscillate.

The damping rate depends on the MBA used - in general the TMA is slowest - and it acts strongest on the perturbed site. Further numerical investigations, not reported here, indicate that the entity of this artificial damping generally decreases with system size. No similar trend was seen for the dependence on the number of particles. Of course, our present results do not give a complete picture of how the large size limit is gradually obtained; nevertheless, we would like to present the reader with some general remarks. In an exact dynamics, a finite system will never fully relax to a stationary steady state. Due to decoherence, a finite but large system will, however, reach a quasi-steady state. As the system grows in size this damping becomes more and more complete. Thus, after long time the system, in an exact time dynamics, would exhibit noise-like fluctuations which never die but which decrease in amplitude with system size, whereas an approximate KBE treatment would lead to a fully stationary state. The difference between exact and approximate time dynamics is therefore expected to become less and less important as the system size grows.

The steady state is not unique for a given final external field, *i.e.* it depends on how the perturbation is switched on. With an adiabatic turn-on we reach the ground state of system with an onsite energy corresponding to the final external final, in accordance with the adiabatic theorem. If the perturbation is switched on suddenly we reach a non-physical steady state with the same energy as at $t = 0^+$. The non uniqueness of the steady state is of great importance

Journal of Physics: Conference Series **220** (2010) 012016 doi:10.1088/1742-6596/220/1/012016

as it implies multiple solutions of the stationary KBE. In other words, given a final external potential, we have in principle infinitely many solutions to the stationary KBE.

The MBAs used here are entirely defined by a generating functionals $\Phi[G]$ and the corresponding set of self-consistent KBE for G: thus, strictly speaking, there is no direct connection to underlying wave-functions. Therefore, there is no built-in guarantee that for instance systems in a finite phase space only can have a finite number of excited states, and that they do not exhibit spurious stationary states in time dynamics.

In finite systems, all self-consistent MBAs, which include partial infinite-order summations, will include diagrams for the self energy which annihilate more holes/particles than those which can be accommodated. In an "exact" theory, all these unphysical contributions will give no net contribution, but this perfect cancellation will in general not occur in approximate schemes like the GWA or the TMA. We saw above that these self-consistent infinite summations give spectral functions with incorrect pole structure, with discrete but infinitely many poles. In the time dynamics these unphysical aspects of our MBAs manifest themselves as a reservoir to which our system becomes attached, and this leads to damping.

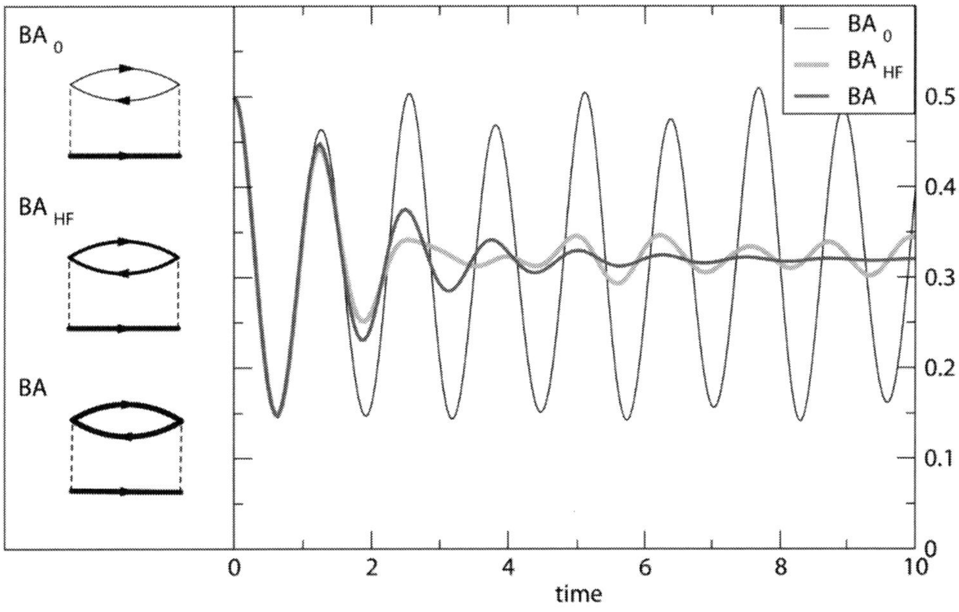

Figure 10. Time dependent densities for Born approximations with different levels of self consistency for $L = 2, n = 1/2, U = 1, w_0 = 5$; BA_0 (black), BA_{HFA} (thick orange) and BA (thick blue).

To show the effect of an increasing level of self consistency on the dynamical behaviour, we discuss in Fig. (10) three particle-conserving versions of the BA [35]. If, in the BA (or GWA), we evaluate the polarisation with ground-state propagators (BA_0), the density does not damp. If we now instead evaluate the polarisation with propagators in the time-dependent HFA approximation (BA_{HFA}), we obtain a partially damped solution. If we allow for full self consistency (BA), finally, the solution damps. In conclusion we see that if all Gs that build up the the self energy move consistently with the external field we reach a steady state [36].

The damping mechanism is inherent to the time propagation scheme and is not consequence of the infinite number of poles in the initial state. To unambiguously illustrate this point, we performed a calculation initialised with the non-interacting propagator G_0, which has a finite

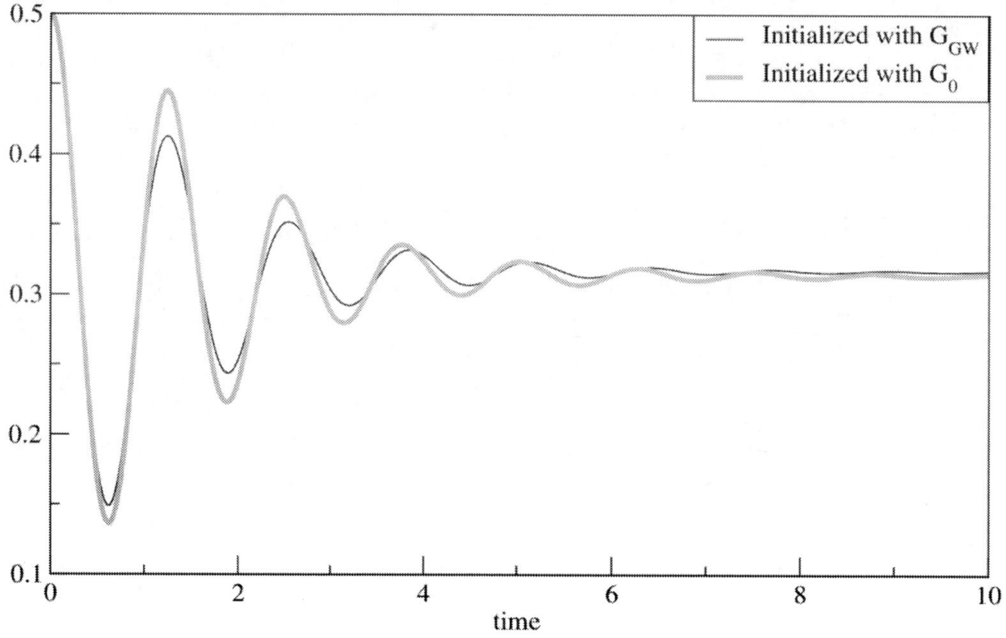

Figure 11. Time dependent densities for the GWA, initialised with the self-consistent GWA ground state (black) and the non-interacting G_0 (thick orange).

number of poles (this calculation corresponds to a sudden switch-on of the interaction as well as of the external field). In Fig. (11) we clearly see that the non-interacting initial state gives rise to a very similar damped density profile. It is worth noting that the curves in Fig. (11), which closely resemble each other, indicate a robustness of the KBE evolution against the initial conditions.

9. Beyond the Hubbard U: long range interactions

An obvious extension of the simple Hamiltonian of Eq. (1) is the inclusion of long range model interactions. These can modify the behaviour of the system in a non trivial way, for example altering/inducing a charge/spin order, affecting the screening response, etc. Most of the features induced by these additional interactions depend considerably on specific details, such as the lattice coordination number, the actual range of the interaction, its functional dependence on the inter-particle distance, etc. However, other aspects of the long range interactions can be rather general in character. For example, let us consider an additional term of the form $\frac{1}{2}\sum_{RR'\sigma\sigma'} v_{RR'} a^{\dagger}_{R\sigma} a^{\dagger}_{R'\sigma'} a_{R'\sigma'} a_{R\sigma}$, where $R \neq R'$. If we examine the effects of this term for a Hubbard dimer with two particles with opposite spin, where $R, R' = 1, 2$, the off-site interaction energy v_{12} leads exactly to an energy shift of $v_{12}/2$ and a renormalisation $U \to |U - v_{12}|$. One can also consider an infinite system with an infinite-range, constant interaction $v_{R \neq R'} = v_{\infty}$ (compensated by a similar uniform jellium-like background). In this case too, the system reduces to one where the non trivial part of the interaction is entirely local, with strength $|U - v_{\infty}|$. Those just discussed are two rather extreme examples; however, some support to the notion that long range interactions may somewhat diminish the role of the local term U comes also from an exact solution for two particles on a lattice with long-range (but not constant) interactions [37]. Results in [37] showed that, in the presence of a local interaction U plus offsite terms, the two particle spectral function can be fairly described by another Hamiltonian with purely on-site interaction $U' < U$. Another result emerging from such a study was that, in the presence of strong enough offsite interaction (schematically, when v/bandwidth $\gg 1$), two-particle repulsive bound stated are formed, which are not purely local in character. Thus, we would expect that in the low density regime, if v/bandwidth $\gg 1$, the TMA would produce a satellite structure in the one-particle spectral density which would correspond to a two-body resonance delocalised in space. This should also affect the non equilibrium behaviour. We argue that the latter two features (reduction of the onsite interaction, and delocalised bound states) could be quite general effects induced by offsite interactions; nevertheless, their detailed analysis is beyond the scope of this paper, and thus deferred to future work.

10. Conclusions and outlook

In this paper, we have studied the non-equilibrium KBE dynamics of clusters with strong correlation within the HFA, the BA, the GWA and the TMA and compared with an exact treatment. We have shown that the TMA performs very well for low fillings and is in general superior to the GWA and the BA. We have also discussed the limitations of conventional MBAs for finite systems. In particular, we observed an unphysical steady state solution of the KBE with approximate self-consistent schemes based upon partial summations. We argue that this is a general result when applying infinite order perturbation theory to finite systems. We have found that the artificial damping is inherent to the self-consistent MBA's time dynamics and that it is insensitive to the initial Green's function, *i.e.*, to the initial correlations. The approximations which damp most strongly are generally those which are also poor in describing the short-time dynamics.

The present work may be extended in several different directions. A first direction is to try to cure the KBE dynamics from the artificial damping without breaking the link with MBPT and the possibility of systematic expansions. A way to do this could be to use the variational properties of the underlying Φ functional and determine the corresponding TDDFT potentials via the Sham-Schlüter equation. At full self-consistency nothing has been gained - the TDDFT dynamics would reproduce the damped KBE dynamics. However, by relaxing the full self-consistency and use optimised one-particle G when constructing the exchange correlation potential, v_{xc} [38, 39] we expect that the artificial damping will be reduced or removed without seriously deteriorating the short-time dynamics. In the BA approximation, this corresponds essentially to use a (time-) optimised independent-particle G for constructing the self-energy and v_{xc}; we saw in section 8 that the damping is indeed reduced. The idea to construct approximate TDDFT potentials from variational MBPT may also be very useful on its own and makes it possible to treat larger and more complex systems.

A second direction is to benchmark approximate TD dynamics also for systems attached to external leads by comparing approximate KBE results with those of time-dependent density-matrix renormalisation group calculations, [40], or other non-perturbative schemes. Work along these lines is currently under way.

As a third line of investigation we note that, at present, time-dependent KBE numerical calculations are relatively heavy. In this work we have used higher order formulae for time propagation and integration in order to speed up the calculations. To treat realistic systems, however, much further work on algorithms is required to make the calculations feasible.

Finally, the KBE are equally suitable for bosonic degrees of freedom. In transport phenomena vibrations often play an crucial role, and it would be interesting to study vibronic coupling (on the model level in the initial phase) to shed more light on these problems.

Acknowledgements

This work was supported by the EU 6th framework Network of Excellence NANOQUANTA (NMP4-CT-2004-500198) and the European Theoretical Spectroscopy Facility (INFRA-2007-

211956).

References

[1] L. P. Kadanoff and G. Baym, *Quantum Statistical Mechanics* (Benjamin, New York, 1962)
[2] L. V. Keldysh, JETP **20**, 1018 (1965)
[3] P. Danielewicz, Ann, Physics **152**, 239 (1984)
[4] General periodicity property for G: $G(t_1, t_2) = -G(t_1, t_2 - i\beta)$ and $G(t_1, t_2) = -G(t_1 - i\beta, t_2)$. The same periodicity is valid for Σ.
[5] See, for example, *Progress in Non-equilibrium Green's Functions III*, J. Phys. Conf. Ser. **35**, edited by M. Bonitz and A. Filinov (2006)
[6] N.-H. Kwong and M. Bonitz, Phys. Rev. Lett. **84**, 1768 (2000)
[7] A. P. Jauho, in Reference [5], p. 313
[8] G. Stefanucci, C.-O. Almbladh, Phys. Rev. B **69**, 195318 (2004)
[9] N. E. Dahlen, R. van Leeuwen, and A. Stan in Reference [5], p. 340
[10] K. Balzer, M. Bonitz, R. van Leeuwen, A. Stan, N. E. Dahlen, Phys. Rev. B **79**, 245306 (2009)
[11] M. Galperin, A. Nitzan, M. A. Ratner, Phys. Rev.
We argue that this is a general result when applying infinite order perturbation theory to finite systB **76**, 035301 (2007)
[12] K. S. Thygesen and A. Rubio, Phys. Rev. B **77**, 115333 (2008)
[13] P. Darancet, A. Ferretti, D. Mayou , V. Olevano Phys. Rev. B **75** , 075102 (2007)
[14] P. Myöhänen, A. Stan, G. Stefanucci and R. van Leeuwen, Eur. Phys. Lett. **84**, 67001 (2008)
[15] J. K. Freericks, Phys. Rev. B **77**, 075109 (2008)
[16] M. Puig von Friesen, C. Verdozzi and C.-O. Almbladh, submitted (2009).
[17] See, for example, T. H. Oosterkamp, T. Fujisawa, W. G. van der Wiel, K. Ishibashi, R. V. Hijman, S. Tarucha and L. P. Kouwenhoven, Nature **395**, 873 (1998)
[18] B. T. Seaman *et al.* Phys. Rev. A **75**, 023615 (2007)
[19] L. Hedin, Phys. Rev. **139**, A796 (1965)
[20] V. Galitzkii, Soviet Phys. JETP **7**, 104 (1958)
[21] D. Semkat, D. Kremp, M. Bonitz, Contributions to Plasma Physics, **42**, 31 (2002)
[22] K. S. Thygesen, Phys. Rev. Lett. **100**, 166804 (2008)
[23] P. Romaniello, D. Sangalli, J. A. Berger, F. Sottile, L. G. Molinari, L. Reining and G. Onida, J. Chem. Phys. **130**, 044108, (2009)
[24] W. Nelson, P. Bokes, Patrick Rinke and R.W. Godby, Phys. Rev. A **75**, 032505 (2007)
[25] E. Runge and E. K. U. Gross, Phys. Rev. Lett. **52**, 997 (1984).
[26] R. van Leeuwen, N. E. Dahlen, G. Stefanucci, C.-O. Almbladh, and U. von Barth, *Time-Dependent Density Functional Theory* (Springer, New York, 2006) and references therein
[27] C. Verdozzi, arXiv:0707.2317 and Phys. Rev. Lett. **101**, 166401 (2008)
[28] H. S. Köhler, N. H. Kwong, and H. A. Yousif Comp. Phys. Comm. **123**, 123 (1999)
[29] A. L. Fetter and J. D. Walecka, *Quantum Theory of Many- Particle Systems*, (McGraw-Hill, New York, 1971)
[30] M. Cini and C. Verdozzi, Nuovo Cimento D **9**, 1 (1987)
[31] C.-O Almbladh, in Reference [5], p. 127
[32] B. Holm, U. von Barth, Phys. Rev. B **57**, 2108 (1998)
[33] Continuous broadened ground state spectral functions have been found for few electron quantum dots [10], in contrast, for the systems considered here our results show only discrete spectral features.
[34] C. Verdozzi, R. W. Godby, S. Holloway, Phys. Rev. Lett. **74**, 2327 (1995)
[35] U. von Barth, unpublished
[36] This conclusion applies to all MBAs here, although partially self-consistent, but particle-conserving schemes can be easily devised for GWA but not for TMA.
[37] C. Verdozzi and M. Cini, Phys.Rev. B **51**, 7412 (1995)
[38] U. von Barth, N. E. Dahlen, R. van Leeuwen, and G. Stefanucci, Phys. Rev. B **72**, 235109 (2005)
[39] C.-O. Almbladh, U. von Barth, and R. van Leeuwen, Int. J. Mod. Phys. B **13**, 535 (1999)
[40] See, for exemple, P. Schmitteckert Phys. Rev. B **70**, 121302 (2004)

Progress in Nonequilibrium Green's Functions IV IOP Publishing
Journal of Physics: Conference Series **220** (2010) 012017 doi:10.1088/1742-6596/220/1/012017

Kadanoff-Baym approach to time-dependent quantum transport in AC and DC fields

Petri Myöhänen[1], Adrian Stan[1], Gianluca Stefanucci[2,3] and Robert van Leeuwen[1,3]

[1]Department of Physics, Nanoscience Center, FIN 40014, University of Jyväskylä, Jyväskylä, Finland
[2]Dipartimento di Fisica, Università di Roma Tor Vergata, Via della Ricerca Scientifica 1, I-00133 Rome, Italy
[3]European Theoretical Spectroscopy Facility (ETSF)

E-mail: `petri.myohanen@jyu.fi`

Abstract. We have developed a method based on the embedded Kadanoff-Baym equations to study the time evolution of open and inhomogeneous systems. The equation of motion for the Green's function on the Keldysh contour is solved using different conserving many-body approximations for the self-energy. Our formulation incorporates basic conservation laws, such as particle conservation, and includes both initial correlations and initial embedding effects, without restrictions on the time-dependence of the external driving field. We present results for the time-dependent density, current and dipole moment for a correlated tight binding chain connected to one-dimensional non-interacting leads exposed to DC and AC biases of various forms. We find that the self-consistent 2B and GW approximations are in extremely good agreement with each other at all times, for the long-range interactions that we consider. In the DC case we show that the oscillations in the transients can be understood from interchain and lead-chain transitions in the system and find that the dominant frequency corresponds to the HOMO-LUMO transition of the central wire. For AC biases with odd inversion symmetry odd harmonics to high harmonic order in the driving frequency are observed in the dipole moment, whereas for asymmetric applied bias also even harmonics have considerable intensity. In both cases we find that the HOMO-LUMO transition strongly mixes with the harmonics leading to harmonic peaks with enhanced intensity at the HOMO-LUMO transition energy.

1. Introduction

The progress in miniaturization of electronic devices has led to a completely new field of research, known as molecular electronics, in which it has become possible to prepare samples and to measure steady state conductivity properties of systems with single molecules squeezed between conducting electrodes [1, 2]. However, in future applications of nanodevices one is not only interested in their steady properties, but one also wants to manipulate these systems in time. For the control and operation of such devices the understanding of the time-dependent properties, such as switch-on times, currents, density fluctuations etc., becomes crucially important. In contrast to the steady-state approach, understanding these phenomena demands truly real-time treatment of the problem which is the topic of this paper.

The study of periodic pulses is of special importance to manipulate and control the electron flow in nanoscale devices. In this paper we present results obtained from the solution of the

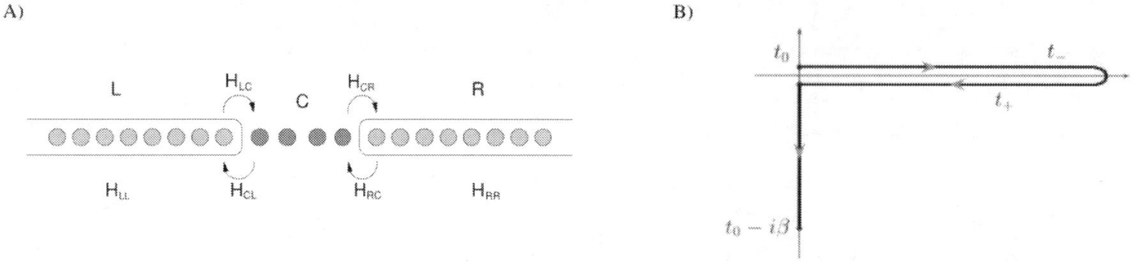

Figure 1. A) The transport setup under consideration: A correlated central region is coupled to semi-infinite left and right noninteracting tight-binding leads. B) The Keldysh contour γ. Times on the upper/lower real-time branches are denoted with $t_{\mp}$. The imaginary (thermal) track runs from t_0 to $t_0 - i\beta$, with β the inverse temperature.

Kadanoff-Baym (KB) equations for periodic voltage pulses. Within this framework also periodic gate voltages or barriers can be included without any additional computational effort. All these kind of signals can be experimentally realized and have been used to pump the electron current in several nanoscale structures ranging from molecules and quantum dots [3, 4] to nanotubes [5, 6]. The AC transport properties of nanoscale systems have been discussed recently [7, 8, 9, 10] and moreover, the advantages of real-time approaches in the context of quantum transport through molecular junctions driven out of equilibrium by periodic fields have been discussed in detail in Ref. [11].

There is another reason for studying AC fields in quantum transport. Usually the assumption is made that the external field in the leads is instantaneously screened. However, in time-dependent transport transient times can be of the same order as the plasma oscillation period (see e.g. Ref.[8]). In such a case it would be reasonable to assume that at Hartree mean-field level the bias in the leads can be described by an AC field that alternates with the plasma frequency.

In this article we use a recently proposed theoretical framework [12, 13] to study time-dependent electron transport through a quantum wire connected to two one-dimensional metallic leads in the presence AC and DC fields. The theoretical approach is based on the real-time propagation of the KB equations [14, 15, 16, 17, 18, 19, 20] for open and interacting systems. The electron-electron interactions are handled perturbatively through the many-body self-energy for which we have implemented the HF (Hartree Fock), 2B (Second Born) and GW conserving approximations while the leads are taken into account through embedding self-energies. We can calculate a wealth of observables such as currents, densities and dipole moments, and study electronic and initial correlations induced by the time nonlocal self-energy.

2. Theory

2.1. The Hamiltonian

We consider a quantum correlated open system (a central region) coupled to noninteracting electron reservoirs (leads) and described by the Hamiltonian (see Fig.1 A)

$$\hat{H}(t) = \hat{H}_C(t) + \sum_{\alpha=L,R} \hat{H}_\alpha(t) + \hat{H}_T - \mu\hat{N}. \tag{1}$$

Here $\hat{H}_C$, $\hat{H}_\alpha$ and $\hat{H}_T$ are the central region, the lead $\alpha = L, R$ and the tunneling Hamiltonians respectively, and $\hat{N}$ is the particle number operator coupled to the chemical potential μ. The explicit expressions for these Hamiltonians read

$$\hat{H}_C(t) = \sum_{ij,\sigma} h_{ij}(t)\hat{d}^\dagger_{i\sigma}\hat{d}_{j\sigma} + \frac{1}{2}\sum_{\substack{ijkl\\ \sigma\sigma'}} v_{ijkl}\hat{d}^\dagger_{i\sigma}\hat{d}^\dagger_{j\sigma'}\hat{d}_{k\sigma'}\hat{d}_{l\sigma}, \tag{2}$$

$$\hat{H}_\alpha(t) = U_\alpha(t)\hat{N}_\alpha + \sum_{ij,\sigma} h^\alpha_{ij}\,\hat{c}^\dagger_{i\sigma\alpha}\hat{c}_{j\sigma\alpha}, \tag{3}$$

$$\hat{H}_T = \sum_{ij,\sigma\alpha} V_{i,j\alpha}[\hat{d}^\dagger_{i\sigma}\hat{c}_{j\sigma\alpha} + \hat{c}^\dagger_{j\sigma\alpha}\hat{d}_{i\sigma}], \tag{4}$$

where σ, σ' are the spin indices, $\hat{d}^\dagger, \hat{d}$ and $\hat{c}^\dagger, \hat{c}$ are the creation and annihilation operators in the device and lead region and i, j label a complete set of one-particle states in the corresponding subspaces of the system. The one-body part $h_{ij}(t)$ of the central region Hamiltonian is in general time-dependent to account for driving fields such as gate voltages or pumping fields. The terms v_{ijkl} in the two-body part are the two-electron integrals of the Coulomb interaction. Furthermore, the one-body part h^α_{ij} of the lead Hamiltonian describes the metallic leads and the tunneling Hamiltonian $\hat{H}_T$ contains the lead couplings to the central region. The system is driven out of equilibrium by a homogeneous time-dependent bias voltage $U_\alpha(t)$ coupled to the number operator $\hat{N}_\alpha = \sum_{i,\sigma}\hat{c}^\dagger_{i\sigma\alpha}\hat{c}_{i\sigma\alpha}$ for lead α. This field can be considered as the sum of the externally applied field and the resulting screening field of the metallic leads which could, for instance, represent an AC field describing a plasmon oscillation.

2.2. Kadanoff-Baym equations for the Keldysh Green's function

We study a system that is in equilibrium at an inverse temperature β at times $t \leq t_0$ and described by a Hamiltonian $\hat{H}_0$. For $t > t_0$ the system is driven out of equilibrium by an external bias voltage and evolves in time according to the Hamiltonian $\hat{H}(t)$. The Keldysh Green's function for this system is defined as the expectation value of the contour-ordered product of the creation and annihilation operators [15, 19, 21, 22]

$$\mathcal{G}_{rs}(z, z') = -i\frac{\mathrm{Tr}\left\{\mathcal{T}[e^{-i\int d\bar{z}\hat{H}(\bar{z})}\hat{a}_r(z)\hat{a}^\dagger_s(z')]\right\}}{\mathrm{Tr}\left\{e^{-\beta\hat{H}_0}\right\}}, \tag{5}$$

where $\hat{a}$ and $\hat{a}^\dagger$ are either lead or central region operators and the indices r and s are collective indices for position and spin. Furthermore, z is a contour time variable and $\mathcal{T}$ orders the operators along the Keldysh contour (see Fig.1 B) by arranging the operators with later contour times to the left. The trace is taken over the many-body states of the system. All the time-dependent one-particle properties can be calculated from the Green's function, for example the time-dependent density matrix is given by

$$n_{rs}(t) = -i\mathcal{G}_{rs}(t_-, t_+), \tag{6}$$

where the time arguments $t_\pm$ are located on the lower/upper branch of the Keldysh contour. Starting from Eq. (5) the equation of motion for the Green's function and its adjoint can easily be derived and are given by

$$i\partial_z\mathcal{G}(z, z') = \delta(z, z')\mathbf{1} + \mathbf{H}(z)\mathcal{G}(z, z') + \int_\gamma d\bar{z}\,\mathbf{\Sigma}^{\mathrm{MB}}(z, \bar{z})\mathcal{G}(\bar{z}, z'), \tag{7}$$

$$-i\partial_{z'}\mathcal{G}(z, z') = \delta(z, z')\mathbf{1} + \mathcal{G}(z, z')\mathbf{H}(z') + \int_\gamma d\bar{z}\,\mathcal{G}(z, \bar{z})\mathbf{\Sigma}^{\mathrm{MB}}(\bar{z}, z'), \tag{8}$$

where $\mathbf{\Sigma}^{\rm MB}$ is the many-body self-energy, $\mathbf{H}(z)$ is the one-body part of the full Hamiltonian and the integration is performed over the Keldysh-contour. These equations of motion need to be solved with the Kubo-Martin-Schwinger (KMS) boundary conditions: $\boldsymbol{\mathcal{G}}(t_0,z') = -\boldsymbol{\mathcal{G}}(t_0-i\beta,z')$ and $\boldsymbol{\mathcal{G}}(z,t_0) = -\boldsymbol{\mathcal{G}}(z,t_0-i\beta)$. All the quantities are considered as block matrices with a structure given by

$$\mathbf{H} = \begin{bmatrix} \mathbf{H}_{\rm LL} & \mathbf{H}_{\rm LC} & \mathbf{0} \\ \mathbf{H}_{\rm CL} & \mathbf{H}_{\rm CC} & \mathbf{H}_{\rm CR} \\ \mathbf{0} & \mathbf{H}_{\rm RC} & \mathbf{H}_{\rm RR} \end{bmatrix}, \quad \mathbf{\Sigma}^{\rm MB} = \begin{bmatrix} \mathbf{0} & \mathbf{0} & \mathbf{0} \\ \mathbf{0} & \mathbf{\Sigma}^{\rm MB}_{\rm CC}[\boldsymbol{\mathcal{G}}_{\rm CC}] & \mathbf{0} \\ \mathbf{0} & \mathbf{0} & \mathbf{0} \end{bmatrix}, \quad \boldsymbol{\mathcal{G}} = \begin{bmatrix} \boldsymbol{\mathcal{G}}_{\rm LL} & \boldsymbol{\mathcal{G}}_{\rm LC} & \boldsymbol{\mathcal{G}}_{\rm LR} \\ \boldsymbol{\mathcal{G}}_{\rm CL} & \boldsymbol{\mathcal{G}}_{\rm CC} & \boldsymbol{\mathcal{G}}_{\rm CR} \\ \boldsymbol{\mathcal{G}}_{\rm RL} & \boldsymbol{\mathcal{G}}_{\rm RC} & \boldsymbol{\mathcal{G}}_{\rm RR} \end{bmatrix}, \tag{9}$$

where the different block matrices describe the projections onto different subregions. We assume that there is no direct coupling between the leads. The many-body self-energy has nonzero elements only in the central region. This follows immediately from the diagrammatic expansion of the self-energy. More specifically, if we expand the self-energy in powers of the two-body interactions we see that the interaction matrix elements v_{ijkl} only connect sites in the central region and therefore all Green function lines in the corresponding Feynman diagrams are of the type $\boldsymbol{\mathcal{G}}_{\rm CC}$. Therefore $\mathbf{\Sigma}^{\rm MB}[\boldsymbol{\mathcal{G}}_{\rm CC}]$ is a functional of $\boldsymbol{\mathcal{G}}_{\rm CC}$ only. Note that we do not expand in the lead-device couplings which are all exactly incorporated in the one-body matrix $\mathbf{H}$. To study the dynamical processes and to extract the observables of interest such as the densities and currents in the whole open and connected system we need the solution for the full Green function in different subregions of the system. In particular we need the Green's functions $\boldsymbol{\mathcal{G}}_{\rm CC}$ projected onto central region C and $\boldsymbol{\mathcal{G}}_{\alpha\alpha}$ projected onto the lead α region. For these we need to extract from the block matrix structure of the Green's function an equation for $\boldsymbol{\mathcal{G}}_{\rm CC}$ and $\boldsymbol{\mathcal{G}}_{\alpha\alpha}$. The projection of the equation of motion (7) onto regions CC and αC gives

$$\Big\{i\partial_z\mathbf{1} - \mathbf{H}_{\rm CC}(z)\Big\}\boldsymbol{\mathcal{G}}_{\rm CC}(z,z') = \delta(z,z')\mathbf{1} + \sum_\alpha \mathbf{H}_{\rm C\alpha}\boldsymbol{\mathcal{G}}_{\alpha\rm C}(z,z') + \int_\gamma d\bar{z}\,\mathbf{\Sigma}^{\rm MB}_{\rm CC}(z,\bar{z})\boldsymbol{\mathcal{G}}_{\rm CC}(\bar{z},z'), \tag{10}$$

for the central region and

$$\Big\{i\partial_z\mathbf{1} - \mathbf{H}_{\alpha\alpha}(z)\Big\}\boldsymbol{\mathcal{G}}_{\alpha\rm C}(z,z') = \mathbf{H}_{\alpha\rm C}\boldsymbol{\mathcal{G}}_{\rm CC}(z,z'), \tag{11}$$

for the projection on the interface αC region. We see that the right hand side of Eq.(11) is of a simple form as a result of the absence of two-body interactions in the leads. The uncontacted and biased lead Green's function $\boldsymbol{g}_{\alpha\alpha}$ satisfies the equation of motion

$$\Big\{i\partial_z\mathbf{1} - \mathbf{H}_{\alpha\alpha}(z)\Big\}\boldsymbol{g}_{\alpha\alpha}(z,z') = \delta(z,z')\mathbf{1}. \tag{12}$$

Using this equation we can construct the solution for $\boldsymbol{\mathcal{G}}_{\alpha\rm C}$ and it reads

$$\boldsymbol{\mathcal{G}}_{\alpha\rm C}(z,z') = \int_\gamma d\bar{z}\,\boldsymbol{g}_{\alpha\alpha}(z,\bar{z})\,\mathbf{H}_{\alpha\rm C}\boldsymbol{\mathcal{G}}_{\rm CC}(\bar{z},z'). \tag{13}$$

Taking into account Eq. (13) the second term on the righthand side of Eq. (10) becomes

$$\sum_\alpha \mathbf{H}_{\rm C\alpha}\boldsymbol{\mathcal{G}}_{\alpha\rm C}(z,z') = \int_\gamma d\bar{z}\,\mathbf{\Sigma}_{\rm em}(z,\bar{z})\boldsymbol{\mathcal{G}}_{\rm CC}(\bar{z},z'), \tag{14}$$

where we have introduced the *embedding* self-energy

$$\mathbf{\Sigma}_{\rm em}(z,z') = \sum_\alpha \mathbf{\Sigma}_{{\rm em},\alpha}(z,z') = \sum_\alpha \mathbf{H}_{\rm C\alpha}\,\boldsymbol{g}_{\alpha\alpha}(z,z')\mathbf{H}_{\alpha\rm C}, \tag{15}$$

accounting for the tunneling of electrons between the central region and the leads. As we see from the definition the embedding self-energy $\mathbf{\Sigma}_{\mathrm{em},\alpha}$ depends only on the coupling Hamiltonians and on the isolated lead Green's function $\boldsymbol{g}_{\alpha\alpha}$. Furthermore, the Green's function is determined once the isolated lead Hamiltonian $\hat{H}_\alpha$ of Eq. (3) is specified. Inserting (14) back to (10) then gives the equation of motion

$$\Big\{i\partial_z\mathbf{1} - \mathbf{H}_{\mathrm{CC}}(z)\Big\}\boldsymbol{\mathcal{G}}_{\mathrm{CC}}(z,z') = \delta(z,z')\mathbf{1} + \int_\gamma d\bar{z}\,\left[\mathbf{\Sigma}^{\mathrm{MB}}_{\mathrm{CC}} + \mathbf{\Sigma}_{\mathrm{em}}\right](z,\bar{z})\,\boldsymbol{\mathcal{G}}_{\mathrm{CC}}(\bar{z},z'). \qquad (16)$$

An adjoint equation can be derived similarly [13]. Equation (16) is an exact equation for the Green's function $\boldsymbol{\mathcal{G}}_{\mathrm{CC}}$ for the class of Hamiltonians of Eq. (1), provided that an exact expression for $\mathbf{\Sigma}^{\mathrm{MB}}_{\mathrm{CC}}[\boldsymbol{\mathcal{G}}_{\mathrm{CC}}]$ as a functional of $\boldsymbol{\mathcal{G}}_{\mathrm{CC}}$ is inserted.

To apply Eq. (16) in practice we need to transform it to real-time equations that we solve by time-propagation. This can be done in Eq. (16) by considering time-arguments of the Green's function and self-energy on different branches of the contour. We therefore have to define these components first. Let us thus consider a function on the Keldysh contour of the general form

$$F(z,z') = F^\delta(z)\delta(z,z') + \theta(z,z')F^>(z,z') + \theta(z';z)F^<(z,z'), \qquad (17)$$

where $\theta(z,z')$ is a contour Heaviside function [19], i.e., $\theta(z,z') = 1$ for z later than z' on the contour and zero otherwise, and $\delta(z,z') = \partial_z\theta(z,z')$ is the contour delta function. By restricting the variables z and z' on different branches of the contour we can define the various components of F as

$$\begin{aligned}
F^{\lessgtr}(t,t') &= F(t_\mp, t'_\pm), &(18)\\
F^{\rceil}(t,\tau) &= F(t_\pm, t_0 - i\tau), &(19)\\
F^{\lceil}(\tau,t) &= F(t_0 - i\tau, t_\pm), &(20)\\
F^M(\tau-\tau') &= -iF(t_0 - i\tau, t_0 - i\tau'), &(21)
\end{aligned}$$

and

$$F^{R/A}(t,t') = F^\delta(t)\delta(t-t') \pm \theta(\pm t \mp t')[F^>(t,t') - F^<(t,t')]. \qquad (22)$$

For the Green's function there is no singular contribution, i.e., $\boldsymbol{\mathcal{G}}^\delta = 0$, but the self-energy has a singular contribution of Hartree-Fock form, i.e., $\mathbf{\Sigma}^\delta = \mathbf{\Sigma}^{\mathrm{HF}}[\boldsymbol{\mathcal{G}}]$ [19]. With these definitions we can now convert Eq. (16) into equations for the separate components. This is conveniently done using the conversion table in Ref. [23]. This leads to a set of coupled real-time differential equations, known as the Kadanoff-Baym equations

$$\begin{aligned}
i\partial_t\boldsymbol{\mathcal{G}}^{\lessgtr}(t,t') &= \mathbf{H}_{\mathrm{CC}}(t)\boldsymbol{\mathcal{G}}^{\lessgtr}(t,t') + \left[\mathbf{\Sigma}^R\cdot\boldsymbol{\mathcal{G}}^{\lessgtr} + \mathbf{\Sigma}^{\lessgtr}\cdot\boldsymbol{\mathcal{G}}^A + \mathbf{\Sigma}^{\rceil}\star\boldsymbol{\mathcal{G}}^{\lceil}\right](t,t'), &(23)\\
-i\partial_{t'}\boldsymbol{\mathcal{G}}^{\lessgtr}(t,t') &= \boldsymbol{\mathcal{G}}^{\lessgtr}(t,t')\mathbf{H}_{\mathrm{CC}}(t') + \left[\boldsymbol{\mathcal{G}}^R\cdot\mathbf{\Sigma}^{\lessgtr} + \boldsymbol{\mathcal{G}}^{\lessgtr}\cdot\mathbf{\Sigma}^A + \boldsymbol{\mathcal{G}}^{\rceil}\star\mathbf{\Sigma}^{\lceil}\right](t,t'), &(24)\\
i\partial_t\boldsymbol{\mathcal{G}}^{\rceil}(t,\tau) &= \mathbf{H}_{\mathrm{CC}}(t)\boldsymbol{\mathcal{G}}^{\rceil}(t,\tau) + \left[\mathbf{\Sigma}^R\cdot\boldsymbol{\mathcal{G}}^{\rceil} + \mathbf{\Sigma}^{\rceil}\star\boldsymbol{\mathcal{G}}^M\right](t,\tau), &(25)\\
-i\partial_t\boldsymbol{\mathcal{G}}^{\lceil}(\tau,t) &= \boldsymbol{\mathcal{G}}^{\lceil}(\tau,t)\mathbf{H}_{\mathrm{CC}}(t) + \left[\boldsymbol{\mathcal{G}}^{\lceil}\cdot\mathbf{\Sigma}^A + \boldsymbol{\mathcal{G}}^M\star\mathbf{\Sigma}^{\lceil}\right](\tau,t), &(26)\\
-\partial_\tau\boldsymbol{\mathcal{G}}^M(\tau-\tau') &= \mathbf{1}\delta(\tau-\tau') + \mathbf{H}_{\mathrm{CC}}\boldsymbol{\mathcal{G}}^M(\tau-\tau') + i\left[\mathbf{\Sigma}^M\star\boldsymbol{\mathcal{G}}^M\right](\tau-\tau'). &(27)
\end{aligned}$$

Here the self-energy is the sum of the many-body and embedding self-energies. The superscript notation for the various components of self-energy and the Green's function, "M",<,>,$\rceil$ and $\lceil$

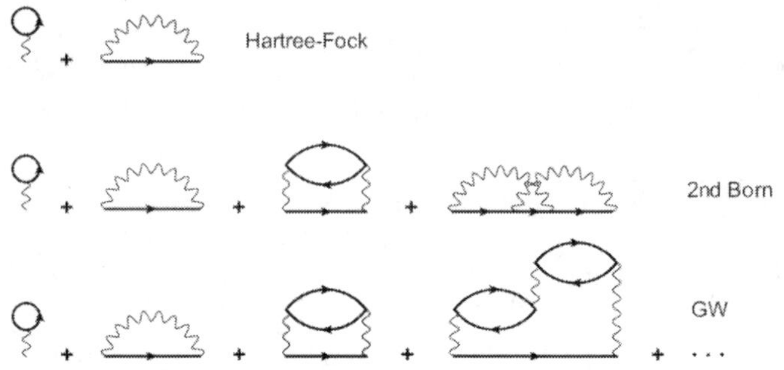

Figure 2. Diagrammatic representation of the self-energies for the conserving many-body approximations used.

are used to identify the time-arguments on different parts of the Keldysh contour [13, 22, 23]. The notations $\cdot$ and $\star$ denote the real-time and imaginary-time convolutions:

$$[a \cdot b]\,(t,t') = \int_0^\infty a(t,\bar{t})b(\bar{t},t')d\bar{t} \quad , \quad [a \star b]\,(t,t') = -i\int_0^\beta a(t,\tau)b(\tau,t')d\tau. \tag{28}$$

In practice, the Matsubara equation (27) is solved first since it is disconnected from the real-time KB equations. The real-time Green's functions are then initialized with the Matsubara Green's function as: $\mathcal{G}^>(0,0) = i\mathcal{G}^M(0^+)$, $\mathcal{G}^<(0,0) = i\mathcal{G}^M(0^-)$, $\mathcal{G}^\rceil(0,\tau) = i\mathcal{G}^M(-\tau)$ and $\mathcal{G}^\lceil(\tau,0) = i\mathcal{G}^M(\tau)$ and then the equations (23)-(26) are solved using the time-propagation method described in Ref. [24].

2.3. Conserving approximations

The importance of particle number conservation in quantum transport transport has been carefully addressed in Refs. [25, 26]. The particle number conservation law in open systems is given by

$$\frac{dN_{\rm C}(t)}{dt} = I_{\rm L}(t) + I_{\rm R}(t), \tag{29}$$

and expresses the fact that the change in the number of particles in the central region sums up to the total current that flows into the leads. Particle number conservation is guaranteed within the KB approach by using expressions for the self-energy that can be obtained as functional derivative with respect to the Green's function of a functional Φ:

$$\mathbf{\Sigma}^{\rm MB}_{{\rm CC},rs}[\mathcal{G}_{\rm CC}](z,z') = \frac{\delta\Phi[\mathcal{G}_{\rm CC}]}{\delta\mathcal{G}_{{\rm CC},sr}(z',z)}. \tag{30}$$

Baym has proven [27] that such Φ-derivable or conserving approximations [28, 27, 29] automatically lead to satisfaction of the conservation laws provided that the equations of motion for the Green's function are solved to full self-consistency. Commonly used conserving approximations for $\mathbf{\Sigma}^{\rm MB}_{\rm CC}[\mathcal{G}_{\rm CC}]$ are the Hartree-Fock, second Born and GW approximations [15, 24, 30, 31] which are displayed diagrammatically in Fig.2.

2.4. Time-dependent current

An equation for the time-dependent current flowing into the lead α can be derived from the time-derivative of the number of particles in lead α using the equation of motion for the lead

Green's function $\mathcal{G}_{\alpha\alpha}$ [13]. This yields

$$I_\alpha(t) = 2\mathrm{Re}\,\mathrm{Tr}_C\left[\mathcal{G}^<_{C\alpha}(t,t)H_{\alpha C}\right]. \tag{31}$$

If we insert the adjoint of Eq. (13) and extract the different components from the resulting equation using the conversion table in Ref.[23], the equation for the current becomes [13]

$$\begin{aligned} I_\alpha(t) &= 2\mathrm{Re}\,\mathrm{Tr}_C\Bigg[\int_{t_0}^{t} d\bar{t}\,\mathcal{G}^<_{CC}(t,\bar{t})\boldsymbol{\Sigma}^A_{\mathrm{em},\alpha}(\bar{t},t) + \int_{t_0}^{t} d\bar{t}\,\mathcal{G}^R_{CC}(t,\bar{t})\boldsymbol{\Sigma}^<_{\mathrm{em},\alpha}(\bar{t},t) \\ &\quad -i\int_0^{\beta} d\tau\,\mathcal{G}^{\rceil}_{CC}(t,\tau)\boldsymbol{\Sigma}^{\lceil}_{\mathrm{em},\alpha}(\tau,t)\Bigg]. \end{aligned} \tag{32}$$

The first two terms in this equation contain integrations over earlier times from t_0 to t and take into account the nontrivial memory effects arising from the time-nonlocality of the embedding self-energy and the Green's functions. Furthermore, the last integral over the imaginary track incorporates the initial many-body and embedding effects. Equation (32) provides a generalization of the widely used Meir-Wingreen formula [32, 33] to the transient time-domain [13].

2.5. Nonequilibrium spectral function

The nonequilibrium spectral function for real times is defined as

$$A(t,t') = i\,\mathrm{Tr}_C[\mathcal{G}^>_{CC} - \mathcal{G}^<_{CC}](t,t') = \sum_{i\in C,\sigma}\langle\left\{\hat{d}_{i\sigma}(t),\hat{d}^\dagger_{i\sigma}(t')\right\}\rangle. \tag{33}$$

where the brackets in the last term denote the anti-commutator. It is common to Fourier transform this function with respect to the relative time-coordinate $\tau = t - t'$ for a given value of the average time $T = (t+t')/2$. The spectral function in the new coordinates is then

$$A(T,\omega) = -\mathrm{Im}\int\frac{d\tau}{2\pi}e^{i\omega\tau}\mathrm{Tr}_C[\mathcal{G}^>_{CC} - \mathcal{G}^<_{CC}](T+\frac{\tau}{2},T-\frac{\tau}{2}) = \mathrm{Re}\int\frac{d\tau}{2\pi}e^{i\omega\tau}A(T+\frac{\tau}{2},T-\frac{\tau}{2}) \tag{34}$$

If the time-dependent external field becomes constant after some switching time, then also the spectral function becomes independent of T after some transient period. In that case we will denote the spectral function by $A(\omega)$. The spectral function for a system in equilibrium has peaks at the addition and removal energies. In the nonequilibrium case this is less clear as one can, in general, not use a Lehmann expansion to prove this fact. However, for a system to which a bias is suddenly applied one can still show that the spectral function will contain peaks at the addition and removal energies from the various excited states of the biased system.

2.6. The density in the leads

The density in the leads can be obtained from the Green's function $\mathcal{G}_{\alpha\alpha}$. This Green's function satisfies an equation of motion that can be obtained by projecting the equation (7) into the $\alpha\alpha$ region:

$$\left\{i\partial_z\mathbf{1} - \mathbf{H}_{\alpha\alpha}(z)\right\}\mathcal{G}_{\alpha\alpha}(z,z') = \delta(z,z')\mathbf{1} + \mathbf{H}_{\alpha C}\mathcal{G}_{C\alpha}(z,z'). \tag{35}$$

Using the adjoint of (13) in (35) we have

$$\left\{i\partial_z - \mathbf{H}_{\alpha\alpha}(z)\right\}\mathcal{G}_{\alpha\alpha}(z,z') = \delta(z,z')\mathbf{1} + \int_\gamma d\bar{z}\boldsymbol{\Sigma}_{\mathrm{in},\alpha}(z,\bar{z})\boldsymbol{g}_{\alpha\alpha}(\bar{z},z'), \tag{36}$$

where we defined the *inbedding* self-energy as

$$\mathbf{\Sigma}_{\text{in},\alpha}(z,z') = \mathbf{H}_{\alpha C}\mathbf{\mathcal{G}}_{CC}(z,z')\mathbf{H}_{C\alpha}. \tag{37}$$

Equation (36) can now be solved using Eq.(12). By taking the time arguments at $t_\pm$ we get

$$\mathbf{\mathcal{G}}_{\alpha\alpha}(t_-,t_+) = \mathbf{g}_{\alpha\alpha}(t_-,t_+) + \int_\gamma d\bar{z}d\bar{\bar{z}}\mathbf{g}_{\alpha\alpha}(t_-,\bar{z})\mathbf{\Sigma}_{\text{in},\alpha}(\bar{z},\bar{\bar{z}})\mathbf{g}_{\alpha\alpha}(\bar{\bar{z}},t_+). \tag{38}$$

We see from Eq.(6) that from this equation we can extract the time-dependent spin occupation of orbital i in lead α by taking $r = s = i\sigma\alpha$. The first term on the r.h.s. of Eq. (38) gives the density of uncontacted biased lead α and the second term gives a correction to the density induced by the presence of the correlated scattering region. In practice we first solve the Kadanoff-Baym equations for $\mathbf{\mathcal{G}}_{CC}$ and then we construct the inbedding self-energy $\mathbf{\Sigma}_{\text{in}}$ which is then used to calculate the time-dependent lead density according to (38).

2.7. Embedding self-energy

In this section we derive the expression for the embedding self-energy for the case of one-dimensional noninteracting tight-binding leads. From equation (15) we see that the embedding self-energy is given by

$$\mathbf{\Sigma}_{\text{em},\alpha,kl}(z,z') = \sum_{ij}^{\mathcal{N}_\alpha} V_{k,i\alpha}\mathbf{g}_{\alpha\alpha,ij}(z,z')V_{j\alpha,l}, \tag{39}$$

where k,l label the sites in the central region and i,j the sites in the lead α. Furthermore, $\mathcal{N}_\alpha$ is the number of sites in the lead α (at the end of the derivation we take $\mathcal{N}_\alpha \to \infty$). We now express the biased lead Green's functions using the eigenbasis of the biased lead Hamiltonian:

$$\mathbf{g}^{\lessgtr}_{\alpha\alpha,ij}(z,z') = \sum_{pq}^{\mathcal{N}_\alpha} U^\alpha_{ip}\tilde{\mathbf{g}}^{\lessgtr}_{\alpha\alpha,pq}(z,z')U^{\alpha\dagger}_{qj}, \tag{40}$$

where

$$\tilde{\mathbf{g}}^{<}_{\alpha\alpha,pq}(z,z') = i\delta_{pq}f(\epsilon_{p\alpha})e^{-i\int_{z'}^{z}(\epsilon_{p\alpha}-\mu+U_\alpha(\bar{z}))d\bar{z}}, \tag{41}$$

$$\tilde{\mathbf{g}}^{>}_{\alpha\alpha,pq}(z,z') = i\delta_{pq}(f(\epsilon_{p\alpha})-1)e^{-i\int_{z'}^{z}(\epsilon_{p\alpha}-\mu+U_\alpha(\bar{z}))d\bar{z}}, \tag{42}$$

are the lead Green's functions in the eigenbasis of the leads with eigenvalues $\epsilon_{p\alpha}$. The columns of matrix $\mathbf{U}^\alpha$ represent these eigenstates in site basis. The function $f(\epsilon) = 1/(e^{\beta(\epsilon-\mu)}+1)$ is the Fermi distribution function. Inserting Eqs.(41) and (42) into the definition of the embedding self-energy (39) and defining the function

$$\Gamma_{kl,\alpha}(\epsilon) = 2\pi\sum_{ijp}^{\mathcal{N}_\alpha} V_{k,i\alpha}U^\alpha_{ip}\delta(\epsilon-\epsilon_{p\alpha})U^{\alpha\dagger}_{pj}V_{j\alpha,l} \tag{43}$$

we get an expression for the one-dimensional lead self-energy

$$\mathbf{\Sigma}^{<}_{\text{em},\alpha,kl}(z,z') = ie^{-i\int_{z'}^{z}U_\alpha(\bar{z})d\bar{z}}\int\frac{d\epsilon}{2\pi}f(\epsilon)\Gamma_{kl,\alpha}(\epsilon)e^{-i(\epsilon-\mu)(z-z')}, \tag{44}$$

$$\mathbf{\Sigma}^{>}_{\text{em},\alpha,kl}(z,z') = ie^{-i\int_{z'}^{z}U_\alpha(\bar{z})d\bar{z}}\int\frac{d\epsilon}{2\pi}(f(\epsilon)-1)\Gamma_{kl,\alpha}(\epsilon)e^{-i(\epsilon-\mu)(z-z')}. \tag{45}$$

The remaining task is to evaluate the function $\Gamma_{kl,\alpha}(\epsilon)$. For a tight-binding (TB) chain we have

$$\epsilon_{p\alpha} = a^\alpha + 2T_\alpha \cos(\phi_p), \tag{46}$$
$$U^\alpha_{np} = \sqrt{\frac{2}{\mathcal{N}_\alpha + 1}} \sin(n\phi_p), \tag{47}$$

where a^α and T_α are the tight binding on-site and hopping parameters for the lead α and $\phi_p = \frac{p\pi}{\mathcal{N}_\alpha+1}$, $p = 1...\mathcal{N}_\alpha$, $n = 1...\mathcal{N}_\alpha$. We consider the case that the coupling Hamiltonian has nonzero elements only at the terminal points adjacent to the endpoints of the central region nanowire, that is when $i = j = 1$. The function $\Gamma_{kl,\alpha}(\epsilon)$ then becomes

$$\Gamma_{kl,\alpha}(\epsilon) = 2\pi V_{k,1\alpha} V_{1\alpha,l} \sum_{p=1}^{\mathcal{N}_\alpha} U^\alpha_{1p} \delta(\epsilon - \epsilon_{p\alpha}) U^{\alpha\dagger}_{p1}. \tag{48}$$

Inserting Eqs. (47) into (48) and taking the $\mathcal{N}_\alpha \to \infty$ limit we obtain

$$\Gamma_{kl,\alpha}(\epsilon) = \frac{2V_{k,1\alpha}V_{1\alpha,l}}{|T_\alpha|} \sqrt{1 - \left(\frac{\epsilon - a^\alpha}{2T_\alpha}\right)^2} \times \theta(2|T_\alpha| - |\epsilon - a^\alpha|). \tag{49}$$

From the definition of $\Gamma_{kl,\alpha}(\epsilon)$ we see that it describes the weighted density of states of the lead α as a function of the energy. Equation (49) then yields the expression for the 1D lead self-energy

$$\boldsymbol{\Sigma}^<_{\text{em},\alpha.kl}(z, z') = i\frac{V_{k,1\alpha}V_{1\alpha,l}}{\pi|T_\alpha|} e^{-i\int_{z'}^{z} U_\alpha(\bar{z})d\bar{z}} \int_{a^\alpha - 2|T_\alpha|}^{a^\alpha + 2|T_\alpha|} d\epsilon f(\epsilon) \sqrt{1 - \left(\frac{\epsilon - a^\alpha}{2T_\alpha}\right)^2} e^{-i(\epsilon-\mu)(z-z')}, \tag{50}$$

and the $\boldsymbol{\Sigma}^>_{\text{em},\alpha,kl}$ component is obtained by simply replacing the $f(\epsilon)$ with $f(\epsilon) - 1$. Furthermore, the $\lceil, \rceil$ and "M" components are similarly obtained by considering the time-arguments on different parts of the Keldysh contour.

3. Numerical results

We now apply the KB approach to study the transport dynamics of a quantum wire consisting of 4 sites connected to left and right one-dimensional TB leads, see Fig. 1 A. The nearest neighbour hoppings in the leads are set to $T_\mathrm{L} = T_\mathrm{R} = -2.0$ and the on-site energy a^α is set equal to the chemical potential, $a^\alpha = \mu$, yielding a half-filling for the lead energy bands. The values for the coupling Hamiltonian are set to $V_{1,1L} = V_{4,1R} = -0.5$ and the central region sites have the on-site energies $h_{ii} = 0$ and hopping parameters $h_{ij} = -1.0$ between nearest neighbour sites i and j. The electron-electron interaction in the central region has the form $v_{ijkl} = v_{ij}\, \delta_{il}\delta_{jk}$ with

$$v_{ij} = \begin{cases} v_{ii} & i = j \\ \frac{v_{ii}}{2|i-j|} & i \neq j \end{cases} \tag{51}$$

where $v_{ii} = 1.5$. The chemical potential is fixed between the highest occupied molecular orbital (HOMO) and lowest unoccupied molecular orbital (LUMO) levels of the isolated central region and has the value $\mu = 2.26$. Furthermore, the inverse temperature is set to $\beta = 90$ corresponding to the zero temperature limit. We use different time-dependent bias voltages. We consider a

Journal of Physics: Conference Series **220** (2010) 012017 doi:10.1088/1742-6596/220/1/012017

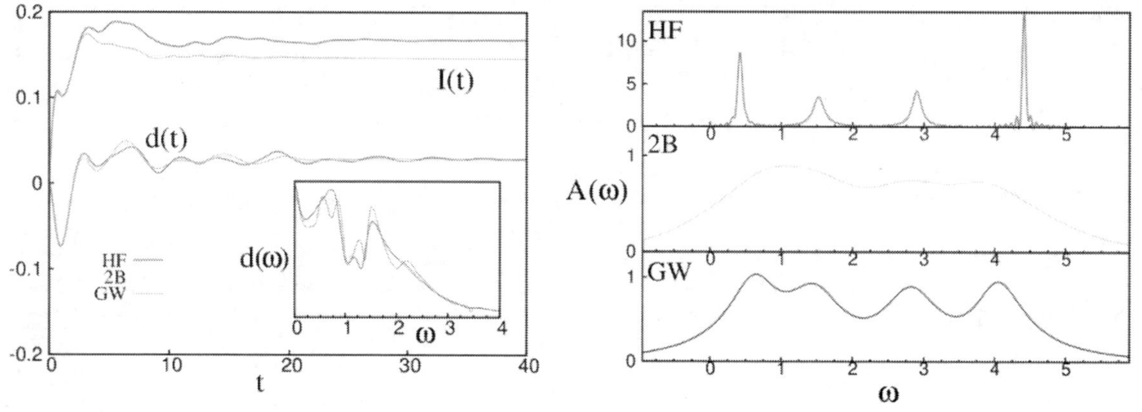

Figure 3. Left panel: Time-dependent right current $I(t)$ and dipole moment $d(t)$ for different self-energy approximations when a DC bias $U_0 = 1.4$ is applied to the leads. The inset shows the absolute value Fourier transform of the dipole moment. Right panel: Steady state spectral function $A(\omega)$.

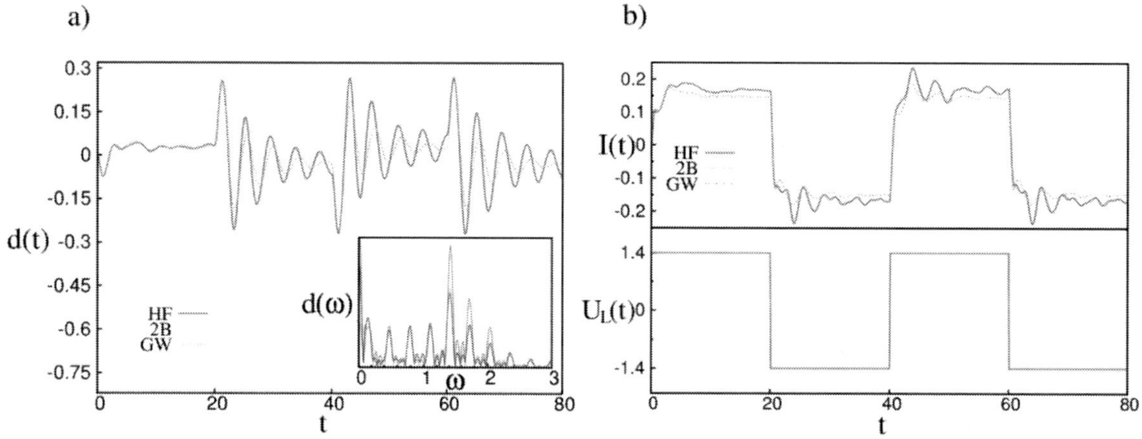

Figure 4. Panel a): Dipole moment $d(t)$ corresponding to bias $U_1(t)$. The inset shows the absolute value of the Fourier transform of the dipole moment. Panel b): Time-dependent current flowing into the right lead (top panel) and the time-dependent bias $U_1(t)$ (bottom panel).

sudden switch on DC case of the form $U_L(t) = -U_R(t) = U_0\theta(t - t_0)$ and periodic AC pulses $U_L(t) = -U_R(t) = U_{1,2}(t)$ of the forms

$$U_1(t) = \begin{cases} U_0 & 0 \le t < \frac{T_0}{2} \\ -U_0 & \frac{T_0}{2} \le t < T_0 \end{cases} \quad \text{(block bias)} \tag{52}$$

$$U_2(t) = \begin{cases} U_0 & 0 \le t < \frac{T_0}{2} \\ 0 & \frac{T_0}{2} \le t < T_0 \end{cases} \quad \text{(on-off bias)} \tag{53}$$

which for $t > T_0$ are periodically extended using $U(t + T_0) = U(t)$. The time T_0 is the period of the complete bias cycle corresponding to oscillation frequency $\omega_0 = 2\pi/T_0$. In all cases we choose the amplitude of the bias voltage to be $U_0 = 1.4$ and for AC biases we set the period to be $T_0 = 40$ a.u. Finally, to study how the charge distributes along the chain after a bias voltage

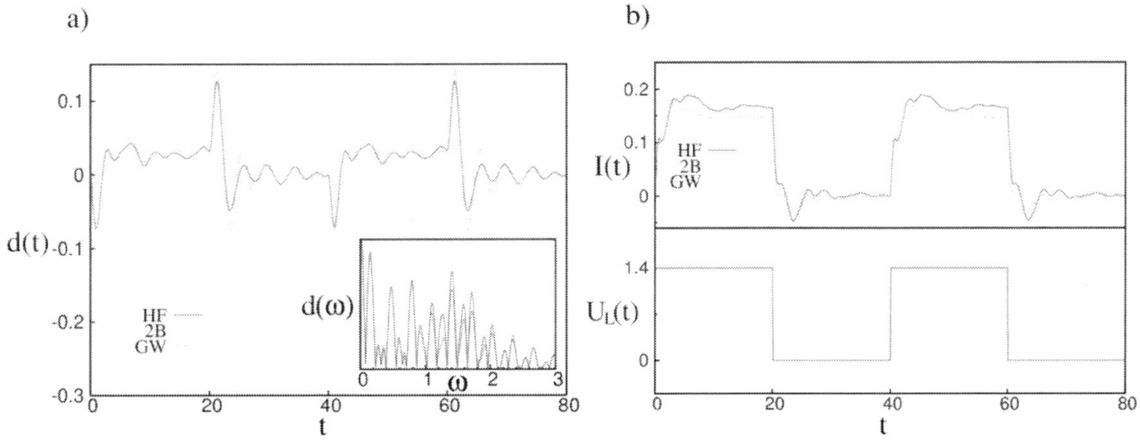

Figure 5. Panel a): Dipole moment $d(t)$ corresponding to bias $U_2(t)$. The inset shows the absolute value of the Fourier transform of the dipole moment. Panel b): Time-dependent current flowing into the right lead (top panel) and the time-dependent bias $U_2(t)$ (bottom panel).

is switched on we calculate the time-dependent dipole moment

$$d(t) = \sum_{i=1}^{4} x_i n_i(t) \tag{54}$$

where the x_i are the coordinates of the sites of the chain (with a lattice spacing of one), with origin halfway between sites 2 and 3, and $n_i(t)$ is the site occupation number.
We start by considering the case of a suddenly switched DC bias. In the left panel of Fig. 3 we compare the time-dependent dipole moments (three lowermost curves) and transient currents (three uppermost curves) for the HF, 2B and GW approximations. The right panel shows the steady-state spectral functions. We see that the Hartree-Fock peaks are relatively sharp with a width approximately given by the imaginary part of the embedding self-energy at the Fermi level of $\Gamma = 2V^2/T_\alpha \approx 0.25$ where $V = V_{1,1L} = V_{4,1R}$. On the other hand the 2B and GW spectral functions are very broadened due to strongly enhanced quasi-particle scattering at finite bias [26, 13]. We see that the Hartree-Fock approximation produces a slightly larger current compared to the 2B and GW approximations. This is in keeping with the fact that the HF spectral function integrated over the bias window leads to a slightly larger value as compared to those from the 2B and GW approximations. We further see that the correlated 2B and GW approximations lead to currents and dipole moments in close agreement with each other. This indicates that the dynamical screening of the Coulomb interaction as described by the first bubble diagram of the self-energy has a leading contribution.

In the inset of the left panel in Fig. 3 we display the absolute value of the Fourier transform of the dipole moment. We see a number of peaks that can be directly related to transitions from the lead electrochemical potentials $\mu + U_\alpha$ to the lead levels and between levels in the central sites themselves. The nature of these transitions can be deduced from the peak positions in the spectral function. Denoting these energies with $\Delta\epsilon_{Li}$, $\Delta\epsilon_{iR}$ and Δ_{ij} we can identify, at HF level, transitions at energies $\Delta\epsilon_{L3,2R} \approx 0.5$, $\Delta\epsilon_{L4} \approx 0.9$, $\Delta_{23} \approx 1.6$, $\Delta_{34} \approx 1.3$, $\Delta\epsilon_{L2,3R} \approx 2.25$ and $\Delta\epsilon_{L1,4R} \approx 3.3$. The largest peaks are due to the HOMO-LUMO transition Δ_{23} and the transitions $\Delta\epsilon_{L4}$, $\Delta\epsilon_{L3,2R}$ and we see that these are the only peaks that are clearly visible for the 2B/GW case, since in the correlated case the transients and dipole moments are strongly

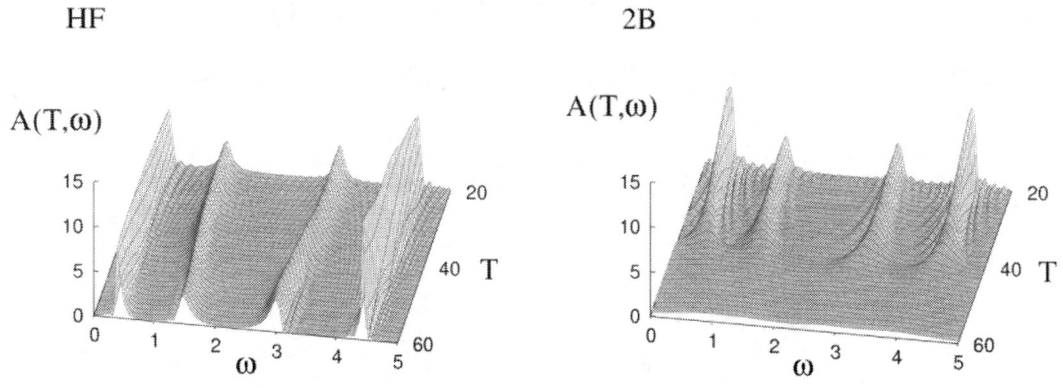

Figure 6. Time-dependent spectral function $A(T,\omega)$ in the HF (left panel) and 2B (right panel) approximation. The system parameters are the same as for Fig. 5.

damped.

We now address the case of the AC block bias $U_1(t)$. In Fig. 4 we show the corresponding dipole moments and transient currents. As in the DC bias case we observe that the GW results are very close to the 2B results. The currents and dipole moments show a similar initial transient behavior as in Fig. 3 up to $t = 20$ after which the bias switches sign. The inset in the Fig. 4 a) contains the Fourier transform of the dipole moment which displays peaks at the odd harmonics $(2n+1)\omega_0$ of the basic driving frequency ω_0 of the AC bias. The appearance of odd harmonics is a consequence of the even inversion symmetry of the unbiased Hamiltonian and the odd inversion symmetry of the applied bias. We see up to 9 harmonic peaks which indicates that we are far from the linear response regime. From the inset of Fig. 4 we also see that the harmonics at energies 1.4 (harmonic order 9) and 1.7 (harmonic order 11) are enhanced. This is due to strong mixing with the HOMO-LUMO transition that occurs at energy 1.6 in the DC case (see fig.3).

High harmonics with even order will develop when we break the symmetry of the unbiased Hamiltonian or of the applied bias. This happens for the case of the on-off AC bias voltage $U_2(t)$ (see Eq.(53)). The dipole moments and transients for this case are displayed in Fig. 5. The applied bias in this case has either even or odd inversion symmetry depending on the time interval considered. Therefore even harmonics at frequencies $2n\omega_0$ are also visible in the Fourier transform of the dipole moment (Fig. 5 inset). Also in this case the higher harmonics mix strongly with the main electronic transition frequencies leading to an increased intensity of the harmonics around the HOMO-LUMO gap frequency $\omega \approx 1.5$. Furthermore there are increased harmonic peaks at frequencies $\omega \approx 0.5, 0.9$ which are due to mixing of the harmonics with the transitions $\Delta\epsilon_{L3,2R}$ and $\Delta\epsilon_{L4}$ (see fig.3).

Now we turn our attention to the spectral functions. In Fig. 6 we show the time-dependent spectral function $A(T,\omega)$ (per spin and shifted with the chemical potential μ) for the HF and 2B approximations for the case of the on-off bias $U_2(t)$. The system is first propagated without the external bias up to 40 a.u. after which the AC bias is turned on. We see that the spectral peaks in Hartree-Fock approximation remain sharp during the whole time-propagation and due to the applied bias a closing of the HOMO-LUMO gap takes place. In contrast to the HF approximation, the spectral peaks in 2B start to broaden and loose intensity rapidly after the bias has been switched on and is caused by enhanced quasiparticle scattering at finite bias. We note that we do not observe any noticeable shift in the position of the spectral peaks as a function of

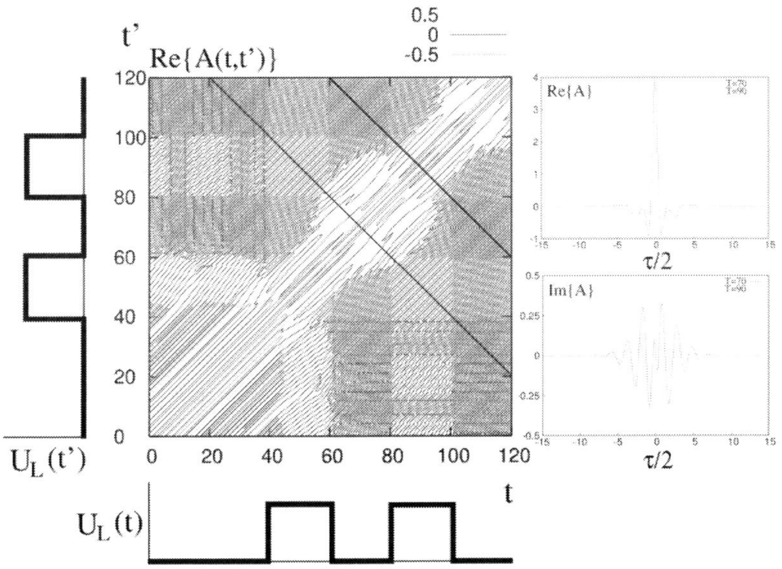

Figure 7. Left panel: Real-time spectral function $A(t,t')$ in the 2B approximation for on-off AC bias voltage. Right panel: The real and imaginary parts of $A(T+\tau/2, T-\tau/2)$ as a function of $\tau/2$ for T=70 (solid line) and T=90 (dashed line). The corresponding cross-sections are indicated with black lines.

the alternating bias. This can be explained from an analysis of the spectral function $A(t,t')$ in real time. We display contour plots of these functions in Figs. 7 and 8. In these figures we also display two center-of-time cross sections $A(T+\tau/2, T-\tau/2)$ for $T=70$ and $T=90$. Also for this case the ground state was first propagated without a bias up to 40 a.u. after which the AC field was switched on. Therefore for $t, t' \leq 40$ the contour plot describes the equilibrium system. For such times the spectral function only depends on the relative time $t-t'$. For $t, t' > 40$ the bias is applied and the spectral function starts to depend on both time-coordinates separately. This is especially visible for the 2B case in Fig.7 where for $40 < t, t' \lesssim 50$ we see a sudden collapse of the spectral function (we have checked this also for other values of the contour lines than displayed in the plot). This time interval coincides with the transient time of the current as shown in Fig.5. Note, however, that the spectral function $A(T,\omega)$ for average times $20 < T \lesssim 30$ is still dominated by the equilibrium state since the Fourier transform over the relative time τ extends mainly over values of t, t' within the equilibrium region of the double time plane. For this reason the collapse of the spectral function $A(T,\omega)$ of Fig.6 occurs within the time interval from $T=20$ to $T=40$. In Fig. 7 we further observe a characteristic checkerboard pattern of the contour line at zero. This reflects the various combinations (on-on, on-off and off-off) of the functions $U_2(t)$ and $U_2(t')$ at times t and t'. If we calculate the spectral function $A(T,\omega)$ by Fourier transforming over the relative time coordinate τ we make an average over these regions. This averaging makes the spectral function less sensitive to the period of the applied AC bias. Nevertheless, some periodic changes in the cross sections $A(T+\tau/2, T-\tau/2)$ can be observed. In the upper and lower right panels of Fig.7 we display these cross sections at average times $T=70$ (center of the off-period) and 90 (center of the on-period) for both the real and the imaginary part of $A(T+\tau/2, T-\tau/2)$ as a function of $\tau/2$. On the time diagonal the real part of A is always equal to 4 since we have four states per spin in our atomic chain. Since

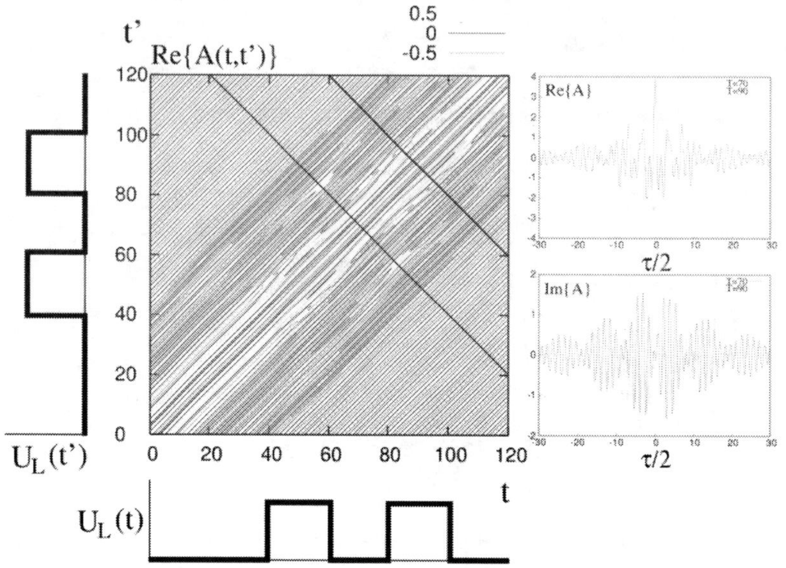

Figure 8. Left panel: Real-time spectral function $A(t, t')$ in the HF approximation for on-off AC bias voltage . Right panel: The real and imaginary parts of $A(T + \tau/2, T - \tau/2)$ as a function of $\tau/2$ for T=70 (solid line) and T=90 (dashed line). The corresponding cross-sections are indicated with black lines.

the spectral function decays fast as a function of τ the main changes in the spectral function occur close to the time diagonal. When we Fourier transform these cross section small shifts in the spectral peaks are observed. In the on-state at $T = 90$ we found a slight closing of the HOMO-LUMO gap as compared to the off-state at $T = 70$.
In Fig.8 we display the spectral function $A(t, t')$ for the HF approximation. It can be seen that the oscillations along the relative time coordinate τ in the spectral functions are much less damped in HF (Fig. 8) than in 2B (Fig. 7). As a consequence we do not see the checkerboard pattern observed in the 2B case since it occurs much farther away from the time diagonal. The functions $A(T + \tau/2, T - \tau/2)$ clearly show a slow and a fast oscillation as a function of $\tau/2$ corresponding to the two outer and inner peaks (at positive and negative frequencies) of the time-dependent spectral function $A(T, \omega)$ displayed in Fig. 6. Note that in this figure we shifted the peaks with the chemical potential $\mu = 2.26$.

We now turn our attention to the lead densities for the case of the AC block bias $U_1(t)$. In right/left panel of Fig. 9 we show the time evolution of the site occupation numbers $n_i(t)$ in the right lead for the AC/DC biases. These are calculated for up to 30 sites deep into the lead from Eq.(38) using $n_i(t) = -i\mathcal{G}_{\alpha\alpha,rr}(t_-, t_+)$ where $r = i\sigma$. For the case of a DC bias shown in panel (a) we can clearly observe a density wave front that propagates into the right lead. Furthermore the electron density displays a zigzag-pattern with an amplitude that decreases away from the first site into the lead. These are the usual Friedel oscillations caused by the presence of the atomic chain in the lattice. Their spatial profile is given by $\delta n(x) \sim \sin(2k_F x)$ where k_F is the Fermi wave vector and x the distance to the impurity site. In the case of the half-filled lead energy band that we consider $k_F = \pi/2$ corresponding to a density variation that alternates on every site. When we switch on the bias in the leads the Friedel oscillations get enhanced by

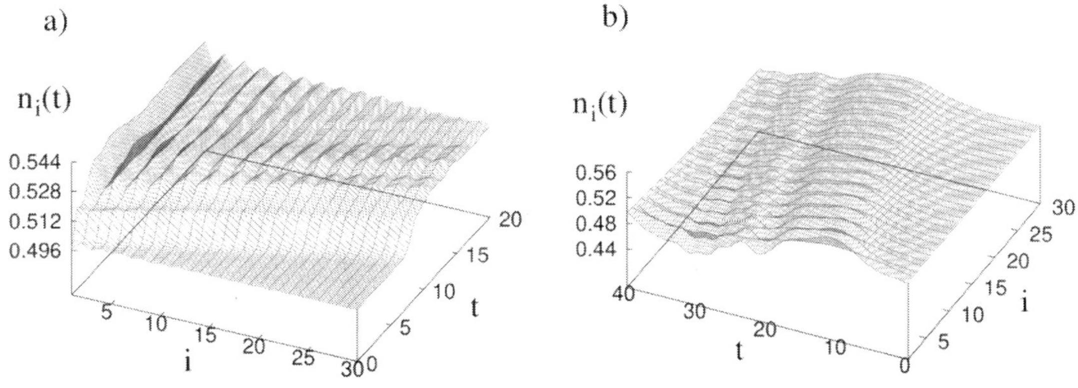

Figure 9. Site occupation number $n_i(t)$ per spin in the right lead as a function of time t and site label i. The results are displayed for (a) DC bias, and (b) AC bias $U_1(t)$ within the HF approximation.

roughly an order of magnitude.

In panel (b) of Fig.9 we shown the lead density profile for the case of the AC bias $U_1(t)$. As in the DC case, we see a wavefront moving into the lead. However, due to the periodic alternation of the bias these wavefronts get superimposed on the returning ones giving rise to interferences causing additional wiggles in the density profile.

4. Conclusions

We proposed a time-dependent many-body approach based on the real time propagation of the KB equations for open and inhomogeneous systems.

The method was applied to study transport dynamics through an atomic chain under AC and DC bias voltages. We calculated the spectral functions and the time-dependent current, density and dipole moment within the HF, second Born and GW conserving approximations. We found that electron correlations beyond mean field have a large impact on time-dependent and steady state properties. Both in the AC and DC case the HF spectral function retains its sharp structures upon applying a bias, while the spectral function calculated within the 2B and GW approximations broadens considerably.

The strong AC biases that we applied lead to highly nonlinear perturbations. As a consequence high-order harmonics of the driving frequency are observed in the time-dependent dipole moment. Depending on the symmetry of the Hamiltonian and of the applied AC voltage the odd and even harmonics are generated. The harmonics with largest intensity are those with energy close to the HOMO-LUMO gap.

References

[1] R. H. M. Smit *et al*, Nature 419, 906-909 (2002).
[2] M. A. Reed *et al*, Science Vol. 278. no. 5336, 252-254 (1997).
[3] L. P. Kouwenhoven, A. T. Johnson, N. C. van der Vaart, C. J. P. M. Harmans, and C. T. Foxon, Phys. Rev. Lett. **67**, 1626 (1991).
[4] M. Switkes, C. M. Marcus, K. Campman, and A. C. Gossard, Science **283**, 1905 (1999).
[5] P. J. Leek, M. R. Buitelaar, V. I. Talyanskii, C. G. Smith, D. Anderson, G. A. C. Jones, J. Wei, and D. H. Cobden, Phys. Rev. Lett. **95**, 256802 (2005).
[6] Z.Zhong, N.M.Gabor, J.E.Sharping, A.L.Gaeta, and P.L. McEuen Nature Nanotechnology **3**, 201 (2008)
[7] A.Tikhonov, R.D.Coalson, and Y.Dahnovsky J.Chem.Phys. **117**, 567 (2002)
[8] R.Baer, T.Seideman, S.Ilani, and D.Neuhauser J.Chem.Phys. **120**, 3387 (2004)

[9] V.Moldoveanu, V.Gudmundsson and A.Manolescu Phys.Rev.B **76**, 165308 (2007)
[10] V. Moldoveanu, A. Manolescu, V.Gudmundsson, cond-mat arXiv:0909.0815.
[11] G. Stefanucci, S. Kurth, A. Rubio and E. K. U. Gross, Phys. Rev. B **77**, 075339 (2008).
[12] P. Myöhänen, A. Stan, G. Stefanucci and R. van Leeuwen, Europhys. Lett. **84**, 67001 (2008).
[13] P. Myöhänen, A. Stan, G. Stefanucci and R. van Leeuwen, Phys. Rev. B **80**, 115107 (2009).
[14] N. E. Dahlen, A.Stan and R.van Leeuwen J. Phys. Conf. Ser. **35**, 324 (2006).
[15] N. E. Dahlen and R. van Leeuwen, Phys. Rev. Lett. **98**, 153004 (2007).
[16] M. Puig von Friessen, C. Verdozzi, and C.-O. Almbladh, condmat arXiv:0905.2061.
[17] K. Balzer, M. Bonitz, R. van Leeuwen, N. E. Dahlen and A. Stan, Phys. Rev. B **79**, 245306 (2009).
[18] L. P. Kadanoff and G. Baym, *Quantum Statistical Mechanics* (Benjamin, New York, 1962).
[19] P. Danielewicz, Ann. Phys. (N.Y.) **152**, 239 (1984).
[20] N.-H. Kwong and M. Bonitz Phys. Rev. Lett. **84**, 1768 (2000).
[21] M. Wagner, Phys. Rev. B **44**, 6104 (1991).
[22] G. Stefanucci and C.-O. Almbladh, Phys. Rev.B **69**, 195318 (2004).
[23] R. van Leeuwen, N. E. Dahlen, G. Stefanucci, C. O. Almbladh, and U. von Barth, *Time-Dependent Density Functional Theory* (Springer, New York, 2006); Lect. Notes Phys. **706**, 33 (2006).
[24] A. Stan, N. E. Dahlen and R. van Leeuwen, J. Chem. Phys. **130**, 224101 (2009).
[25] K. S. Thygesen and A. Rubio Phys. Rev. B**77**, 115333 (2008).
[26] K. S. Thygesen, Phys. Rev. Lett. **100**, 166804 (2008).
[27] G. Baym Phys. Rev. **127**, 1391 (1962).
[28] G. Baym and L. P. Kadanoff, Phys. Rev. **124**, 287 (1961).
[29] U. von Barth, N. E. Dahlen, R. van Leeuwen and G. Stefanucci, Phys. Rev. B **72**, 235109 (2005).
[30] A. Stan, N. E. Dahlen and R. van Leeuwen, J. Chem. Phys. **130**, 114105 (2009).
[31] N. E. Dahlen and R. van Leeuwen, J. Chem. Phys. **122**, 164102 (2005).
[32] Y. Meir and N. S. Wingreen, Phys. Rev. Lett. **68**, 2512 (1992).
[33] A.-P. Jauho, N. S. Wingreen and Y. Meir, Phys. Rev. B **50**, 5528 (1994).

Real-time switching between multiple steady-states in quantum transport

A.-M. Uimonen[1], E. Khosravi[2,5,6], G. Stefanucci[3,5], S. Kurth[4,5], R. van Leeuwen[1,5], E.K.U. Gross[2,5,6]

[1]Department of Physics, Nanoscience Center, FIN 40014, University of Jyväskylä, Jyväskylä, Finland
[2]Institut für Theoretische Physik, Freie Universität Berlin, Arnimallee 14, 14195, Berlin, Germany
[3]Dipartimento di Fisica, Università di Roma Tor Vergata, Via della Ricerca Scientifica 1, I-00133 Rome, Italy
[4]Nano-Bio Spectroscopy Group, Dpto. de Física de Materiales, Universidad del País Vasco UPV/EHU, Centro Mixto CSIC-UPV/EHU, Av. Tolosa 72, E-20018 San Sebastián, Spain
IKERBASQUE, Basque Foundation for Science, E-48011 Bilbao, Spain
[5]European Theoretical Spectroscopy Facility (ETSF)
[6]Max Planck Institute of Microstructure Physics, Weinberg 2, 06120, Halle, Germany

E-mail: anna-maija.uimonen@jyu.fi

Abstract. We study transport through an interacting model system consisting of a central correlated site coupled to finite bandwidth tight-binding leads, which are considered as effectively noninteracting. Its nonequilibrium properties are determined by real-time propagation of the Kadanoff-Baym equations after applying a bias voltage to the system. The electronic interactions on the central site are incorporated by means of self-energy approximations at Hartree-Fock, second Born and GW level. We investigate the conditions under which multiple steady-state solutions occur within different self-energy approximations, and analyze in detail the nature of these states from an analysis of their spectral functions. At the Hartree-Fock level at least two stable steady-state solutions with different densities and currents can be found. By applying a gate voltage-pulse at a given time we are able to switch between these solutions. With the same parameters we find only one steady-state solution when the self-consistent second Born and GW approximations are considered. We therefore conclude that treatment of many-body interactions beyond mean-field can destroy bistability and lead to qualitatively different results as compared those at mean-field level.

1. Introduction

The experimental observation [1, 2] of a hysteresis loop in the I/V characteristic of double-barrier resonant tunneling structures prompted intense theoretical activities to gain a microscopic understanding of this phenomenon. Several authors have been able to reproduce the hysteresis behavior by treating the Coulomb interaction at a mean field level [3, 4, 5, 6]. Self-consistent calculations have revealed the presence of bistable solutions, one of the solution being characterized by a considerable accumulation of charge in the potential well. Subsequent experimental work on double-barrier resonant tunneling diodes has, however, demonstrated that hysteresis loops do not always occur [7]. The suppression of the intrinsic bistability has been attributed to exchange-correlation effects [8].

With the advent of molecular electronics [9] the study of intrinsic bistability in nanoscale devices has gained attention due to the possibility of developing, for example, molecular diodes. So far, most of the work has focused on the steady-state I/V curve of molecular junctions attached to metallic leads. Up to date calculations are performed within the one-particle scheme of time-independent density functional theory (DFT). At the Hartree level bistability was reported by Negre *et. al* for a double quantum dot structure [10].

The mechanism of bistability and the calculation of switching times between two different states are mostly unexplored and the question how correlations affect the bistability has received very scarce attention. The purpose of the present exploratory paper is twofold: to extend the analysis to the time-domain and to study the role of memory effects in a bistable interacting resonant level model (IRLM).

Two complementary theoretical approaches will be used for calculating the time-dependent current and density, namely Time-Dependent (TD) DFT and Many-Body Perturbation Theory (MBPT). TDDFT [11] provide an exact framework to account for correlation effects both in the leads and the central region [12]. Within TDDFT the basic quantities that are propagated in time are the one-particle orbitals which depend on only one time variable. This property renders the implementation computationally favorable [13]. Most approximations to the TDDFT potential, however, do not include memory effects and the exchange-correlation part is approximated by local or semi-local functionals of the density. The lack of more sophisticated approximations represents, at present, a major obstacle for an accurate first principle description of TD quantum transport through interacting regions. MBPT has the advantage over TDDFT of allowing for a systematic inclusion of relevant physical processes through a selection of Feynman diagrams. Thus, MBPT provides an important tool to proceed beyond the commonly used adiabatic approximations and to quantify the importance of memory effects through advanced approximations to the self-energy. We recently proposed [14, 15] a time dependent MBPT formulation of quantum transport, based on the real-time propagation of the Green function [16, 17, 18] for open and interacting systems. First the Dyson equation for the connected system is solved to self-consistency on the imaginary axis. After that the Green function is propagated with the Kadanoff-Baym equations using different level of conserving approximations. As the Green function depends on two time variables the implementation of the MBPT scheme is more demanding than that of the TDDFT scheme. We expect that the interplay between MBPT and TDDFT will be essential to develop accurate approximation for systems more complex than the one studied here.

2. Interacting Resonant Level Model

We study an Anderson-type of system [19] where the impurity is an interacting site coupled to the infinite one dimensional tight-binding leads of finite band width. The Hamiltonian of the system reads

$$\begin{aligned}\hat{H}(t) &= \sum_{\sigma}[\varepsilon_0 + V_g(t)]\hat{d}^\dagger_{0\,\sigma}\hat{d}_{0,\sigma} + \frac{1}{2}\sum_{\sigma,\sigma'}\mathcal{U}\,\hat{d}^\dagger_{0,\sigma}\hat{d}^\dagger_{0,\sigma'}\hat{d}_{0,\sigma'}\hat{d}_{0,\sigma} + \sum_{i,\alpha,\sigma}[a + U_\alpha(t)]\hat{c}^\dagger_{i\alpha,\sigma}\hat{c}_{i\alpha,\sigma} \\ &+ \sum_{i,\alpha,\sigma}[b\hat{c}^\dagger_{i\alpha,\sigma}\hat{c}_{i+1\alpha,\sigma} + h.c.] + \sum_{\alpha,\sigma}[V_{0,1\alpha}\hat{d}^\dagger_{0,\sigma}\hat{c}_{1\alpha,\sigma} + h.c.], \end{aligned} \tag{1}$$

where i denotes the site indices and σ is the spin index, ε_0 is the on-site energy of the localized site, $\mathcal{U}$ is the strength of the two-particle interaction on the central site, b is the hopping parameter between lead sites, $U_\alpha(t)$ is time-dependent bias voltage in the leads ($\alpha = L/R$), a is the on-site parameter in the leads and $V_{0,1\alpha}$ denotes coupling between the lead and the localized site. The fermionic creation- and annihilation operators in the leads α are denoted as $\hat{c}^\dagger$, $\hat{c}$ whereas for the localized site they are denoted as $\hat{d}^\dagger$, $\hat{d}$. The quantity $V_g(t)$ denotes a time-dependent gate voltage.

The main reason for studying the IRLM is that for this system the multiple steady-state solutions are easily found from a fixed-point equation for the density on the localized site. The IRLM is the simplest system in which bistability occurs and hence allows for a clear interpretation of its multiple steady-state solutions. We study the system at a finite temperature and under a finite bias, such that we are out of the Kondo regime in which the IRLM is often used.

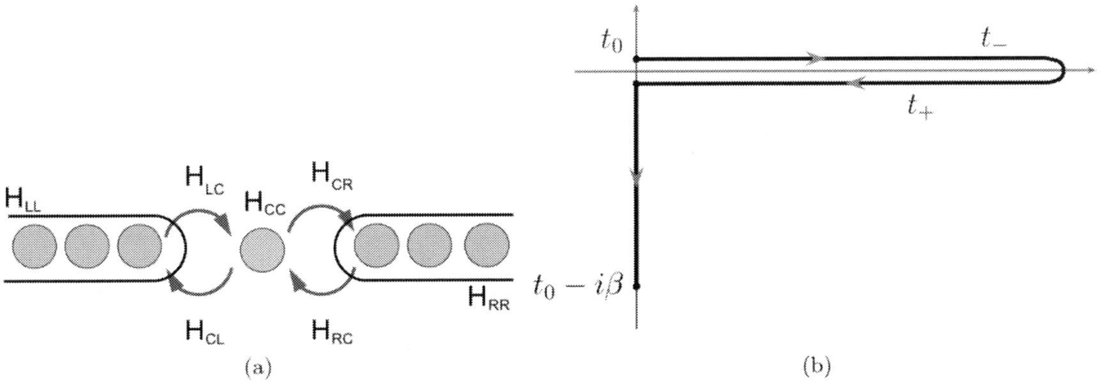

Figure 1: a) A schematic representation of the studied system. The correlated central region (C) is coupled to infinite one dimensional left (L) and right (R) tight-binding leads via a coupling Hamiltonian. b) The Keldysh contour c. Times on the lower branch t_+ are later than times on the upper branch t_-. The imaginary track extends up to the inverse temperature β.

3. Theoretical approach

3.1. Kadanoff-Baym equations

We study the nonequilibrium properties of the IRLM by means of time-propagation of the Kadanoff-Baym equations for the nonequilibrium Green function [20]. We assume the system to be contacted and in equilibrium at a chemical potential μ and at inverse temperature β before the time $t = t_0$. For $t > t_0$ the system is driven out of equilibrium by an external bias and we aim to study the time-evolution of the electron density and current. We give here a brief description of the approach described in more detail in Ref. [14].

The Keldysh Green function is defined as the expectation value of the contour ordered product [17]

$$G(1,2) = -i\frac{\mathrm{Tr}\{\hat{U}(t_0 - i\beta, t_0)\mathcal{T}_C[\hat{\psi}_H(1)\hat{\psi}_H^\dagger(2)]\}}{\mathrm{Tr}\{\hat{U}(t_0 - i\beta, t_0)\}}, \tag{2}$$

where $\mathcal{T}_C$ denotes the time ordering operator along the Keldysh contour c (see Fig.1b) and $\hat{U}$ is the time evolution operator. The average is taken over the grand canonical ensemble. We use the compact notation $1 = (\mathbf{x}_1, t_1)$ and $2 = (\mathbf{x}_2, t_2)$ where $\mathbf{x} = (\mathbf{r}, \sigma)$ is a collective space-spin index. From the Green function it is possible to calculate any one particle property of the system. For example the time-dependent density is given as

$$\langle \hat{n}(1) \rangle = -iG^<(1, 1^+), \tag{3}$$

where t^+ approaches t from an infinitesimally later time $t^+ = t + \delta$. The equation of motion for the full system can be derived from the definition of the Green function Eq. (2) and reads

$$[i\partial_{t_1} - H(1)]G(1,2) = \delta(1,2) + \int_c \mathrm{d}3\, \Sigma^{MB}[G](1,3)G(3,2), \tag{4}$$

where H is the one-body part of the Hamiltonian. The self-energy $\Sigma[G]$ incorporates all of the effects of exchange and correlation in the central region and is a functional of the Green function [17, 16]. In a localized basis the one-body part of the Hamiltonian (1) and the many-body self-energy can be written in a block-matrix form

$$H = \begin{pmatrix} H_{LL} & H_{LC} & 0 \\ H_{CL} & H_{CC} & H_{CR} \\ 0 & H_{RC} & H_{RR} \end{pmatrix}, \qquad \Sigma^{MB} = \begin{pmatrix} 0 & 0 & 0 \\ 0 & \Sigma^{MB}_{CC}[G_{CC}] & 0 \\ 0 & 0 & 0 \end{pmatrix}, \tag{5}$$

where the $H_{\alpha\alpha}$ and H_{CC} components describe the leads and the central system respectively, whereas the off-diagonal components describe the hopping between them [14]. We only consider the central region as interacting whereas the leads are effictively noninteracting. As a consequence, the many-body self-energy in Eq. (5) has non-vanishing elements only for the central region because the diagrammatic expansion starts and ends with an interaction line. In this work the electronic interactions are incorporated in $\Sigma^{MB}[G]$ at HF, 2B and GW level [14].

Solving the problem of an open infinite system is equivalent to solving the problem of a closed system with an equation of motion which considers the leads through an embedding term [14]. In our case this reads (in the reminder of this paper we will suppress the spatial indices of the objects involved)

$$[i\partial_t - H_{CC}(t)]\, G_{CC}(t,t') = \delta(t,t') + \int_c \mathrm{d}\bar{t}\, \left\{ \left[\Sigma_{em}(t,\bar{t}) + \Sigma^{MB}_{CC}[G_{CC}](t,\bar{t})\right] G_{CC}(\bar{t},t') \right\}, \tag{6}$$

where the embedding self-energy $\Sigma_{em}(t,t')$ accounts for the tunneling of electrons between leads and central region. In its general form, the embedding self-energy reads [14, 15, 21]

$$\Sigma_{em}(t,t') = \sum_\alpha \Sigma_{em,\alpha}(t,t') = \sum_\alpha H_{C\alpha} g_{\alpha\alpha}(t,t') H_{\alpha C}, \tag{7}$$

where $g_{\alpha\alpha}(t,t')$ is the lead Green function and $H_{C\alpha}, H_{\alpha C}$ is the coupling Hamiltonian between the central site and the leads.

The current through the lead α can be expressed in terms of Keldysh Green functions as [14, 15]

$$\begin{aligned} \mathrm{I}_\alpha(t) \;=\; & 2\mathrm{Re}\Big\{ \mathrm{Tr}_C \Big[\int_{t_0}^{t} d\bar{t} [G^<_{CC}(t,\bar{t})\Sigma^A_{em,\alpha}(\bar{t},t) + \int_{t_0}^{t} d\bar{t} G^R_{CC}(t,\bar{t})\Sigma^<_{em,\alpha}(\bar{t},t)] \\ & -i \int_0^\beta d\bar{\tau} G^{\rceil}_{CC}(t,\bar{\tau})\Sigma^{\lceil}_{em,\alpha}(\bar{\tau},t) \Big] \Big\}, \end{aligned} \tag{8}$$

where we integrated on the Keldysh contour and where the superscripts $A, R, <$ refer to advanced/retarded/lesser component of Green function/self-energy and $\rceil, \lceil$ are the mixed components having one time argument on a imaginary axis and the other on the real axis [20, 18, 14]. The trace is taken over the central region. The current accounts for the initial many-body and embedding effects through the last term in equation (8) which is an integral over the vertical track. Equation (8) is a generalization of the Meir-Wingreen formula [22]. We further define the nonequilibrium spectral function

$$A(T,\omega) = -\mathrm{Im}\,\mathrm{Tr}\int \frac{\mathrm{d}\tau}{2\pi} e^{i\omega\tau}[G^{>} - G^{<}](T+\frac{\tau}{2}, T-\frac{\tau}{2}), \tag{9}$$

where $\tau = t-t'$ is a relative time and $T = (t+t')/2$ is an average time coordinate. In equilibrium, this function is independent of T and has peaks below the Fermi level at the electron removal energies of the system, while above the Fermi level it has peaks at the electron addition energies. If the time-dependent external field becomes constant after some switching time, then also the spectral function becomes independent of T after some transient period and has peaks at the addition and removal energies of the biased system.

3.2. Time-dependent density functional theory

TDDFT provides an alternative framework to describe electron transport through interacting systems. In TDDFT [23] the time-dependent density of the interacting system is obtained through time-propagation of a Kohn-Sham system in an effective local potential. While in MBPT the correlation level is determined by the choice of the many-body self-energy, in TDDFT the main approximation is the functional used for the effective potential. The difficulty of TDDFT, compared to MBPT is the current lack of sufficient accurate approximations to the time-dependent exchange-correlation potential. However, the computational effort is much lower compared to a MBPT propagation.

The problem brought forward by considering an open system can be resolved in a very similar manner as in MBPT, with the help of an embedding self-energy. The equation of motion for the k-th single-particle orbital projected onto the central region, $\psi_{k,C}(t)$, reads

$$[i\partial_t - \mathrm{H}_{CC}(t)]\,\psi_{k,C}(t) = \int_0^t \mathrm{d}\bar{t}\,\Sigma^R_{em}(t,\bar{t})\psi_{k,C}(\bar{t}) + \sum_\alpha H_{C\alpha} g^R_{\alpha\alpha}(t,0)\psi_{k,\alpha}(0), \tag{10}$$

where $\Sigma^R_{em}(t,\bar{t})$ is the retarded embedding self-energy (see Eq. (7)) and $g^R_{\alpha\alpha}$ is the retarded lead Green function. The time-dependent density in the central region is obtained by

$$n(t) = \sum_k^{occ} |\psi_{k,C}(t)|^2, \tag{11}$$

where the summation is taken over all occupied orbitals in the time-dependent Slater determinant [24]. The technique to propagate Eq. (10) is described in detail in Ref. [13]. In this work we used this approach at a level, in which, for the system studied, the local exchange potential is equal to half the Hartree potential of the localized site. The results were found, as expected, to be identical to those obtained from the Kadanoff-Baym approach at HF level.

3.3. Steady-state density

We begin our study of the bistable regime by deriving an equation for the density on the localized site. This quantity is given by the lesser Green function at equal times

$$\langle\hat{n}(t)\rangle = -iG^{<}(t,t^{+}). \tag{12}$$

Since we consider the steady-state, we can assume that in the long-time limit the Green functions depend only on the relative time coordinate $t-t'$. In that case the Green function can be Fourier

Journal of Physics: Conference Series **220** (2010) 012018 doi:10.1088/1742-6596/220/1/012018

transformed with respect to the relative time variable and the expression for the steady-state becomes

$$n = \int \frac{d\omega}{2\pi i} G^<(\omega). \tag{13}$$

The Green function appearing in this expression satisfies the equation

$$G^<(\omega) = G^R(\omega)\Sigma^<_{tot}(\omega)G^A(\omega), \tag{14}$$

where $\Sigma^<_{tot}(\omega) = \Sigma^<_{em}(\omega) + \Sigma^{MB,<}_{CC}(\omega)$ and where $G^A = [G^R]^*$. The retarded Green function has the expression

$$G^R(\omega) = \left[\omega - \varepsilon_0 - \mathrm{Re}[\Sigma^R_{tot}(\omega)] - i\mathrm{Im}[\Sigma^R_{tot}(\omega)]\right]^{-1}. \tag{15}$$

For the tight-binding leads, that we consider, the retarded embedding self-energy is given by

$$\Sigma^R_{em,\alpha}(\omega) = \Lambda_\alpha(\omega) - \frac{i}{2}\Gamma_\alpha(\omega) = \frac{V_{1\alpha,0}V_{0,1\alpha}}{2b^2}\begin{cases} \omega_\alpha - \sqrt{\omega_\alpha^2 - 4b^2} & ,\omega_\alpha > 2|b| \\ \omega_\alpha + \sqrt{\omega_\alpha^2 - 4b^2} & ,\omega_\alpha < -2|b| \\ \omega_\alpha - i\sqrt{4b^2 - \omega_\alpha^2} & ,|\omega_\alpha| < 2|b| \end{cases} \tag{16}$$

where $\omega_\alpha = \omega - a + \mu - U_\alpha$, with the lead-on-site parameter a and the hopping parameter between the lead sites b. The chemical potential is denoted by μ, the applied bias for the lead α by U_α, and $V_{1\alpha,0}, V_{0,1\alpha}$ are the left/right couplings between leads and the central site. The lesser component of the embedding self-energy can be expressed as $\Sigma^<_{em,\alpha}(\omega) = if_\alpha(\omega)\Gamma_\alpha(\omega)$ where f_α is the Fermi distribution of lead α and Γ_α is defined in Eq. (16).

If we integrate the left hand side of Eq. (14) over all frequencies then according to Eq. (13) we obtain an expression for the density per spin n on the localized site

$$n = \int_{-\infty}^{\infty} \frac{d\omega}{2\pi} \frac{\Gamma_L(\omega)f_L(\omega) + \Gamma_R(\omega)f_R(\omega) - i\Sigma^{MB,<}_{CC}(\omega)}{(\omega - \mathrm{Re}[\Sigma^{MB,R}_{CC}(\omega)] - \Lambda(\omega))^2 + (\mathrm{Im}[\Sigma^{MB,R}_{CC}(\omega)] - \Gamma(\omega)/2)^2}, \tag{17}$$

where $\Lambda = \Lambda_R + \Lambda_L$ and $\Gamma = \Gamma_R + \Gamma_L$. This is a Meir-Wingreen-type equation for the density [25, 22] and is valid for transport through interacting systems. Within the HF approximation for the IRLM one has $\Sigma^{MB,<} = \mathrm{Im}[\Sigma^{MB,R}] = 0$ and $\mathrm{Re}[\Sigma^{MB,R}] = \mathcal{U}n$. Note that we include the time-singular part of the self-energy in the definition of the retarded/advanced component, see *e.g.*, [18]. Within this approximation, Eq. (17) now becomes a self-consistent fixed-point equation for the density n, since the value of the integral (17) depends on the density via the term $\mathcal{U}\,n$ in the denominator.

4. Results and discussion

We consider a biased system with the following parameters: $V_{0,1R} = V_{0,1L} = V = -0.35$, $U_L = 1.5$, $U_R = 0.0$, $\mathcal{U} = 2.0$, $b = -0.5$, $a = \mu$ and $\mu = 0.3$, $\beta = 90$. The leads are half-filled such that the Fermi level of lead α is positioned at $\mu + U_\alpha$ and the width of the lead band is $[\mu + U_\alpha - 2b, \mu + U_\alpha + 2b]$. With these parameters, within the HF approximation, the Meir-Wingreen approach [25, 22] predicts three solutions for the steady-state density n in Eq. (17). The three fixed points are shown in the inset of Fig. 2a, where we display the left and right hand side of Eq. (17). The corresponding densities are $n_1 = 0.33$, $n_2 = 0.58$ and $n_3 = 0.66$, which should be compared to the density of $n_0 = 0.28$ of the unbiased equilibrium state.

In 2a we show the steady-state spectral functions corresponding to these three solutions, which can be obtained directly from the retarded Green function $G^R(\omega)$. The peak of the spectral function corresponding to density $n_1 = 0.33$ is positioned at the energy 0.5. We thus see that this spectral function is located in an energy range within the right lead energy-band

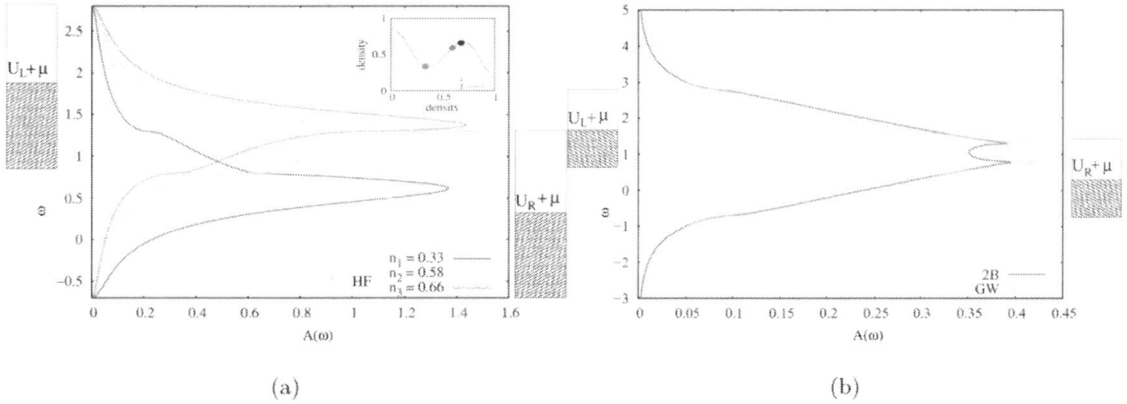

Figure 2: a) Spectral functions for the HF approximation corresponding to different steady-state solutions. *Inset*: The graphical solution of the integral in Eq. (17): (1) is the left hand side of the equation whereas (2) is the value of the integral on the right hand side with given density n. b) Steady-state spectral functions for the 2nd Born and GW approximations.

(see right side of Fig. 2a). The spectral function corresponding to density $n_3 = 0.66$ is peaked approximately at energy 1.5 and is located in an energy range within the left lead energy-band. The HF spectral function corresponding second solution with density ($n_2 = 0.58$) is peaked on the top edge of the right lead energy-band and located in between the spectral functions corresponding to the densities n_1 and n_3. By time-propagation (see below), we find that the two solutions corresponding to densities n_1 and n_3 lead to stable steady-states, *i.e*, states that are reachable by time-propagation after applying a bias. On the other hand, the state corresponding to density n_2 is unstable and cannot be reached by time-propagation. The spectral function corresponding to the state with density n_2 has a large overlap with the one of density n_3. This indicates that (for sufficiently slow gate switching, see below) during the time-propagation more charge will flow onto the central site resulting in a stable steady-state with density n_3. From the analysis of the spectral functions we thus conclude that the density bistability in this system occurs when the spectral functions of the two stable solutions are localized well within one of the lead energy-bands and are well-separated. This happens exactly when the leads have a small overlap and the system is within the region of negative differential resistance (NDR).

Let us now consider the situation when we go beyond the HF approximation. In Fig. 2b we show the steady-state spectral functions for the 2B and GW approximations, as obtained from time-propagation of the Kadanoff-Baym equations. With these approximations we found only one steady-state solution, with very broad spectral function due to enhanced quasi-particle scattering at finite bias [14]. The spectral weight of the 2B and GW states is spread almost uniformly over the energy range from the bottom of the right lead energy band to the top of left lead energy band and extends well outside the lead bands. We also observe two small peaks in these spectral functions which occur approximately at 0.8 and 1.3, corresponding to the edges of the lead energy bands.

We now go to the time domain and consider how the steady-states at HF level, corresponding to densities $n_1 = 0.33$ and $n_3 = 0.66$, can be reached by time-propagation starting from the initially unbiased equilibrium state. We generate these steady-states by applying time-dependent gate pulses and we use the same technique to switch between them. In this work we have used an exponentially decaying gate voltage of the form

$$V_g(t) = \begin{cases} V_g e^{-\gamma t} & , \text{ if } 0 < t < T_g \\ -V_g e^{-\gamma(t-T_g)} & , \text{ if } T_g < t < 2T_g \\ V_g e^{-\gamma(t-2T_g)} & , \text{ if } t > 2T_g \end{cases} . \qquad (18)$$

The steady-state with density n_1 is obtained by time-propagation after applying a sudden constant bias $U_L(t) = U_L\theta(t)$ in the left lead, without applying the gate voltage. This is displayed in Fig. 3a. The steady-state of highest density n_3 is obtained (in addition to switching on the sudden bias in the leads) by applying an exponentially decaying gate voltage to the impurity site (with amplitude $V_g = 1.5$, decay rate $\gamma = 0.1$ and $T_g = \infty$).

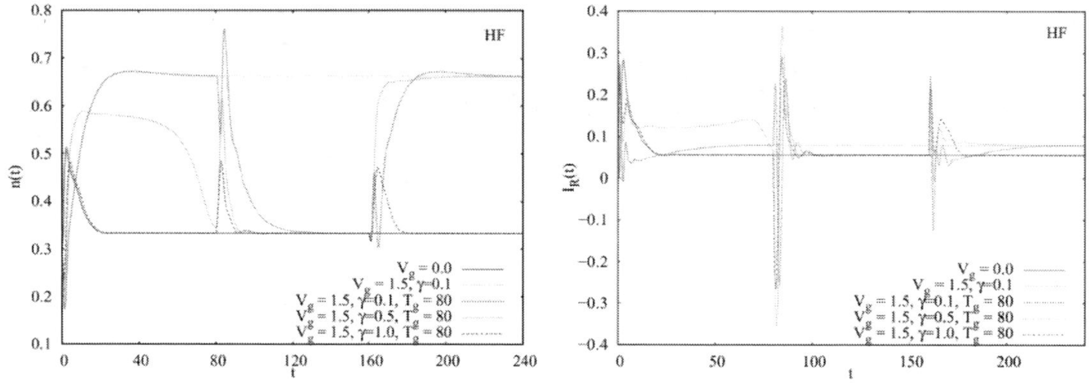

(a) (b)

Figure 3: a) Two different steady-state densities and the switching between the solutions by applying a gate voltage in the form of the equation (18). b) Two different steady-state currents through right lead and the switch between the solutions with a gate voltage of the form (18).

In Fig. 3a we show the time evolution of the density for various switchings between the steady-states. After we apply the first (sufficiently slow) gate voltage of equation (18) we let the system reach the steady-state of density n_3. At a time T_g we apply a second gate voltage in the opposite direction. The system shows a transient behavior after which, the system reaches the steady-state of density n_1. If we apply the third gate voltage at time $2T_g$ (see equation (18)), the system exhibits a short transient behavior and attains again the original steady-state of density n_3.

Corresponding to the densities n_1 and n_3 there are two distinguishable solutions for the currents $I_R(t)$ flowing into the right lead. These are shown in Fig. 3b where the lower value of the current corresponds to the state of density n_1 and the higher value of the current corresponds to the state of density n_3. The various transients observed in the density correspond directly to the transients in current shown in Fig. 3b. The frequency of the strongly damped oscillations of the transients (observed upon switching between the states in Fig.3b) is approximately given by the gate voltage, which causes a temporary change of V_g in the level position. As one can see from the Fig. 3a, if the decay rate $\gamma > 0.5$, *i.e.*, when the switching is too fast, the steady-state of density $n_3 = 0.66$ cannot be reached from the initial ground state of density $n_0 = 0.28$, because the system does not have enough time to acquire sufficient density.

In Fig. 4*a* we show, within the HF approximation, the nonequilibrium spectral function $A(T, \omega)$ of Eq.(9) for a switch (with $T_g = 60$, $V_g = 1.5$, and $\gamma = 0.1$) from the steady-state with density n_1 to the state with density n_3. When the gate is applied, the upper side of the spectral peak, starts to oscillate. The spectral peak undergoes a transition and overshoots before it settles to a new value of 1.5 within the right lead band. During this transition we can observe the appearance of a sharp peak in the spectral function located approximately at energy 1.3 corresponding to the top edge of the right lead band. This is shown in the inset of Fig. 4a, where we display a snapshot of $A(T,\omega)$ at time $T = 70$ at which the central site has density $n(T) = 0.53$, which is very close to the density of unstable state n_2. Therefore, although this state does not lead to a steady-state, it can still be observed in the nonequilibrium spectral function.

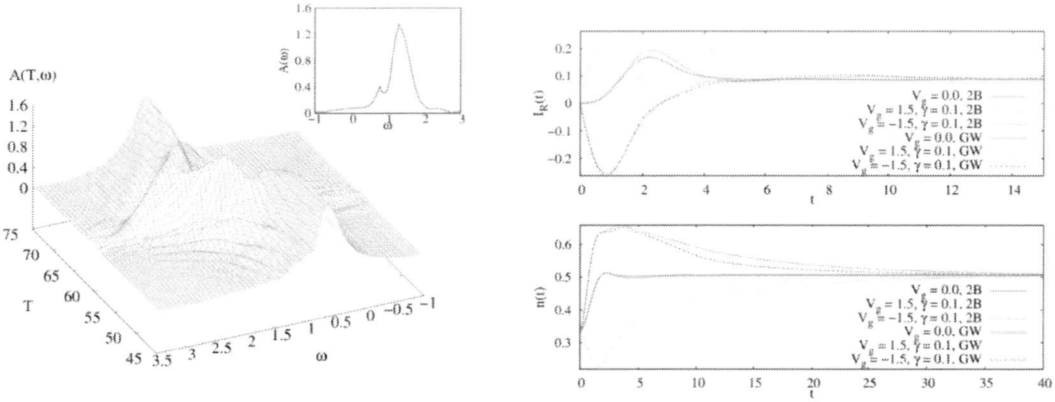

(a) (b)

Figure 4: a) Nonequilibrium spectral function $A(T,\omega)$ within the HF-approximation for the switch from the density n_1 to density n_3. *Inset*: Snapshot of spectral function corresponding to density n_2. b) The densities and currents calculated within the 2B and the GW approximations with and without an exponentially decaying gate voltage.

In Fig. 4b we show the densities and the currents obtained within the 2B and the GW self-energy approximations. For these cases only one steady-state is obtained with a density of about 0.5 on the central site. We have applied a time-dependent gate voltage of the form $V_g(t) = V_g e^{-\gamma t}$ for $t > 0$. If no gate is applied the transient regime is shorter and the density attains its steady-state value faster, compared to the case where the gate is applied. One can observe that the steady-state values of the currents and the densities for both GW and 2B are close to each other implying that for this system the single-bubble diagram, common to both approximations, plays a crucial role [14].

We conclude that for both the 2B and the GW approximation the spectral functions are very broad (see Fig. 2b) which makes it impossible to locate these spectral functions within an energy range in one of the lead energy-bands leading to two well-separated states. As a consequence the bistable regime is lost in the 2B and GW approximations, at least for the parameters considered.

5. Conclusions

In this paper we have investigated the switching between the steady-states of an interacting resonant level model connected to leads. For a given set of parameters we used a Meir-Wingreen type of approach to identify three different steady-state values for the density within the HF approximation. We showed that by applying an exponentially decaying gate voltage pulse during the time-propagation, at Hartree-Fock level, we can reach two different solutions and switch between them. However, due to a strong quasi-particle broadening of the spectral function, bistability is lost when the second Born and GW approximations are considered. Hence treatment of many-body interactions beyond mean-field can destroy bistability and lead to qualitatively different results as compared to those at mean-field level.

References

[1] V. J. Goldman and D. C. Tsui. *Phys. Rev. Lett.*, 58(12), 1987.
[2] F. W. Sheard and G. A. Toombs. *Appl. Phys. Lett.*, 52(15), 1988.
[3] J. O. Sofo and A. Balsero. *Phys. Rev. B*, 42(7292), 1990.
[4] T. Fiig and A. P. Jauho. *Surface Science*, 267(392), 1992.
[5] P. L. Pernas and F. Florens. *Phys. Rev. B*, 47(4779), 1993.
[6] N.C. Kluksdahl, A.M. Kriman, D.K. Ferry, and C. Ringhofer. *Phys.Rev.B*, 39(7720), 1989.
[7] L. Eaves, G. A. Toombs, F. W. Sheard, C. A. Payling, M. L Leadbeater, E. S. Alves, T. J. Foster, P. E. Simmonds, M. Henini, O. H. Hughes, J. C. Portal, G. Hill, and M. A. Pate. *Appl. Phys. Lett.*, 52(212), 1997.
[8] N. Zou, M. Willander, I. Linnerud, U. Hanke, K. A. Chao, and Y. M. Galperin. *Phys. Rev. B*, 49(2193), 1994.
[9] M. Di Ventra. *Electrical Transport in Nanoscale Systems.* Cambridge University Press, 2008.
[10] C. F. A. Negre, P. A. Gallay, and C. G. Sánchez. *Chem. Phys. Lett.*, 460:220–224, 2008.
[11] R. M. Dreizler and E. K. U Gross. *Density Functional Theory: An Approach to the Quantum Many-Body Problem.* Springer-Verlag, New York, 1990.
[12] G. Stefanucci and C.-O. Almbladh. *Phys. Rev. B*, 69(195328), 2004.
[13] S. Kurth, G. Stefanucci, C.-O. Almbladh, A. Rubio, and E. K. U. Gross. *Phys. Rev. B*, 72(035308), 2005.
[14] P. Myöhänen, A. Stan, R. van Leeuwen, and G. Stefanucci. *Phys. Rev. B*, (accepted), 2009.
[15] P. Myöhänen, A. Stan, G. Stefanucci, and R. van Leeuwen. *Europhys. Lett.*, 84(67001), 2008.
[16] L. P. Kadanoff and G. Baym. *Quantum Statistical Mechanics.* Benjamin, New York, 1962.
[17] P. Danielewicz. *Ann. Phys. (NY)*, 152(239), 1984.
[18] N. E. Dahlen and R. van Leeuwen. *Phys. Rev. Lett.*, 98(153004), 2007.
[19] P. W. Anderson. *Phys. Rev.*, 124(1):41–53, 1961.
[20] R. van Leeuwen, N. E. Dahlen, G. Stefanucci, C.-O. Almbladh, and U. von Barth. Time-Dependent Density Functional Theory. *Lect. Notes Phys.*, 706:33–59, 2006.
[21] P. Myöhänen, A. Stan, R. van Leeuwen, and G. Stefanucci. Kadanoff-Baym approach to time-dependent quantum transport in AC and DC fields. *(these proceedings)*, 2009.
[22] Y. Meir and N. S. Wingreen. *Phys. Rev. Lett.*, 68(2512), 1992.
[23] M. A. L. Marques, C. A. Ullrich, F. Nogueira, A. Rubio, K. Burke, and E. K. U. Gross. *Time-Dependent Density Functional Theory.* Springer Heidelberg, 2006.
[24] A. Szabo and N. S. Ostlund. *Modern Quantum Chemistry: Introduction to Advanced Electronic Structure Theory.* Dover Publications, 1996.
[25] A. P. Jauho, N. S. Wingreen, and Y. Meir. *Phys. Rev. B*, 50(5528), 1994.
[26] A. L. Fetter and J. D. Walecka. *Quantum Theory of Many-Particle Systems.* Dover Publications, 1971.

Journal of Physics: Conference Series **220** (2010) 012019 doi:10.1088/1742-6596/220/1/012019

Multiconfigurational time-dependent Hartree-Fock calculations for photoionization of one-dimensional Helium

David Hochstuhl, Sebastian Bauch, Michael Bonitz

Institut für Theoretische Physik und Astrophysik, Christian-Albrechts-Universität Kiel, Leibnizstrasse 15, 24098 Kiel, Germany

E-mail: `hochstuhl@theo-physik.uni-kiel.de`

Abstract. The multiconfigurational time-dependent Hartree-Fock equations are discussed and solved for a one-dimensional model of the Helium atom. Results for the ground state energy and two-particle density as well as the absorption spectrum are presented and compared to direct solutions of the time-dependent Schrödinger equation.

1. Introduction

In recent years new powerful radiation sources became available for the precise investigation of photoionization processes of matter. New methods have made it possible to observe the electronic motion in atoms and molecules in a time-resolved fashion on a scale of attoseconds [1; 2]. With this, several effects, such as strong-field tunneling [3] or time-resolved Auger decay [4] could be studied in detail. Similarly, sub-femtosecond processes in solids are becoming increasingly relevant [5; 6]. The explanation of the arising effects is a challenge for theoretical physicists, which need to face an old problem, namely the solution of the electronic Schrödinger equation with its exponentially growing effort with increasing number of the degrees of freedom.

Several methods have been developed to circumvent this fundamental limitation. Among them are time-dependent density functional theory (TDDFT), see e.g. [7], the method of nonequilibrium Green functions (NEGF), e.g. [8–10], or time-dependent reduced density-matrix theory (TDRDM), e.g. [11–13], which all aim at projecting the Schrödinger equation on a more convenient set of equations requiring only a polynomially growing effort in solution. Despite the indisputable successes of these methods, they lack a systematical and practically feasible way to achieve convergence to the exact result.[1]

In this paper, we apply a method which provides this mentioned feature, namely time-dependent Multiconfigurational Hartree-Fock (MCTDHF) [14]. It can either be seen as an extension of Hartree-Fock, which includes several Slater determinants (or permanents in the case of Bosons [15]) instead of a single one, or as an extension of Configuration Interaction, that

[1] In TDDFT, this is actually the main problem, since the result crucially depends on the choice of the exchange-correlation functional. NEGF (and TDRDM) are principally exact, if all equations in a hierarchy of equations were taken into account, resp. if all self-energy diagrams were summed up. In practice, however, the hierarchy is decoupled already on a low level.

Journal of Physics: Conference Series **220** (2010) 012019 doi:10.1088/1742-6596/220/1/012019

employs time-dependent single-particle orbitals instead of a fixed basis. In the case of infinitely many determinants, the results essentially become exact. In MCTDHF, the exponential problem is not really avoided, but it is shifted to much larger systems than in direct solutions of the Schrödinger equation. Thus, MCTDHF is applicable to few particle systems of roughly ten particles. To become familiar with the method, in this work we consider a directly solvable one-dimensional model of the Helium atom, and compare the MCTDHF results with those from the time-dependent Schrödinger equation (TDSE).

The outline of the paper is as follows: After this introduction, we give an overview on the MCTDHF formalism, recapitulate the working equations and provide the main ideas of our implementation. Subsequently, groundstate as well as time-dependent results for one-dimensional Helium are presented. In this paper, we apply the notation of Ref. [16].

2. The MCTDHF method

The system of our interest are few electron atoms in an external electromagnetic field described by the Hamiltonian (in atomic units)

$$\hat{H} = \sum_{k=1}^{N} \left\{ \frac{\mathbf{p}_k{}^2}{2} - \frac{Z}{|\mathbf{r}_i|} + \mathcal{E}(t) \cdot \mathbf{r}_k \right\} + \frac{1}{2} \sum_{k \neq l} \frac{1}{|\mathbf{r}_k - \mathbf{r}_l|}, \tag{1}$$

where the motion of the nucleus is neglected. In this paper, we focus on the one-dimensional Helium atom ($Z = 2$), for which the singularities in the Coulomb potential are softened by a cutoff, see Sec. 3.1. However, since the following theoretical considerations are completely general, we use the three-dimensional formulation. The studied systems are considered at zero temperature, so a wavefunction treatment is appropriate.

2.1. Overview

Standard wavepacket propagation methods, e.g. time-dependent Configuration Interaction, typically approximate the many-body wavefunction as a linear expansion in a set of basis vectors of the subspace $\mathcal{H}(2M, N)$ of the N-particle Hilbert space, that is, the subspace spanned by all N-fold anti-symmetrized products of $2M$ single-particle orbitals (the factor 2 is due to the two possible spin-projections). Commonly, the many-body basis states are taken to be either Slater determinants (SD) or configuration state functions (CSF). The latter are special linear combinations of Slater determinants, which are not only eigenfunctions of the projected-spin operator $\hat{S}_z$ – as Slater determinants are – but also eigenfunctions of the total-spin operator $\hat{S}^2$ [16]. This ansatz for the wavefunction is inserted into the Schrödinger equation to obtain a linear equation of motion for the time-dependent expansion coefficients, which may be solved by a couple of methods (e.g. short iterative Lanczos- [17; 18], Chebycheff- [19], split operator- [20] methods, etc. [21]). However, this simple approach suffers from a main drawback, namely that the size of the N-particle Hilbert space basis grows exponentially with the number of particles N and orbitals $2M$, what restricts the applicability to rather small systems. A common way around this problem is to drop several Slater-determinants, which are believed to be physically less important, as it is done for instance in the time-dependent Configuration Interaction singles (TDCIS) method [22]. This enables the treatment of higher particle numbers resp. the inclusion of a larger single-particle basis at the cost of a reduced description of correlation effects.

The multiconfigurational time-dependent Hartree-Fock method uses an alternative strategy. It also approximates the wavefunction by a linear expansion of basis states of $\mathcal{H}(2M, N)$, which are, however, allowed to vary in time. This is achieved by assuming the $2M$ orbitals $\{|\,\phi_k\,\rangle\}$ to be explicitly time-dependent, and being expressed by an expansion in a set of N_b time-independent orbitals $\{|\chi_l\rangle\}$:

$$|\phi_k\rangle(t) = \sum_{l=1}^{N_b} b_{kl}(t)\,|\chi_l\rangle\,, \quad k = 1,\cdots,2M\,. \tag{2}$$

By using Slater determinants built with such time-dependent orbitals, it is likewise only possible to represent states in a subspace $\mathcal{H}(2M, N)$. However, now this space is allowed to vary freely in the much larger subspace $\mathcal{H}(2N_b, N)$. In this way it is possible to arrive at an accurate description of the wavefunction and to defer the problem of the exponential growth of the Hilbert basis size to larger systems.
In the following, we give the MCTDHF equations for systems described by the standard electronic Hamiltonian (1), which in second quantization reads

$$\hat{H}(t) = \sum_{pq} h_{pq}(t)\,\hat{E}_{pq} + \frac{1}{2}\sum_{pqrs} g_{pqrs}\,\hat{e}_{pqrs}\,, \tag{3}$$

with the one- and two-particle excitation operators [16]

$$\hat{E}_{pq} = \sum_{\sigma} \hat{a}^\dagger_{p\sigma}\hat{a}_{q\sigma}\,, \tag{4}$$

$$\hat{e}_{pqrs} = \sum_{\sigma\tau} \hat{a}^\dagger_{p\sigma}\hat{a}^\dagger_{r\tau}\hat{a}_{s\tau}\hat{a}_{q\sigma}\,, \tag{5}$$

acting in the time-dependent basis $\{|\phi_l\rangle\}$. The electron integrals are given by

$$h_{pq}(t) = \int d\mathbf{r}\,\phi_p^*(\mathbf{r})\left\{-\frac{1}{2}\Delta + V(\mathbf{r}) + \boldsymbol{\mathcal{E}}(t)\cdot\mathbf{r}\right\}\phi_q(\mathbf{r})\,, \tag{6}$$

$$g_{pqrs} = \iint d\mathbf{r}\,d\bar{\mathbf{r}}\;\phi_p^*(\mathbf{r})\phi_q(\mathbf{r})\,\frac{1}{|\mathbf{r}-\bar{\mathbf{r}}|}\,\phi_r^*(\bar{\mathbf{r}})\phi_s(\bar{\mathbf{r}})\,. \tag{7}$$

The single particle Hamiltonian $\hat{h}(t)$ includes the action of an external electromagnetic field $\boldsymbol{\mathcal{E}}(t)$ in dipole approximation (length gauge) and thus carries the only explicit time-dependence.

2.2. The MCTDHF equations

As mentioned above, the multiconfigurational time-dependent Hartree-Fock ansatz approximates the wavefunction by a linear combination of time-dependent basis states of the N-particle Hilbert space, which are in the following assumed to be Slater determinants:

$$|\Psi\rangle = \sum_{\mathbf{n}} C_{\mathbf{n}}(t)\,|n_{1\alpha}, n_{1\beta}, n_{2\alpha}\cdots, n_{M\beta}; t\rangle\,. \tag{8}$$

The Slater determinants are written in occupation number representation specifying the occupation of the M time-dependent spatial orbitals $\{|\phi_k\rangle\}$ with an electron with spin-projection α (spin-up) or β (spin-down), and $\sum_k n_{k\alpha} + n_{k\beta} \equiv N$. This corresponds to a *spin-restricted* treatment, i.e. we assume that α- and β-electrons share a common spatial orbital [23]. For the derivation of the equations of motion we follow Refs. [24; 25] and employ the Lagrange formulation of the time-dependent variational principle, in which the action functional

$$S\Big[\{C_{\mathbf{n}}(t)\}, \{|\phi_k\rangle(t)\}\Big] = \int dt\,\left\{\left\langle\Psi\left|\hat{H} - i\frac{\partial}{\partial t}\right|\Psi\right\rangle - \sum_{kl}\mu_{kl}(t)\Big(\langle\phi_k|\phi_l\rangle - \delta_{kl}\Big)\right\} \tag{9}$$

Journal of Physics: Conference Series **220** (2010) 012019 doi:10.1088/1742-6596/220/1/012019

is minimized with respect to the variational parameters. The time-dependent Lagrange multipliers are introduced to ensure the orbitals to remain orthonormal during the temporal evolution.

We first derive the equations of motion for the orbitals by requiring the variation to be stationary,

$$\frac{\delta}{\delta\langle\,\phi_n\,|}\;S\Big[\{C_{\mathbf{n}}(t)\},\{|\,\phi_k\,\rangle(t)\}\Big]\;\stackrel{!}{=}\;0\,. \tag{10}$$

After a few steps we arrive at the following nonlinear equation:

$$\hat{\mathbf{P}}\,i\frac{\partial}{\partial t}\,|\,\phi_n\,\rangle\;=\hat{\mathbf{P}}\left\{\hat{h}(t)\,|\,\phi_n\,\rangle\;+\;\sum_{pqrs}\;(\mathbf{D}^{-1})_{np}\;d_{pqrs}\,\hat{g}_{rs}\,|\,\phi_q\,\rangle\right\}, \tag{11}$$

where we introduced the (spin-restricted) one- and two-particle density matrices

$$D_{pq}\;=\;\langle\,\Psi\,|\,\hat{E}_{pq}\,|\,\Psi\,\rangle\,, \tag{12}$$

$$d_{pqrs}\;=\;\langle\,\Psi\,|\,\hat{e}_{pqrs}\,|\,\Psi\,\rangle\,, \tag{13}$$

as well as the mean-field operator $\hat{g}_{rs}$, which in coordinate representation reads

$$g_{rs}(\mathbf{r})\;=\;\int d\mathbf{r}'\;\phi_r^*(\mathbf{r}')\,\frac{1}{|\mathbf{r}-\mathbf{r}'|}\,\phi_s(\mathbf{r}')\,. \tag{14}$$

Further, the elimination of the Lagrange multipliers led to a projection operator

$$\hat{\mathbf{P}}\;=\;1-\sum_m\,|\,\phi_m\,\rangle\langle\,\phi_m\,|\,, \tag{15}$$

which projects on the orthogonal complement of the span of the orbitals. In order to remove the projection operator on the lhs. of Eq. (11) and obtain explicit equations, a unitary transformation among the orbitals is applied, which ensures

$$\left\langle\,\phi_k\,\middle|\,\frac{\partial}{\partial t}\,\middle|\,\phi_l\,\right\rangle\;=\;0\,, \tag{16}$$

i.e. the change of an orbital is orthogonal to the subspace spanned by all orbitals. After insertion into Eq. (11), we obtain the MCTDH orbital equations

$$i\frac{\partial}{\partial t}\,|\,\phi_n\,\rangle\;=\hat{\mathbf{P}}\left\{\hat{h}(t)\,|\,\phi_n\,\rangle\;+\;\sum_{pqrs}\;(\mathbf{D}^{-1})_{np}\;d_{pqrs}\,\hat{g}_{rs}\,|\,\phi_q\,\rangle\right\}. \tag{17}$$

The minimization with respect to the coefficients, i.e.

$$\frac{\delta}{\delta C^*_{\mathbf{n}}}\;S\Big[\{C_{\mathbf{n}}(t)\},\{|\,\phi_k\,\rangle(t)\}\Big]\;\stackrel{!}{=}\;0\,, \tag{18}$$

then straightforwardly leads to a Schrödinger equation in matrix representation in the SD-basis:

$$i\frac{\partial}{\partial t}\,C_{\mathbf{n}}(t)\;=\;\sum_{\mathbf{m}}\langle\,\mathbf{n}\,|\,\hat{H}(t)\,|\,\mathbf{m}\,\rangle\;C_{\mathbf{m}}(t)\,. \tag{19}$$

Note, that again Eq. (16) has been used, which also causes the matrix element of the time-derivative operator between Slater determinants to vanish,

$$\left\langle\,\mathbf{n}\,\middle|\,\frac{\partial}{\partial t}\,\middle|\,\mathbf{m}\,\right\rangle\;=\;0\,. \tag{20}$$

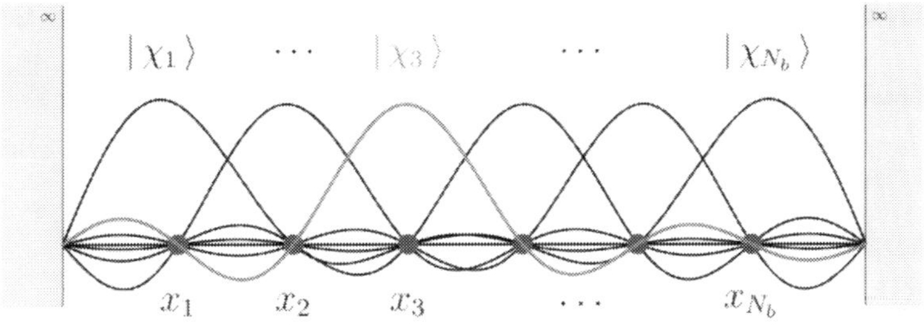

Figure 1. Schematic view of the sine DVR functions, for a basis size of $N_b = 6$. The basisfunctions are constructed over an equidistant grid x_k in such a way, that $\chi_i(x_k) = \delta_{ik}/\sqrt{w_k}$ holds, with a set of integration weights w_k. Matrix elements may then be evaluated approximately by a summation, i.e. $\int dx\, \chi_i^*(x) f(x) \chi_j(x) \longrightarrow \sum_k w_k\, \chi_i^*(x_k) f(x_k) \chi_j(x_k)$. This leads directly to diagonal spatial matrix elements, $\langle\, \chi_i \,|\, f(\hat{x}) \,|\, \chi_j \,\rangle = f(x_i)\, \delta_{ij}$.

2.3. Numerical implementation

We give some notes on our numerical solution of the coupled set of MCTDHF equations, Eqs. (19) and (17).

Single-particle basis. First of all, like in Eq. (2), an appropriate time-independent single particle basis $\{|\,\chi_l\,\rangle\}$ is chosen, which inserted into the orbital equation (17) yields an equation for the time-dependent expansion coefficients $b_{kl}(t)$ [23]. In this work we use a sine discrete variable representation (DVR) basis [26; 27], see Fig. 1. As is common to all DVR bases, matrix elements of spatial operators are diagonal, and they are simply given by the function values on a grid $\mathbf{r}_i$ of Gaussian integration points,

$$\langle\, \chi_p \,|\, f(\mathbf{r}) \,|\, \chi_q \,\rangle \;=\; \delta_{pq}\, f(\mathbf{r}_q)\,, \tag{21}$$

$$\langle\, \chi_p\, \chi_r \,|\, g(\mathbf{r}, \mathbf{r}') \,|\, \chi_s\, \chi_q \,\rangle \;=\; \delta_{pq}\, \delta_{rs}\, g(\mathbf{r}_i, \mathbf{r}_j)\,. \tag{22}$$

For the sine DVR, the grid consists of equally spaced nodes and the non-diagonal kinetic energy matrix can be evaluated analytically.[2]

Density matrices and Hamiltonian. For the evaluation of the density matrices and the Hamiltonian, the matrix elements of one-and two-particle excitation operators in the basis of Slater determinants, $\langle\, \mathbf{n} \,|\, \hat{E}_{pq} \,|\, \mathbf{m} \,\rangle$ and $\langle\, \mathbf{n} \,|\, \hat{e}_{pqrs} \,|\, \mathbf{m} \,\rangle$, have to be evaluated, both of which may attain either zero, plus or minus one. Since these quantities are needed very frequently, all non-zero contributions are determined and their sign is stored in memory before the actual time propagation. The Hamiltonian can then be easily calculated using the Slater-Condon rules [16].

Time evolution. After writing the wavefunction coefficients $\mathbf{C}$ and the single-particle basis expansion coefficients $\mathbf{b}$ in a single vector, the MCTDHF equations may be casted to the general form

$$i\,\frac{\partial}{\partial t}\,\boldsymbol{\mathcal{V}} \;=\; \boldsymbol{\mathcal{F}}(\boldsymbol{\mathcal{V}})\,, \qquad \boldsymbol{\mathcal{V}} \;=\; \begin{pmatrix} \mathbf{b} \\ \mathbf{C} \end{pmatrix}. \tag{23}$$

We solve this coupled set of equations by means of general purpose integrators, such as Runge-Kutta or Burlisch-Stoer methods [30]. Other possible techniques particularly well suited for the MCTDHF scheme are given in Ref. [27].

Solution effort. As mentioned above, the basis of Slater-determinants grows with $\binom{2M}{N}$, where M is the number of time-dependent spatial orbitals and N the number of particles, leading

[2] Instead of a normal DVR, one could also employ a finite-element DVR, see e.g. [28; 29].

to the typical exponential problem of Configuration Interaction. However, for the two-particle model we consider here, the SD-basis only grows with $\mathcal{O}(M^2)$ and thus poses no difficulties. Here, the effort is rather determined by the number N_b of orbitals, which may become large in order to adequately describe the continuum. For each evaluation of the rhs. $\boldsymbol{\mathcal{F}}(\boldsymbol{\mathcal{V}})$ of Eq. (23), the electron integrals in the time-dependent basis are needed, which are formed through a transformation from the time-independent basis using the orbital coefficients. The time-consuming part is the transformation of the two-particle interaction matrix. Assuming $N_b \gg M$, the leading term for a DVR basis is given by $\mathcal{O}(MN_b^2)$. To further diminish the effort and obtain an almost linear scaling with the size of the underlying time-independent basis, low-rank approximations of the interaction potential can be used [27; 31]. We also note, that for a fixed basis size N_b the effort grows as $\mathcal{O}(M^4N_b)$.

3. Numerical results

3.1. One-dimensional Helium model

The here considered one-dimensional model of Helium is given by a potential

$$V(x) = -\frac{2}{\sqrt{x^2+1}}, \qquad (24)$$

and has been well tested for roughly 30 years. Compared to the real three-dimensional Helium atom, in this model the electron movement is restricted only to the laser polarization axis. Its usefulness derives on the one hand from the fact, that electronic correlation effects of Helium and also larger atoms can be qualitatively well explained, most prominently the nonsequential double ionization [32; 33]. On the other hand, it is exactly solvable by viewing the one-dimensional two-particle problem as a Schrödinger equation of one-particle moving in the two-dimensional external potential

$$V(x,y) = -\frac{2}{\sqrt{x^2+1}} - \frac{2}{\sqrt{y^2+1}} + \frac{1}{\sqrt{(x-y)^2+1}}. \qquad (25)$$

In this paper, we use it as a benchmark for the approximate MCTDHF simulations. For the details of the employed solution method of the two-dimensional TDSE, see [34].

3.2. Groundstate results

In the following we present the results for the groundstate of the Helium model obtained with MCTDHF. A similar investigation has already been given in [35], however with a slightly different atomic potential. The groundstate is obtained by propagation of the MCTDHF equations in imaginary time (ITP), starting from the orbitals of the noninteracting system. The system is

M	1	2	3	4	5	exact
Energy [Hartree]	−2.2242	−2.2365	−2.2381	−2.2382	−2.23825	−2.23826
% of corr. energy	–	87%	98.9%	99.6%	99.9%	100%

Table 1. Total energies of the groundstate for different numbers M of time-dependent orbitals, plus the fraction of the included correlation energy (the entire correlation energy is defined as the difference between the exact and the Hartree-Fock result). With increasing M, the energy rapidly approaches the exact energy as obtained by a direct solution of the Schrödinger equation. For $M > 5$, the results equal the exact result for the given number of digits.

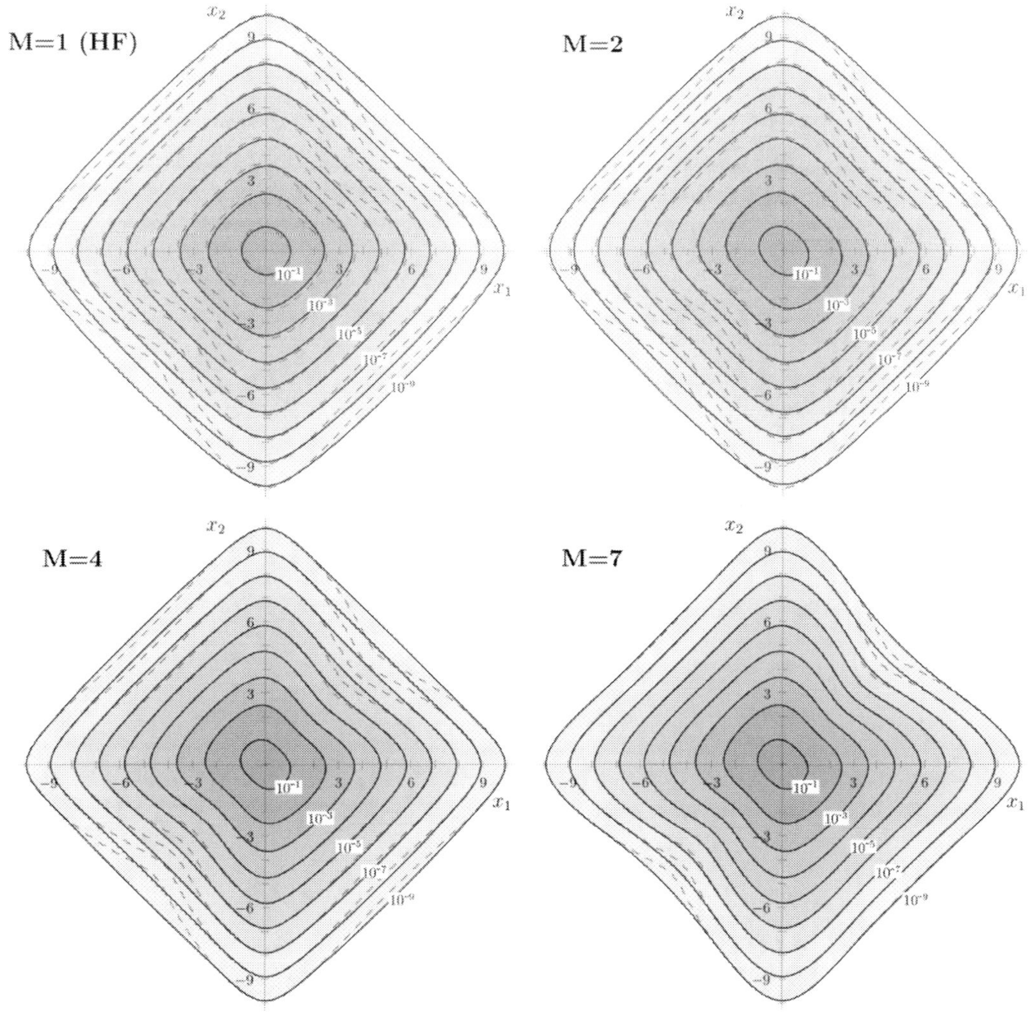

Figure 2. Logarithmic contour plot of the groundstate two-particle densities for different numbers M of MCTDHF orbitals. x_1 and x_2 are the coordinates of the two electrons. In each plot, the red dashed curves show the result from a direct solution of the Schrödinger equation.

evolved until the difference of the total energy between two steps falls below a certain limit (here 10^{-10} a.u.). The results are compared to direct solutions of the Schrödinger equation, which were obtained by ITP as well. In Tab. 1 we display the groundstate energies for different MCTDHF approximations, and the corresponding fraction of the correlation energy as defined by the difference between the exact and the Hartree-Fock ($M = 1$) energy. We observe a rapid convergence with the number of orbitals, already the first correction to Hartree-Fock, $M = 2$, is able to account for 87% of the entire correlation energy. Note that $M = 2$ only needs roughly an eight-fold effort as compared to Hartree-Fock, which is due to the CPU time growth with $\mathcal{O}(M^4 N_b)$ (see section 2.3).

In Fig. 2 we plot the two-particle (tp) densities, that for a two-electron system equal the absolute square of the wavefunction, again for different approximations. In each plot, the (red) dashed curve depicts the tp-density from the exact solution. From the figure it is obvious, that the

Journal of Physics: Conference Series **220** (2010) 012019 doi:10.1088/1742-6596/220/1/012019

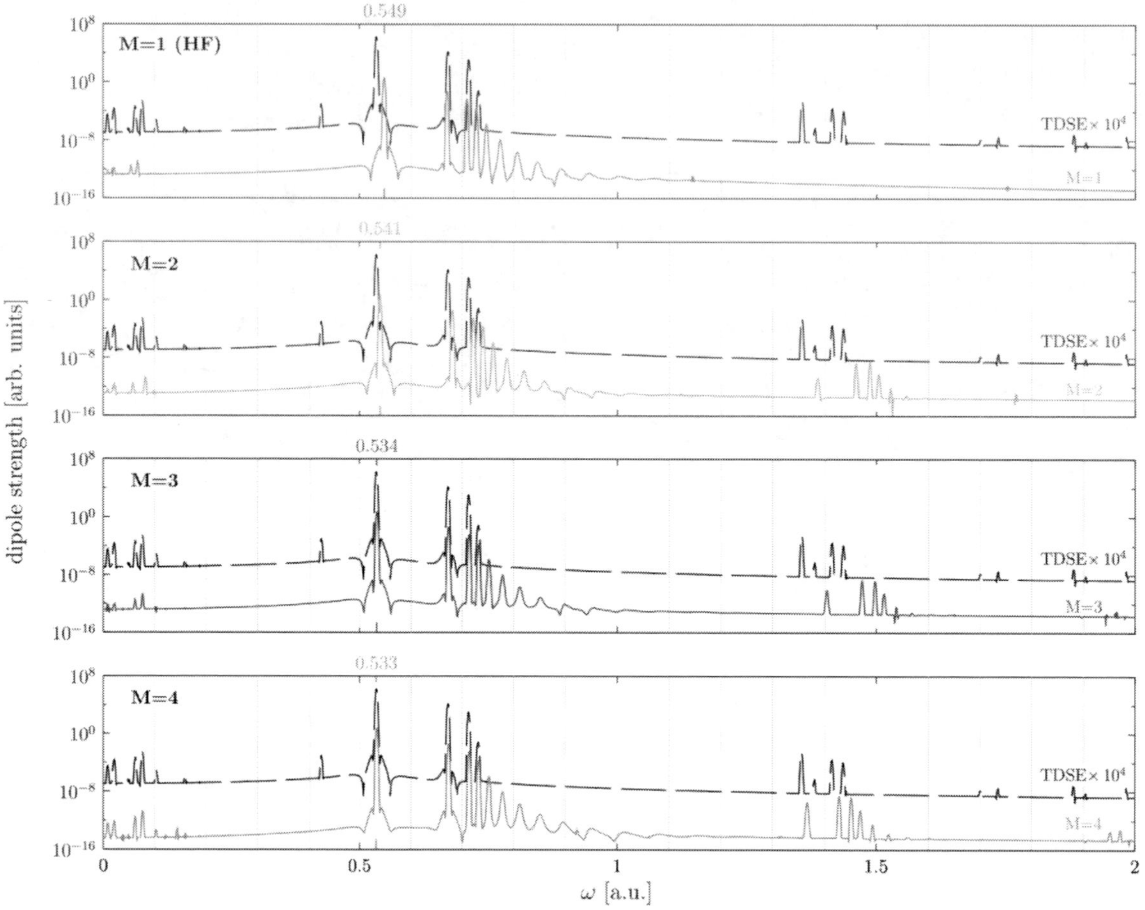

Figure 3. Linear response of the one-dimensional Helium model atom as obtained from MCTDHF and the TDSE (shifted by 10^4). The pictures show results for different numbers M of time-dependent orbitals. The peaks in the tail of the spectrum are caused by correlations. The numbers on the upper x-axes denote the positions of the first peak, which with increasing M converge towards the exact position $\omega = 0.533$ a.u.

Hartree-Fock approximation is not able to reproduce the correct butterfly shape caused by the Coulomb repulsion (which is largest for $x_1 = x_2$). With increasing number of orbitals, the tp-density approaches the exact result, until it attains an accuracy of seven digits for $M = 7$.

3.3. Time-dependent results

As an example for time-dependent calculations, we investigate the reaction of the model atom to an external perturbation in linear response[3]. To this end, we disturb the groundstate by a dipole kick, which is sufficiently short in order to provide a homogeneous spectral density and sufficiently weak to avoid non-linear effects. In this work, we chose a kick with duration 0.01 a.u. and an amplitude of $\mathcal{E}_0 = 0.01$ a.u. The disturbed system was propagated for 2000 a.u.

[3] The linear response can also be obtained from time-independent calculations by diagonalization of the Hamiltonian. In fact, the following results present an alternative for the diagonalization called "filter-diagonalization", see e.g. [36].

($\sim$ 50 fs) and the expectation value of position has subsequently been Fourier-transformed (using a Blackman-Harris window). This yields the dipole spectrum, from which the excitation frequencies can be observed. In Fig. 3 we plot the obtained spectra for different approximations and compare them with the result from the TDSE. First let us consider the energy region below the first ionization threshold $\omega \leq 0.75$ a.u. [37], which corresponds to the one-electron excitations. These excitations are rather well reproduced already within Hartree-Fock, only the position of the peaks and thus the excitation energies deviate slightly. However, for Hartree-Fock ($M = 1$), there are no peaks in the tail of the curve, which consequently are caused by correlations. Already in the first correlation correction, $M = 2$, the main peaks exist, but their positions deviate from the TDSE result. For $M = 4$, the exact spectrum is well reproduced. The values on the upper x-axes indicate the position of the first peak, which determine the difference between the groundstate energy and the energy of the first excited singlet state. We observe a similar trend as before, namely that $M = 4$ is able to reproduce the exact result of $\omega = 0.533$ a.u.

4. Summary

We have given an introduction to the time-dependent Multiconfiguration Hartree-Fock formalism and discussed the main ideas of our numerical implementation. In order to investigate the characteristics and capabilities of the method, it has been applied to a well-known one-dimensional model of the Helium model. We calculated the groundstate energies and two-particle densities, as well as the linear response of the system, each one for different MCTDHF approximations, i.e. for different numbers M of time-dependent spatial orbitals. All results were compared to direct solutions of the time-dependent Schrödinger equation. Our investigation showed, that MCTDHF is well suited for the time-resolved study of this two-electron system. We are very confident, that a comparable level of accuracy can also be obtained for larger system (up to, say, $N = 10$).
Our future work therefore will concentrate on time-dependent electronic correlations in larger model atoms, which are not accessible by direct solutions of the Schrödinger equation.

Acknowledgements

This work is supported by the Deutsche Forschungsgemeinschaft via SFB-TR 24 and the U.S. Department of Energy award DE-FG02-07ER54946.

References

[1] Krausz F and Ivanov M 2009 *Reviews of Modern Physics* **81** 163
[2] Brabec T and Krausz F 2000 *Reviews of Modern Physics* **72** 545
[3] Uiberacker M, Uphues T, Schultze M, Verhoef A J, Yakovlev V, Kling M F, Rauschenberger J, Kabachnik N M, Schröder H, Lezius M, Kompa K L, Muller H G, Vrakking M J J, Hendel S, Kleineberg U, Heinzmann U, Drescher M and Krausz F 2007 *Nature* **446** 627
[4] Drescher M, Hentschel M, Kienberger R, Uiberacker M, Yakovlev V, Scrinzi A, Westerwalbesloh T, Kleineberg U, Heinzmann U and Krausz F 2002 *Nature* **419** 803
[5] Krasovskii E E and Bonitz M 2007 *Physical Review Letters* **99** 247601
[6] Krasovskii E E and Bonitz M 2009 *Physical Review A* **80** 053421
[7] Runge E and Gross E K U 1984 *Physical Review Letters* **52** 997
[8] Dahlen N E and van Leeuwen R 2007 *Physical Review Letters* **98** 153004
[9] Kadanoff L P and Baym G 1962 *Quantum Statistical Mechanics.* (W. Benjamin, New York)
[10] Keldysh L 1964 *Zh. Eksp. Teor. Fiz.* **47** 1515
[11] Bonitz M 1998 *Quantum Kinetic Theory* (B.G. Teubner)

[12] Kremp D, Bonitz M, Kraeft W D and Schlanges M 1997 *Annals of Physics* **52** 320

[13] Viktor K, Axt V and Stahl A 1995 *Physical Review B* **51** 14164

[14] Bonitz M, Hochstuhl D, Bauch S and Balzer K 2010 *Contributions to Plasma Physics* **50** 54

[15] Heimsoth M and Bonitz M 2010 *Physica E* **42** 420

[16] Helgaker T, Joergensen P and Olsen J 2000 *Molecular electronic structure theory* (Wiley (New York))

[17] Park T J and Light J C 1986 *Journal of Chemical Physics* **85** 5870

[18] Jie Q and Liu D 2006 *Journal of Physics A* **39** 1691

[19] Tal-Ezer H and Kosloff R 1984 *Journal of Chemical Physics* **81** 3967

[20] Chin S A and Chen C R 2001 *Journal of Chemical Physics* **114** 7338

[21] Leforestier C, Bisseling R H, Cerjan C, Feit M D, Friesner R, Guldberg A, Hammerich A, Jolicard G, Karrlein W, Meyer H D, Lipkin N, Roncero O and Kosloff R 1991 *Journal of Computational Physics* **94** 59

[22] Rohringer N, Gordon A and Santra R 2006 *Physical Review A* **74** 043420

[23] Hochstuhl D and Bonitz M 2010 *to be published*

[24] Alon O E, Streltsov A I and Cederbaum L S 2008 *Physical Review A* **77** 033613

[25] Alon O E, Streltsov A I and Cederbaum L S 2007 *Journal of Chemical Physics* **127** 154103

[26] Colbert D T and Miller W H 1992 *Journal of Chemical Physics* **96** 1982

[27] Beck M and Meyer H 1997 *Zeitschrift für Physik D Atoms, Molecules and Clusters* **42** 113

[28] Balzer K, Bauch S and Bonitz M 2010 *Physical Review A* **81** 022510

[29] Balzer K, Bauch S and Bonitz M 2009 *this volume*

[30] Press W H, Teukolsky S A, Vetterling W T and Flannery B P 2007 *Numerical Recipes 3rd Edition: The Art of Scientific Computing* (New York: Cambridge University Press)

[31] Caillat J, Zanghellini J, Kitzler M, Koch O, Kreuzer W and Scrinzi A 2005 *Physical Review A* **71** 012712

[32] Liu W C, Eberly J H, Haan S L and Grobe R 1999 *Physical Review Letters* **83** 520

[33] Dahlen N E and van Leeuwen R 2001 *Physical Review A* **64** 023405

[34] Bauch S and Bonitz M 2008 *Physical Review A* **78** 043403

[35] Zanghellini J 2004 *Multielectron dynamics in the ionization of molecules by strong laser pulses.* (PhD Thesis. University of Vienna)

[36] Mandelshtam V A and Taylor H S 1992 *Journal of Chemical Physics* **107** 6756

[37] Haan S L, Grobe R and Eberly J H 1994 *Physical Review A* **50** 378

Journal of Physics: Conference Series **220** (2010) 012020 doi:10.1088/1742-6596/220/1/012020

Finite elements and the discrete variable representation in nonequilibrium Green's function calculations. Atomic and molecular models

Karsten Balzer, Sebastian Bauch, and Michael Bonitz
Institut für Theoretische Physik und Astrophysik, Christian-Albrechts-Universität Kiel, Leibnizstrasse 15, 24098 Kiel, Germany

E-mail: balzer@theo-physik.uni-kiel.de

Abstract. In this contribution, we discuss the finite-element discrete variable representation (FE-DVR) of the nonequilibrium Green's function and its implications on the description of strongly inhomogeneous quantum systems. In detail, we show that the complementary features of FEs and the DVR allow for a notably more efficient solution of the two-time Schwinger/Keldysh/Kadanoff-Baym equations compared to a general basis approach. Particularly, the use of the FE-DVR leads to an essential speedup in computing the self-energies.

As atomic and molecular examples we consider the He atom and the linear version of H_3^+ in one spatial dimension. For these closed-shell models we, in Hartree-Fock and second Born approximation, compute the ground-state properties and compare with the exact findings obtained from the solution of the few-particle time-dependent Schrödinger equation.

1. Introduction

In the last decade, the application of the nonequilibrium Green's function (NEGF) to describe strongly inhomogeneous quantum systems has started to become an actively considered subject. Thereby, various finite and localized systems have challenged attention and different many-body approximations have been applied. Recent state-of-the-art approaches discuss small atoms and molecules [1, 2], few-electron quantum dots [3, 4] and quantum dots coupled to leads [5], molecular junctions [6], and Hubbard nanoclusters [7]. In part, these works also include the monitoring of the system's temporal evolution which, accounting for correlation and memory effects, requires to solve the two-time Schwinger/Keldysh/Kadanoff-Baym equations [8, 9, 10] (SKKBE).

To solve the SKKBE for homogeneous quantum systems [11, 12, 13, 14, 15, 16, 17, 18, 19, 20, 21, 22] has become routine. However, this still does not hold for finite, localized and inhomogeneous systems. The reason is, that in contrast to homogeneous systems where one spatial coordinate or momentum drops out of the NEGF, the inhomogeneity claims adequate resolution in both coordinates or momenta. To meet these requirements all above mentioned works rely on an expansion of the NEGF in one-particle orbitals: for atomic and molecular systems, linear combinations of Slater-type or Gauss-type orbitals are being used (also in connection with tight-binding models), whereas for other classes of problems e.g. potential-eigenstate basis functions are being utilized. Beyond a basis ansatz, alternatives such as direct grid (finite-difference) methods do not exist as they are computationally very expensive.

Nevertheless, also basis approaches are so far limited to relatively small basis sets. Generally, the numerical complexity involved in the description of the binary interactions does not permit to extend nonequilibrium calculations to much larger basis dimensions than a guide number of $n_b < 50$ orbitals. This explains the low resolution in the description of photoionization processes of model atoms [23, 24] using nonequilibrium Green's functions.

In this contribution, we—for the first time—develop a grid-based approach in the frame of the finite-element discrete variable representation (FE-DVR), e.g. Refs. [25, 26, 27, 28] and references therein. This method leads to specially-designed, flexible basis sets which are capable to combine the advantages of pure grid and standard basis approaches, see Sec. 2 and also Ref. [29]. In particular, it allows for a very efficient treatment of the binary interactions and, in turn, a drastic simplification of higher-order self-energy expressions, which, generally, require the main effort in all NEGF calculations. As a result, only $\mathcal{O}(n_b^2)$ semi-analytical matrix elements of the interaction energy operator are required to compose the second Born self-energy. This has to be compared to a general basis representation of the NEGF: There, $\mathcal{O}(n_b^4)$ matrix elements are involved, cf. [4]. As a consequence, the use of the FE-DVR enables more efficient calculations at less storage memory and computing time and provides the basis to consider also spatially extended hamiltonians, where particles may occupy large domains in coordinate space.

After the outline of the method, we, in Sec. 3, apply the FE-DVR representation of the NEGF to compute the ground-state properties of atomic and molecular models. First, in Sec. 3.1, we discuss the one-dimensional (1D) helium atom [30, 31, 32, 33, 34, 35, 36, 37], where we focus on technical details such as grid size, DVR basis size, and convergence in the case of the Hartree-Fock and second Born approximation. The benchmarking results shown are also of relevance for time-dependent calculations, as they define and border the requirements to resolve explicit correlation effects (within a NEGF approach) e.g. the two-electron resonances [38] which are embedded in the one-electron continuum of the atom and are going to be occupied during laser-atom interactions—see also [39] in the present volume. Finally, we consider the linear molecular ion H_3^+ in the symmetric 1D singlet configuration [40, 41] as an example of a two-electron molecule and vary the interatomic distance to record the binding-energy curves in Hartree-Fock and second Born approximation, see Sec. 3.2. From this we can extract the minimum ground-state energies and the respective bond-lengths, which are compared to the findings from the few-particle time-dependent Schrödinger equation.

2. The nonequilibrium Green's function in FE-DVR representation

For the description of the NEGF using finite elements together with the discrete variable representation [42] (DVR), we consider the general N-electron Hamiltonian

$$\hat{h} = \sum_{i=1}^{N} (\hat{t}_i + \hat{v}_i) + \sum_{i<j} \hat{u}_{ij} \,, \tag{1}$$

with the kinetic energy $\hat{t}_i = -\nabla_i^2/2$, the time-dependent potential energy $\hat{v}_i = v(x_i, t)$, and the pair interaction energy $\hat{u}_{ij} = u(|x_i - x_j|)$ formulated in atomic units. The one-particle nonequilibrium Green's function with space-time arguments $1 = (x, t)$ and $1' = (x', t')$ reads

$$G(1, 1') = -i \left\langle T_{\mathcal{C}} \psi(1) \psi^\dagger(1') \right\rangle \,, \tag{2}$$

where spin is omitted, $\psi(1)$ and $\psi^\dagger(1')$ are electron field operators, and $T_{\mathcal{C}}$ denotes time-ordering on the full Keldysh contour [9] $\mathcal{C}$. According to system (1), $G(1, 1')$ obeys the SKKBE [8, 9, 10]

$$\{\mathrm{i}\,\partial_t - H(1)\}\, G(1, 1') = \delta_{\mathcal{C}}(1 - 1') + \int_{\mathcal{C}} \mathrm{d}2\, \Sigma[G](1, 2)\, G(2, 1') \,, \tag{3}$$

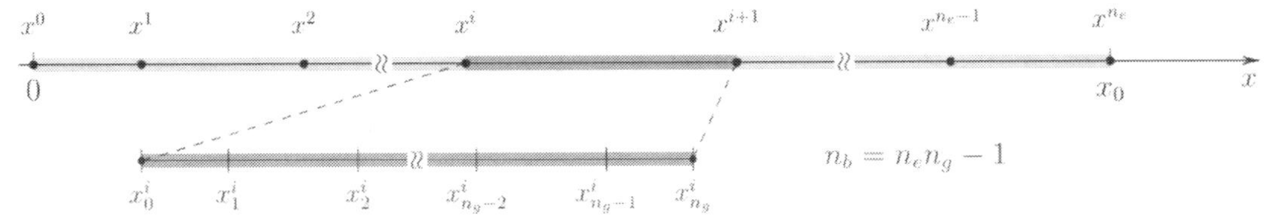

Figure 1. Interval $[0, x_0]$, as discretized in FE-DVR representation with a number of n_e finite elements and n_g generalized Gauss-Lobatto points (per element). n_b denotes the dimensionality of the extended basis covering the whole interval.

with addition of its adjoint equation. Further, $H(1)$ is the one-electron (kinetic plus potential) energy, $\Sigma(1, 1')$ denotes the self-energy, and equilibrium initial correlations are treat in the mixed Green's function approach [43, 44, 45], cf. Sec. 2.3.

The favorable aspects of the FE-DVR representation have been successfully used for the time-dependent Schrödinger equation (TDSE) by Rescigno [25] and others, e.g. [28]. There, the accuracy of the DVR [46] on the one hand and the sparse character of FEs on the other led to an efficient TDSE code which is well parallelizable. In our case, this hybrid approach allows us to rewrite the SKKBE (3) in a highly effective matrix notation using optimal combinations of a grid and a local basis.

2.1. Grid-based ansatz

On a predefined spatial 1D interval $\mathcal{I} = [0, x_0]$, we expand the nonequilibrium Green's function of hamiltonian (1) as

$$G(1, 1') = \sum_{i_1 m_1} \sum_{i_2 m_2} \chi^{i_1}_{m_1}(x)\, \chi^{i_2}_{m_2}(x')\, g^{i_1 i_2}_{m_1 m_2}(t, t') , \qquad x, x' \in \mathcal{I} , \tag{4}$$

with time-dependent complex coefficients $g^{ii'}_{mm'}(t, t')$ and real basis functions $\chi^i_m(x)$. Outside the interval $\mathcal{I}$, we assume the NEGF to vanish. All indices i (superscripts), ranging $0, 1, \ldots, n_e - 1$ in Eq. (4), are linked to a grid of length x_0 composed of n_e finite elements, see Fig. 1. All indices m (subscripts), ranging $0, 1, \ldots, n_g - 1$, are connected to locally defined basis functions $\chi^i_m(x)$ which are constructed as follows: First, we divide the interval $\mathcal{I}$ into finite elements with boundaries $x^0 = 0 < x^1 < x^2 < \ldots < x^{n_e-1}, x^{n_e} = x_0$. To each FE $[x^i, x^{i+1}]$, we then attach a local DVR basis using the generalized Gauss-Lobatto (GGL) points [25] x^i_m and weights w^i_m:

$$x^i_m = \frac{1}{2}\left\{\left(x^{i+1} - x^i\right) x_m + \left(x^{i+1} + x^i\right)\right\} , \tag{5}$$

$$w^i_m = \frac{w_m}{2}\left(x^{i+1} - x^i\right) , \tag{6}$$

with the standard Gauss-Lobatto points x_m (and weights w_m). For the special case of Legendre interpolating functions, the points x_m are defined as roots of the first derivative of Legendre polynomials $P_n(x)$ according to

$$\left.\frac{\mathrm{d}}{\mathrm{d}x} P_{n_g}(x)\right|_{x=x_m} = 0 , \tag{7}$$

and the weights w_m are given by

$$w_m = \frac{2}{n_g(n_g+1)[P_{n_g}(x_m)]^2} , \tag{8}$$

Journal of Physics: Conference Series **220** (2010) 012020 doi:10.1088/1742-6596/220/1/012020

where n_g denotes the total number of basis functions per element. In our approach, we keep the number of basis functions constant in each FE [47]. However, in order to combine the locally defined DVR basis functions to a continuous basis set, the FE-DVR space is being spanned by two classes of functions—'bridge' and 'element' functions, see also Ref. [29]. The bridge function (the case $m = 0$ in Eq. (9)) extends over two adjacent FEs and ensures *communication* between the grid domains i and $i+1$. In particular, it guarantees the continuity of the NEGF. The 'element' functions are zero at and outside the respective element boundaries. Using definitions (5) and (6), the basis functions have the explicit form

$$\chi^i_m(x) = \begin{cases} \dfrac{f^i_{n_g-1}(x) + f^{i+1}_0(x)}{\sqrt{w^i_{n_g-1} + w^{i+1}_0}} & , m = 0 \quad (\text{'bridge' function}) \\[2ex] \dfrac{f^i_m(x)}{\sqrt{w^i_m}} & , \text{else} \quad (\text{'element' function}) \end{cases} , \tag{9}$$

and are orthonormal in the sense of the generalized Gauss-Lobatto quadrature. In Eq. (9), the so-called Lobatto shape functions[25, 48] $f^i_m(x)$ are defined as

$$f^i_m(x) = \begin{cases} \prod_{\bar{m}\neq m} \dfrac{x - x^i_{\bar{m}}}{x^i_m - x^i_{\bar{m}}} & , x^i \le x \le x^{i+1} \\[2ex] 0 & , x < x^i \quad \text{and} \quad x > x^{i+1} \end{cases} , \tag{10}$$

and obey $f^i_m(x^{i'}_{m'}) = \delta_{ii'}\delta_{mm'}$. As in the last finite element $n_e - 1$ (i.e. in the element $[x^{n_g-1}, x_0]$) no bridge function is needed due to the boundary condition of vanishing $G(1,1')$ outside the interval I, the total FE-DVR set consists of

$$n_b = n_e n_g - 1 \,. \tag{11}$$

basis functions. Finally, we note, that a generalization of Eq. (9) to higher dimensions (2D and 3D) is, in principle, possible by using a product ansatz for the coordinate functions [27].

2.2. Matrix elements of the kinetic, potential and interaction energy operators

With representation (4) of the NEGF, the SKKBE will transform into an equation of motion for the matrix $\mathbf{g}(t,t')$ with elements $g^{i_1 i_2}_{m_1 m_2}(t,t')$, cf. Sec. 2.3. Obviously, this equation involves also the kinetic, potential and interaction energy operator of Eq. (1) in matrix form, which we specify in the following. Thereby, integrations over coordinate space are performed by using the generalized Gauss-Lobatto quadrature rule, and case differentiations arise from the basis functions $\chi^i_m(x)$ being split into element and bridge functions.

First, let us consider the potential and the kinetic energy in FE-DVR representation: The potential-energy matrix is given by

$$\begin{aligned} v^{i_1 i_2}_{m_1 m_2}(t) &= \int_0^{x_0} \mathrm{d}x\, \chi^{i_1}_{m_1}(x)\, v(x,t)\, \chi^{i_2}_{m_2}(x) \\ &= \delta_{i_1 i_2}\, \delta_{m_1 m_2}\, \tilde{v}^{i_1}_{m_1}(t) \,, \end{aligned} \tag{12}$$

with

$$\tilde{v}^i_m(t) = \begin{cases} v(x^i_m, t) & , m > 0 \\[2ex] \dfrac{v(x^i_{n_g-1}, t)\, w^i_{n_g-1} + v(x^{i+1}_0, t)\, w^{i+1}_0}{w^i_{n_g-1} + w^0_{i+1}} & , m = 0 \end{cases} . \tag{13}$$

This implies that the potential energy is diagonal with respect to elements i and local DVR basis indices m, and that, consequently, it can be represented by a vector of dimension n_b. Moreover, Eq. (12) holds true also for any other local operator. As the operator of the kinetic energy is non-local in coordinate space, the matrix elements $t^{i_1 i_2}_{m_1 m_2}$ are not diagonal. We follow the derivation of Ref. [25] and obtain the kinetic-energy matrix as

$$
\begin{aligned}
t^{i_1 i_2}_{m_1 m_2} &= -\frac{1}{2}\int_0^{x_0} \mathrm{d}x\, \chi^{i_1}_{m_1}(x)\, \nabla^2\, \chi^{i_2}_{m_2}(x) \qquad (14)\\
&= \begin{cases}
\frac{1}{2}\,\delta_{i_1 i_2}\,\tilde{t}^{i_1}_{m_1 m_2}\left[w^{i_1}_{m_1} w^{i_1}_{m_2}\right]^{-1/2} & ,\ m_1>0,\ m_2>0\\[2ex]
\frac{1}{2}\left(\delta_{i_1 i_2}\,\tilde{t}^{i_1}_{n_g-1,m_2} + \delta_{i_1 i_2-1}\,\tilde{t}^{i_2}_{0 m_2}\right)\left[w^{i_1}_{n_g-1} + w^{i_1+1}_0\right]^{-1/2} & ,\ m_1=0,\ m_2>0\\[2ex]
\frac{1}{2}\left(\delta_{i_1 i_2}\,\tilde{t}^{i_1}_{m_1 n_g-1} + \delta_{i_1 i_2+1}\,\tilde{t}^{i_1}_{m_1 0}\right)\left[w^{i_1}_{m_1}\left(w^{i_2}_{n_g-1} + w^{i_2+1}_0\right)\right]^{-1/2} & ,\ m_1>0,\ m_2=0\\[2ex]
\dfrac{\delta_{i_1 i_2}\left(\tilde{t}^{i_1}_{n_g-1,n_g-1} + \tilde{t}^{i_1+1}_{00}\right) + \delta_{i_1 i_2-1}\,\tilde{t}^{i_2}_{0,n_g-1} + \delta_{i_1 i_2+1}\,\tilde{t}^{i_1}_{n_g-1,0}}{2\left[\left(w^{i_1}_{n_g-1} + w^{i_1+1}_0\right)\left(w^{i_2}_{n_g-1} + w^{i_2+1}_0\right)\right]^{1/2}} & ,\ m_1=m_2=0
\end{cases}
\end{aligned}
$$

Here, the case differentiations lead to the matrix having a block-diagonal form [27], and the quantity $\tilde{t}^{i}_{m_1 m_2}$ is given by

$$
\tilde{t}^{i}_{m_1 m_2} = \sum_m \frac{\mathrm{d}f^i_{m_1}(x^i_m)}{\mathrm{d}x}\,\frac{\mathrm{d}f^i_{m_2}(x^i_m)}{\mathrm{d}x}\, w^i_m\,, \qquad (15)
$$

which involves the first derivatives of the Lobatto shape functions at the GGL points, see also Ref. [25].

Next, let us focus on the matrix elements of the binary-interaction operator $\hat{u}$ (the two-electron integrals) which are carrying a set of four index-pairs (i, m) and are defined by

$$
u^{i_1 i_2, i_3 i_4}_{m_1 m_2, m_3 m_4} = \int_0^{x_0}\mathrm{d}x \int_0^{x_0}\mathrm{d}x'\, \chi^{i_1}_{m_1}(x)\,\chi^{i_3}_{m_3}(x')\, u(|x-x'|)\, \chi^{i_2}_{m_2}(x)\,\chi^{i_4}_{m_4}(x')\,. \qquad (16)
$$

In a general basis approach, the two-electron integrals [Eq. (16) with all index-pairs replaced by single indices] often require a careful analysis, as they are not analytically accessible and have to be numerically precomputed for all combination of indices, e.g. [4]. Although symmetry relations [49] help to restrict oneself to a subset of indices, the effort scales with $\mathcal{O}(n_b^4)$ and thus can be huge for larger basis sets. On the contrary, using the FE-DVR, the evaluation of the two-electron integrals turns out to be much simpler. In particular, the integrals can be performed in a semi-analytical way such that Eq. (16) reduces to

$$
u^{i_1 i_2, i_3 i_4}_{m_1 m_2, m_3 m_4} = \delta_{i_1 i_2}\delta_{i_3 i_4}\delta_{m_1 m_2}\delta_{m_3 m_4}\,\tilde{u}^{i_1 i_3}_{m_1 m_3}\,. \qquad (17)
$$

where the kernel matrix $\tilde{\mathbf{u}}$ is symmetric and follows as

$$
\tilde{u}^{i_1 i_2}_{m_1 m_2} = \sum_{i_3 m_3} \alpha^{i_3}_{m_3}\,\beta^{i_1 i_3}_{m_1 m_3}\,\beta^{i_2 i_3}_{m_2 m_3}\,. \qquad (18)
$$

To obtain Eq. (18), we have used the separable form of the discretized interaction potential $u(|x-x'|)$, and, correspondingly, the quantities α^i_m denote the eigenvalues of the matrix

$$
U_{(im)(i'm')} = u(|x^i_m - x^{i'}_{m'}|) = \sum_{i_3 m_3} \alpha^{i_3}_{m_3}\,\tilde{\beta}^{m_3 m}_{i_3 i}\,\tilde{\beta}^{m_3 m'}_{i_3 i'}\,, \qquad (19)
$$

and $\beta^{ii'}_{mm'}$ are related to the eigenvectors $\tilde{\beta}^{ii'}_{mm'}$:

$$\beta^{ii'}_{mm'} = \begin{cases} \tilde{\beta}^{i'i}_{m'm} & , m > 0 \\ \dfrac{\tilde{\beta}^{i'i}_{m'(n_g-1)} w^i_{n_g-1} + \tilde{\beta}^{i'(i+1)}_{m'0} w^{i+1}_0}{w^i_{n_g-1} + w^0_{i+1}} & , m = 0 \end{cases} . \tag{20}$$

In comparison with any single-electron matrix element (such as the kinetic or potential energy), in FE-DVR, the calculation of the binary-interaction matrix elements involves just an additional but numerically elementary matrix diagonalization. Furthermore, besides the fact that with Eq. (17) the two-electron integrals attain a very simple form independent of the specific pair-interaction potential, only a single matrix of dimension $n_b \times n_b$ needs to be stored in the code. This memory-friendly property is based on the high degree of diagonality determined by the product of Kronecker deltas and represents a main attractive feature of the FE-DVR representation. Particularly, this aspect opens the way towards efficient NEGF calculations, since Eq. (17) has direct consequences for the structure of the self-energies, see the following Section.

2.3. Equations of motion

Once all relevant matrix elements are known with respect to the chosen FE-DVR basis (9), we can start to solve the Schwinger/Keldysh/Kadanoff-Baym equations (3) for the one-particle Green's function $G(1,1')$ expanded in the form of Eq. (4). This implies, though, the SKKBE in the matrix form of the finite-element discrete variable representation:

$$\sum_{im} \left\{ \mathrm{i}\partial_t \delta^{i_1 i}_{m_1 m} - h^{i_1 i}_{m_1 m}(t) \right\} g^{i i_2}_{m m_2}(t,t') = \delta_{\mathcal{C}}(t-t') + \sum_{im} \int_{\mathcal{C}} \mathrm{d}t_2\, \Sigma^{i_1 i}_{m_1 m}(t,t_2)\, g^{i i_2}_{m m_2}(t_2,t') , \tag{21}$$

$$\Sigma^{i_1 i_2}_{m_1 m_2}(t,t') = \delta_{\mathcal{C}}(t-t')\, \Sigma^{\mathrm{HF},i_1 i_2}_{m_1 m_2}(t) + \Sigma^{\mathrm{corr},i_1 i_2}_{m_1 m_2}(t,t') , \tag{22}$$

where we have denoted $\delta^{ii'}_{mm'} = \delta_{ii'}\delta_{mm'}$, $\mathbf{h}(t) = \mathbf{t} + \mathbf{v}(t)$, and Eq. (21) has to be supplied with its adjoint equation. Further, Eq. (22) separates the self-energy matrix $\Sigma^{i_1 i_2}_{m_1 m_2}(t,t')$ into a time-local Hartree-Fock part ($\mathbf{\Sigma}^{\mathrm{HF}}$) and a contribution $\mathbf{\Sigma}^{\mathrm{corr}}$ that accounts for electron-electron (e-e) correlation and memory effects. However, as an exact treatment of e-e correlations is impractical, we have to apply many-body approximations for which the second Born diagrams provides one of the most basic models. Hence, besides the general form of the HF self-energy

$$\Sigma^{\mathrm{HF},i_1 i_2}_{m_1 m_2}(t) = -\mathrm{i} \left\{ \sigma\, \delta^{i_1 i_2}_{m_1 m_2} \sum_{i_3 m_3} g^{i_3 i_3}_{m_3 m_3}(t,t^+)\, \tilde{u}^{i_1 i_3}_{m_1 m_3} - g^{i_2 i_1}_{m_2 m_1}(t,t^+)\, \tilde{u}^{i_2 i_1}_{m_2 m_1} \right\} , \tag{23}$$

where t^+ denotes $t \to t + \epsilon_{\geq 0}$ we, in second Born approximation, have

$$\Sigma^{\mathrm{corr},i_1 i_2}_{m_1 m_2}(t,t') = \sum_{i_3 m_3} \sum_{i_4 m_4} \left\{ \sigma\, g^{i_1 i_2}_{m_1 m_2}(t,t')\, g^{i_3 i_4}_{m_3 m_4}(t,t') - g^{i_1 i_4}_{m_1 m_4}(t,t')\, g^{i_3 i_2}_{m_3 m_2}(t,t') \right\} \times$$
$$\times\; g^{i_4 i_3}_{m_4 m_3}(t',t)\, \tilde{u}^{i_1 i_4}_{m_1 m_4}\, \tilde{u}^{i_2 i_3}_{m_2 m_3} . \tag{24}$$

Eqs. (23) and (24) involve the spin degeneracy factor $\sigma \in \{1,2\}$, the matrix elements $\tilde{u}^{i_1 i_2}_{m_1 m_2}$ of Eq. (17) and show a very simple form compared to the situation when a general basis is applied, e.g. [4]. The reason for this is the subtle structure of the FE-DVR. In detail, the Hartree term is completely diagonal [including a single sum over n_b elements] and the exchange term involves only a product of two matrix elements. For the second-order Born terms, the degree of simplification is even more drastic: In a general basis representation, two sums are required for each full vertex point in the second-order self-energy diagrams and, additionally, a single sum is needed for the start- and the end-point. This leads to a scaling with $\mathcal{O}(n_b^6)$. However, in our case, due to the diagonality of the two-electron integrals, cf. Eq. (17), the evaluation reduces remarkably to a scaling with $\mathcal{O}(n_b^2)$ per matrix element.

For the atomic and molecular model calculations to be outlined as first benchmarks of the method in Sec. 3, we, in this contribution, restrict ourselves to the (equilibrium) ground-state properties. For completeness, we give a short summary of the main computational steps. First, in the FE-DVR picture, the Hartree-Fock equilibrium Green's function, denoted $\mathbf{g}^0(\tau)$, follows from

$$h^{0,i_1i_2}_{m_1m_2}[\mathbf{g}^0(0^-)] = t^{i_1i_2}_{m_1m_2} + v^{i_1i_2}_{m_1m_2} + \Sigma^{0,i_1i_2}_{m_1m_2}[\mathbf{g}^0(0^-)]\,, \tag{25}$$

where $g^{0,i_1i_2}_{m_1m_2}(\tau)$ is the HF approximation of $g^{i_1i_2}_{m_1m_2}(t,t')|_{t-t'=0+\mathrm{i}\tau}$ with $\tau \in [-\beta, 0]$ and β is the inverse temperature, and Σ^0 is defined via Eq. (23), but with $\mathbf{g}$ being replaced by matrix $\mathbf{g}^0$. Thereby, Eq. (25) has to be solved to self-consistency by iteration and as result we obtain

$$g^{0,i_1i_2}_{m_1m_2}(\tau) = \sum_{im} c^{ii_1}_{mm_1}\, f_\beta(\epsilon^i_m - \mu)\, e^{-\tau(\epsilon^i_m-\mu)} c^{ii_2}_{mm_2}\,. \tag{26}$$

Here, the vector ϵ^i_m contains the energy eigenvalues of $\mathbf{h}^0$, the matrix $c^{ii'}_{mm'}$ summarizes the corresponding eigenvectors, and the chemical potential μ is determined by normalization of the Fermi distribution: $N = \sum_{im} f_\beta(\epsilon^i_m - \mu)$. Eq. (26) solves a simple differential equation, see e.g. [4], and corrections due to e-e correlations in second Born approximation are obtained by insertion into the Dyson equation [43] (the SKKBE in the limit $t - t' = 0 + \mathrm{i}\tau$) for the full Matsubara Green's function $\mathbf{g}^M(\tau)$:

$$\begin{aligned}
g^{M,i_1i_2}_{m_1m_2}(\tau) &= g^{0,i_1i_2}_{m_1m_2}(\tau) + I^{(2)\,i_1i_2}_{m_1m_2}(\tau)\,, \\
I^{(2)\,i_1i_2}_{m_1m_2}(\tau) &= \sum_{im}\int_0^\beta \mathrm{d}\bar\tau\, g^{0,i_1,i}_{m_1m}(\tau-\bar\tau)\, I^{(1)\,ii_2}_{mm_2}(\bar\tau)\,, \\
I^{(1)\,i_1i_2}_{m_1m_2}(\tau) &= \sum_{im}\int_0^\beta \mathrm{d}\bar\tau\, \{\Sigma^{M,i_1i}_{m_1m}(\tau-\bar\tau) - \delta(\tau-\bar\tau)\,\Sigma^{0,i_1i}_{m_1m}\}\, g^{M,ii_2}_{mm_2}(\bar\tau)\,,
\end{aligned} \tag{27}$$

where $\mathbf{\Sigma}^M(\tau) = \delta(\tau)\mathbf{\Sigma}^{\mathrm{HF}}(\tau) + \mathbf{\Sigma}^{\mathrm{corr}}(\tau)$ with $\mathbf{g}^M$ instead of $\mathbf{g}$ in Eqs. (23) and (24). The convolution integrals in Eq. (27) are performed in sequence by direct integration. This has been found to be more stable and controllable than the method applied before in Refs. [2, 4]. Once the self-consistent $\mathbf{g}^M(\tau)$ is computed, we have direct access to many observables, e.g. the one-electron density is obtained as $n(x) = \sum_{i_1m_1}\sum_{i_2m_2} \chi^{i_1}_{m_1}(x)\chi^{i_2}_{m_2}(x) g^{M,i_1i_2}_{m_1m_2}(0^-)$. The total energy is computed similar as in Refs. [43, 4]. Overall, in order to ensure the atoms and molecules being in the ground state, we set the inverse temperature $\beta = 100$.

3. Performance for atomic and molecular models

As first benchmarks and preparatory work for the investigation of the temporal evolution of small atoms and/or molecules following an external (e.g. laser-induced) perturbation, we here consider their equilibrium-(initial-)state preparation within the FE-DVR context. As examples, we focus on the He atom and the linear molecular ion H_3^+, modeled in one spatial dimension. For both two-electron systems, the Coulomb potential is considered in the regularized form $u(|x - x'|) = [(x - x')^2 + 1]^{-1/2}$ which, from the physical point of view, allows for a transverse extension of the few-particle wave function. Furthermore, e-e correlations are treated in second Born approximation.

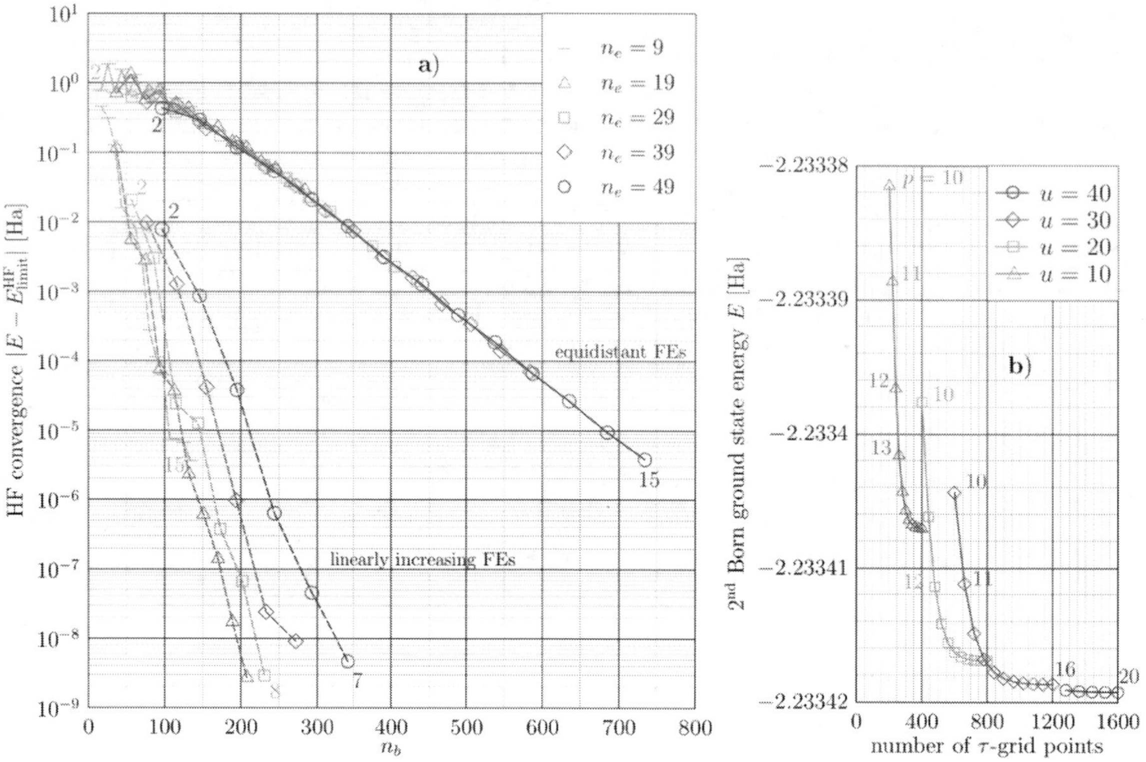

Figure 2. **a**) Convergence of the Hartree-Fock ground-state energy for the 1D helium atom against basis dimension n_b. As indicated, each curve corresponds to a different arrangement of the FEs within a total interval of 200 a.u. length. For the upper set of lines the FEs are equally distributed, whereas for the lower set the central FE is one atomic unit wide, and the width of neighboring elements is linearly increasing towards the boundaries. In addition, the number of local DVR basis functions has been varied between $n_g = 2$ and 15, see numbers on the curves. **b**) He ground-state energy in second Born approximation ($n_b = 202$) with respect to different τ-grid parameters u and p.

3.1. The 1D helium atom

The helium atom is the most elementary closed-shell two-electron system. In one spatial dimension, it is well modeled by the nucleus potential $v(x) = -Z\,[(x - x_0/2)^2 + \rho]^{-1/2}$, where the atomic number is $Z = 2$, $x \in [0, x_0]$, and ρ is a regularization parameter. In this setup, the 1D helium atom serves as a fundamental 'testing ground' for multi-electron calculations [36, 37, 39] and provides many features of the single- and double-ionization dynamics including the so-called knee structure [34, 35]. Considering the singlet state, we refine the model and set $\sigma = 2$ in Eqs. (23) and (24), and used $\rho = 1$.

Further, for the equilibrium calculations, we have used a FE-DVR basis that covers a total interval of 200 a.u. length. This corresponds to a domain that is about 100 times larger than the characteristic extension of the ground-state wave function or density, cf. Ref. [29]. Such a grid extension is more than adequate to resolve the ground-state features of the model (to be discussed here) but will become crucial, when the helium atom is perturbed by external fields and, in turn, electrons start to occupy highly excited or continuum states. In this sense, our results are benchmarks also with relevance for the computation of the system's temporal evolution. In particular, a grid with an extension of about 200 a.u. should be well capable to resolve the two-electron resonance states embedded within the one-electron continuum of dipole spectra. This follows from the performance of the few-particle time-dependent Schrödinger equation (TDSE) using absorbing potentials, e.g. [50] and references therein. We note, that also in the FE-DVR approach, such imaginary one-electron potentials that damp reflections at the interval boundaries are easily implemented, just allowing the matrix elements $v^{i_1 i_2}_{m_1 m_2}$ in Eq. (12) to be complex.

The explicit partitioning of the interval into finite elements has been organized as follows: (case I) the interval is divided into equidistant segments, (case II) the central FE has a width of one atomic unit and the width of the surrounding elements is linearly increasing towards the interval boundaries. The effect of these segmentations on the Hartree-Fock ground-state energy convergence of the helium atom is displayed in Fig. 2 **a**) as function of the total basis size. In case I, for equidistant FEs, the convergence is relatively slow with n_b and strongly depends on the number of elements as well as on the number of local DVR basis functions used (see the different symbols and lines). Consequently, more than $n_b > 550$ functions are needed for the HF energy to deviate by less than 10^{-4} Ha from the converged HF result $E^{\mathrm{HF}}_{\mathrm{limit}} = -2.2242096$ Ha. In case II, the situation is completely different as essentially more basis functions are available in the center region of the interval. This enables a superior representation of the Matsubara Green's function $G^M(x, x'; \tau)$ in coordinate space, and leads to adequate convergence at 200 to 300 FE-DVR basis functions with an error reduced by several orders of magnitude compared to case I. Furthermore, the ground-state energy depends less on the number of elements. For completeness, we note that, besides the total energy, an additional indicator for the basis quality, is to look at how well the potential $v(x)$ can be expanded into the chosen FE-DVR basis.

For case II with $n_e = 29$ and $n_g = 7$, we have computed the ground-state energy in second Born approximation. Thereby, all quantities are found to be well converged with respect to the basis size, compare with the HF case in Fig. 2 **a**). The time-argument τ in the Matsubara Green's function ($\tau \in [-\beta, 0]$) has been discretized using a uniform power mesh [51] with parameters u and p—for definition see e.g. Refs. [4, 43]. Fig. 2 **b**) indicates the convergence with respect to these parameters, where the total number of τ-grid points is given by $2up + 1$. At $\beta = 100$, a mesh parameter $p \geq 10$ ensures the particle number $N = \sum_{im} g^{M,ii}_{mm}(0^-)$ being sufficiently stable during iteration of the Dyson equation (27). Particularly, with more than 1000 grid points it is possible to compute the ground-state energy to relatively high precision, $E^{\mathrm{2ndB}} = -2.233419$ Ha. As result, the electron-electron correlations lower the total energy accounting for 66% of the correlation energy and, hence (improving the HF result), approaches the exact ground-state energy which is -2.2382578 Ha. For the discussion of other observables such as the one-electron density in HF and second Born approximation, the reader is referred to Ref. [29].

3.2. The linear molecule H_3^+

With more than two nuclei, the molecular ion H_3^+ can, in 1D, only be realized in its linear version [52], where the H-H bonds are oriented parallel to each other. Hence, the electrons move along the molecular axis and the one-electron potential is modeled as

$$v_d(x) = \left[(x-(x_0-d)/2)^2+1\right]^{-1/2} + \left[(x-(x_0+d)/2)^2+1\right]^{-1/2} + \left[x^2+1\right]^{-1/2} + \frac{5}{d}\,, \quad (28)$$

where $d/2$ denotes the interatomic distance, and the last term (offset) collects all internuclei interaction energy contributions. For the ground-state NEGF calculations, the coordinate space has been constrained to an interval of 50 a.u. with a grid of $n_e = 13$ FE being linearly increasing, starting from a one atomic unit wide central element. In total, $n_b = 142$ FE-DVR basis functions have been used.

The total binding energy E_b of the singlet state is shown in Fig. 3 **a**) against distance d for the exact solution of the TDSE (dotted [and triangles]), the Hartree-Fock (dashed) and the

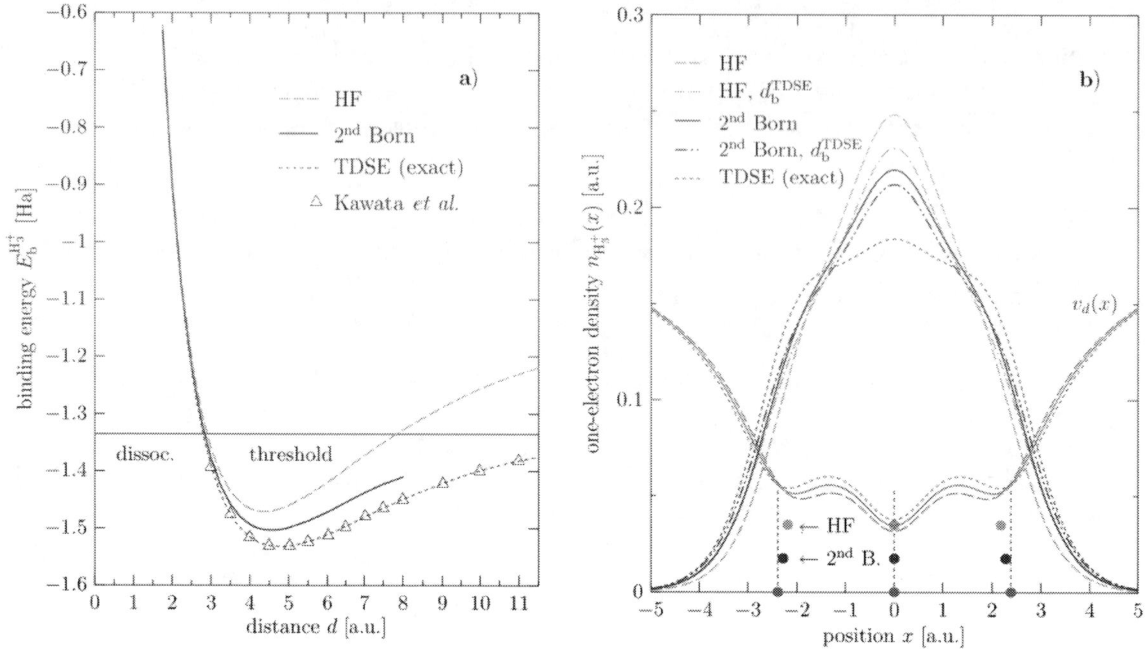

Figure 3. **a**) Binding-energy curves for the linear H_3^+ molecule in one spatial dimension for different approximations. The dotted line corresponds to the exact solution obtained from the few-particle time-dependent Schrödinger equation (TDSE). The triangles denote the binding energy of Ref. [40]. **b**) One-electron density at the equilibrium bond-lengths $2d_b$: 4.3654 (HF), 4.5579 (second Born), and 4.7698 (exact). The dots indicate the self-consistent positions of the nuclei. The gray curves indicate the corresponding potential $v_d(x)$ [scaled by 0.35 and shifted].

second Born approximation (solid). Over a broad range of internuclear distances, the second Born approximation, thereby, accounts for about 60-70% of the correlation energy, and—leading overall to a larger bond—essentially improves the HF result.

Furthermore, with a value of -1.3396 Ha, H_3^+ has the same dissociation threshold as the 1D hydrogen molecule [29], but, in addition, leaves behind a positively charged hydrogen ion. However, we note, that the Hartree-Fock and the second Born approximation, cannot resolve this threshold. The reason is that the H_3^+ molecule dissociates into open-shell fragments—two hydrogen atoms and a single hydrogen ion. These cannot be represented within a spin-restricted calculation (with $\sigma = 2$), and, hence, lead to a strong deviation of E_b in the limit of large d. Regardless of this failure of the ansatz, the NEGF calculations are well capable to describe the behavior of E_b around the minima in the binding energy curves. Also, the equilibrium positions are consistently reproduced, and the correct trend is observed when e-e correlations are being included: the bond-length d_b shifts to larger nuclear separations, for the specific values obtained see caption of Fig. 3.

The electron ground-state density of the molecular ion is displayed in Fig. 3 **b**) together with schematic curves for the spatial potentials $v_d(x)$, where d is twice the equilibrium internuclear distance, cf. Eq. (28). In HF approximation (dashed curve), the density shows a pronounced maximum in the central region of the molecule, whereas the exact density (dotted curve) is essentially less peaked. However, also the exact result does not indicate onset of electron localization, i.e. does not show separated maxima in $n_{H_3^+}(x)$. In second Born approximation, corresponding to a lower total energy (cf. Fig. 3 **a**)), we resolve the correct trend of this density reduction. Moreover, replacing the self-consistent bond-lengths by the corresponding length obtained from the TDSE (dash-dotted curves) only slightly improves the results for the Hartree-Fock and the second Born approximation. This explains that the substantial differences in the density profiles are unambiguously due to e-e correlation effects.

4. Conclusion

The FE-DVR ansatz (4) provides an elegant and very efficient way to treat binary interactions in NEGF calculations for inhomogeneous quantum systems. To this end, the method uses a flexible combination of grid (FE) and basis (DVR) strategies, which allows for simple, (semi)-analytical formulas for the matrix elements of the kinetic-, potential- and, especially, the interaction-energy operator, cf. Sec. 2.2. Further, for the most basic model that accounts for particle-particle correlations—the second Born approximation—the use of a FE-DVR basis enables remarkable scaling properties: Only $\mathcal{O}(n_b^2)$ summations are required for the computation of a single matrix element of the second-order self-energy instead of $\mathcal{O}(n_b^6)$ summations that are needed in a general basis approach.

Also, we emphasize that the FE-DVR space can, e.g. via finite-element variations, be well adjusted to the problem considered and, thus, allows for an efficient expansion of the NEGF [also for spatially extended hamiltonians] and quick convergence of the observables of interest. This has been exemplified for the He atom and the triatomic molecule H_3^+ in Sec. 3.1 and 3.2. In addition, the method is found to be stable also for large grids [large basis sets with $n_b > 500$] and also for larger particle numbers $N \leq 20$, considering interacting fermions in a harmonic trap potential.

Finally, we believe that the FE-DVR method is attractive also for other classes of many-body approximations, such as GW or T-matrix calculations, as it will likewise simplify the computation of self-energy contributions of higher than second order. Moreover, though we, in this contribution, focused on the solution of the Dyson equation, the formalism presented is, in particular, well applicable in nonequilibrium situations solving the full two-time SKKBEs which will be demonstrated in a forthcoming publication.

References

[1] A. Stan etal, Europhys. Lett. **76**, 298 (2006).
[2] N.E. Dahlen, and R. van Leeuwen, Phys. Rev. Lett. **98**, 153004 (2007).
[3] K. Balzer, and M. Bonitz, J. Phys. A: Math. Theor. **42**, 214020 (2009).
[4] K. Balzer, M. Bonitz, R. van Leeuwen, A. Stan, and N.E. Dahlen, Phys. Rev. B **79**, 245306 (2009).
[5] P. Myöhänen, A. Stan, G. Stefanucci, and R. van Leeuwen, Europhys. Lett. **84**, 67001 (2008).
[6] K.S. Thygesen, Phys. Rev. Lett. **100**, 166804 (2008).
[7] M. Puig von Friesen, C. Verdozzi, and C.-O. Almbladh, Phys. Rev. Lett. **103**, 245306 (2009).
[8] P.C. Martin, and J. Schwinger, Phys. Rev. **115**, 1342 (1959).
[9] L.V. Keldysh, Zh. Eksp. Teor. Fiz. **47**, 1515 (1964) [Sov. Phys. JETP **20**, 1018 (1965)].
[10] L.P. Kadanoff, and G. Baym, *Quantum Statistical Mechanics* (Benjamin, New York, 1962).
[11] P. Danielewicz, Ann. Phys. (N.Y.) **152**, 305 (1984).
[12] H.S. Köhler, Phys. Rev. C **51**, 3232 (1995).
[13] P. Bożek, Phys. Rev. C **56**, 1452 (1997).
[14] N.H. Kwong, and M. Bonitz, Phys. Rev. Lett. **84**, 1768 (2000).
[15] M. Bonitz, D. Kremp, D.C. Scott, R. Binder, W.D. Kraeft, and H.S. Köhler, J. Phys.: Cond. Mat. **8**, 6057 (1996).
[16] D. Semkat, D. Kremp, and M. Bonitz, Phys. Rev. E **59** 1557 (1999).
[17] D. Semkat, D. Kremp, and M. Bonitz, J. Math. Phys. **41**, 7458 (2000).
[18] R. Binder, H.S. Köhler, and M. Bonitz, Phys. Rev. B **55**, 5110 (1997).
[19] N.H. Kwong, M. Bonitz, R. Binder, and S. Köhler, phys. stat. sol. (b) **206**, 197 (1998).
[20] W. Schäfer, J. Opt. Soc. Am. B **13**, 1291 (1996).
[21] L. Banyai, H. Haug, and P. Gartner, Eur. Phys. J. B **1**, 209 (1998).
[22] P. Gartner, J. Seebeck, and F. Jahnke, Phys. Rev. B **73**, 115307 (2006).
[23] D. Hochstuhl, K. Balzer, S. Bauch, and M. Bonitz, Physica E **42**, 513 (2010).

[24] M. Bonitz, D. Hochstuhl, S. Bauch, and K. Balzer, Contrib. Plasma Phys. **50**, 54 (2010).
[25] T.N. Rescigno, and C.W. McCurdy, Phys. Rev. A **62**, 032706 (2000).
[26] L.A. Collins, S. Mazevet, J.D. Kress, B.I. Schneider, and D.L. Feder, Physica Scripta **T110**, 408 (2004).
[27] B.I. Schneider, L.A. Collins, and S.X. Hu, Phys. Rev. E **73**, 036708 (2006).
[28] J. Feist, R. Pazourek, S. Nagele, E. Persson, B.I. Schneider, L.A. Collins, and J. Burgdörfer, J. Phys. B: At. Mol. Opt. Phys. **42**, 134014 (2009).
[29] K. Balzer, S. Bauch, and M. Bonitz, Phys. Rev. A **81**, 022510 (2010).
[30] M.S. Pindzola, D.C. Griffin, and C. Bottcher, Phys. Rev. Lett. **66**, 2305 (1991).
[31] R. Grobe and J.H. Eberly, Phys. Rev. A **48**, 4664 (1993); S.L. Haan, R. Grobe, and J.H. Eberly, Phys. Rev. A **50**, 378 (1994).
[32] D. Bauer, Phys. Rev. A **56**, 3028 (1997).
[33] W.-C. Liu, J.H. Eberly, S.L. Haan, and R. Grobe, Phys. Rev. Lett. **83**, 520 (1999).
[34] N.E. Dahlen, and R. van Leeuwen, Phys. Rev. A **64**, 023405 (2001).
[35] M. Lein, E.K.U. Gross, and V. Engel, Phys. Rev. Lett. **85**, 4707 (2000).
[36] J. Zanghellini, M. Kitzler, T. Brabec, and A. Scrinzi, J. Phys. B: At. Mol. Opt. Phys. **37**, 763 (2004).
[37] M. Ruggenthaler, D. Bauer, Phys. Rev. Lett. **102**, 233001 (2009).
[38] G. Tanner, K. Richter, and J.-M. Rost, Rev. Mod. Phys. **72**, 497 (2000).
[39] D. Hochstuhl, and M. Bonitz, proceedings of conference "PNGF IV", this issue (2009).
[40] I. Kawata, H. Kono, A.D. Bandrauk, Phys. Rev. A **64** 043411 (2001).
[41] N. Suzuki, I. Kawata, and K. Yamashita, Chemical Physics **338**, 348-353 (2007).
[42] J.C. Light, I.P.Hamilton, and J.V. Lill, J. Chem. Phys. **82**, 1400 (1985).
[43] N.E. Dahlen, and R. van Leeuwen, J. Chem. Phys. **122**, 164102 (2005).
[44] N.E. Dahlen, A. Stan, and R. van Leeuwen, J. Phys: Conf. Ser. **35**, 324 (2006); N.E. Dahlen, R. van Leeuwen, and A. Stan, J. Phys: Conf. Ser. **35**, 340 (2006).
[45] A. Stan, N.E. Dahlen, and R. van Leeuwen, J. Chem. Phys. **130**, 224101 (2009).
[46] J.C. Light, and T. Carrington Jr., Advances in Chemical Physics **114**, 263 (2007)
[47] The generalization to different numbers of basis functions per element is straightforward, and only slightly alters the matrix elements involved.
[48] D.E. Manolopoulos and R.E. Wyatt, Chem. Phys. Lett. **152**, 23 (1988).
[49] The two-electron integrals $u_{ij,\bar{i}\bar{j}}$ are symmetric with respect to interchange of $i \leftrightarrow j$, $\bar{i} \leftrightarrow \bar{j}$ and tupels $(i,j) \leftrightarrow (\bar{i},\bar{j})$.
[50] S. Bauch, and M. Bonitz, Phys. Rev. A **78**, 043403 (2008).
[51] W. Ku and A.G. Eguiluz, Phys. Rev. Lett. **89**, 126401 (2002).
[52] J.D. Alexander, C.R. Calvert, R.B. King, O. Kelly, L. Graham, W.A. Bryan, G.R.A.J. Nemeth, W.R. Newell, C.A. Froud, I.C.E. Turcu, E. Springate, I.D. Williams, and J.B. Greenwood, J. Phys. B: At. Mol. Opt. Phys. **42**, 141004 (2009).

Quantum transport through a quantum dot: combining the scattering-states numerical renormalization group with nonequilibrium Green functions

Frithjof B. Anders and Sebastian Schmitt

Lehrstuhl für Theoretische Physik II, Technische Universität Dortmund, Otto-Hahn-Str. 4, 44221 Dortmund, Germany

E-mail: `frithjof.anders@tu-dortmund.de`

Abstract.
Scattering states fulfill the correct boundary conditions of a current carrying open quantum system. Discretizing the energy continuum of these states allows for employing Wilson's numerical renormalization group approach without violating the boundary conditions by using a finite size system. We evolve the analytically known steady-state density operator for a non-interacting quantum-system at finite bias to the full interacting problem by the time-dependent numerical renormalization group after switching on the local charging energy. Using a newly developed algorithm for steady-state nonequilibrium Green functions, we can calculate the current I as function of bias voltage V for arbitrary temperature and magnetic field. A comparison with second-order and GW Kadanoff-Baym-Keldysh results shows excellent agreement for weak interaction strength U.

1. Introduction

Quantum dots have been considered as possible building blocks for nano-electronics and for quantum information [1, 2]. Their physical properties are dominated by many-body effects at temperatures below the charging energy due to the quantization of the charge. Furthermore, such devices are fully controllable by external gate electrodes. Therefore, their investigation is of fundamental importance for our understanding of open quantum systems out of equilibrium.

The quantization of the electric charge dominates the electrical response of such quantum dots at low temperature leading to Coulomb blockade physics: for voltages smaller than the charging energy $U = E_c = e^2/2C$, where C is the classical capacitance, transport can only occur via tunneling of single-electrons coining the name single-electron transistor (SET) for such devices – see the review by Kastner [1]. This blockade of charge transport is lifted in certain parameter regimes due to the Kondo effect [3]. Even when real charge fluctuations are suppressed, the remaining spin degree of freedom in the Coulomb blockade valley can fluctuate due to virtual spin-flip transitions at odd fillings of a quantum dot mediating charge transfer between the leads. As a consequence, a pinned many-body resonance at the chemical potential opens a new transport channel below a characteristic temperature T_K which is exponentially dependent on the charging energy.

Journal of Physics: Conference Series **220** (2010) 012021 doi:10.1088/1742-6596/220/1/012021

By now the properties of such nano-devices are well understood at zero bias in the crossover regime from high to low temperatures using non-perturbative approaches such as Wilson's numerical renormalization group (NRG) [4]. In contrast to equilibrium conditions, the understanding of the real-time evolution and the steady-state nonequilibrium of many-particle quantum systems is still at its infancy. Over the past 40 years, the Keldysh technique [5] has proven to be the most successful approach to nonequilibrium dynamics as it furbishes a perturbative expansion for correlation functions. It is applicable to problems having a well defined ground state in equilibrium such as in semi-conductor quantum optics [6] where the band gaps act as a cutoff for any divergence in perturbation theory.

Quantum impurity models are used in the theory of quantum transport on the molecular level. Since most quantum impurity models exhibit infrared divergencies in the perturbation theory – for instance a local spin coupled anti-ferromagnetically to the spin of a metallic host [7] – the Keldysh diagrammatic approaches are only applicable in certain parameter regimes usually at high temperature, large bias or weak interactions. In these regimes, the infrared divergencies are cut off by temperature or voltage, both causing decoherence and a finite memory time. This is true, in particular, for the single impurity Anderson model [7]: the minimal model to describe Coulomb blockade physics and its lifting via a many-body resonance. Simple equation of motion techniques also fail for these type of problems [7] since they are not able to capture the correct infrared physics which develops through correlations in all orders.

Summing cumulants used in Hubbard decoupling schemes [8, 9, 10] to infinite order yields conserving approximations such as the Non-Crossing-Approximation (NCA). The NCA [11, 12, 13] and their extensions [14] based on skeleton expansion become unreliable at low temperatures and large magnetic fields [15] since the logarithmically divergence terms are only partially included. However, the combination of NCA and Keldysh techniques pioneered by Meir and Wingreen [16] has been a standard approach for calculating quantum transport in strongly correlated systems with ballistic leads in the 90ths [17]: for large values of U it is better suitable than recent GW approaches [18] since it contains (i) exactly the atomic/molecular limit with arbitrary Coulomb repulsions and (ii) also predicts the correct low-energy scale [19].

For a certain topology of the couplings, the current is fully determined by the nonequilibrium (NEQ) spectral functions of the impurity [20, 21, 16]. For large voltage V ($V/T_K \gg 1$) scaling functions in V/T_K have been derived for the Kondo model [22, 23]. Such generalizations of the NCA or the "poor-man's scaling" [7] to nonequilibrium can be backed in certain limits with field theoretical techniques [24]. In the Toulouse limit [25] the Kondo model has been solved exactly for finite bias [26] as well as in an AC field [27]. Using bosonisation and refermionisation, this model is mapped exactly onto a resonant level model. The Keldysh NCA solution for pulsed bias [28] agrees well with the exact solution at the Toulouse point [29].

Recently, a large number of different methods have been developed to tackle this difficult problem for arbitrary parameter regimes. Using scattering states, perturbative methods [30], Quantum Monte Carlo methods [31] and the Bethe Ansatz approach [32] were extended to finite bias along the line of Hershfield's expression of a steady-state density operator [33]. Schmitteckert applied the time-dependent density matrix renormalization group (DMRG) [34] to different interacting 1D wires [35] which agrees well with a field theoretical solution [36]. Iterative propagation of the density matrix [37] along the Keldysh contour allows the calculation of the current in the weakly correlate regime governed by short memory times. Wegner's flow equation approach [38, 39] as well as the real-time renormalization group approach [40, 41] has been extended to a spin coupled anti-ferromagnetically to two leads.

In this paper, we report on the application of a scattering-states approach to quantum-transport though a single-electron transistor [1]. Oguri has pointed out [42] that discretizing the Lippmann-Schwinger states of a noninteracting quantum dot can serve as a single-particle basis which is current carrying. The scattering-state NRG (SNRG) approach consists of the following steps to calculate the current through a SET for arbitrary temperature, bias, magnetic field and interaction strength U:

(i) construct the single-particle scattering states of the noninteracting Hamiltonian which contain the correct boundary conditions of the current-carrying open quantum system,

(ii) use the analytically known steady-state density operator [33] of a noninteracting quantum-dot for an initial NRG run at finite bias at a given temperature and coupling to the leads,

(iii) employ the time-dependent numerical renormalization group [43, 44] to propagate the steady-state density operator to the analytically unknown density operator at finite bias and charging energy, i.e. $\hat{\rho}_0(V, U = 0, t = 0) \to \hat{\rho}(V, U, t \to \infty)$,

(iv) use $\hat{\rho}(V, U)$ to calculate the steady-state nonequilibrium spectral functions [45] which enters the Meir-Wingreen formula [20, 21, 16] for the current.

The paper is organized as follows. In Sec. 2 we introduce the model and discuss briefly the two approaches used: the scattering-states numerical renormalization group in Sec. 2.2 and the Keldysh Green function technique in Sec. 2.7. The comparison between the SNRG and the Keldysh Green functions for small values of U are discussed in Sec. 3.1. The IV characteristics of a quantum dot in the Kondo regime is presented in Sec. 3.2. While the zero-bias conductance remains independent of U for a symmetric junction for $T \to 0$, the IV curves are strongly U dependent for finite bias.

2. Theory

Quantum impurity models (QIM) are used to describe quantum transport on the molecular level. Their Hamiltonian $\mathcal{H}$

$$\mathcal{H} = \mathcal{H}_{imp} + \mathcal{H}_{bath} + \mathcal{H}_I \tag{1}$$

consists of three parts: an impurity part $\mathcal{H}_{imp}$ modelling the interacting device with a finite number of degrees of freedom, one or more Bosonic or Fermionic baths assigned to $\mathcal{H}_{bath}$ and the coupling of these subsystems by $\mathcal{H}_I$. Examples of such quantum impurities are a simple two-level system, a localized spin, a superconducting qubit, a molecule or a quantum dot. The leads provide two such fermionic baths at different chemical potentials, while dephasing of isolated spins or qubits is usually driven by bosonic fluctuations such as fluctuating electromagnetic fields or molecular vibrations [46].

2.1. Model

In this work, we restrict ourselves to a junction with a single spinful orbital coupled to a left (L) and a right (R) lead

$$\begin{aligned}\mathcal{H} &= \sum_{\sigma\alpha=L,R} \int d\epsilon\, (\epsilon - \mu_\alpha)\, c^\dagger_{\epsilon,\sigma\alpha} c_{\epsilon,\sigma\alpha} + \sum_{\sigma=\pm 1} \left[E_d + \frac{U}{2} - \frac{\sigma}{2} H\right] \hat{n}^d_\sigma \\ &\quad + \frac{U}{2}\left(\sum_\sigma \hat{n}^d_\sigma - 1\right)^2 + \sum_{\alpha\sigma} V_{\alpha\sigma} \int d\epsilon\, \sqrt{\rho_\alpha(\epsilon)} \left(d^\dagger_\sigma c_{\epsilon\sigma\alpha} + c^\dagger_{\epsilon\sigma\alpha} d_\sigma\right).\end{aligned} \tag{2}$$

Here $\hat{n}^d_\sigma = d^\dagger_\sigma d_\sigma$ measures the orbital occupancy, H denotes the external magnetic field and U the local charging energy if the quantum dot level is doubly occupied. The Hartree term has been partially absorbed into the single-particle energy $E_d \to E_d + U/2$. The different chemical potentials μ_α in both leads appear as a shift of the band centers and are functions of the external voltage $V = \mu_R - \mu_L$. For simplicity, we assume that both leads have the same density of states $\rho(\epsilon)$ characterized by the same band width D but different band centers. This Hamiltonian is commonly used to model a single Coulomb-blockade resonance in ultra-small quantum dots [3, 16].

Journal of Physics: Conference Series **220** (2010) 012021 doi:10.1088/1742-6596/220/1/012021

2.2. *Definition of the scattering states*

In the absence of the local Coulomb repulsion H_U

$$H_U = \frac{U}{2}\left(\sum_\sigma \hat{n}^d_\sigma - 1\right)^2 , \tag{3}$$

the Hamiltonian (2) is diagonalized exactly [30, 33, 47, 31, 42, 48, 49] in the continuum limit by scattering states. These scattering-states are created by the operator $\gamma^\dagger_{\epsilon\sigma\alpha}$

$$\gamma^\dagger_{\epsilon\sigma\alpha} = c^\dagger_{\epsilon\sigma\alpha} + V_\alpha\sqrt{\rho_\alpha(\epsilon)}G^r_{0\sigma}(\epsilon)\left[d^\dagger_\sigma + \sum_{\alpha'}\int d\epsilon' \frac{V_{\alpha'}\sqrt{\rho_{\alpha'}(\epsilon')}}{\epsilon + i\delta - \epsilon'} c^\dagger_{\epsilon'\sigma\alpha'}\right] . \tag{4}$$

The local retarded resonant level Green function

$$G^r_{0\sigma}(\omega) = \left[\omega + i\delta - (E_d + \frac{U}{2} - \sigma\frac{H}{2}) - \sum_\alpha V^2_\alpha \int d\epsilon \frac{\rho_\alpha(\epsilon)}{\omega + i\delta - \epsilon}\right]^{-1} \tag{5}$$

enters as expansion coefficient. Defining $\bar{V} = \sqrt{V_L^2 + V_R^2}$, we will use $r_{R(L)} = V_{R(L)}/\bar{V}$ and

$$\Delta(\omega) = \bar{V}^2 \sum_\alpha r^2_\alpha \int d\epsilon \frac{\rho_\alpha(\epsilon)}{\omega + i\delta - \epsilon} = \Re e[\Delta(\omega)] - i\Gamma(\omega) \tag{6}$$

in the rest of the paper, where $\Gamma(0) = \Gamma = \sum_\alpha \pi V^2_\alpha \rho(0)$.

For infinitely large leads, these scattering states diagonalize the Hamiltonian (2) for $U = 0$:

$$\mathcal{H}^i_0 = \mathcal{H}(U=0) = \sum_{\alpha=L,R;\sigma} \int d\epsilon\, \epsilon \gamma^\dagger_{\epsilon\sigma\alpha}\gamma_{\epsilon\sigma\alpha} . \tag{7}$$

Note that these scattering states are complex and current carrying. This is embedded in the complex expansion coefficients $G^r_{0\sigma}(\epsilon)$ reflecting the continuum limit. Time-reversal symmetry manifests itself in the symmetric spectrum for left and right movers. To avoid any contribution from bound states, we will implicitly assume a wide band limit: $D \gg \max\{|E_d|, \Gamma, |V|, |H|\}$, where $\Gamma_\alpha = \pi V^2_\alpha \rho(0)$ and $\Gamma = \Gamma_L + \Gamma_R$.

Hershfield has shown that the density operator for such a noninteracting current-carrying quantum system retains its Boltzmannian form [33]

$$\hat{\rho}_0 = \frac{e^{-\beta(\mathcal{H}^i_0 - \hat{Y}_0)}}{\mathrm{Tr}\left[e^{-\beta(\mathcal{H}^i_0 - \hat{Y}_0)}\right]} , \hat{Y}_0 = \sum_{\alpha\sigma} \mu_\alpha \int d\epsilon\, \gamma^\dagger_{\epsilon\sigma\alpha}\gamma_{\epsilon\sigma\alpha} \tag{8}$$

even for finite bias. The $\hat{Y}_0$ operator accounts for the occupation of the left- and right-moving scattering states, and μ_α for the different chemical potentials of the leads. Therefore, all steady-state expectation values of operators can be calculated using $\hat{\rho}_0$ which includes the finite bias. In the absence of a Coulomb repulsion U, the problem is well understood. The current expectation value [42] calculated with this density-operator $\hat{\rho}_0$ reproduces the standard result [20, 21, 16] for noninteracting devices. The knowledge of the analytical form of $\hat{\rho}_0$, however, makes this steady-state model accessible to a NRG approach [4, 49].

The expansion coefficients in the definition of $\gamma^\dagger_{\epsilon\sigma\alpha}$ in Eq. (4) contain the complex single-particle Green function $G^r_{0\sigma}(\epsilon)$ which we separate in modulus and phase

$$G^r_{0\sigma}(\epsilon) = |G^r_{0\sigma}(\epsilon)|e^{-i\Phi_\sigma(\epsilon)} . \tag{9}$$

This phase $\Phi_\sigma(\epsilon)$ is absorbed into the new scattering states $\gamma^\dagger_{\epsilon\sigma\alpha} \to \tilde\gamma^\dagger_{\epsilon\sigma\alpha} = \gamma^\dagger_{\epsilon\sigma\alpha} e^{i\Phi_\sigma(\epsilon)}$ by a gauge transformation. The impurity operator $d^\dagger_\sigma$ is expanded into left and right mover contributions

$$d^\dagger_\sigma = r_R d^\dagger_{\sigma R} + r_L d^\dagger_{\sigma L} \tag{10}$$

using the inversion of Eq. (4). These two new operators are defined as

$$d^\dagger_{\sigma\alpha} = \bar V \int d\epsilon \sqrt{\rho(\epsilon)} |G^r_{0\sigma}(\epsilon)| \tilde\gamma^\dagger_{\epsilon\sigma\alpha} \; , \tag{11}$$

and obey the anti-commutator relation $\{d_{\sigma\alpha}, d^\dagger_{\sigma'\alpha'}\} = \delta_{\alpha\alpha'}\delta_{\sigma\sigma'}$.

2.3. Discretization of the scattering states

The scattering-states numerical renormalization group (SNRG) approach [49] starts from a logarithmic discretization of the scattering-states continuum $\tilde\gamma_{\epsilon\sigma\alpha}$ controlled by the parameter [50, 4] $\Lambda > 1$. The discretized version of the noninteracting Hamiltonian (7) is mapped [50, 4] onto a semi-infinite Wilson chain Hamiltonian $H_0(\Lambda)$

$$\begin{aligned} H_0(\Lambda) &= \lim_{m\to\infty} H^0_m \\ H^0_m &= \sum_{\sigma\alpha}\sum_{n=0}^{m} w_{n\sigma\alpha} f^\dagger_{n\sigma\alpha} f_{n\sigma\alpha} + \sum_{\sigma\alpha}\sum_{n=0}^{m-1} \left(t_{n\sigma\alpha} f^\dagger_{n\sigma\alpha} f_{n+1\sigma\alpha} + t^*_{n\sigma\alpha} f^\dagger_{n+1\sigma\alpha} f_{n\sigma\alpha} \right) \end{aligned} \tag{12}$$

whose tight-binding matrix elements $t_{n\sigma\alpha}$ decay exponentially $t_{n\sigma\alpha} \propto \Lambda^{-n/2}$. We use $d_{\sigma\alpha}$ defined in Eq. (11) as starting vector $f_{0\sigma\alpha} = d_{\sigma\alpha}$ for the Householder transformation [4] and obtain the tight-binding coefficients [4] of the Wilson chain (12). It is straight forward to show that the energy of the first chain link corresponds to the energy of the original quantum-dot orbital: $w_{0\sigma\alpha} = E_d + U/2 - \sigma H/2$.

In contrast to the standard NRG [50, 4], the impurity degree of freedom has been included into $H_0(\Lambda)$ since not the leads but the full scattering states have been discretized. Any complex phase of the tight-binding parameters t_n can be absorbed into the creation (annihilation) operators $f^\dagger_{n\sigma\alpha}(f_{n\sigma\alpha})$ of an electron on the chain link n with spin σ and α (=left/right) mover by a local gauge transformation.

The NRG solves $H_0(\Lambda)$ by iteratively diagonalizing the series of Hamiltonians H^0_m. Hereby, the discarding of the high-energy states at the end of each iteration circumvents the exponentially increasing dimension of the many-body Fock space: the Fock space of the next Hamiltonian H_{m+1} is spanned only by the lowest N_s eigenstates of H_m plus the degrees of freedom of the additional chain link $m+1$. The difference $\Delta H_m = H_{m+1} - H_m$ combined with rescaling of the energy scale provides the renormalization group equation of the problem [4] which is the core of the NRG iteration procedure.

In the SNRG, however, the current operator acquires energy dependent phases $\Phi_\sigma(\epsilon)$ from the gauge transformation when expanded in the scattering states. Since it remains an open question how to set up a Wilson chain for which the current operator is local, we calculate [21, 33, 42] the IV curve using the bias-dependent spectrum of the retarded Green function $\rho_\sigma(\omega, V) = -\Im m[G^r_{d\sigma}(\omega, V)]/\pi$

$$I(V) = \frac{G_0}{e}\sum_\sigma \int_{-\infty}^{\infty} d\omega \, [f(\omega-\mu_L) - f(\omega-\mu_R)] \, \pi\Gamma\rho_\sigma(\omega, V) \tag{13}$$

where $f(\omega)$ denotes the Fermi function, and $G_0 = (e^2/h)4\Gamma_L\Gamma_R/\Gamma^2$. This expression remains valid even for finite U [20, 21, 16].

2.4. Local Coulomb interaction

In order to include the local Coulomb interaction, the number operator $\hat{n}^d_\sigma = d^\dagger_\sigma d_\sigma$ must be expanded in the new orbitals $d_{\sigma\alpha}$. It consists of two contributions: a density term and a backscattering term $\hat{n}^d_\sigma = \hat{n}^0_\sigma + \hat{O}^{back}_\sigma$, where

$$\hat{n}^0_\sigma = \sum_\alpha r^2_\alpha d^\dagger_{\sigma\alpha} d_{\sigma\alpha} \tag{14}$$

and the backscattering term is defined as

$$\hat{O}^{back}_\sigma = r_L r_R \left(d^\dagger_{\sigma R} d_{\sigma L} + d^\dagger_{\sigma L} d_{\sigma R} \right). \tag{15}$$

The local Coulomb interaction term

$$U \hat{n}^d_\uparrow \hat{n}^d_\downarrow = U \left(\hat{n}^0_\uparrow \hat{n}^0_\downarrow + \sum_\sigma \hat{O}^{back}_\sigma \hat{n}^0_{-\sigma} + \hat{O}^{back}_\uparrow \hat{O}^{back}_\downarrow \right) \tag{16}$$

leads to a mixing of left and right movers, since $\hat{O}^{back}_\sigma$ does not commute with Y_0. However, the term H^0_U,

$$H^0_U = \frac{U}{2} \left(\sum_\sigma \hat{n}^0_\sigma - 1 \right)^2, \tag{17}$$

commutes with $\hat{Y}_0$ and can be absorbed into the steady-state density operator $\hat{\rho}_0 \to \tilde{\rho}_0 = \exp[-\beta(\mathcal{H}^i - \hat{Y}_0)]/Z$ with $\mathcal{H}^i = \mathcal{H}^i_0 + H^0_U$ using the arguments given in Ref. [24].

2.5. Time-dependent numerical renormalization group (TD-NRG) approach

Starting from an equilibrated system for times $t \leq 0$, the initial Hamiltonian $\mathcal{H}_i$ is changed to $\mathcal{H}_f$ by a sudden quench at $t = 0$. Then, the density matrix $\hat{\rho}(t)$ evolves as

$$\hat{\rho}(t) = e^{-i\mathcal{H}_f t} \hat{\rho}_0 e^{i\mathcal{H}_f t} \tag{18}$$

starting from its initial value $\hat{\rho}_0$ at $t = 0$. If $\mathcal{H}_{i(f)}$ describes a quantum impurity problem and $\hat{O}$ is an impurity operator, it was recently shown that the real-time dynamics of the expectation value of $O(t) = \langle \hat{O}(t) \rangle$ can be assessed [43, 44] by evaluating

$$O(t) = \sum_m \sum_{r,s}^{dis} \rho^{red}_{r,s}(m) O_{s,r} e^{-i(E^m_r - E^m_s)t} \tag{19}$$

where $O_{s,r}(m)$ denotes the matrix elements of the operator $\hat{O}$ at NRG iteration m, E^m_r the NRG eigenenergy of state $|r\rangle$ and $\rho^{red}_{r,s}$ is the reduced density matrix

$$\rho^{red}_{r,s}(m) = \sum_e \langle r, e; m | \hat{\rho}_0 | s, e; m \rangle \tag{20}$$

tracing out all degrees of freedom e of the Wilson chain links m', where $m < m' \leq N$. The sum restriction $\sum^{dis}_{r,s}$ indicates that at least one of the states $|r\rangle, |s\rangle$ must be a discarded state at iteration m: excitations between two retained states will be refined in the next iterations.

The derivation of Eq. (19) requires a complete basis set of the Wilson chain. It was shown that the set of all discarded states of the NRG truncation procedure [43, 44] provides such a

Journal of Physics: Conference Series **220** (2010) 012021 doi:10.1088/1742-6596/220/1/012021

complete basis set. The full Fock-space of the Wilson-chain of length N is spanned by this complete basis set $\{|l,e;m\rangle\}$

$$\hat{1} \;=\; \sum_m^N \sum_{l,e} |l,e;m\rangle\langle l,e;m| \tag{21}$$

where l labels a discarded state at iteration m, and e denotes the set of quantum numbers characterizing the chain links from $m+1$ to N. At each iteration m, we can also partition the Fock-space into two parts

$$\hat{1} \;=\; \underbrace{\sum_{m'}^{m-1}\sum_{l,e} |l,e;m'\rangle\langle l,e;m'|}_{=\mathcal{I}_m^-} + \underbrace{\sum_{r,e} |r,e;m\rangle\langle r,e;m|}_{=\mathcal{I}_m^+} = \mathcal{I}_m^- + \mathcal{I}_m^+ \tag{22}$$

where r accounts for all states which are present at iteration m. The derivation of Eq. (19) makes heavy use of the different possibilities of partitioning of the complete Fock-space of the Wilson chain.

When applying the time-dependent NRG, we have to iteratively diagonalize not only the initial Hamiltonian $K_i = \mathcal{H}_i - \hat{Y}_0$ but also the final Hamiltonian $\mathcal{H}_f$ (here $\mathcal{H}_f = \mathcal{H}_i + U\hat{n}^d_\uparrow \hat{n}^d_\downarrow$) using their corresponding Wilson chains (12) and the RG equation described above. The approximate eigenstates of the non-interacting $\mathcal{H}_i(N)$ obtained from the initial NRG run are used for a faithful representation of the steady-state density operator $\hat{\rho}_0$ at the inverse temperature $1/T = \beta_N \propto \Lambda^{N/2}$ and the finite bias V. The iteratively obtained eigenstates of $\mathcal{H}_f(m)$ and their overlap matrix elements with the eigenstates of $\mathcal{H}_i(m)$ allows to construct a matrix representation of $\hat{\rho}(t)$.

In contrast to the standard RG procedure which eliminates high energy degrees of freedom, the TD-NRG sums up the contributions from all discarded states via Eq. (19): the usage of a complete basis set ensures that the initial condition $O(t \to 0^+) = \mathrm{Tr}\left[\rho_0 \hat{O}\right]$ is always fulfilled up to machine precision of 10^{-15} accuracy.

2.6. Steady-state Green function

In Sec. 2.2 we have argued that the analytic form of the steady-state nonequilibrium density operator is only known explicitly [33] for $U = 0$. This allows for applying the NRG approach to construct a faithful representation of $\hat{\rho}_0(V, U = 0)$. Since switching on the local Coulomb interaction is a $1/N$ effect in the coupled system of the two leads and quantum dot, we can assume that for infinitely large leads (i) a steady-state is reached after some characteristic but finite time which is (ii) unique and independent of the initial condition. Then, the time averaging of the density operator

$$\hat{\rho}_\infty \;=\; \lim_{T\to\infty} \frac{1}{T}\int_0^T dt\hat{\rho}(t) \tag{23}$$

projects out the steady-state contributions to the density operator $\hat{\rho}(t) = \exp(-i\mathcal{H}_f t)\,\hat{\rho}_0\,\exp(i\mathcal{H}_f t)$ even in a finite size system: only the energy diagonal terms contribute in accordance with the condition $[\mathcal{H}_f, \hat{\rho}_\infty] = 0$. Even though $\hat{\rho}_\infty$ remains unknown analytically, we can construct it systematically using the TD-NRG described above.

The steady-state retarded Green function (GF) is defined as

$$G^r_{A,B}(t) \;=\; -i\mathrm{Tr}\left[\hat{\rho}_\infty [\hat{A}(t), \hat{B}]_s\right]\Theta(t), \tag{24}$$

where $[\hat{A}(t), \hat{B}]_s$ denotes the commutator for Bosonic, and the anti-commutator for Fermion correlation functions. The GF can be calculated [45] using the TD-NRG [43, 44] and extending ideas developed for equilibrium Green functions [51] which utilized the partitioning properties (22) of the complete basis set. In contrast to the equilibrium, however, two reduced-density matrices, $\rho^{red}_{r,s}(m)$ and $\tilde{\rho}^{red}_{r,s}(m)$ must be introduced [45]. One of the contributions to the (anti)-commutator can be written as

$$\begin{aligned} \mathrm{Tr}\left[\hat{\rho}_\infty \hat{A}(t) B\right] &= \sum_{m=m_{min}}^{N} \sum_{r,s}^{dis} \sum_{k} \delta_{E^m_r, E^m_s} A^m_{r,k} e^{i(E^m_r - E^m_k)t} B^m_{k,s} \rho^{red}_{s,r}(m) \\ &+ \sum_{m=m_{min}}^{N-1} \sum_{l'}^{dis} \sum_{k_1} \sum_{k_2} A^m_{k_1,l'} e^{i(E^m_{k_1} - E^m_{l'})t} B^m_{l',k_2} \tilde{\rho}^{red}_{k_2,k_1}(m) \ , \end{aligned} \tag{25}$$

where the recursion relation for the second reduced density matrix $\tilde{\rho}^{red}_{k_2,k_1}(m)$ reads

$$\begin{aligned} \tilde{\rho}^{red}_{k_2,k_1}(m) &= \sum_{r,s}^{dis} \sum_{\alpha_{m+1}} \langle k_2, \alpha_{m+1}; m | s; m+1 \rangle \rho^{red}_{s,r}(m+1) \delta_{E^{m+1}_r, E^{m+1}_s} \langle r; m+1 | k_1, \alpha_{m+1}; m \rangle \\ &+ \sum_{k',k''}^{dis} \sum_{\alpha_{m+1}} \langle k_2, \alpha_{m+1}; m | k'; m+1 \rangle \tilde{\rho}_{k',k''}(m+1) \langle k''; m+1 | k_1, \alpha_{m+1}; m \rangle \end{aligned} \tag{26}$$

which is very similar to the recursion relation obeyed by reduced density matrix $\rho^{red}_{r,s}(m)$. The matrix elements of the type $\langle k''; m+1 | k_1, \alpha_{m+1}; m \rangle$ are generated in the standard NRG algorithm. The Kronecker $\delta_{E,E'}$ projects out the energy diagonal parts of the reduced density matrix since only those contribute to $\hat{\rho}_\infty$.

Details of its derivation are found in Ref. [45]. It was shown that the algorithm is identical to the equilibrium algorithm [51] if $\mathcal{H}_i = \mathcal{H}_f$. Combining the Laplace transformation of both contributions (25) of the anti-commutator to $G^r_{A,B}(z)$ yields the steady-state spectral function for the retarded GF which is used to calculate the current via Eq. (13).

2.7. *The Kadanoff-Baym-Keldysh approach*

In order to gauge the quality of the results for the SNRG for small values of U where perturbative approaches are applicable, we employ the nonequilibrium perturbation theory as formulated by Kadanoff, Baym [52] and Keldysh [5] on the usual Keldysh time contour, for example, see Refs. [53, 54, 55]. Since we are only interested in the steady-state properties, the information and correlations of the initial conditions are assumed to be lost. This is achieved by sending the initial time $t_0 \to -\infty$ and dropping all correlation functions which involve the initial state. It is assumed, that the system reaches a steady state which is translational invariant in time. Therefore the single-particle Green function only depends on the difference between the two formerly independent times of particle creation and annihilation. The Fourier transform of the time difference leads to the formulation in frequency space for all Green functions of the steady state.

In the nonequilibrium steady-state formulation, two independent components of the contour ordered Green functions survive which are chosen to be the retarded and lesser Green functions, $G^r(\omega)$ and $G^<(\omega)$ respectively. The advanced and greater functions are related via

$$G^a(\omega) = G^r(\omega)^\dagger \tag{27}$$

$$G^>(\omega) = G^<(\omega) + G^r(\omega) - G^a(\omega) \ . \tag{28}$$

The solution of the retarded Green function at finite bias enters Eq. (13) to obtain the steady-state current.

The two relevant Green functions can be expressed as [56, 57]

$$G^r_\sigma(\omega) = \frac{1}{\omega + i\delta - E_d - \Sigma^H_\sigma - \Delta(\omega) - \Sigma^r_\sigma(\omega)} \tag{29}$$

$$G^<_\sigma(\omega) = G^r_\sigma(\omega)\Big[2if_L(\omega)\Gamma_L(\omega) + 2if_R(\omega)\Gamma_R(\omega) + \Sigma^<_\sigma(\omega)\Big]G^a_\sigma(\omega) \quad , \tag{30}$$

where, again, $\Delta(\omega) = \Delta_R(\omega) + \Delta_L(\omega)$ are the hybridization functions of the leads, $\Gamma_\alpha(\omega)$ their imaginary parts (see Eq. (6)) and $f_\alpha(\omega) = 1/\{\exp[\beta(\omega - \mu_\alpha)] + 1\}$ are the Fermi functions of the corresponding leads. The retarded and lesser self-energies, Σ^r and $\Sigma^<$ respectively, include all correlation effects induced by the Coulomb interaction U. Σ^H_σ accounts for the frequency independent Hartree energy shift.

We focus on three different approximations for the self-energy: (A) The direct expansion up to second order in U, where noninteracting $U = 0$ propagators are used as internal lines, labeled as $2^{nd}U$, (B) full self-consistent evaluation of the second order skeleton diagram, and (C) the GW approximation (GWA). In the GW approximation (GWA) the bare Coulomb interaction U is screened by an infinite series of particle-hole excitations. The approximation (B) is called second Born approximation (2BA). In contrast to the usual 2BA no exchange contribution exists for our model with only one spinful orbital. Since the 2BA and the GWA are both evaluated self-consistently, the self-energies can be derived from a Luttinger-Ward functional [58] and are conserving approximations [59].

The analytic expressions for the self-energies read

$$\Sigma^H_\sigma = U \int \frac{d\omega}{2\pi i} G^<_{\bar\sigma}(\omega) \tag{31}$$

$$\Sigma^r_\sigma(\omega) = i\int \frac{dx}{2\pi} G^<_\sigma(x)\, W^r_\sigma(\omega - x) + i\int \frac{dx}{2\pi} G^r_\sigma(x)\, W^>_\sigma(\omega - x) \tag{32}$$

$$\Sigma^<_\sigma(\omega) = i\int \frac{dx}{2\pi} G^<_\sigma(x)\, W^<_\sigma(\omega - x) \quad , \tag{33}$$

where the effective interactions are given by

$$W^r_\sigma(\omega) = U^2 P^r_{\bar\sigma}(\omega) \qquad (2BA) \tag{34}$$

$$W^r_\sigma(\omega) = \frac{U^2 P^r_{\bar\sigma}(\omega)}{1 - U^2 P^r_\sigma(\omega)\, P^r_{\bar\sigma}(\omega)} \qquad (GW) \tag{35}$$

$$W^<_\sigma(\omega) = W^r_\sigma(\omega)\, P^<_\sigma(\omega)\, W^a_\sigma(\omega) \tag{36}$$

$$W^>_\sigma(\omega) = W^r_\sigma(\omega)\, P^>_\sigma(\omega)\, W^a_\sigma(\omega) \quad , \tag{37}$$

and the particle-hole bubbles are

$$P^r_\sigma(\omega) = -i\int \frac{dx}{2\pi}\, G^r_\sigma(x)\, G^<_\sigma(x - \omega) - i\int \frac{dx}{2\pi}\, G^<_\sigma(x)\, G^a_\sigma(x - \omega) \tag{38}$$

$$P^a_\sigma(\omega) = -i\int \frac{dx}{2\pi}\, G^a_\sigma(x)\, G^<_\sigma(x - \omega) - i\int \frac{dx}{2\pi}\, G^<_\sigma(x)\, G^r_\sigma(x - \omega) \tag{39}$$

$$P^<_\sigma(\omega) = -i\int \frac{dx}{2\pi}\, G^<_\sigma(x)\, G^>_\sigma(x - \omega) \quad . \tag{40}$$

In the above expressions, the advanced and greater Green functions are determined via equations (27) and (28) and $\bar\sigma = -\sigma$ denotes the opposite spin of σ.

Equations (29)-(40) form a closed set solved self-consistently for the 2BA and GWA. For the $2^{nd}U$-approximation all particle-hole propagators (38)-(40) are evaluated only once with the

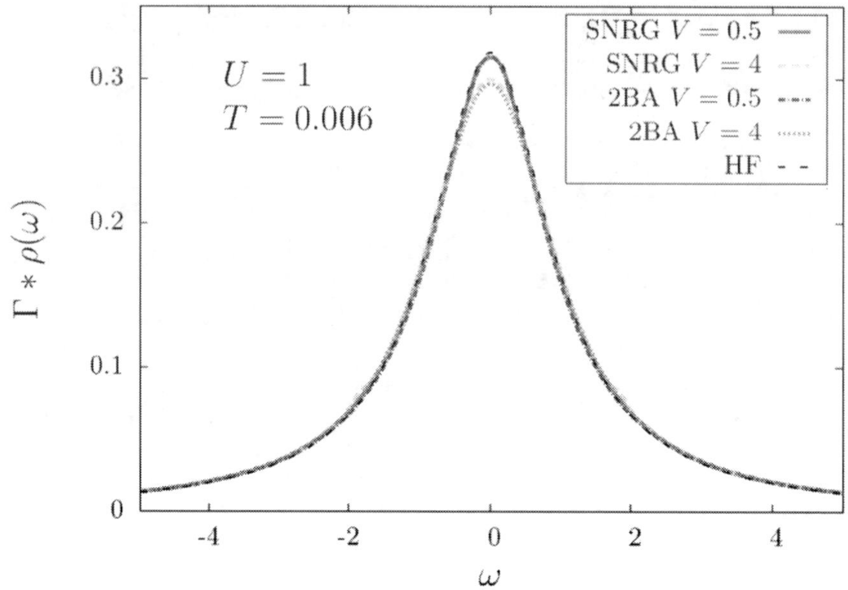

Figure 1. Taken from Ref. [60]: Comparison of the SNRG and the Keldsyh spectral functions obtained in second Born approximation (2BA) for $U/\Gamma = 1, E_d = -U/2$ for two values of the bias voltage at $T = 0.006$ and symmetric coupling $R = 1$. We add $\rho(\omega)$ obtained in Hartree-Fock (HF) approximation as comparison (dashed line).

bare Green functions

$$g^r_\sigma(\omega) = \frac{1}{\omega + i\delta - E_d - \Sigma^H_\sigma - \Delta(\omega)} \tag{41}$$

$$g^<_\sigma(\omega) = g^r_\sigma(\omega)\Big[2if_L(\omega)\Gamma_L(\omega) + 2if_R(\omega)\Gamma_R(\omega)\Big]g^a_\sigma(\omega) \quad . \tag{42}$$

Only the Hartree shift is included in order to determine the desired filling, and equation (34) is used as the effective interaction.

The GW approximation [61, 62] has been successfully applied to overcome some shortcomings of local-density calculations and estimate the screening of the Coulomb interactions in solid state physics. Recently, the method has been employed to calculate quantum transport through nanoscale devices. It furbished a conserving approximation and captures some essential many-body effects [63, 64, 57, 18]. It was shown to describe accurately the equilibrium properties in the weakly interacting regime and in asymmetric situations with a nearly empty or nearly full impurity orbital [65, 66]. In the strongly interacting Kondo regime $\Gamma - U < E_d < -\Gamma$, the GW approximation produces an exponentially narrow peak in the spectral function at the Fermi level, which could be interpreted as remnants of the expected many-body resonance [66]. However, the line shape of this low energy resonance as well as the high-energy Hubbard peaks at $\omega \approx E_d$ and $\omega \approx E_d + U$ are not reproduced by this approximation [65, 66]. Additionally, for very large interaction strength $U/\Gamma > 8$, all those perturbative approaches favor unphysical magnetic solutions: these sets of equations are close to bifurcation points which might yield unphysical hysteretic response. This indicates restricted applicability of such approach to small values of U/Γ.

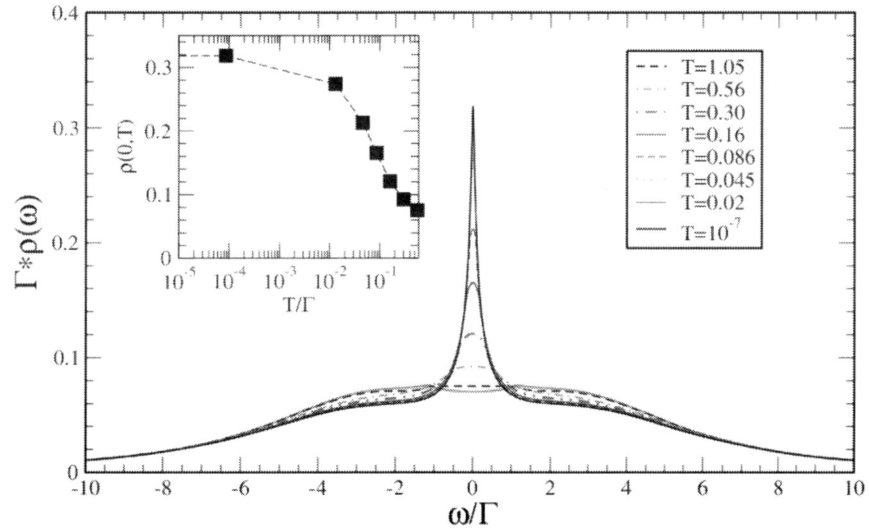

Figure 2. Temperature dependent equilibrium spectral function for a symmetric junction ($E_d = -U/2$) and a charging energy of $U/\Gamma = 8$. The inset shows the temperature dependence evolution of the many-body resonance at $\omega = 0$. NRG parameters: $\Lambda = 3.5$, $N_s = 1500$.

3. Results

3.1. Spectral functions for small charging energies U

All energies as well as the bias voltage are measured here in units of the single-particle charge fluctuation scale Γ. We used symmetric structureless leads characterized by a constant DOS $\rho_0 = 1/(2D)$ on the interval $[-D : D]$ where $D/\Gamma = 20$ is chosen.

For $U/\Gamma < 1$, the nonequilibrium spectral functions obtained with the SNRG and all three Keldysh approximations are identical [60]. In all cases, the major contribution stems from the Hartree-term Σ^H, which causes a level shift from $E_d \to E_d + \Sigma^H$. The line shape remains nearly identical to the $U = 0$ solution as can seen in comparison with the $U = 0$ spectrum: the additional scattering introduced by the finite value of U is much smaller than the resonant level width Γ. Furthermore, the spectrum – not shown here – remains almost bias independent.

A comparison between different spectral functions is depicted in Fig. 1 for moderate values of $U/\Gamma = 1$ and particle-hole symmetry. While for zero bias all curves agree, slight deviations between the SNRG solution and the Keldysh approach can been seen at small frequencies and large voltages. The perturbative approaches predict a slightly larger dephasing than the SNRG, leading to a stronger decrease of the spectral function around $\omega = 0$. Nevertheless, the overall agreement of all approaches is remarkable. Again, the line shape is dominated by the Hartree-shift and resonant level with Γ.

3.2. Transport in the Kondo regime

We have established the quality of the SNRG by comparing the SNRG spectral functions with the Keldysh spectral functions for small charging energy. Theses perturbative and conserving diagrammatic approximations only yield reliable results for small values of U since higher order diagrams have been omitted.

The spectral functions look quite different for larger values of U. A temperature dependent many-body resonance at $\omega = 0$ emerges when lowering T below the characteristic low-energy scale T_K as depicted in Fig. 2. The single lead calculation reduced the necessary number of NRG states N_s keeps after each iteration significantly. T_K depends exponentially on the charging

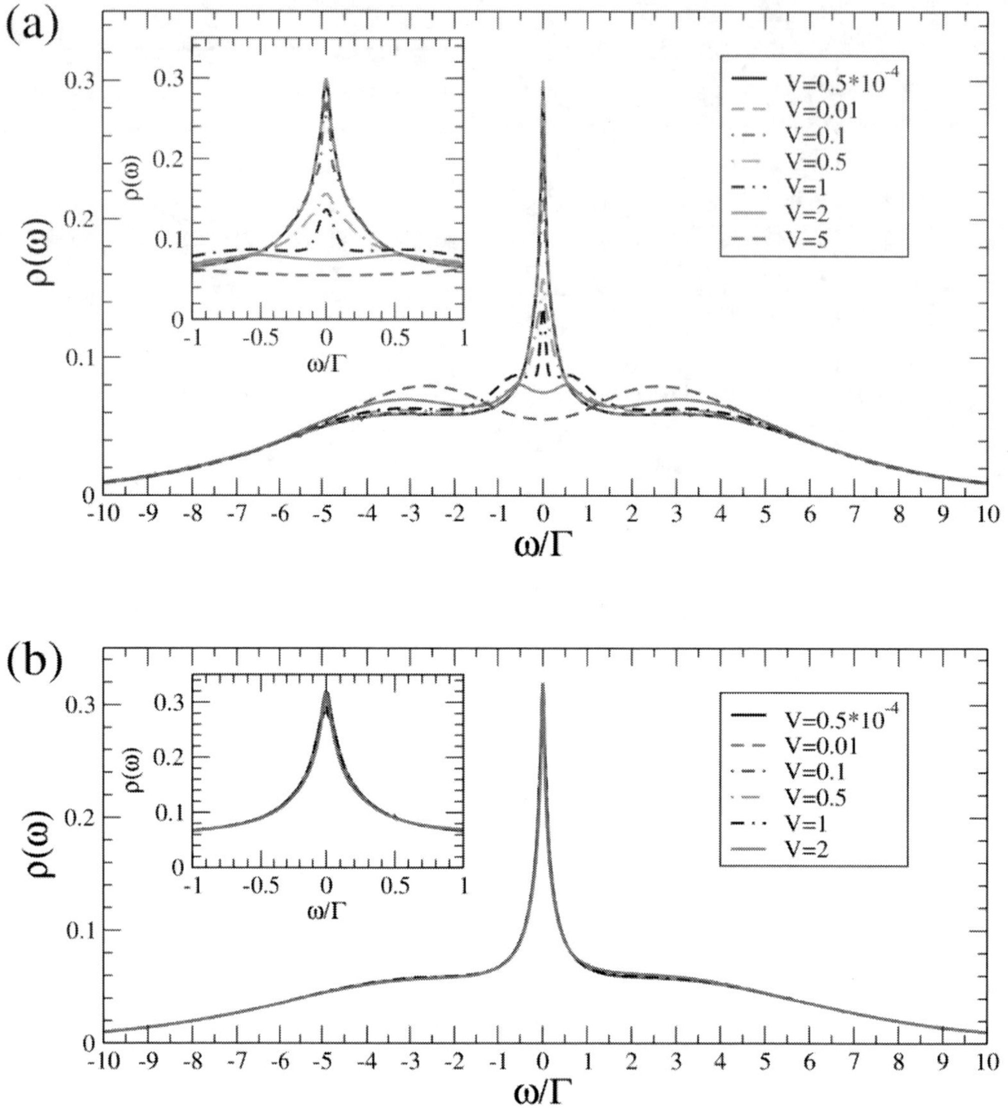

Figure 3. Taken from Ref. [49]: nonequilibrium spectral functions for (a) a symmetric junction $R = 1$ and different value of the bias V, and (b) for a strongly asymmetric junction $R = 1000$. The charging energy is set to $U/\Gamma = 8$ and the single-particle level at $E_d = -U/2$ and $T \to 0$. The insets show the evolution of the Kondo-resonance. NRG Parameter: $\Lambda = 4$, $N_s = 2200$.

energy U [67], i.e. $T_K \propto \exp(-\pi U/8\Gamma)$, and cannot be accessed by the diagrammatic methods described in Sec. 2.7 [60]. In addition, the single-particle resonance of the noninteracting problem splits into an upper and lower part with fractal weights which are the broaden atomic peaks at E_d and $E_d + U$. The inset of Fig. 2 shows the evolution of the central many-body peak – $\rho(\omega = 0)$ – as a function of temperature. The logarithmic increase of $\rho(0)$ around $T \approx T_K$ as well as the saturation to the unitary limit $1/(\Gamma\pi)$ for $T \to 0$ is characteristic for the Kondo effect.

We have investigated [49] the bias dependence of the spectral functions for $U/\Gamma = 8$ in the quantum-point contact regime ($R = 1$) and in the tunneling regime ($R = 1000 \gg 1$).

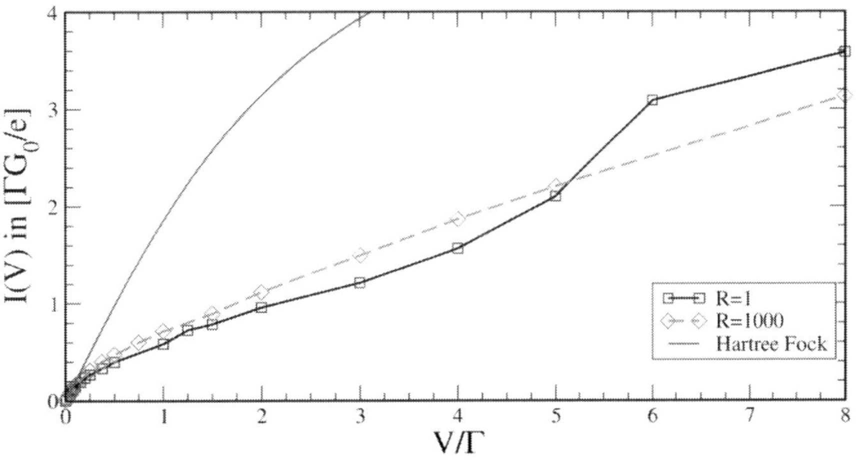

Figure 4. Current versus voltage for the point contact ($R = 1$) and the tunneling junction ($R = 1000$) calculated using the spectral functions of Fig. 3. Note that the current has been normalized to $G_0 = (e^2/h)(4\Gamma_L\Gamma_R/\Gamma^2) = (e^2/h)4R/(1+R)^2$. The blue curve indicates the Hartree-Fock solution.

The spectral function changes significantly with V in the point-contact junction as depicted in Fig. 3(a): the many-body resonance at $\omega = 0$ decreases with increasing voltage on the scale of T_K. The onset of a splitting of the resonance is also observed for $|V| > \Gamma$.

As expected for the tunneling regime, the spectrum remains almost unaltered as shown in Fig. 3(b). A Kondo effect develops due to the interaction between the quantum dot and the lead which strongly coupled to the device; the weakly coupled lead only serves as probe.

The IV characteristic for a large charging energy $U/\Gamma = 8$ was calculated by inserting the spectra of Fig. 3 into Eq. (13). The results are shown for $R = 1$ and $R = 1000$ in Fig. 4. For better comparison, the current $I(V)$ has been normalized by G_0 which contains the prefactor $4R/(1+R)^2$. For $R = 1000$, $G_0 = (e^2/h)4 \times 10^{-4}$: the current is strongly suppressed in the tunneling regime. In the quantum-point contact regime, the many-body resonance is destroyed with increasing voltage, and spectral weight is transfered to higher frequencies. As a consequence, the normalized current is initially lower than in the tunneling regime. This spectral weight loss accumulates at intermediate frequencies in the vicinity of the original atomic charge peaks. Hence, the current is enhanced once the voltage approaches the charging energy U.

The normalized current will saturate at 2π for very large voltages independent of U due to the normalization of the Green function. At zero bias and odd integer fillings of the quantum-dot, the differential conductance is always given by values close to G_0 for $T \to 0$. For larger Coulomb interaction U, however, the current increases with bias significantly slower than predicted by the Hartree-Fock solution. Therefore, effective single-particle theories such as Hartree-Fock are unsuitable to describe interacting nano-devices and molecular junctions in the coherent regime at low temperature since they predict incorrect IV curves.

4. Conclusion

We have presented a novel non-perturbative approach to quantum-transport which extends the Wilson's numerical renormalization group to open quantum systems. It includes the crossover from the Coulomb-blockade regime at high temperature to coherent transport at temperatures smaller than T_K. Since T_K is exponentially dependent on the charging energy U, diagrammatic techniques fail to predict the correct energy scale and are only valid for small values of the charging energy U.

Our approach uses a scattering-state description for the single-particle states and allows for simulation of a current-carrying open quantum system on a finite size system by including the correct boundary condition. Knowing the analytical form of the steady-state density operator for a noninteracting device at arbitrary bias [33], makes it possible to employ Wilson's numerical renormalization to steady-state problems. By letting the system evolve from the Hartree-Fock solution to a full many-body problem at a finite value U, the time-dependent TD-NRG provides all information on the steady-state density matrix of the interacting system needed to calculate the steady-state spectral function at finite bias for arbitrary temperatures.

The scattering-states numerical renormalization group nonequilibrium spectra agree excellently with the diagrammatic Keldysh spectra for small values of U, establishing the reliability of the approach. The SNRG is suitable to access small and large interaction strength at arbitrary temperature in contrast to the Keldysh approaches such as second order Born or GW approximation which are limited to small values of U.

The scattering-states numerical renormalization group descriptions includes (i) Hubbard decoupling schemes [8, 9, 10] which are only applicable at high temperature as well as (ii) diagrammatic expansions in U suitable for small values of the interaction strength. In addition, the SNRG does not suffer from artificial bifurcation points or unphysical magnetic solutions for large values of U as seen in Hartree-Fock, second-order or GW approximation. Therefore, the method is suitable to investigate the IV characteristics of single-electron transistors in a finite magnetic field. A detailed study on the magneto-transport and the spin-currents relevant for spintronics is under way.

5. Acknowledgments

We would like to acknowledge financial support by the Deutsche Forschungsgemeinschaft under AN 275/6-1 and supercomputer support of the NIC, Forschungszentrum Jülich under project no. HHB000.

[1] Kastner M A 1992 *Rev. Mod. Phys.* **64** 849
[2] Elzerman J M, Hanson R, van Beveeren L H W, Witkamp B, Vandersypen L M K and Kouvenhoven L P 2004 *Nature* **430** 431
[3] Goldhaber-Gordon D, Shtrikman H, Mahalu D, Abusch-Magder D, Meirav U and Kastner M 1998 *Nature* **391** 156
[4] Bulla R, Costi T A and Pruschke T 2008 *Rev. Mod. Phys.* **80** 395
[5] Keldysh L V 1965 *Sov. Phys. JETP* **20** 1018
[6] Khitrova G, Gibbs H M, Jahnke F, Kira M and Koch S W 1999 *Rev. Mod. Phys.* **71** 1591
[7] Hewson A C 1993 *The Kondo Problem to Heavy Fermions* (Cambridge UK: Cambridge Press)
[8] Fransson J, Eriksson O and Sandalov I 2002 *Phys. Rev. Lett.* **88** 226601
[9] Fransson J, Eriksson O and Sandalov I 2002 *Phys. Rev. B* **66** 195319
[10] Esposito M and Galperin M 2009 *Phys. Rev. B* **79** 205303
[11] Grewe N 1983 *Z. Phys. B* **52** 193
[12] Kuramoto Y 1983 *Z. Phys. B* **53** 37
[13] Bickers N E 1987 *Rev. Mod. Phys.* **59** 845
[14] Anders F B 1995 *J. Phys.: Condens. Matter* **7** 2801
[15] Lorek A, Grewe N and Anders F B 1991 *Solid State Commun.* **78** 167
[16] Wingreen N S and Meir Y 1994 *Phys. Rev. B* **49** 11040
[17] Hettler M H, Kroha J and Hershfield S 1998 *Phys. Rev. B* **58** 005649
[18] Myohanen P, Stan A, Stefanucci G and van Leeuwen R 2009, *Kadanoff-Baym approach to quantum transport through interacting nanoscale systems: From the transient to the steady-state regime* arXiv:0906.2136
[19] Pruschke T and Grewe N 1989 *Z. Phys. B* **74** 439
[20] Hershfield S, Davies J H and Wilkins J W 1991 *Phys. Rev. Lett.* **67** 003720
[21] Meir Y and Wingreen N S 1992 *Phys. Rev. Lett.* **68** 2512
[22] Rosch A, Kroha J and Wölfle P 2001 *Phys. Rev. Lett.* **87** 156802
[23] Rosch A, Paaske J, Kroha J and Wölfle P 2003 *Phys. Rev. Lett.* **90** 076804
[24] Doyon B and Andrei N 2006 *Phys. Rev. B* **73** 245326

[25] Toulouse G 1969 *Phys. Rev. B* **2** 270
[26] Schiller A and Hershfield S 1995 *Phys. Rev. B* **51** 12896
[27] Schiller A and Hershfield S 1996 *Phys. Rev. Lett.* **77** 1821
[28] Plihal M, Langreth D C and Nordlander P 2000 *Phys. Rev. B* **61** R13341
[29] Schiller A and Hershfield S 2000 *Phys. Rev. B* **62** R16271
[30] Han J E 2006 *Phys. Rev. B* **73** 125319
[31] Han J E and Heary R J 2007 *Phys. Rev. Lett.* **99** 236808
[32] Mehta P and Andrei N 2006 *Phys. Rev. Lett.* **96** 216802
[33] Hershfield S 1993 *Phys. Rev. Lett.* **70** 2134
[34] Schollwöck U 2005 *Rev. Mod. Phys.* **77** 259
[35] Schmitteckert P 2004 *Phys. Rev. B* **70** 121302(R)
[36] Boulat E, Saleur H and Schmitteckert P 2008 *Phys. Rev. Lett.* **101** 140601
[37] Weiss S, Eckel J, Thorwart M and Egger R 2008 *Phys. Rev. B* **77** 195316
[38] Wegner F 1994 *Ann. Physik* **3** 77
[39] Kehrein S 2005 *Phys. Rev. Lett.* **95** 056602
[40] Schoeller H and Koenig J 2000 *Phys. Rev. Lett.* **84** 3686
[41] Schoeller H 2009 *Eur. Phys. J. Special Topics* **168** 179–266
[42] Oguri A 2007 *Phys. Rev. B* **75** 035302
[43] Anders F B and Schiller A 2005 *Phys. Rev. Lett.* **95** 196801
[44] Anders F B and Schiller A 2006 *Phys. Rev. B* **74** 245113
[45] Anders F B 2008 *J. Phys.: Condens. Matter* **20** 195216
[46] Leggett A J, Chakravarty S, Dorsey A T and Fisher M P A 1987 *Rev. Mod. Phys.* **59** 1
[47] Enss T, Meden V, Andergassen S, Barnabe-Theriault X, Metzner W and Schoenhammer K 2005 *Phys. Rev. B* **71** 155401
[48] Lebanon E, Schiller A and Anders F B 2003 *Phys. Rev. B* **68** 041311(R)
[49] Anders F B 2008 *Phys. Rev. Lett.* **101** 066804
[50] Wilson K G 1975 *Rev. Mod. Phys.* **47** 773
[51] Peters R, Pruschke T and Anders F B 2006 *Phys. Rev. B* **74** 245114
[52] Kadanoff L P and Baym G 1962 *Quantum Statistical Mechanics* (Benjamin)
[53] Danielewicz P 1984 *Ann. Phys.* **152** 239–304
[54] Wagner M 1991 *Phys. Rev. B* **44** 6104–6117
[55] van Leeuwen R, Dahlen N E, Stefanucci G, Almbladh C O and von Barth U 2006 *Lect. Notes Phys.* **706** 33–59
[56] Haug H and Jauho A P 1996 *Quantum Kinetics in Transport and Optics of Semiconductors* (Springer, Berlin)
[57] Spataru C D, Hybertsen M S, Louie S G and Millis A J 2009 *Phys. Rev. B* **79** 155110
[58] Luttinger J M and Ward J C 1960 *Phys. Rev.* **118** 1417–1427
[59] Baym G 1962 *Phys. Rev.* **127** 1391–1401
[60] Schmitt S and Anders F B 2010 *Phys. Rev. B* **81** 165106
[61] Aryasetiawan F and Gunnarsson O 1998 *Rep. Prog. Phys.* **61** 237–312 and references therein
[62] Aulbur W G, Jönsson L and Wilkins J W 1999 *Solid State Physics* **54** and references therein
[63] Thygesen K S and Rubio A 2007 *J. Chem. Phys.* **126** 091101
[64] Darancet P, Ferretti A, Mayou D and Olevano V 2007 *Phys. Rev. B* **75** 075102
[65] Wang X, Spataru C D, Hybertsen M S and Millis A J 2008 *Phys. Rev. B* **77** 045119
[66] Thygesen K S and Rubio A 2008 *Phys. Rev. B* **77** 115333
[67] Krishna-murthy H R, Wilkins J W and Wilson K G 1980 *Phys. Rev. B* **21** 1003

Journal of Physics: Conference Series **220** (2010) 012022 doi:10.1088/1742-6596/220/1/012022

Calculating Green Functions from Finite Systems

Peter Schmitteckert

Institute of Nanotechnology, Karlsruhe Institute of Technology, 76344 Eggenstein-Leopoldshafen, Germany

E-mail: Peter.Schmitteckert@kit.edu

Abstract. In calculating Green functions for interacting quantum systems numerically one often has to resort to finite systems which introduces a finite size level spacing. In order to describe the limit of system size going to infinity correctly, one has to introduce an artificial broadening larger than the finite size level discretization. In this work we compare various discretization schemes for impurity problems, i.e. a small system coupled to leads. Starting from a naive linear discretization we will then discuss the logarithmic discretization of the Wilson NRG, compare it to damped boundary conditions and arbitrary discretization in energy space. We then discuss the importance of choosing the right single particle basis when calculating bulk spectral functions. Finally we show the influence of damped boundary conditions on the time evolution of wave packets leading to a NRG-tsunami.

1. Introduction

Correlation, or Green, functions are a fundamental concept of condensed matter theory, for an introduction see e.g. [1–3]. However, for interacting systems exact solutions are rare and one often has to resort to perturbative or numerical approaches. In calculating Green functions for infinite systems, let it be a bulk Green function, or an impurity problem, numerically one hast to resort to a discretized, finite system. In this work we discuss the influence of various discretization schemes on the spectral function, i.e. the imaginary part of the retarded Green function.

To this end we restrict ourselves to non-interacting Fermi systems where numerics can be performed without any approximation and the resulting errors can be traced back to the discretization scheme.

We start with the definition of Green functions in time domain, and derive resolvent equations in frequency domain, which demonstrates that for the calculation of spectral functions one does not need the spectrum explicitly. We then discuss the case of an impurity problem, namely the resonant level model, where a single level is coupled to a non interacting lead / bath. It turns out that it is non-trivial to treat this very simple model accurately on a finite lattice. We then proceed with the problem of an energy resolved spectral function of a non-interacting tight binding chain. It turns out, that one can reconstruct the δ-function of the spectral function in the continuum. However, care has to be taken in order to avoid discretization errors. Finally we show that damped, or Numerical Renormalization Group- (NRG) like, boundary conditions lead to a phenomenon we call the NRG tsunami.

1.1. Green Functions in Time Domain

The lesser (greater) Green functions $G^<$ ($G^>$) and the retarded (advanced) Green functions G^r (G^a) are defined [1–3] by

$$G^>_{\hat{A},\hat{B}}(t,t') = -\mathrm{i}\,\langle \hat{A}(t)\,\hat{B}(t')\rangle \tag{1}$$

$$G^<_{\hat{A},\hat{B}}(t,t') = -\mathrm{i}\zeta\,\langle \hat{B}(t')\,\hat{A}(t)\rangle \tag{2}$$

$$G^r_{\hat{A},\hat{B}}(t,t') = -\mathrm{i}\,\Theta(t-t')\langle\Big[\hat{A}(t),\,\hat{B}(t')\Big]_{-\zeta}\rangle = \Theta(t-t')(G^>_{\hat{A},\hat{B}}(t,t') - G^<_{\hat{A},\hat{B}}(t,t')) \tag{3}$$

$$G^a_{\hat{A},\hat{B}}(t,t') = \mathrm{i}\,\Theta(t'-t)\langle\Big[\hat{A}(t),\,\hat{B}(t')\Big]_{-\zeta}\rangle = \Theta(t'-t)(G^<_{\hat{A},\hat{B}}(t,t') - G^>_{\hat{A},\hat{B}}(t,t')) \tag{4}$$

where $\hat{A}$ and $\hat{B}$ denote two arbitrary operators. For fermionic operators, $\zeta = -1$, $[A,B]_+ = AB + BA$ denotes the anti-commutator of operators A and B, and for bosonic operators, $\zeta = 1$, the anti-commutator is replaced by a commutator $[A,B]_- = AB - BA$. Throughout this work $\langle\cdots\rangle$ denotes the zero temperature ground state average.

1.2. Resolvent Representation in Frequency Domain

In order to simplify notation when switching to frequency domain we assume translational invariance in time and introduce the following Green functions of two arbitrary operators $\hat{A},\hat{B}$:

$$G^+_{\hat{A},\hat{B}}(t) = -\mathrm{i}\,\Theta(t)\,\langle \hat{A}(t)\,\hat{B}(0)\rangle \tag{5}$$

$$G^-_{\hat{A},\hat{B}}(t) = \mathrm{i}\,\Theta(t)\,\langle \hat{A}(0)\,\hat{B}(t)\rangle\,, \tag{6}$$

leading to

$$G^>_{\hat{A},\hat{B}}(t,0) = G^+_{\hat{A},\hat{B}}(t) - G^-_{\hat{A},\hat{B}}(-t) \tag{7}$$

$$G^<_{\hat{A},\hat{B}}(t,0) = \zeta G^+_{\hat{B},\hat{A}}(-t) - \zeta G^-_{\hat{B},\hat{A}}(t) \tag{8}$$

$$G^r_{\hat{A},\hat{B}}(t,0) = G^+_{\hat{A},\hat{B}}(t) + \zeta G^-_{\hat{B},\hat{A}}(t) \tag{9}$$

$$G^a_{\hat{A},\hat{B}}(t,0) = \zeta G^+_{\hat{B},\hat{A}}(-t) + G^-_{\hat{A},\hat{B}}(-t)\,. \tag{10}$$

The Fourier transformed Green function $\mathcal{G}^+_{\hat{A},\hat{B}}(\omega)$ is defined by

$$\mathcal{G}^+_{\hat{A},\hat{B}}(\omega) = \int_{-\infty}^{+\infty} \mathrm{d}t\, \mathrm{e}^{\mathrm{i}\omega t}\, G^+_{\hat{A},\hat{B}}(t) = -\mathrm{i}\int_0^{+\infty} \mathrm{d}t\, \mathrm{e}^{\mathrm{i}\omega t}\,\langle \hat{A}(t)\,\hat{B}(0)\rangle \tag{11}$$

$$= -\mathrm{i}\int_0^{+\infty} \mathrm{d}t\, \mathrm{e}^{\mathrm{i}\omega t}\,\langle\Psi_0|\mathrm{e}^{\mathrm{i}\mathcal{H}t}\,\hat{A}\,\mathrm{e}^{-\mathrm{i}\mathcal{H}t}\hat{B}|\Psi_0\rangle \tag{12}$$

$$= -\mathrm{i}\int_0^{+\infty} \mathrm{d}t\,\langle\Psi_0|\hat{A}\,\mathrm{e}^{\mathrm{i}(E_0-\mathcal{H}+\omega+\mathrm{i}\eta)t}\,\hat{B}|\Psi_0\rangle \tag{13}$$

$$= -\mathrm{i}\,\langle\Psi_0|\hat{A}\left.\frac{\mathrm{e}^{\mathrm{i}(E_0-\mathcal{H}+\omega+\mathrm{i}\eta)t}}{\mathrm{i}(E_0-\mathcal{H}+\omega+\mathrm{i}\eta)}\right|_0^\infty \hat{B}|\Psi_0\rangle \tag{14}$$

$$= \langle\Psi_0|\hat{A}\,\frac{1}{E_0-\mathcal{H}+\omega+\mathrm{i}\eta}\,\hat{B}|\Psi_0\rangle \tag{15}$$

and similarly

$$\mathcal{G}^-_{\hat{A},\hat{B}}(\omega) = \langle\Psi_0|\hat{A}\,\frac{1}{E_0-\mathcal{H}-\omega-\mathrm{i}\eta}\,\hat{B}|\Psi_0\rangle\,, \tag{16}$$

where a convergence generating $\eta = 0^+$ has been introduced to ensure convergence. Finally, we obtain

$$\mathcal{G}^{>}_{\hat{A},\hat{B}}(\omega) = \mathcal{G}^{+}_{\hat{A},\hat{B}}(\omega) - \mathcal{G}^{-}_{\hat{A},\hat{B}}(-\omega) \tag{17}$$

$$\mathcal{G}^{<}_{\hat{A},\hat{B}}(\omega) = \zeta\mathcal{G}^{+}_{\hat{B},\hat{A}}(-\omega) - \zeta\mathcal{G}^{-}_{\hat{B},\hat{A}}(\omega) \tag{18}$$

$$\mathcal{G}^{r}_{\hat{A},\hat{B}}(\omega) = \mathcal{G}^{+}_{\hat{A},\hat{B}}(\omega) + \zeta\mathcal{G}^{-}_{\hat{B},\hat{A}}(\omega) \tag{19}$$

$$\mathcal{G}^{a}_{\hat{A},\hat{B}}(\omega) = \zeta\mathrm{G}^{+}_{\hat{B},\hat{A}}(-\omega) + \mathrm{G}^{-}_{\hat{A},\hat{B}}(-\omega) \tag{20}$$

1.3. Correction Vector Approach

It is interesting to note that the resolvent representation Eqs. (15, 16) allows for calculating the Green function without a complete knowledge of the spectrum via the correction vector approach [4]. Starting from a general resolvent expression

$$\mathcal{G}(E) = \langle\Psi|\hat{A}\underbrace{\frac{1}{\mathcal{H} - E + \mathrm{i}\eta}\hat{B}|\Psi\rangle}_{|\xi\rangle} = \langle\Psi|\hat{A}|\xi\rangle \tag{21}$$

we obtain the correction vector $|\xi\rangle$ from the linear system

$$(\mathcal{H} - E + \mathrm{i}\eta)\,|\xi\rangle = \hat{B}|\Psi\rangle\,, \tag{22}$$

which can be solved by standard solvers. However, note that in this approach one has to solve a different linear system for each desired energy E. An alternative approach consists in expanding the resolvent in orthogonal polynomials, for a review see [5].

1.4. Single Particle Propagator

The single particle Green functions for fermionic systems are defined by

$$\mathrm{G}^{>}(x,t;y,t') = -\mathrm{i}\,\langle c_x(t)\,c_y^{+}(t')\rangle \tag{23}$$

$$\mathrm{G}^{<}(x,t;y,t') = \mathrm{i}\,\langle c_y^{+}(t')\,c_x(t)\rangle \tag{24}$$

$$\mathrm{G}^{r}(x,t;y,t') = -\mathrm{i}\,\Theta(t-t')\langle\big[c_x(t),\,c_y^{+}(t')\big]_{+}\rangle \tag{25}$$

$$\mathrm{G}^{a}(x,t;y,t') = \mathrm{i}\,\Theta(t'-t)\langle\big[c_x(t),\,c_y^{+}(t')\big]_{+}\rangle \tag{26}$$

where we use x and y to denote the position in the lattice, $c_x(t)$ and $c_x^{+}(t)$ are the fermionic annihilation and creation operators at site x and time t. For spinful calculations x denotes a super index of the spatial coordinate and the spin orbital. From Eqs. (15, 16) we obtain for the retarded and advanced Green functions

$$\mathcal{G}^{r}(x,y,\omega) = \mathcal{G}^{+}_{\hat{c}_x,\hat{c}_y^{+}}(\omega) - \mathcal{G}^{-}_{\hat{c}_y^{+},\hat{c}_x}(\omega) \tag{27}$$

$$\mathcal{G}^{a}(x,y,\omega) = \mathcal{G}^{r}(x,y,\omega)^{*}\,. \tag{28}$$

1.5. Free Fermions

Up to now the description for Green functions was completely general. In the following we restrict ourselves to the description of non-interacting Fermi systems since our goal is to describe the problem induced by evaluating Green functions on finite systems. In result we are able to perform approximation free numerics. Nevertheless, our findings are applicable for interacting system and can be exploited by other methods.

Journal of Physics: Conference Series **220** (2010) 012022 doi:10.1088/1742-6596/220/1/012022

For non-interacting fermions we start with a general Hamiltonian

$$\mathcal{H} = \vec{c}^{+} \cdot H \cdot \vec{c} = \sum_{x,y} \hat{c}_x^{+} H_{x,y} \hat{c}_y \,. \tag{29}$$

We can now switch to a diagonal basis by diagonalizing the matrix H:

$$\mathrm{diag}(\vec{\varepsilon}) = U \cdot H \cdot U^{+} \tag{30}$$
$$\mathbf{1} = U \cdot U^{+} \tag{31}$$
$$\vec{\tilde{c}} = U \cdot \vec{\hat{c}}, \tag{32}$$

where ε_ℓ are the single particle levels. If the ground state $|\Psi_0\rangle$ is non degenerate it is given by

$$|\Psi_0\rangle = \prod_{\varepsilon_\ell < \varepsilon_{\mathrm{F}}} \tilde{c}_\ell^{+} |-\rangle \,, \tag{33}$$

where ε_{F} is the Fermi energy and $|-\rangle$ is the vacuum state. However, on finite systems at zero temperature this definition of the Fermi energy is ambiguous since ε_{F} can sit anywhere between the highest occupied and the lowest unoccupied level. We set ε_{F} in the middle of those two levels to ensure numerical stability. For degenerate ground states one has to take care of the different possibilities of filling the highest level. When evaluating expectation values one then has to average degenerate levels at the Fermi energy by taking the zero temperature limit of the finite temperature result, e.g. the number of particles N is then given by

$$N = \lim_{\beta \to \infty} \sum_\ell \mathrm{f}\left(\beta(\varepsilon_\ell - \varepsilon_{\mathrm{F}})\right) U_{\ell,x}^{*} U_{\ell,x} \,, \tag{34}$$

where f() is the Fermi function and one should work at a small, but non-vanishing temperature.

Evaluating the retarded single particle Green functions $\mathrm{G}^r(x,t;y,t')$ of Eq. (27) in frequency domain using the formulae of section 1.2 we obtain

$$\mathcal{G}^{+}_{\hat{c}_x,\hat{c}_y^{+}}(\omega) = \langle \Psi_0 | \hat{c}_x \frac{1}{E_0 - \mathcal{H} + \omega + \mathrm{i}\eta} \hat{c}_y^{+} | \Psi_0 \rangle = \sum_\ell \left(1 - \langle \Psi_0 | \tilde{n}_\ell | \Psi_0 \rangle\right) \frac{U_{\ell,x}^{*} U_{\ell,y}}{\omega - \varepsilon_\ell + \mathrm{i}\eta} \tag{35}$$

and

$$\mathcal{G}^{-}_{\hat{c}_y^{+},\hat{c}_x}(\omega) = -\sum_\ell \langle \Psi_0 | \tilde{n}_\ell | \Psi_0 \rangle \frac{U_{\ell,x}^{*} U_{\ell,y}}{\omega - \varepsilon_\ell + \mathrm{i}\eta} \,. \tag{36}$$

Finally we obtain from Eqs. (27, 28)

$$\mathcal{G}^r(x,y,\omega) = \mathcal{G}^{+}_{c_x,c_y^{+}}(\omega) - \mathcal{G}^{-}_{c_y^{+},c_x}(\omega) = \sum_\ell \frac{U_{\ell,x}^{*} U_{\ell,y}}{\omega - \varepsilon_\ell + \mathrm{i}\eta} \tag{37}$$

2. Resonant Level Model

Following the rather general introduction on calculating Green functions we will now concentrate on the spectral function

$$\mathcal{A} = \frac{-1}{\pi} \Im \mathcal{G}^r(x,x,\omega) \tag{38}$$

of a single resonant level $\epsilon_d \hat{n}_d$ coupled via a hybridization t' to a one-dimensional lead , where we set the hopping element to $t = 1$. Ignoring the finite width and the cosine dispersion of the lead band results in the wide band limit solution of an area normalized Lorentzian

$$\mathcal{A} = \frac{1}{\pi} \frac{w}{\omega^2 + w^2} \qquad w = t'^2 \tag{39}$$

Throughout this section we apply Eq. (37) to evaluate the resolvent equation (27). A problem that arises is that the convergence factor η of (27), which is 0^+ for continuum leads has to be larger than the finite size level splitting of the leads while it has to be much smaller than any physical scale of interest as it also acts as a broadening. In addition, in finite systems the existence of a boundary influences the ground state result typically on a scale $\omega_{\mathrm{BC}} \sim \Delta_{\mathrm{F}}$, where Δ_{F} is the level spacing at the Fermi surface. Note, that in the case of a single level coupled to a lead with finite width there may exist bound states outside the conduction band. While these states are not accessible by single particle of the conduction band due to energy conservation, they can be access by few- or many- particle processes.[6].

The goal of this section is to provide an overview of various discretization schemes of the leads and their impact on the spectral function. We start with the natural choice of a finite nearest neighbour tight binding chain. It turns out that their resolution is quiet limited. Next we discuss discretization schemes used in the Numerical Renormalization Group (NRG) approach[7; 8] and a variation of it called Smooth[9] or Damped[10] Boundary conditions, which give good results for low frequencies, but the high frequency tails are spoiled. Freyn and Florens[11] used an effective increase of the density of states to lower the broadening close to a resonance within standard NRG, to obtain a larger accuracy. Finally we provide a discretization scheme which is able to provide high resolution on all energy scales, where we generalize the variable discretization approach of Nishimoto and Jeckelmann.[12] Žitkoa[13] has applied a variable discretization within the context of NRG.

2.1. Linear Leads

In Figure 2 we show the numerical evaluation of the spectral function of a resonant level coupled to a lead via a hybridization of $t' = 0.1$ evaluated for a total number of 700 sites and 350 fermions and hard wall boundary conditions (HWBC). The corresponding Lorentzian (39) has a half halfwidth of

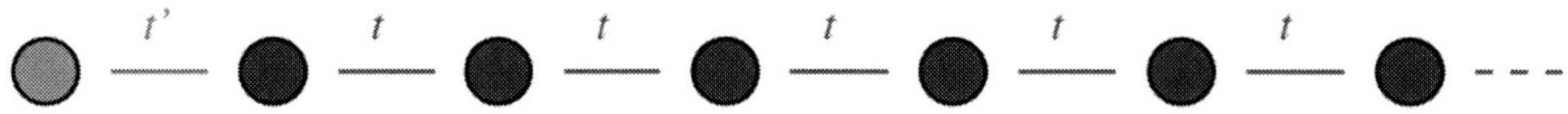

Figure 1. Sketch of a single impurity coupled to a lead via t' and a lead hopping of t.

$w = t'^2$=0.01 which is already larger than the finite size level splitting of $\Delta_F \approx 0.00352$. Nevertheless the figure demonstrates that the discretization of the leads is too coarse to reproduce the Lorentzian although the level spacing is already significantly smaller than the resonance width. For an η larger than the finite size level spacing the resonance is artificially broadened while for smaller η the discrete nature of leads appears in the correlation function as can be seen by the spikes of the $\eta = 0.0025$ result. This observation is explained by looking more closely at the resolvent equation (27). In the thermodynamic limit of continuum leads η is 0^+ and the imaginary part of the resolvent $\mathcal{G}^+_{c_x, c_y^+}(\omega)$

$$\Im \frac{1/\pi}{E_0 - \mathcal{H} + \omega + 0^+} = \pi\delta(\omega - (\mathcal{H} - E_0)) \tag{40}$$

gives a contribution only at $\omega = E_n - E_0$, where E_n are the eigenenergies of the Hamiltonian $\mathcal{H}$. By switching to a finite η the δ function gets replaced by a Lorentzian of half half-width η. This corresponds

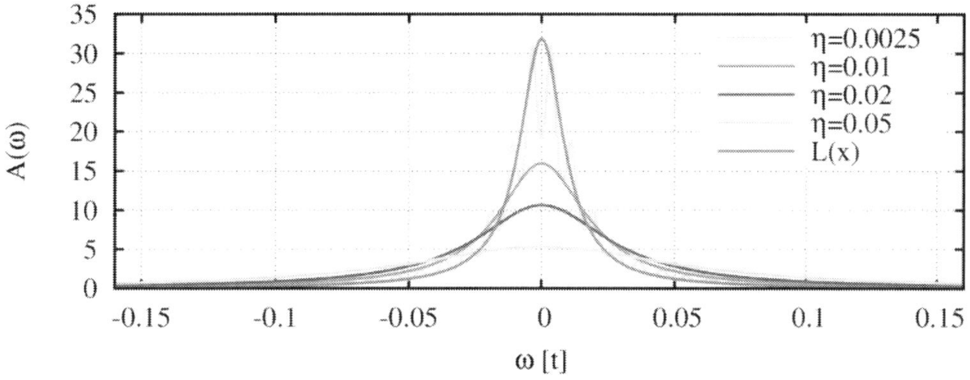

Figure 2. Spectral function of a resonant level, $t' = 0.1$, $M = 700$, $N = 350$, $\epsilon_d = 0$, HWBC, and $\eta = 0.05$, 0.02, 0.01, 0.0025. $L(\omega)$ shows the Lorentzian of half halfwidth $w = 0.01$ corresponding to the spectral function in the wide band limit.

to a convolution of the original spectral function with a Lorentzian of the same width and our result is replaced by

$$L_\eta(\omega) = \int_{-\infty}^{\infty} d\varepsilon \, \frac{w/\pi}{(\omega-\varepsilon)^2 + w^2} \frac{\eta/\pi}{\varepsilon^2+\eta^2} = \frac{1}{\pi} \frac{w+\eta}{\omega^2 + (w+\eta)^2} . \tag{41}$$

By taking into account the broadening induced by η we can fit Eq. (41) to our data as shown in Figure 3

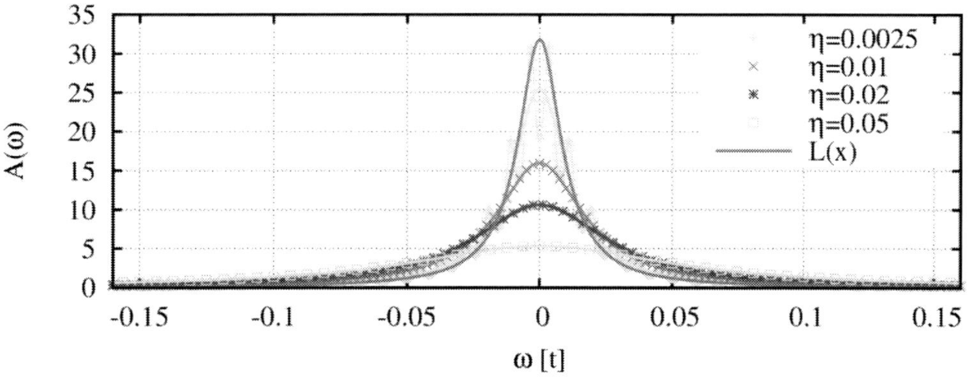

Figure 3. Lorentzian fits of the spectral function of a resonant level, $t' = 0.1$, $M = 700$, $N = 350$, $\epsilon_d = 0$, HWBC, and $\eta = 0.05, 0.02, 0.01, 0.0025$. The level spacing at the Fermi energy is $\Delta_F \approx 0.00352$. The full line shows the Lorentzian of half half-width $w = 0.01$ corresponding to the spectral function in the wide band limit.

and Table 1. The rather good results for even large η suggest that the results may be strongly improved by unfolding the η broadening by a deconvolution.

2.2. Logarithmic Discretization

In order to increase the energy resolution of our lead we can replace the tight binding chain with constant hopping by a tight binding chain similar to the one used in NRG where the hopping element is

Journal of Physics: Conference Series **220** (2010) 012022 doi:10.1088/1742-6596/220/1/012022

η	0.1	0.05	0.02	0.01	0.0025	0.001
w_L	0.00942	0.00973	0.00990	0.00995	0.00999	0.0170
w_{Log}	0.00855	0.00876	0.00887	0.00891	0.00894	0.00895
w_{DBC}	0.00945	0.00974	0.00987	0.00979	0.00942	0.00929

Table 1. Fits of a Lorentzian (41) to the spectral function of a single impurity coupled to a single lead. All results are in units of the lead hopping element t. w_L corresponds to the linear leads used in Figure 2 and fits are shown in Figure 3. w_{Log} corresponds to the logarithmic discretization used in Figure 4 and w_{DBC} corresponds to the damped boundary conditions of Figure 6 The analytical result is $w = 0.01t$.

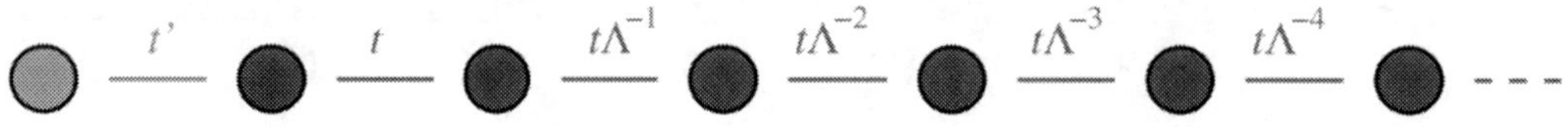

Figure 4. Sketch of a single impurity coupled to a lead via t' and a NRG like lead hopping of $t\Lambda^{-n}$.

exponentially decreased by a factor $\Lambda^{-n/2}$, with $\Lambda > 1$ and n the index of the NRG iteration.[7] Here we use a chain where the hopping is reduced by a factor of Λ on each bond as displayed in Figure 4.

In Figure 5 we show the results for a single level coupled to 31 NRG like lead sites where the hopping $t = 1$ is reduced by $\Lambda^{-1} = 0.8$ on each bond as sketched in Figure 4. For these parameters we get a level spacing at the Fermi surface of $\Delta_F \approx 0.000739t$. Correspondingly the resolution for $\omega = 0$ is now much higher and the $\eta = 0.0025$ curve is well resolved at small frequencies. However, for large frequencies the resolution drops exponentially leading to spikes even in the $\eta = 0.01t$ curve. As a consequence the Lorentzian fits do not work as well as with the (albeit much larger) linear lead, compare Table (1). A general feature of the logarithmic discretization consists in the overly broad tails, compare also the the section on frequency dependent broadening.

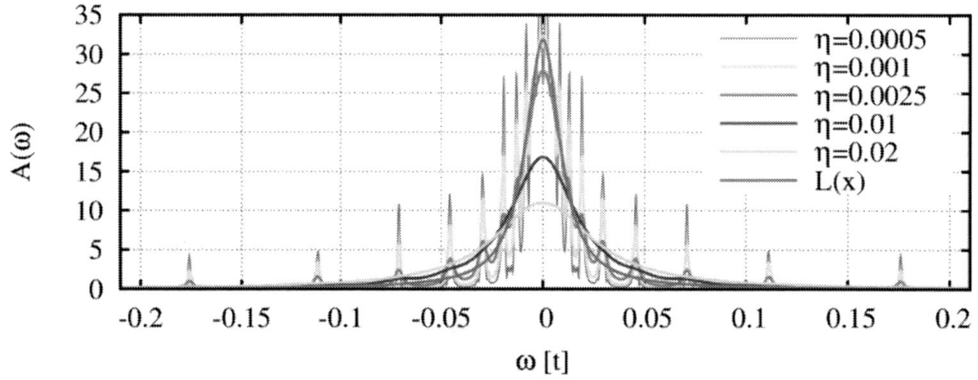

Figure 5. Spectral function of a resonant level, $\epsilon_d = 0$, coupled to 31 NRG like lead sites via $t' = 0.1t$ using $t = 1$, $\Lambda^{-1} = 0.8$. The finite size level spacing is $\Delta_F \approx 0.000739t$. $L(\omega)$ shows the Lorentzian of half half-width $w = 0.01$ corresponding to the spectral function in the wide band limit.

2.3. Damped Boundary Conditions

In this section we combine the linear and exponentially decaying leads by first attaching the impurity to M_{RS} real space sites with constant hopping element t and then coupling to an exponentially decaying

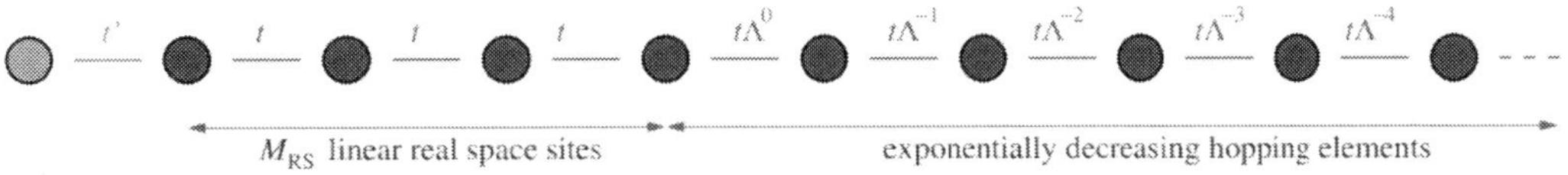

Figure 6. Sketch of a single impurity coupled to a lead via t' and a NRG like lead hopping of $t\Lambda^{-n}$.

lead. In Figure 7 we show the result for a setup that corresponds to Figure 5 where a hundred additional sites with hopping element t have been inserted between the impurity and exponential decaying lead. A similar kind of boundary condition has been originally introduced by Vekić and White[9] to mimic the thermodynamic limit in a bulk system. The version employed here was introduced by Bohr, Schmitteckert, and Wï¿½lfle [14] to tackle the finite size effects in the evaluation of the Kubo formula for linear transport. At a first glance the additional lead sites result in a slight reduction of the spike of the

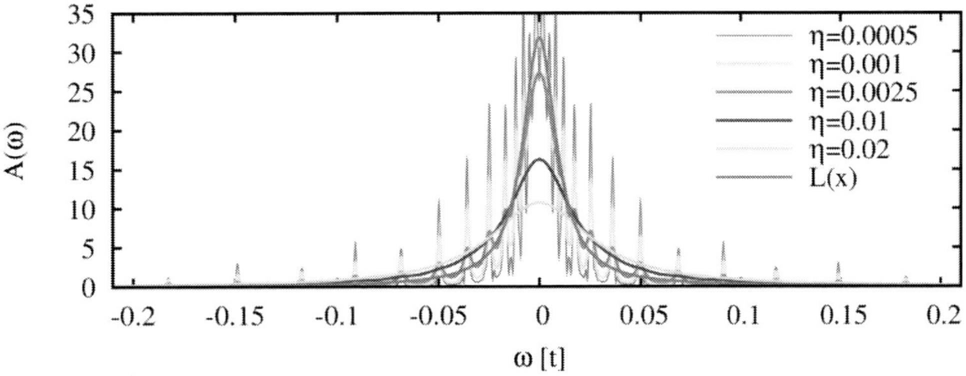

Figure 7. Spectral function of a resonant level, $\epsilon_d = 0$, coupled to an NRG like Hamiltonian via $t' = 0.1$ as in Figure 5, $\Lambda^{-1} = 0.8$, $M_\Lambda = 30$, plus 100 additional sites with a constant hopping of t inserted between the impurity and the NRG like chain leading to a level spacing at the Fermi surface of $\Delta_F \approx 0.00144$. The fits of Lorentzian (41) are shown in Table (1). The full red line shows the Lorentzian of half half-width $w = 0.01$ corresponding to the spectral function in the wide band limit.

logarithmic discretization only. However, one obtains significantly better Lorentzian (41) fits compared to the Logarithmic discretization alone. Nevertheless, the result is still quite disappointing.

2.4. Frequency Dependent Broadening

In order to remove the spikes in the logarithmic or DBC discretization one has to employ frequency dependent broadening. In Figure 8 we present the results for the same system as in Figure 7. However, this time we employ a broadening η that is proportional to the level spacing at energy ω. Since the energy spectrum is discrete we take a linear weighted average of the level spacing of the level below and above ω. Let us define the level spacing Δ_ω at energy ω as

$$\Delta_\omega = (\varepsilon_n - \varepsilon_{n-1}) + (\varepsilon_{n+1} - 2\varepsilon_n + \varepsilon_{n-1}) \frac{\omega - 0.5\,(\varepsilon_n + \varepsilon_{n-1})}{0.5\,(\varepsilon_{n+1} - \varepsilon_{n-1})} \tag{42}$$

where n is the level index with

$$0.5\,(\varepsilon_{n-1} + \varepsilon_n) \;\leq\; \omega \;<\; 0.5\,(\varepsilon_n + \varepsilon_{n+1})\,. \tag{43}$$

Note, that in the case of degenerate levels one should only count distinct energy levels to avoid a vanishing distance. For an ω outside the energy range where the corresponding n exists we use the level spacing of the corresponding first or last level distance. We then define a relative η scaling

$$\eta = \eta_\omega * \Delta_\omega . \tag{44}$$

For an $\eta_\omega = 1$ we obtain a broadening η that is of the order of the level spacing at energy ω. In Figure 8 we compare spectral functions for the same system as in Figure 6, only the constant η is replaced by a relative η broadening. With this approach one can eliminate the spikes by using an $\eta_\omega \geq 1.0$. However, the tails are still too broad and Eq. (41) can not be used to fit the result, as η is now energy dependent. In order to filter the η induced broadening one would now have to resort to an energy dependent deconvolution.

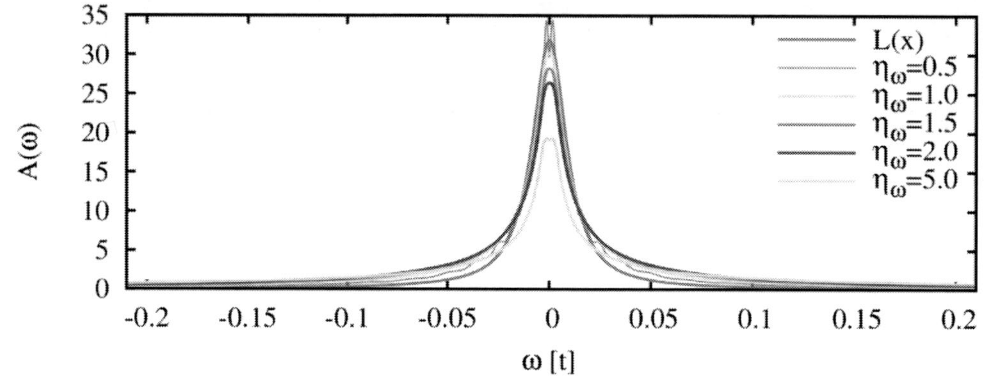

Figure 8. Spectral function of a resonant level using DBC as in Figure 7, $\epsilon_d = 0$, $t' = 0.1$, $\Lambda^{-1} = 0.8$, $M_\Lambda = 30$, $M_{\mathrm{RS}} = 30$, and $\Delta_{\mathrm{F}} \approx 0.00144$. Here we used a broadening $\eta = \eta_\omega * \Delta_\omega$, where Δ_ω corresponds to the level spacing at energy ω, see Eq. (42). The full red line shows the Lorentzian of half half-width $w = 0.01$ corresponding to the spectral function in the wide band limit.

2.5. *Frequency Adapted Grids*

In the previous sections we showed different lattice schemes to evaluate a spectral function. None of the schemes gave actually satisfying results. Either the resolution was poor or the tails were not represented correctly. A solution to this problem consists in adapting the lattice for each frequency ω.

We would like to note that this is a generalization to the variable discretization approach of Nishimoto and Jeckelmann[12] in the sense that we can apply a constant broadening for the complete frequency range due to our recipe for constructing discretizations.

2.5.1. Momentum Space Leads This goal can be achieved by switching to leads in momentum (or energy) space. Let us start with the infinite chain[1]

$$\sum_{x=0}^{M} t_x \hat{c}_x^+ \hat{c}_{x-1} + t_x^* \hat{c}_{x-1}^+ \hat{c}_x \rightarrow \sum_{x=-\infty}^{\infty} t_x \hat{c}_x^+ \hat{c}_{x-1} + t_x^* \hat{c}_{x-1}^+ \hat{c}_x . \tag{45}$$

[1] Strictly speaking, we should replace the finite lead by a semi-infinite chain. However, this could be incorporated into t_x.

We now switch to momentum representations

$$\sum_{x=-\infty}^{\infty} t_x \hat{c}_x^+ \hat{c}_{x-1} + t_x^* \hat{c}_{x-1}^+ \hat{c}_x = \frac{1}{2\pi} \int_{-\pi}^{\pi} \mathrm{d}k\, \epsilon_k\, \tilde{c}_k^+ \tilde{c}_k\,. \tag{46}$$

For a nearest neighbour chain one obtains $\epsilon(k) = -2t\cos(k)$, however one can now use any desired band $\epsilon(k)$. Motivated by these considerations we use the general form of a lead in 'momentum space'

$$\frac{1}{2\pi} \int_{D_\mathrm{l}}^{D_\mathrm{u}} \mathrm{d}k\, \mathcal{N}(k)\, \epsilon_k\, \tilde{c}_k^+ \tilde{c}_k\,, \tag{47}$$

where $\mathcal{N}(k)$ is the momentum density of states, and D_l (D_u) the lower (upper) momentum cut off. Note, here we name 'k' momentum although it can be any labelling. We avoid using an energy density of states in this work as we might be interested in transport properties and keep the flexibility to describe left and right movers by negative and positive momenta.

2.5.2. Rediscretization In order to use the leads in momentum space for our numerics we have to rediscretize the leads,

$$\mathcal{H}_\mathrm{Lead} = \sum_{\ell=1}^{M_\mathrm{L}} \epsilon_{k_\ell}\, \check{c}_\ell^+ \check{c}_\ell\,, \tag{48}$$

where k_ℓ are the M_L discretization points between D_l and $D_\mathrm{u}$with $k_{\ell-1} < k_\ell$, which could be used in a scheme as displayed in Figure 9, and $\check{c}_\ell$ are the fermionic annihilation operators in the new discretization scheme. For convenience we define the interval edges

$$d_\ell = \begin{cases} D_\mathrm{l} & \ell = 0 \\ (k_\ell + k_{\ell+1})/2 & 1 \le \ell < M \\ D_\mathrm{u} & \ell = M \end{cases}\,, \tag{49}$$

the level spacing

$$\Delta_\ell = d_\ell - d_{\ell-1}\,, \tag{50}$$

and density of state weights

$$\mathcal{N}_\ell = \int_{d_{\ell-1}}^{d_\ell} \mathrm{d}k\, \mathcal{N}(k)\,. \tag{51}$$

In order to preserve the density of states and to ensure canonical commutation relations we have to use the following rule:

$$\frac{1}{\sqrt{2\pi}} \int_{d_{\ell-1}}^{d_\ell} \mathrm{d}k\, \sqrt{\mathcal{N}(k)}\, \tilde{c}(k) \to \sqrt{\frac{\mathcal{N}_\ell}{2\pi}}\, \check{c}_\ell\,, \tag{52}$$

which leads to

$$\left[\check{c}_\ell^+, \check{c}_{\ell'}\right]_+ = \frac{2\pi}{\sqrt{\mathcal{N}_\ell \sqrt{\mathcal{N}_{\ell'}}}} \int_{d_{\ell-1}}^{d_\ell} \frac{\mathrm{d}k}{\sqrt{2\pi}} \sqrt{\mathcal{N}(k)} \int_{d_{\ell'-1}}^{d'_\ell} \frac{\mathrm{d}q}{\sqrt{2\pi}} q\sqrt{\mathcal{N}(q)}\, \left[\hat{c}_k^+, \hat{c}_q\right]_+ \tag{53}$$

$$= \delta_{\ell,\ell'} \mathcal{N}_\ell^{-1} \int_{d_{\ell-1}}^{d_\ell} \mathrm{d}k\, \sqrt{\mathcal{N}(k)} \int_{d_{\ell-1}}^{d_\ell} \mathrm{d}q\, \sqrt{\mathcal{N}(q)} \delta_{k,q} \tag{54}$$

$$= \delta_{\ell,\ell'} \mathcal{N}_\ell^{-1} \int_{d_{\ell-1}}^{d_\ell} \mathrm{d}k\, \mathcal{N}(k) \tag{55}$$

$$= \delta_{\ell,\ell'}\,. \tag{56}$$

Note that in the discretization of a single one-dimensional lead we have $\mathcal{N}(k) = 1$ and therefore $\mathcal{N}_\ell = \Delta_\ell$.

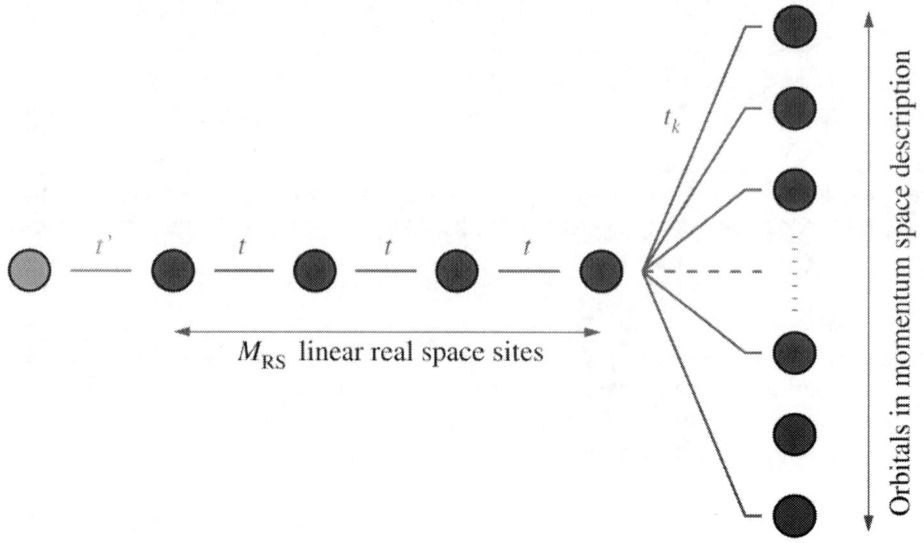

Figure 9. Sketch of a single impurity coupled to M_{RS} real space sites which are then coupled to a lead in momentum space representation.

2.5.3. Level Distribution Function In order to obtain a distribution similar to the logarithmic distribution in section 2.2 we use an integrated distribution function for the levels and discretize the interval $[D_\mathrm{l}, D_\mathrm{u}]$ in equal area sections. Here we use a regularized $1/x$ function

$$\tilde{P}_{\log}(p, w_1, w_2)) = \begin{cases} \frac{1}{\sqrt{(\omega_p+w_2)^2+w_1^2}} & : \omega_p < -w_2 \\ \frac{1}{w_1} & : -w_2 \le \omega_p \le w_2 \\ \frac{1}{\sqrt{(\omega_p-w_2)^2+w_1^2}} & : \omega_p > w_2 \end{cases} \tag{57}$$

$$P_{\log}(p, w_1, w_2, D_\mathrm{l}, D_\mathrm{u})) = \frac{\tilde{P}_{\log}(p, w_1, w_2))}{\int_{D_\mathrm{l}}^{D_\mathrm{u}} \mathrm{d}k\, \tilde{P}_{\log}(k, w_1, w_2))} . \tag{58}$$

to generate the levels. Setting $\omega_p = \epsilon_p$ one can focus the distribution at the energy associated to the momentum. For $w_2 = 0$ one obtains a logarithmic distribution for the levels at ω_p from $P_{\log}(p, w_1, 0, D_\mathrm{l}, D_\mathrm{u}))$ which is slightly smoothed on a scale w_1. The inset of the constant part in the centre enables a linear spacing in the high resolution region. This is important for obtaining accurate results in the calculation of linear transport from the Kubo formula.[10] In addition, by adding distribution functions centred at different ω_p one can obtain a discretization with several accumulation points.

2.5.4. RLM Spectral Function For a symmetric dispersion $\epsilon(p) = \epsilon(-p)$ we can now employ another transformation by taking a symmetric level distribution for left ($p < 0$) and right ($p > 0$) movers and combining them to

$$\check{c}_{\pm,p} = (\check{c}_p \pm \check{c}_{-p})/\sqrt{2} \qquad p > 0 \tag{59}$$

leading to the lead Hamiltonians

$$\mathcal{H}_\pm = \sum_{p>0} \epsilon(p)\, \check{c}^\dagger_{\pm,p} \check{c}_{\pm,p} , \tag{60}$$

where only $\mathcal{H}_+$ couples to the impurity. Therefore we can ignore the $\mathcal{H}_-$ part of the Hamiltonian. If we denote the last real space site with n, then the coupling to the momentum leads is given by

$$-t\,\hat{c}_n^+\hat{c}_{n+1}^+ \rightarrow -\sqrt{2}t\,\hat{c}_n^+\sum_\ell \sqrt{\mathcal{N}_\ell}\,\tilde{c}_{+,\ell}\,. \tag{61}$$

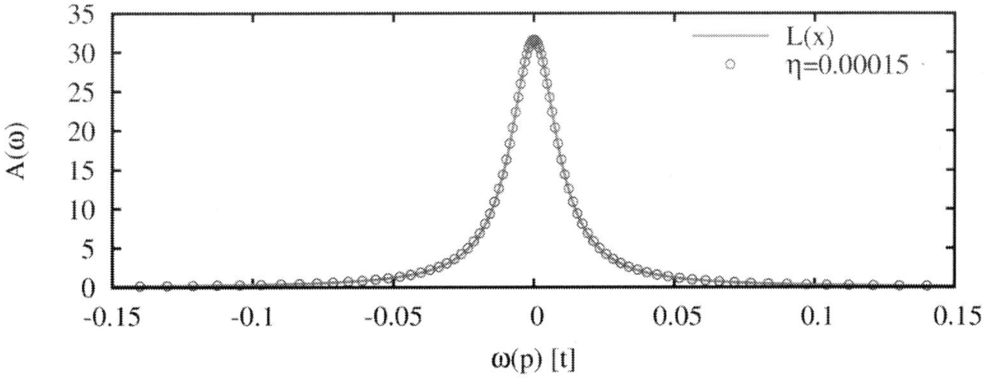

Figure 10. Spectral function of a resonant level using momentum leads and frequency adapted grids. In the numerics we employed a fine grained resolution at ω using $P_{\log}(p = \omega/2, 0.0001, 0.0002, -1, 1)$, a linearised dispersion $\varepsilon_k = 2kt$, and a constant broadening $\eta = 0.00015t$. The momentum leads consist of 50 sites, and the impurity is first coupled to 3 real space sites. The full red line shows the Lorentzian of half half-width $w = 0.01$ corresponding to the spectral function in the wide band limit.

The main advantage of momentum leads is that one can adapt the discretization to the frequency ω that should be resolved. We test this idea by generating a small level distance at frequency $\omega_p = -2t\cos(p)$ by slicing $P_{\log}(p, w_1, 2w_1, D_l, D_u)$ and plot the result in Figure 10. Clearly, this approach gives an excellent result which reproduces the central peak and the tails. In addition, we can even go back to use a fixed η, as the level spacing at ω is now always of the same order of magnitude. This allows us to fit Eq. (41) leading to a bare width of $w_{\mathrm{ML}} = 0.00991$. In addition, the momentum lead approach allows us to restrict the bandwidth to the relevant region. Keeping the parameter of Figure 10 and only changing the momentum cut-off to ± 0.1 we obtain a bare width of $w_{\mathrm{ML}} = 0.00997$.

Finally we demonstrate in Figure 11 that within this scheme one can even obtain accurate derivatives of Green functions with respect to the frequency. The derivative of $\mathcal{G}^\pm$ is given by

$$\frac{\mathrm{d}}{\mathrm{d}\omega}\mathcal{G}^+_{\hat{A},\hat{B}}(\omega) = \langle\Psi_0|\hat{A}\,\frac{-1}{(E_0 - \mathcal{H} + \omega + \mathrm{i}\eta)^2}\,\hat{B}|\Psi_0\rangle \tag{62}$$

$$\frac{\mathrm{d}}{\mathrm{d}\omega}\mathcal{G}^-_{\hat{A},\hat{B}}(\omega) = \langle\Psi_0|\hat{A}\,\frac{1}{(E_0 - \mathcal{H} - \omega - \mathrm{i}\eta)^2}\,\hat{B}|\Psi_0\rangle\,. \tag{63}$$

It turns out that the numerical evaluation of Eqs. (62, 63) is more sensitive to the discretization used. We switch to a linear band, $\omega = 2kt$, with cutoffs of $D_{\mathrm{l,u}} = \pm 0.1$, 100 momentum lead sites, the distribution function $P_{\log}(p, w_1, 2w_1)$, $w_1 = 0.0005$, and $\eta = 0.0005$. As a comparison we plot the exact result and the exact result convoluted with a Lorentzian of half half-width $\eta = 0.0005$,

$$L'_\eta(\omega, w) = \frac{\mathrm{d}}{\mathrm{d}\omega}L_\eta(\omega, w) = \frac{-2\omega}{\pi}\,\frac{w+\eta}{(\omega^2 + (w+\eta)^2)^2}\,. \tag{64}$$

By fitting $L'_{0.0005}(\omega, w)$ of Eq. (41) to the numerical result we obtain $w = 0.01004$. For the same discretization a fit Eq. (41) of the spectral function gives $w = 0.0099996$.

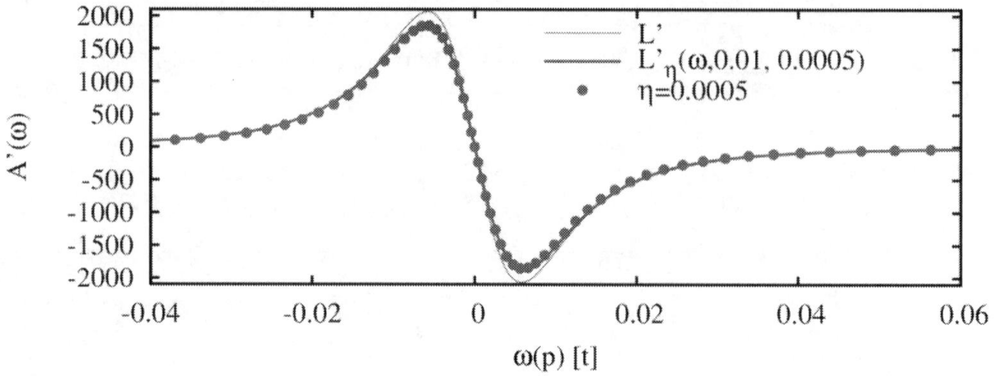

Figure 11. Derivative of the spectral function of a resonant level using momentum leads. The red line is the analytical result in the wide band limit, the blue line is the analytical result convoluted with a Lorentzian of half half-width $\eta = 0.0005$. The circles are obtained from momentum leads using a distribution centred around $\pm\omega$ with $P_{\log}(p, w_1, 2w_2)$, $w_1 = 0.0005$, $w_2 = 0.0001$, a linear lead $\omega(k) = 2kt$ with cutoffs $D_{\mathrm{l,u}} = \pm 0.1t$, and 3 real space sites sandwiched between the impurity and the momentum leads consisting of 100 sites.

3. Bulk Green Functions

In the previous sections we discussed the spectral properties of an impurity, namely a single resonant level, coupled to a non-interacting lead. In this case it was natural to look at the local spectral functions of the impurity. We now turn to the momentum resolved spectral function of a bulk system. As an example we look at the retarded Green function of a tight binding chain

$$\mathcal{H} = -t \sum_x \hat{c}_x^+ \hat{c}_{x-1} + \hat{c}_{x-1}^+ \hat{c}_x = -2t \int_{-\pi}^{\pi} \mathrm{d}k\, \cos(k)\, \hat{f}_k^+ \hat{f}_k$$

$$\hat{f}_k = \frac{1}{\sqrt{2\pi}} \sum_x \mathrm{e}^{\mathrm{i}kx}\, \hat{c}_x$$

At $k = \pi/2$ the spectral function is simply given by $A(\omega) = \delta(\omega)$.

3.1. Choosing Correct Single Particle States

For simplicity let us start with a sine function ansatz for the single particle states on M sites

$$\hat{f}_n = \frac{1}{\sqrt{M}} \sum_{x=1}^{M} \sin(k_n x)\, \hat{c}_x \tag{65}$$

$$k_n = \frac{2\pi n}{M} \tag{66}$$

As can be seen in Figure 12 this ansatz leads to a fictitious double peak structure and an oscillatory part which does not resemble the δ-peak of the system in the thermodynamic limit. The reason for this is that by using the single particle states of Eqs. (65, 66) we used single particle states which are not eigenstates of the system with periodic boundary conditions (PBC).

Therefore we repeat this calculation for a tight binding chain employing hard wall boundary conditions (HWBC). The graph in Figure 13 shows that the result looks much better now, but we still obtain the fictitious double peak structure. Again, the reason for this lies in a wrong single particle basis.

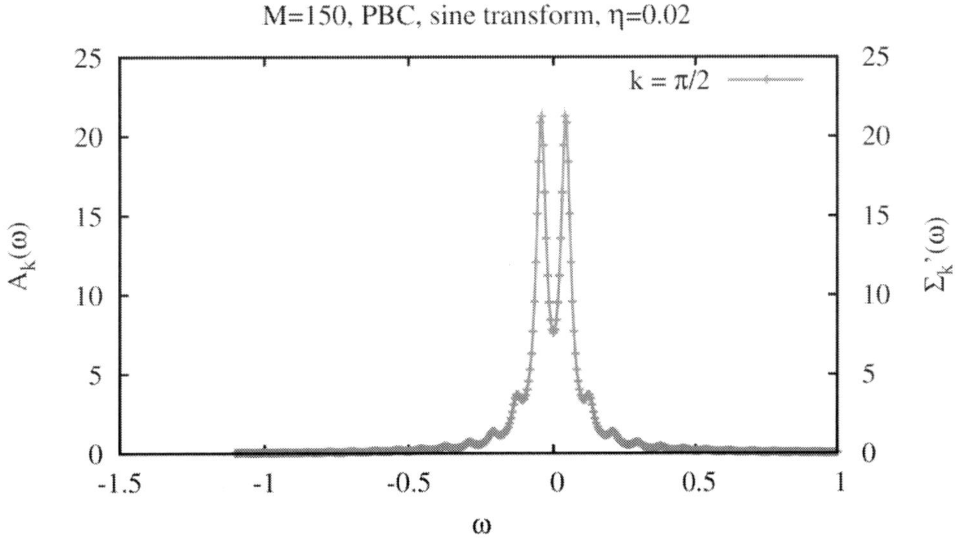

Figure 12. Spectral function of the tight binding chain at $k = \pi/2$ using the single particle states of Eqs. (65, 66) and a tight binding chain with $M = 150$ lattice sites and periodic boundary conditions (PBC) and sine function as the basis set.

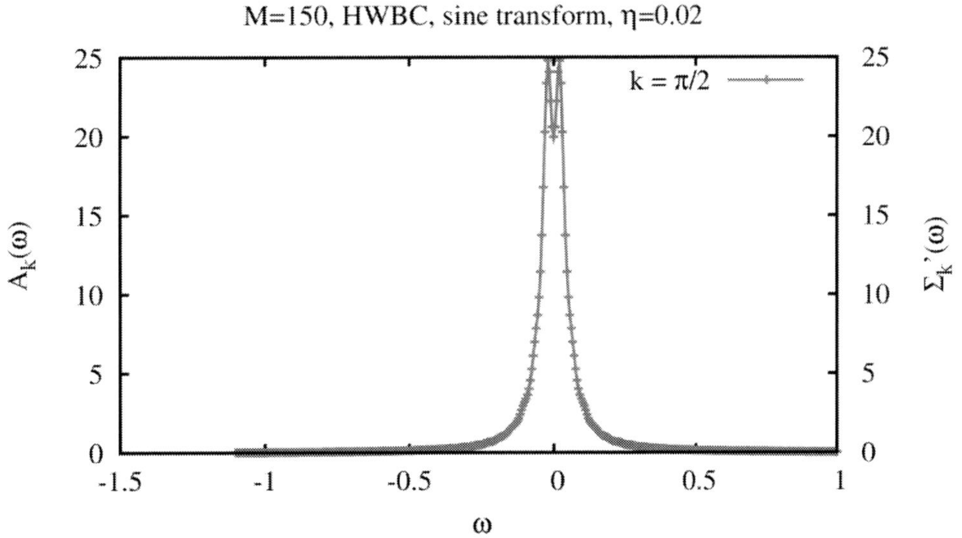

Figure 13. Spectral function of the tight binding chain at $k = \pi/2$ using the single particle states of Eqs. (65, 66) and a tight binding chain with $M = 150$ lattice sites and periodic boundary conditions (PBC) and sine function as the basis set.

However, this time the reason for its failure it is more subtle. The problem is that Eq. (66) gives the wrong eigenstates for HWBC and the correct single particle basis is given by

$$\hat{f}_n = \sqrt{\frac{2}{M+1}} \sum_{x=1}^{M} \sin(k_n x)\, \hat{c}_x \qquad k_n = \frac{n\pi}{M+1} \quad n = 1, 2, \cdots, M \tag{67}$$

leading to the result of Figure 14 which finally resembles the δ-peak broadened by η. In summary we

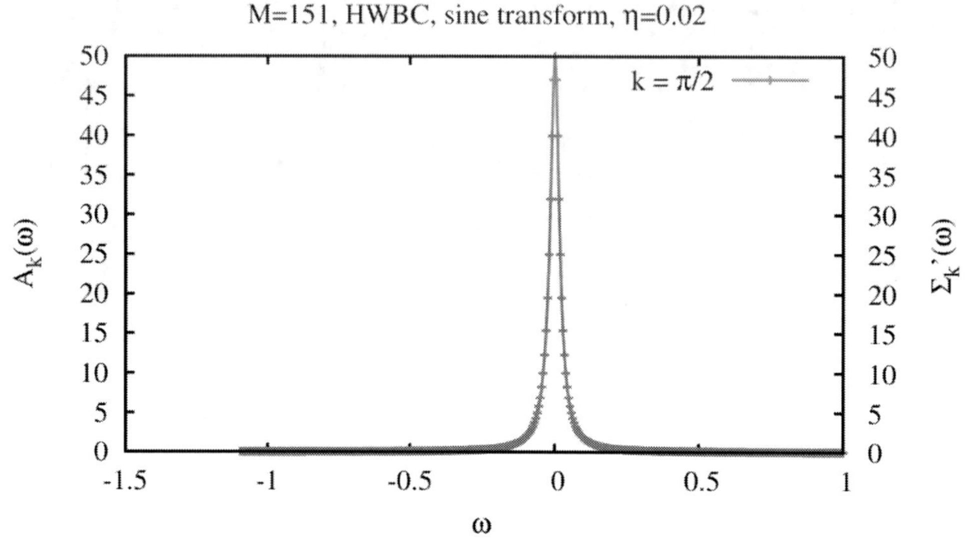

Figure 14. Spectral function of the tight binding chain at $k = \pi/2$ using the single particle states of Eq. (67) and a tight binding chain with $M = 150$ lattice sites and periodic boundary conditions (PBC) and sine function as the basis set.

would like to point out that one can obtain nice spectral functions for the bulk system from finite systems, however care has to be taken to choose the correct representation. Otherwise spurious structures may appear. The need for using the sine solution for HWBC has been pointed out by Benthien, Gebhard, and Jeckelmann[15]. It was shown by Ulbricht and Schmitteckert that for interacting particle in a harmonic trap one can obtain spectral functions from finite systems by resorting to hermite polynomials.[16]

4. Poor Man's Deconvolution: Self Energy Sharpening

In this section we discuss a simple strategy to "sharpen" Green functions obtained in the previous sections. It was first applied to improve the spectral function of a spin polarized one-dimensional Hubbard model.[16] As an example we improve the result for the spectral function of the tight binding chain of the previous section. A straight forward strategy to improve the result consists in a deconvolution to remove the broadening introduced by the finite η in the denominator of the resolvents, i.e. perform the inverse operation of Eq. (41). While this procedure is mathematically well defined, it is highly unstable numerically and very sensitive to numerical noise.[17] Here we want to introduce a simple method to improve the results, which has the advantage of being stable and allows to preserve analytical properties of spectral functions, e.g. $\mathcal{A}(\omega) \leq 0$.

We start by defining the self energy $\Sigma(\omega)$ for the infinite system

$$\mathcal{G}^r(\omega) = \frac{1}{\omega - \Sigma(\omega) + \mathrm{i}0^+} \tag{68}$$

and the self energy $\Sigma_\eta(\omega)$ for the finite system which includes the finite broadening η

$$\mathcal{G}^r_\eta(\omega) = \frac{1}{\omega - \Sigma_\eta(\omega) + \mathrm{i}0^+} . \tag{69}$$

When switching from the infinite system to the finite system we should trace out the discarded part which would lead to a contribution to the self energy of the reduced system. However, we do not know this part. Therefore we make the ansatz that the discarded part can be modelled by the $\mathrm{i}\eta$ term we already have in the resolvents to enable the mixing of excited states. In this way we obtain our "poor man's deconvolution" ansatz $\Sigma_\eta = \Sigma + \mathrm{i}\eta$:

$$\Sigma_\eta(\omega) = \omega - \frac{1}{\mathcal{G}^r_\eta(\omega)} + \mathrm{i}0^+ \tag{70}$$

$$\Sigma(\omega) = \omega - \frac{1}{\mathcal{G}^r_\eta(\omega)} - \mathrm{i}\eta + \mathrm{i}0^+ \tag{71}$$

where $\mathcal{G}^r_\eta(\omega)$ is the Green function we obtain from our numerics on a finite lattice and η is the broadening explicitly used in the numerics. The result of this sharpening applied to the data of Figure 14 is shown in Figure 15. Since real and imaginary part are vanishing up to numerical precision we have rediscovered the δ-peak structure of the infinite system from the numerics on a finite lattice.

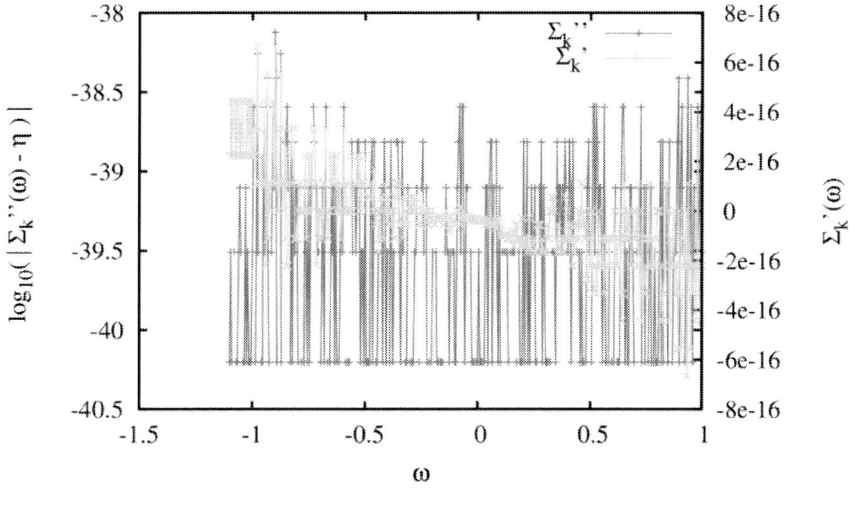

Figure 15. Imaginary (Σ'') and real (Σ') part of the self energy of tight binding chain obtained by sharpening the results Figure 14 for a 151 site tight binding chain at $k = \pi/2$ using the single particle states of Eq. (67), hard wall boundary conditions (HWBC) and sine function as the basis set.

5. The NRG Tsunami

Finally we would like to discuss the effect of damped boundary conditions on the dynamics of wave packets to show that boundary conditions are not only important in frequency space, but that they can also change the results in time domain dramatically. Besides the extraction of excitation velocities,[18–21] it has recently gained interested for the calculation of finite bias conductance[19; 20; 22–29] In this section we follow the route of [18] adapted for free fermions using exact diagnalization of the quadratic form. In Figure 16 we show the initial states of a 201 site tight binding chain with periodic boundary

Journal of Physics: Conference Series **220** (2010) 012022 doi:10.1088/1742-6596/220/1/012022

conditions at half filling, where half a fermion was trapped in the middle of the system by applying a Gaussian potential with a width of $\sigma = 2.5$. In addition we show the response to the same perturbation for a system where the PBC are replaced by damped boundary conditions, compare Figure 6.

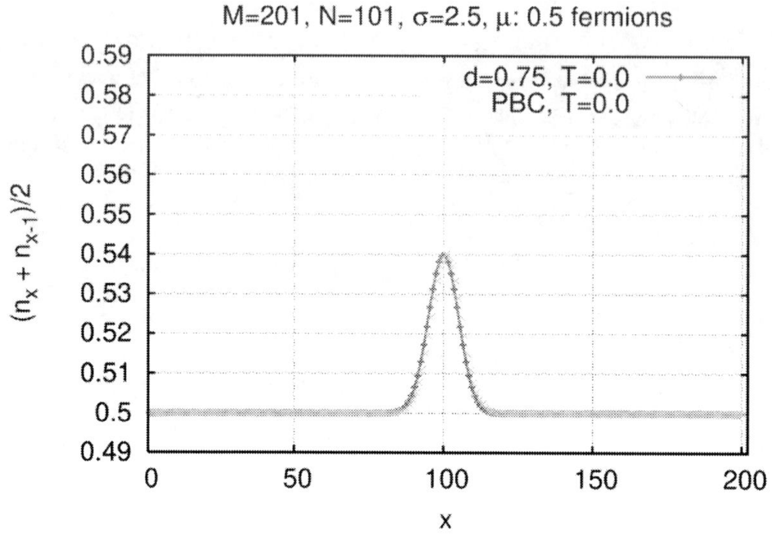

Figure 16. Density of a $M = 201$ site tight binding chain with $N = 101$ fermions, where a Gaussian potential of width $\sigma = 2.5$ was applied to trap an additional half fermion in the centre region in addition to an average density of $\rho = 0.5$. In the results for periodic boundary conditions (PBC) are given by the crosses. The result for Damped Boundary conditions as displayed in Figure 6 applied to the left and right end of the chain is displayed by the line with plusses.

In Figure 17 we show the system after performing a time evolution up to time $T = 90$, in which the homogeneous system moves a distance of $T{*}\mathrm{v_F} = 180$ sites. Due to the periodic boundary conditions the wave packets appear now at position $x \approx 100 \pm 20$. Note that by applying a low energy perturbation we create excitations around $\pm\mathrm{k_F}$ and our initial wave packet splits into a right and left moving wave packet travelling at $\pm\mathrm{v_F}$, where $\mathrm{k_F}$ is the Fermi momentum and $\mathrm{v_F}$ is the excitation velocity. However, once the wave packet hits the region of damped boundary conditions (DBC) it sees an exponentially decreasing hopping element which results in an exponentially decreasing excitation velocity. In the similar way as long water waves hitting the shore, where the excitation velocity is decreased, our wave packet has to pile up like a tsunami, since the front is moving much slower than the back. In addition each changed hopping element creates a small back reflection leading to an additional wiggling.

In using the DPC for conductance calculation from extended wave packets, e.g. see [30; 31], one can profit from the enhanced density of states at the Fermi surface, however one has to take the inverse tsunami into account, namely particle trapped in the DBC region are exponentially slow and can therefore not sustain a steady state.

6. Summary

In summary we have shown that one can extract sound results for the thermodynamic limit in the context of calculating Green functions from a finite lattice. However, care has to be taken to choose the correct representation of states. In order to calculate spectral functions at finite frequencies for impurity problems one should apply frequency adapted grids to achieve high resolution. A sharpening procedure based on the self energy can be used as a replacement for deconvoluting the broadening introduced by a finite η.

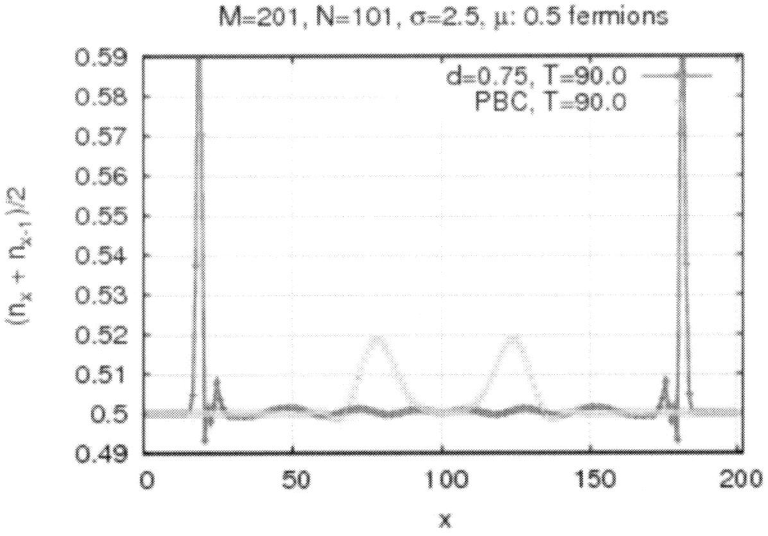

Figure 17. Density of a $M = 201$ site tight binding chain with $N = 101$ fermions, where a Gaussian potential of width $\sigma = 2.5$ as in Figure 16 after performing a time evolution of time $T = 90$ with the Hamiltonian without the perturbation. The results for periodic boundary conditions (PBC) are given by the crosses. The result for damped boundary conditions as displayed in Figure 6 applied to the left and right end of the chain is shown by the line with plusses.

Acknowledgments
The author wishes to acknowledge assistance and clarifying discussions with Frithjof Anders, Alexander Branschï¿½del, Dan Bohr, Eric Jeckelmann, Stefan Kremer, Christina Stawiarsky, and Tobias Ulbricht.

References

[1] Mahan G D 2000 *Many particle physics* 3rd ed (New York: Kluwer Academics / Plenum Publishers) ISBN 0-306-46338-5
[2] Orland H and Negele J W 1998 *Quantum Many Particle Systems* (Perseus Books)
[3] Bruus H and Flensberg K 2004 *Many-Body Quantum Theory in Condensed Matter Physics: An Introduction.* (Oxford University Press) ISBN 0-19-856633-6
[4] Ramasesha S 1990 *J. Comp. Chem.* 545
[5] Weiße A, Wellein G, Alvermann A and Fehske H 2006 *Rev. Mod. Phys.* **78** 275–306
[6] Longo P, Schmitteckert P and Busch K 2010 *Phys. Rev. Lett.* **104** 023602
[7] Krishna-murthy H R, Wilkins J W and Wilson K G 1980 *Phys. Rev. B* **21** 1003–1043
[8] Bulla R, Costi T A and Pruschke T 2008 *Rev. Mod. Phys.* **80** 395–450
[9] Vekić M and White S R 1993 *Phys. Rev. Lett.* **71** 4283–4286
[10] Bohr D and Schmitteckert P 2007 *Phys. Rev.* B **75** 241103(R)
[11] Freyn A and Florens S 2009 *Phys. Rev. B* **79** 121102
[12] Nishimoto S and Jeckelmann E 2004 *J.Phys.:Condens. Matter* **16** 613
[13] Žitkoa R 2009 *Comput. Phys. Comm.* **180** 1271
[14] Bohr D, Schmitteckert P and Wölfle P 2006 *Europhys. Lett.* **73** 246
[15] Benthien H, Gebhard F and Jeckelmann E 2004 *Phys. Rev. Lett.* **92** 256401

[16] Ulbricht T and Schmitteckert P 2010 *EPL* **89** 47001

[17] Raas C and Uhrig G 2005 *EPJ* B **45** 293

[18] Schmitteckert P 2004 *Phys. Rev.* B **70** 121302(R)

[19] Schmitteckert P and Schneider G 2006 *High Performance Computing in Science and Engineering '06* ed Nagel W E, Jäger W and Resch M (Berlin: Springer) pp 113–126

[20] Branschädel A, Ulbricht T and Schmitteckert P 2009 *High Performance Computing in Science and Engineering '09* ed Nagel W E, Kröner D B and Resch M (Berlin: Springer) p 123 ISBN 978-3-642-04664-3

[21] Ulbricht T and Schmitteckert P 2009 *EPL* **86** 57006

[22] Di Ventra M and Todorov T N 2004 *J. Phys.: Condens. Matter* **16** 8025

[23] Stefanucci G and Almbladh C O 2004 *Europhys. Lett.* **67** 14

[24] Stefanucci G and Almbladh C O 2004 *Phys. Rev. B* **69** 195318

[25] Bushong N, Sai N and Di Ventra M 2005 *Nano Letters* **5** 2569

[26] Schneider G and Schmitteckert P 2006 Conductance in strongly correlated 1d systems: Real-time dynamics in dmrg (unpublished) (*Preprint* `cond-mat/0601389`)

[27] Al-Hassanieh K A, Feiguin A E, Riera J A, Büsser C A and Dagotto E 2006 *Physical Review B (Condensed Matter and Materials Physics)* **73** 195304

[28] Boulat E, Saleur H and Schmitteckert P 2008 *Physical Review Letters* **101** 140601

[29] Ulbricht T and Schmitteckert P 2008 *High Performance Computing in Science and Engineering '08* ed Nagel W E, Kröner D B and Resch M (Berlin: Springer) pp 71–82 ISBN 978-3-540-88301-2

[30] Anders F B 2008 *Phys. Rev. Lett.* **101** 066804

[31] Dias da Silva L G G V, Heidrich-Meisner F, Feiguin A E, Büsser C A, Martins G B, Anda E V and Dagotto E 2008 *Phys. Rev. B* **78** 195317

Journal of Physics: Conference Series **220** (2010) 012023 doi:10.1088/1742-6596/220/1/012023

Analytical calculation of the Green's function and Drude weight for a correlated fermion-boson system

A Alvermann[1], **D M Edwards**[2] **and H Fehske**[1]

[1]Institute of Physics, Ernst-Moritz-Arndt University, Greifswald, Germany

[2] Department of Mathematics, Imperial College London, London SW7 2BZ, United Kingdom

E-mail: `alvermann@physik.uni-greifswald.de`

Abstract. In classical Drude theory the conductivity is determined by the mass of the propagating particles and the mean free path between two scattering events. For a quantum particle this simple picture of diffusive transport loses relevance if strong correlations dominate the particle motion. We study a situation where the propagation of a fermionic particle is possible only through creation and annihilation of local bosonic excitations. This correlated quantum transport process is outside the Drude picture, since one cannot distinguish between free propagation and intermittent scattering. The characterization of transport is possible using the Drude weight obtained from the f-sum rule, although its interpretation in terms of free mass and mean free path breaks down. For the situation studied we calculate the Green's function and Drude weight using a Green's functions expansion technique, and discuss their physical meaning.

1. Motivation

The motion of a fermionic particle, such as an electron or hole, that interacts strongly with some background medium is a constantly recurring theme in condensed matter physics. The present contribution is motivated partly by the study of the motion of a hole in an antiferromagnetic spin background, as in a doped Mott insulator. Mott insulators occur for electrons on a lattice subject to strong Coulomb repulsion [1], and are commonly studied in the context of the Hubbard model. Since putting two electrons at the same lattice site costs a large energy if Coulomb repulsion is strong, the number of doubly occupied lattice sites is negligibly small. At half-filling we can therefore assume that every lattice site is occupied by a single electron. The system is an insulator because electron motion, which creates doubly occupied sites, is energetically forbidden.

If electron motion is suppressed, the electron spin is the only relevant degree of freedom. Perturbation theory in the kinetic energy shows that spins couple antiferromagnetically in a Mott insulator. Starting from the Hubbard model, spin dynamics is described by the antiferromagnetic Heisenberg model with complex properties arising from the interplay of strong correlations and quantum spin fluctuations. In a very simple picture, we may alternatively consider the Néel state of a classical antiferromagnet (see Fig. 1, upper panel).

If an electron is removed from a two-dimensional (2D) Mott insulator in a Néel state, the hole can move without creating doubly occupied sites. But the moving hole distorts the antiferromagnetic spin background, creating misaligned spins with substantial energy (see Fig. 1, lower panel). The 'string' of misaligned spins that forms along the path of the hole strongly

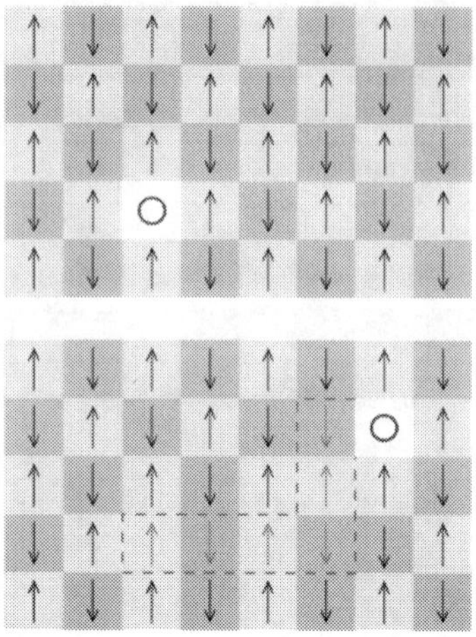

Figure 1. Upper panel: Hole in a Néel antiferromagnet on a square lattice. Lower panel: When the hole moves by nearest-neighbour hopping, it creates misaligned spins that do not match the Néel order. These distortions of the spin background lead to parallel spins on neighbouring sites – indicated with a dashed line – with a substantial increase in energy. To 'unwind' the string of background distortions, the hole in principle has to retrace the path it moved, and is therefore bound to its origin. The existence of (quantum) spin fluctuations, and spin-order restoring processes (see Fig. 2) weakens, or completely destroys, this 'string effect'.

restricts propagation, and tends to bind the hole to its origin. Therefore the hole does not move as a free particle. In a quantum antiferromagnet, which does not possess the strict order of the Néel antiferromagnet, the spin distortions can 'heal out' or relax by quantum spin fluctuations. It is then the relaxation rate of spin distortions that determines how fast the hole can move: The hole is continuously creating a string of distortions but can move on 'freely' at a speed which gives the distortions time to decay.

In the absence of quantum spin fluctuations free-particle like motion of a hole is completely suppressed. Surprisingly enough, a hole can propagate through so-called 'Trugman paths' [2], which realize a certain combinatorial rearrangement of spins compatible with the spin order. The simplest of these processes, where the hole moves by two sites and restores the antiferromagnetic spin alignment, is shown in Fig. 2. In contrast to the situation for an almost free particle propagation through such processes depends entirely on the successive rearrangement of spins in precise coordination with the hole motion, just like executing a complicated dance pattern. With respect to the correlations between spin and hole motion we call this a correlated transport process.

The focus of the present contribution is on similar correlated transport processes in a simpler fermion-boson model. This model is introduced in the next section (Sec. 2). In Sec. 3 we define the Drude weight as the appropriate measure of the efficiency of a correlated transport process. In Sec. 4 we derive the one-fermion Green's function, which is discussed together with the Drude weight in Sec. 5, before we conclude in Sec. 6.

2. The model

The picture sketched here — how the motion of a particle depends on the creation of background distortions and their decay — is very general, and applies e.g. to the polaron problem of electron

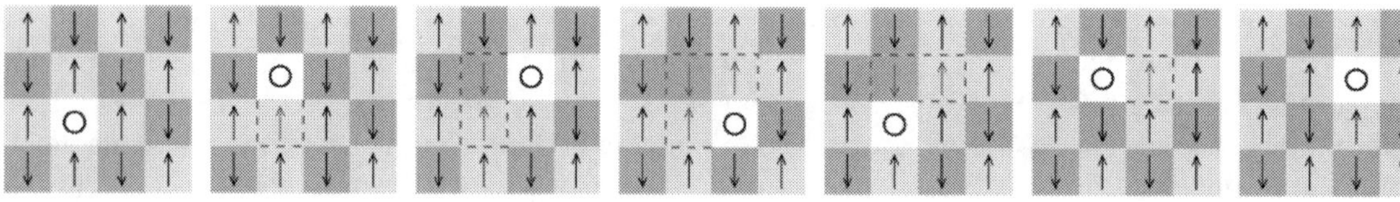

Figure 2. The simplest 'Trugman path' that allows for propagation of a hole in a Néel antiferromagnet [2]. After six steps, the hole has moved by two sites to a next-nearest neighbour and the Néel spin order is restored.

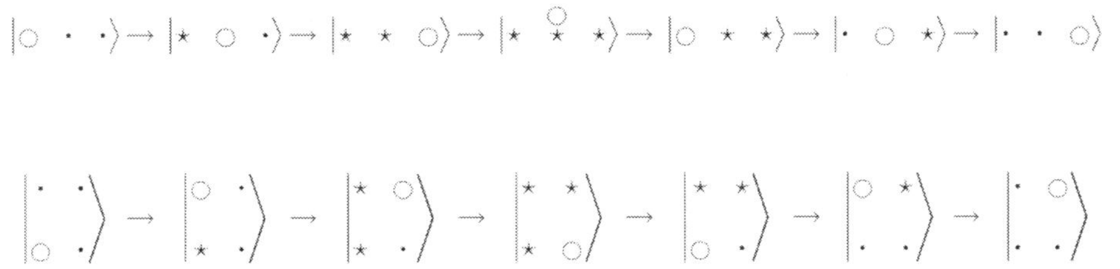

Figure 3. Lowest order vacuum-restoring process that allows for propagation of a fermion in the model Eq. (1), on a 1D chain (upper row) and 2D square lattice (lower row). In any dimension, it comprises six steps and propagates the fermion by two sites. In steps 1–3, three bosons are excited, which are consumed in steps 4–6. The efficiency of this process consequently scales as t_b^6/ω_0^5 (see Sec. 5.1). Note that the 2D process is the exact translation of the Trugman path shown in Fig. 2, where the correlated many-particle background of the spin model is replaced by the bosonic vacuum. For our model, such processes play a role already in 1D.

motion in a deformable lattice studied using the Holstein model [3,4], and to charge transport in the presence of strong electronic correlations, such as for hole-doped Mott insulators studied using the t-J-model [5,6]. For that reason, we recently introduced a simple but general fermion-boson model that captures the fundamental aspects without modelling any very specific details. This model was originally proposed by Edwards [7] and subsequently used to study single-particle quantum transport [8] and metal-insulator transitions [9–11].

We here consider the Hamiltonian

$$H = -t_b \sum_{\langle i,j \rangle} c_j^\dagger c_i (b_i^\dagger + b_j) + \omega_0 \sum_i b_i^\dagger b_i \tag{1}$$

for a spinless fermion (created by $c_i^\dagger$) coupled to bosons (created by $b_i^\dagger$) of energy ω_0. The first term in the Hamiltonian specifies how bosons are created (and annihilated) when the fermion hops to a nearest neighbour. Any hop of the fermion changes the number of bosons by one. In this sense, fermion motion is boson-controlled.

Fermion hopping creates a boson only on the site the fermion leaves, and destroys a boson only on the site the fermion enters. As a consequence motion of the fermion creates, in a lattice of arbitrary dimension, a 'string' of bosonic excitations: In each hop the fermion creates additional bosons that can be destroyed only by returning to sites already visited. This 'string effect' restricts propagation just as for the hole in the 2D antiferromagnet.

The original model (in Refs. [7–9]) contains an additional term $-\lambda \sum_i (b_i^\dagger + b_i)$ that accounts for boson relaxation. Bosons can decay when $\lambda \neq 0$, so that in this case the string effect is weakened. In the present situation, the string effect is not overcome by boson relaxation. But just as for the Trugman paths, there exist processes whereby the fermion propagates, while the boson vacuum is restored [8]. The simplest of these vacuum-restoring processes comprises six steps, promoting the fermion by two sites (see Fig. 3). If we compare to Fig. 2, we see that the 2D process in our model is an exact representation of a Trugman path. Here, such processes occur already in 1D, and can be considered as a projection of the corresponding 2D process. In our model such correlated transport processes emerge quite naturally from some basic assumptions. Our general model now allows us to study the implications of the existence of such processes for the motion of a fermionic particle. An analytical calculation of the one-fermion Green's function, which is only possible in the $\lambda = 0$-case considered here, enables us to characterize correlated transport using the Drude weight. We restrict ourselves to the study of a single fermion in 1D at zero temperature.

$$
\begin{aligned}
|\underline{\circ}\rangle &= c_k^\dagger|\mathrm{vac}\rangle = e^{\mathrm{i}k}\,|\underline{\cdot}\ \circ\rangle = e^{-\mathrm{i}k}\,|\circ\ \underline{\cdot}\rangle \\
|\underline{\star}\rangle &= b_k^\dagger|\mathrm{vac}\rangle \\
|\overset{\circ}{\underline{\star}}\rangle &= \sum_n e^{\mathrm{i}kn} c_n^\dagger b_n^\dagger|\mathrm{vac}\rangle \\
|\underline{\circ}\ \star\rangle &= \sum_n e^{\mathrm{i}kn} c_n^\dagger b_{n+1}^\dagger|\mathrm{vac}\rangle = e^{-\mathrm{i}k}\,|\circ\ \underline{\star}\rangle \\
|\underline{\star}\ \circ\rangle &= \sum_n e^{\mathrm{i}kn} c_{n+1}^\dagger b_n^\dagger|\mathrm{vac}\rangle = e^{-\mathrm{i}k}\,|\star\ \underline{\circ}\rangle \\
|\star\ \overset{\circ}{\underline{\star}}\ \star\rangle &= \sum_n e^{\mathrm{i}kn} c_n^\dagger b_{n-1}^\dagger b_n^\dagger b_{n+1}^\dagger|\mathrm{vac}\rangle
\end{aligned}
$$

Figure 4. Examples of the graphical notation for fermion-boson states. An empty site is represented by a dot ($\cdot$), the fermion by a circle ($\circ$), and a boson by a star ($\star$). The origin is indicated with an underscore (_). All states are Bloch states with momentum k. The last row shows the symmetric state which appears in the 6-step-process (cf. Fig. 5).

3. Drude weight

Transport of almost free particles is conveniently characterized through the particle mass and an effective mean free path. How, then, can transport be characterized in a situation where it differs entirely from free particle motion? As stressed by Kohn [12] one should consider the Drude weight D, which can be defined through its relation

$$
D = 1/(2m^*)\,, \quad \frac{1}{m^*} = \frac{\partial^2 E(k)}{\partial k^2}\Big|_{k=0} \tag{2}
$$

to the quasiparticle mass m^*, where $E(k)$ is the quasiparticle dispersion. Alternatively, the Drude weight can be introduced through the relation $\mathrm{Re}\,\sigma(\omega) = 2\pi D\,\delta(\omega) + \sigma_{\mathrm{reg}}(\omega)$ for the real part of the conductivity $\sigma(\omega)$, or be obtained from the f-sum rule, which relates the integrated conductivity to the kinetic energy. Note that for our model with boson-controlled hopping, the current operator is more complicated than usual [8].

The Drude weight generalizes the original Drude theory with its concept of an effective mean free path between two scattering events, which is valid only for an almost free particle. Note that just through the above relation between Drude weight and quasiparticle mass the latter is identified as a relevant quantity in a rigorous quantum transport theory. Indeed, the quasiparticle mass itself is a rather complicated quantity, which can be strongly renormalized as a consequence of correlated many-particle dynamics.

4. Derivation of the Green's function with a 3-boson constraint

Even for the simple-looking Hamiltonian in Eq. (1) an exact calculation of Green's functions is not possible. The situation changes if an additional constraint is imposed, namely a bound on the number of simultaneously excited bosons. Since the vacuum restoring 6-step-process in Fig. 3 involves 3 bosons, a non-trivial result is obtained allowing for at most 3 bosons in the system (we will justify this restriction further in the next section). Then, only a finite number of states contribute to the one-fermion Green's function

$$
G(k,z) = \langle \mathrm{vac}|c_k[z-H]^{-1}c_k^\dagger|\mathrm{vac}\rangle\,. \tag{3}
$$

This enables us to derive a closed expression for this Green's function.

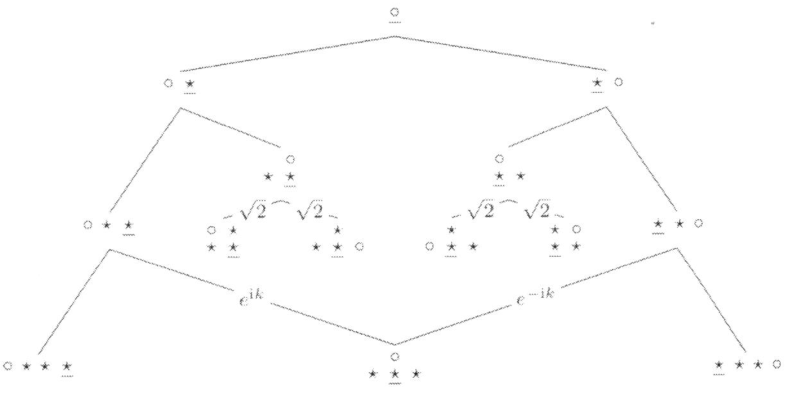

Figure 5. Graph structure for the calculation of $G(k, z)$. Each unlabelled edge corresponds to a matrix element t_b, the labelled edges to t_b times the given factor. The $\sqrt{2}$ arises from matrix elements of bosonic operators.

4.1. The tree of states

As in Fig. 3, we use a graphical notation for states, explained in Fig. 4. The calculation requires the use of Bloch states with given momentum k. As an example, $|\circ\rangle = \sum_n e^{ikn} c_n^\dagger |\text{vac}\rangle$. The graph in Fig. 5 shows all states with at most 3 bosons that are accessible from $|\circ\rangle$ by repeated application of H, while the edges of the graph correspond to non-zero matrix elements of H.

The graph of states contains a loop corresponding to the 6-step-process. To obtain a simple expression for the Green's function with a recursive calculation we must get rid of this loop. To this end we perform a change of basis that exploits the reflection symmetry of the Hamiltonian (1), and rearrange all states in the tree structure shown in Fig. 6. Note that the states in this tree are not eigenstates of the parity (or reflection) operator, since they all have finite momentum k. Note further that it is important to choose the position of the origin correctly. The origin in the two states below the tree root is shifted by one site. Then the symmetric state in the last line of Fig. 4, which appears at the bottom of the loop in Fig. 5, appears only in the left subtree. The corresponding linear combination with a minus sign, which would appear in the right subtree, is zero due to reflection symmetry. Therefore Fig. 6 has no loop. The shift of the origin accounts for the two sites the fermion propagates in the 6-step-process.

4.2. Recursive calculation of the Green's function

The tree in Fig. 6 allows for the recursive calculation of $G(k, z)$. The basic elements in this calculation are Green's functions $G_s(z)$ associated to a state $|s\rangle$ and the subtree with $|s\rangle$ at its top. Such a Green's function is defined through the usual equation $G_s(z) = \langle s|[z - H_s]^{-1}|s\rangle$ (similar to Eq. (3)), where H_s is the projection of the Hamiltonian onto the subspace spanned by all states in the subtree. Note that these projected Hamiltonians commute for disjoint subtrees, so that the corresponding Green's functions can be calculated independently. It is for this reason that we had to remove the loop in the original graph of states (Fig. 5).

At each step of the calculation, a Green's function $G_s(z)$ associated to a subtree is written in the form

$$G_s(z) = [z - a - b_l^2 G_l(z) - b_r^2 G_r(z)]^{-1} , \tag{4}$$

where $G_l(z)$ and $G_r(z)$ are the Green's functions associated to the left and right subtrees one step further down the tree, b_l and b_r are the matrix elements of H along the corresponding edges,

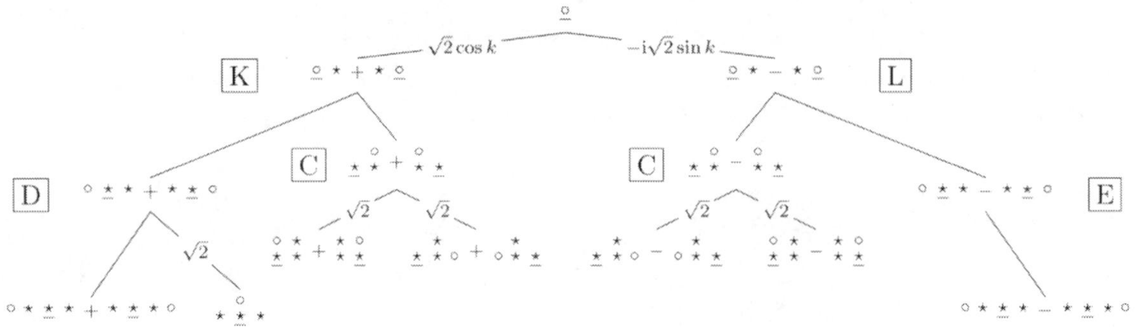

Figure 6. Rearranged tree structure for the calculation of $G(k,z)$ after a basis change. Note that the matrix element $\sqrt{2}t_b$ at the edge in the D-subtree results from the scalar product of the two corresponding states, and not from bosonic operators.

and $a = \langle s|H|s\rangle$ is the diagonal matrix element of H in the state $|s\rangle$. Note that when $|s\rangle$ is at the bottom of the tree, the associated Green's function is simply given by $G_s(z) = [z-a]^{-1}$.

To justify Eq. (4) we proceed in a standard way and split H_s as

$$H_s = a|s\rangle\langle s| + b_l V_l + H_l + b_r V_r + H_r \;, \tag{5}$$

where $|l\rangle$, $|r\rangle$ correspond to the states associated with $G_l(z)$, $G_r(z)$, and $V_l = |s\rangle\langle l| + |l\rangle\langle s|$, $V_r = |s\rangle\langle r| + |r\rangle\langle s|$. Multiplying both sides of the operator identity

$$1 = [z - H_r - H_l]^{-1}[z - H_s \, + a|s\rangle\langle s| + b_l V_l + b_r V_r] \tag{6}$$

with $[z-H_s]^{-1}$ from the right, we get the (generalized) Dyson equation

$$[z-H_s]^{-1} = [z-H_r-H_l]^{-1} + [z-H_r-H_l]^{-1}\left[a|s\rangle\langle s| + b_l V_l + b_r V_r\right][z-H_s]^{-1} \,. \tag{7}$$

Taking scalar products with the state $|s\rangle$ leads to

$$G_s(z) = z^{-1} + z^{-1}\Big(aG_s(z) + b_l\langle l|[z-H_s]^{-1}|s\rangle + b_r\langle r|[z-H_s]^{-1}|s\rangle\Big) \,, \tag{8}$$

and in a second step, with $|l\rangle$, $|r\rangle$, to

$$\begin{aligned} \langle l|[z-H_s]^{-1}|s\rangle &= G_l b_l G_s(z) \,, \\ \langle r|[z-H_s]^{-1}|s\rangle &= G_r b_r G_s(z) \,. \end{aligned} \tag{9}$$

Inserting Eq. (9) into Eq. (8) and solving for $G(s)$ yields our result Eq. (4).

We now apply Eq. (4) to the tree of states in Fig. 6. Working from the top of the tree of states to the bottom, one obtains

$$\begin{aligned} G(k,z) &= \Big[z - 2t_b^2(K(z)\cos^2 k + L(z)\sin^2 k)\Big]^{-1} \\ &= \Big[z - t_b^2(K(z)+L(z)) - t_b^2(K(z)-L(z))\cos 2k\Big]^{-1} \,, \end{aligned} \tag{10}$$

where the intermediate results to the subtrees labelled in Fig. 6 are

$$\begin{aligned} C(z) &= \left[z - 2\omega_0 - \frac{4t_b^2}{z - 3\omega_0} \right]^{-1}, \\ D(z) &= \left[z - 2\omega_0 - \frac{3t_b^2}{z - 3\omega_0} \right]^{-1}, \\ E(z) &= \left[z - 2\omega_0 - \frac{t_b^2}{z - 3\omega_0} \right]^{-1}, \\ K(z) &= \Big[z - \omega_0 - t_b^2 (C(z) + D(z)) \Big]^{-1}, \\ L(z) &= \Big[z - \omega_0 - t_b^2 (C(z) + E(z)) \Big]^{-1}. \end{aligned} \tag{11}$$

4.3. Discussion

Our derivation of the Green's function is closely related to well-known methods such as the renormalized perturbation expansion [13], the (Lanczos) recursion method [14] or the Mori-Zwanzig formalism [15]. For our problem the most important aspect is the appearance of two Green's functions in Eq. (4), corresponding to the two branches of each node in the binary tree of states in Fig. 6. This generalizes the standard recursive construction leading, e.g. in Lanczos recursion, to a continued fraction of the form

$$G(z) = \cfrac{1}{z - a_0 - \cfrac{b_1^2}{z - a_1 - \cfrac{b_2^2}{z - a_2 - \dots}}}. \tag{12}$$

In the present case this construction would result in very complicated expressions, since arbitrary linear combinations of the states shown in Figs. 5, 6 occur in the recursion, e.g. of states with different boson number. The additional freedom provided by Eq. (4) is important to obtain simple results for $G(k, z)$ in closed form (obviously, our result is not given in the form of Eq. (12)). The success depends on our ability to find a suitable arrangement of states in a tree as in Fig. 6. This task can hardly be automatized, and such an arrangement is difficult to find for more complicated situations. Nevertheless, the idea presented here can be generalized so that it may be useful also in further calculations.

5. Evaluation of the Green's function and Drude weight

In Fig. 7 we show the spectral function

$$A(k, \omega) = -(1/\pi)\,\mathrm{Im}\, G(k, z)\,. \tag{13}$$

We see that the Green's function is periodic in k with period π. The reason is that the vacuum-restoring 6-step process (Fig. 3) propagates the fermion by two sites, resulting in the term $\propto \cos 2k$ in Eq. (10). The spectral function is dominated by a coherent quasiparticle band at low energy, which is almost dispersionless and separated by a large gap (of the order ω_0) from other (coherent) signatures at higher energies. Note that the dispersion of the quasiparticle band is related to the Drude weight, as discussed in the next subsection. As Fig. 7 further shows, $G(k, z)$ compares quite well to the numerical result without the 3-boson-constraint [8]. The reason is that, for all but very small $\omega_0/t_b \simeq 1$, hole motion is dominated by the 6-step process that involves only 3 bosons. We will next discuss how this fact shows up in the Drude weight.

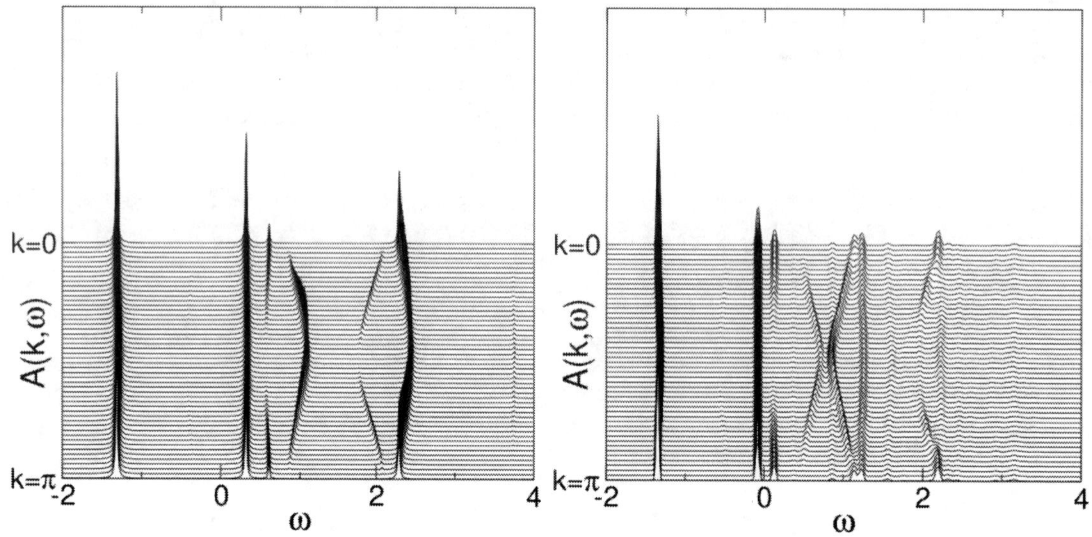

Figure 7. Spectral function $A(k,\omega)$ for $t_b = 1.0$ and $\omega_0 = 1.0$ with the 3-boson constraint from Eq. (10) (left panel) compared to the numerical result (right panel, from Ref. [8]) without the constraint.

5.1. The Drude weight

Following Eq. (2) the Drude weight D is obtained from the ground-state energy

$$E(k) = E_0 + (D/2)k^2 + O(k^4) , \tag{14}$$

which is given by lowest pole of $G(k,z)$. With decreasing ω_0, the Drude weight grows as boson-controlled transport becomes more efficient, but remains finite for $\omega_0 \to 0$ (see Fig. 8). Since the 6-step-process involves 5 intermediate states with excited bosons, the leading order of D/t_b in an expansion in t_b/ω_0 is $(t_b/\omega_0)^5$. Any process contributing to D/t_b that has an intermediate state with more than 3 bosons, must be of order $(t_b/\omega_0)^7$ or higher since at least 4 bosons have to be excited and de-excited afterwards, which requires 8 steps. Therefore the asymptotic behaviour of D for large ω_0/t_b can be obtained from $G(k,z)$ in Eq. (10), with the result $D \simeq t_b^6/(3\omega_0^5)$ shown in Fig. 8.

Once again, we compare our analytical D to the numerical result without the 3-boson constraint. Both results fit surprisingly well although the restriction to a small number of bosons might appear to be a crude approximation for small ω_0/t_b. But the inset in Fig. 8 shows that in numerical calculations without the 3-boson constraint the number of bosons in the ground-state remains small over a large range of values of ω_0/t_b. This indicates, together with the agreement between the two curves for D, that the main contribution to fermion propagation comes from the 6-step-process even when ω_0/t_b is small, although higher order vacuum-restoring processes involving more than three bosons might become energetically favourable in this case. One explanation for that behaviour is that correlated transport through vacuum-restoring processes relies on the correlated excitation and de-excitation of bosons. Higher order processes involving more than 3 bosons cannot give a large contribution to D since correlations between many bosons are inevitably reduced. This also explains the saturation of D as $\omega_0 \to 0$: The contribution of the 6-step-process to D remains finite for $\omega_0 \to 0$ (restricting to 3 bosons, the Hamiltonian is a bounded operator, and $E(k)$ and D remain finite). Then, if the main contribution to D comes from the 6-step-process, D itself remains finite for $\omega_0 \to 0$ even without the 3-boson constraint.

Journal of Physics: Conference Series **220** (2010) 012023 — doi:10.1088/1742-6596/220/1/012023

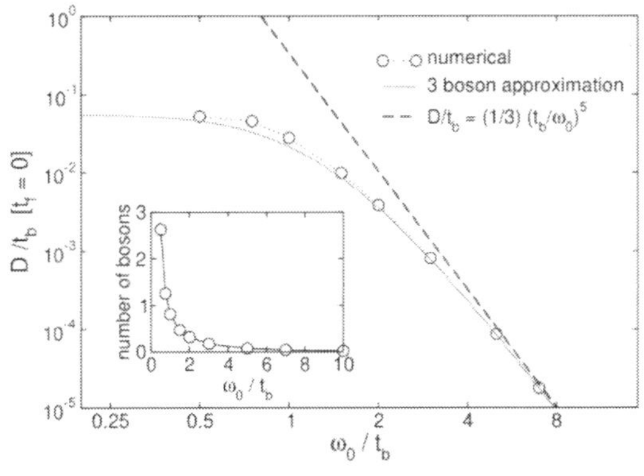

Figure 8. Drude weight D with 3-boson constraint, and from exact numerics without the constraint [8]. The dashed line gives the asymptotic result $D = t_b^6/(3\omega_0^5)$ for large ω_0/t_b. The inset displays the average number of bosons in the ground-state without the 3-boson constraint.

6. Conclusions and Outlook

In this contribution we discuss correlated transport through vacuum-restoring processes in a general fermion-boson model. We obtain a simple analytical expression for the one-fermion Green's function. From the Green's function we deduce the Drude weight which characterizes transport even when the description in terms of a mean-free path between two scattering events, as in Drude's original theory, breaks down. The dependence of the Drude weight on the boson energy ω_0/t_b reveals two opposing tendencies in the correlated transport process: On the one hand, transport is assisted by bosons, and therefore D increases with decreasing ω_0/t_b. On the other hand, it is the involvement of bosonic excitation in the transport process itself that limits its efficiency, and D remains finite in the limit of small ω_0/t_b. This competition between correlations and fluctuations is a general physical phenomenon, which can be further studied in the full model [8].

The correlated transport process studied here is also important at finite fermion density, when the formation of a charge-density wave at half-filling triggers a metal-insulator transition [9]. The charge-density-wave insulator is peculiar in that it involves few bosons, in contrast to the common Peierls scenario, and is caused by strong boson-fermion correlations. A hole doped into this insulator propagates through the 6-step-process discussed here, while an additional electron can propagate through a more efficient 2-step-process. The simple results obtained here for a single fermion are important to understand the more complicated physics in this case. Our considerations may also allow for the development of an analytical description of the correlated charge-density-wave insulator.

Acknowledgments

The authors are grateful for discussions with S. Ejima, T. Koch, and G. Wellein. This work was supported by Deutsche Forschungsgemeinschaft through SFB 652.

References

[1] Mott N F 1990 *Metal-Insulator Transitions* (Taylor & Francis)
[2] Trugman S A 1988 *Phys. Rev. B* **37** 1597
[3] Holstein T 1959 *Ann. Phys. (N.Y.)* **8** 325
[4] Holstein T 1959 *Ann. Phys. (N.Y.)* **8** 343
[5] Brinkman W F and Rice T M 1970 *Phys. Rev. B* **2**
[6] Kane C L, Lee P A and Read N 1989 *Phys. Rev. B* **39** 6880
[7] Edwards D M 2006 *Physica B* **378-380** 133
[8] Alvermann A, Edwards D M and Fehske H 2007 *Phys. Rev. Lett.* **98** 056602
[9] Wellein G, Fehske H, Alvermann A and Edwards D M 2008 *Phys. Rev. Lett.* **101** 136402
[10] Ejima S, Hager G and Fehske H 2009 *Phys. Rev. Lett.* **102** 106404
[11] Ejima S and Fehske H 2009 *Phys. Rev. B* **80** 155101

Journal of Physics: Conference Series **220** (2010) 012023 doi:10.1088/1742-6596/220/1/012023

[12] Kohn W 1964 *Physical Review* **133** A171
[13] Economou E N 1983 *Green's Functions in Quantum Physics* (Springer)
[14] Haydock R, Heine V and Kelly M J 1972 *J. Phys. C* **5** 2845–2858
[15] Fulde P 1991 *Electron Correlation in Molecules and Solids* (Springer)

AUTHOR INDEX